Integrated Circuits and Semiconductor Devices: Theory and Application

Dedicated to

Carol, Mary, Florence,
Jeff, and Christopher

Integrated Circuits and Semiconductor Devices: Theory and Application

second edition

GORDON J. DEBOO
CLIFFORD N. BURROUS
NASA/Ames Research Center, California
West Valley College, Saratoga, California

McGraw-Hill Book Company
Gregg Division
New York St. Louis Dallas San Francisco Auckland
Bogotá Düsseldorf Johannesburg London Madrid
Mexico Montreal New Delhi Panama Paris São Paulo
Singapore Sydney Tokyo Toronto

Library of Congress Cataloging in Publication Data
Deboo, Gordon J
Integrated circuits and semiconductor devices.

Includes bibliographical references and index.
1. Semiconductors. 2. Integrated circuits.
3. Optoelectronic devices. I. Burrous,
Clifford N., joint author. II. Title.
TK7871.85.D43 1977 621.381'73 76-45746
ISBN 0-07-016246-8

INTEGRATED CIRCUITS AND SEMICONDUCTOR DEVICES:
Theory and Application

234567890 DODO 783210987

The editors for this book were George J. Horesta
and Mark Haas,
the designer was Charles A. Carson,
the art supervisor was George T. Resch,
and the production supervisor was Regina R. Malone.
It was set in Times Roman by Santype Ltd. (overseas).
Printed and bound by R. R. Donnelley & Sons Company.

Contents

THYRISTORS

MISCELLANEOUS SEMICONDUCTOR DEVICES

Preface

The organization of this second edition is substantially the same as that of the first, and the changes reflect the results of a questionnaire completed by users of the first edition. There are two main differences between the two editions. The first is that many new topics have been introduced. For example, in response to the increasing use of digital technology in all areas of electronics, Chapter 5 now includes sections on microprocessors, CMOS, I^2L, three-state logic, RAM/ROM memories, error detection and correction, encoders and code converters, multiplexers and demultiplexers, and additional LSI examples such as calculators and digital watches. Other chapters have been updated or expanded so that there are new sections on phase-locked loops, IC transducers, displays, and opto-isolators. The second difference is that many more worked examples have been added within the text.

The book covers virtually all semiconductor devices and ICs and is intended to serve two groups. The first group consists of electronics students at a number of different levels, including community and junior college, engineering technology, industrial training, and extension courses. The second group consists of practicing engineers, engineering technicians, physicists, and those in related fields who need to update their knowledge of electronic techniques involving semiconductor devices. For both groups, the book is a review of the basic theory and application of integrated circuits and semiconductor devices.

The material is based on an extremely popular course of a similar nature and purpose taught in the San Francisco Bay area by the authors. The text is frequently used in one of the last courses in a two- or four-year electronics-technology curriculum to complete the student's background in integrated circuits and semiconductor devices other than bipolar transistors. It is the authors' hope that the text will help close the gap between academic instruction and industrial practice, thereby better preparing students for their role in industry. The prerequisites for a course using this book include basic semiconductor physics, basic circuit analysis, and an introduction to transistors. A brief review of the hybrid-π approach to bipolar-transistor analysis is given in the appendix.

The book has five parts. First, junction and metal oxide semiconductor field effect transistors are covered. The second part introduces integrated circuits, including operational amplifiers, other linear integrated circuits, digital integrated circuits, and integrated-circuit fabrication. Optoelectronic devices are dealt with in the third part, with subsections on light sources, displays, photodetectors, and source-detector combinations. The fourth part contains a chapter on thyristors such as silicon-controlled rectifiers and triacs. The last section covers miscellaneous semiconductors such as diodes and unijunction transistors.

A feature of the text is that the level of mathematics required is generally not above elementary algebra. Useful and valid approximations are used wherever possible to avoid obscuring principles with unwieldy mathematics. Examples are practical and often use data from actual data sheets. The book is up to date and comprehensive, yet written for the Community College level. A paperback instructor's guide is available to supplement the text. It contains problem solutions, instruction hints, and suggested experiments which can be performed with listed inexpensive versions of semiconductor devices, ICs, and conventional test equipment.

Many thanks are due Mr. Don Billings of NASA/Ames and Mr. Hans Sorenson of Hewlett-Packard, for their constructive criticism and suggestions regarding the second edition, and to the very helpful reviewers of the first edition.

Gordon J. Deboo
Clifford N. Burrous

Integrated Circuits and Semiconductor Devices: Theory and Application

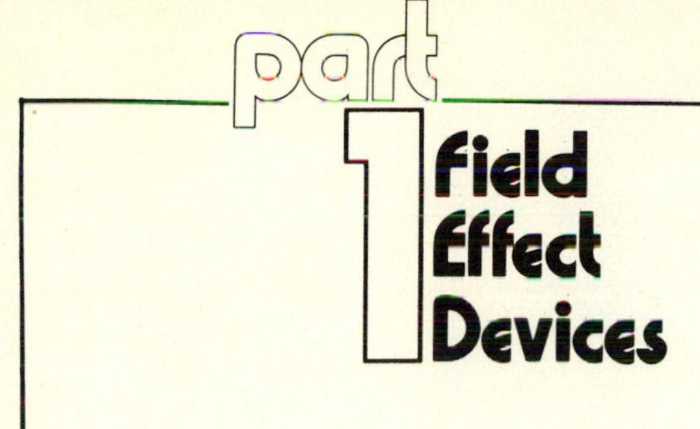

Part 1 Field Effect Devices

chapter 1

Junction Field Effect Transistors

1-1 INTRODUCTION

The *field effect transistor* (*FET*) is a semiconductor device that combines the small size and low power requirements of the bipolar transistor with the high input impedance of the vacuum tube.

There are two kinds of FET, although within each classification there are many variations. In this chapter the abbreviation JFET will be used to denote the *junction field effect transistor*. The abbreviation MOSFET will be used to denote the *metal oxide semiconductor* (*MOS*) *FET*, which is discussed in Chap. 2. The differences between these two field effect devices will become apparent as the discussion proceeds.

In contrast to bipolar transistors, in which the controlling effect is a *current* (the base current), both JFETs and MOS devices have very small input currents, which are due to leakage effects. The controlling *effect* is a *field* produced by an input voltage, which is why JFETs and MOS devices are called *field effect* transistors.

Figure 1-1 is a comparison of the symbols for a JFET, a bipolar transistor, and a vacuum tube. Also included is the symbol for a diode. Whereas there is only one polarity for vacuum tubes, there are two polarities of JFET available, just as there are in the case of bipolar transistors. Figure 1-1*a* shows an *n*-channel JFET, a term that will be explained later. Consider the biasing of the input diode of each of the three devices. For the *n*-channel JFET shown, the gate (or input) is marked with an arrow, which indicates the polarity of the input diode. (For a *p*-channel device, the arrow would have been drawn in the opposite direction.) The symbol for a diode is given in Fig. 1-1*d*. If the anode is sufficiently positive with respect to the cathode, the diode will conduct. If the anode is not sufficiently positive or is negative with respect to the cathode, virtually no current will flow. The arrows on JFETs and bipolar transistors follow the same convention. For a silicon diode, about 0.6 V of positive bias is required for conduction.

From the typical voltages shown in Fig. 1-1*a*, it can be seen that the JFET input diode is reversed-biased, which produces a high input impedance. The situation is similar for the vacuum tube for which the grid is the anode of the input diode. In the case of the bipolar transistor, however, the arrow shows that the base-emitter diode is forward-biased, thereby giving rise to a relatively low

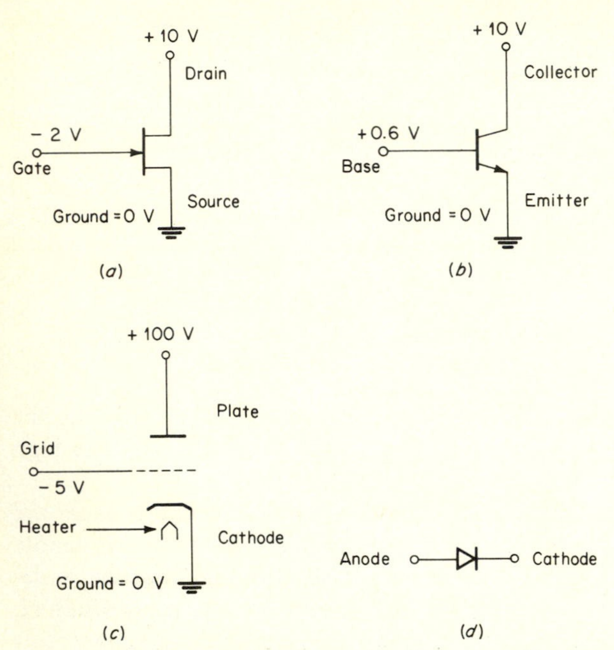

Fig. 1-1 JFET symbol compared with those for a bipolar transistor, a vacuum tube and a diode: (*a*) JFET (*n*-channel); (*b*) bipolar transistor; (*c*) vacuum tube; (*d*) diode.

input impedance. The input resistance of a JFET ranges from 10^8 to 10^{12} Ω, and for the bipolar transistor from about 10^2 to 10^6 Ω.

In a broad sense the drain, gate, and source of a JFET correspond to the collector, base, and emitter of a bipolar transistor and to the plate, grid, and cathode of a vacuum tube. When biased as a voltage amplifier, there is current flow in a JFET from source to drain, and this current flow is controlled by the gate voltage. In a bipolar-transistor, which is a current-controlled device, the flow of current from emitter to collector is controlled by the base current. While it is true that a base voltage exists, the relationship between base current and collector current is more linear than that between base voltage and collector current. The JFET, like the vacuum tube, can be thought of as a voltage-controlled device because, when it is used as an amplifier, the gate or grid currents are for all practical purposes zero.

Although first described as early as 1954, the JFET was slow to come into general use, partly because of manufacturing difficulties and partly because of the rapid development of the bipolar transistor. The JFET has more than fulfilled its early promise and now finds application in virtually every phase of circuit design.

In addition to being a high-input-impedance, low-bias current amplifier, the JFET may be used as an analog switch, as a voltage-variable resistor, as a very high frequency amplifier with low cross-modulation distortion, and as a device with an adjustable temperature coefficient, which can be set to be positive, negative, or nearly zero.

1-2 THEORY OF OPERATION

JFETs operate by means of the creation and control of the *depletion layer* that exists in all reverse-biased *pn* junctions. Figure 1-2*a* represents an unbiased *pn* junction. The left-hand or *p* region has excess holes, which are available for conduction. The right-hand or *n* region has excess electrons, which are also available for conduction.

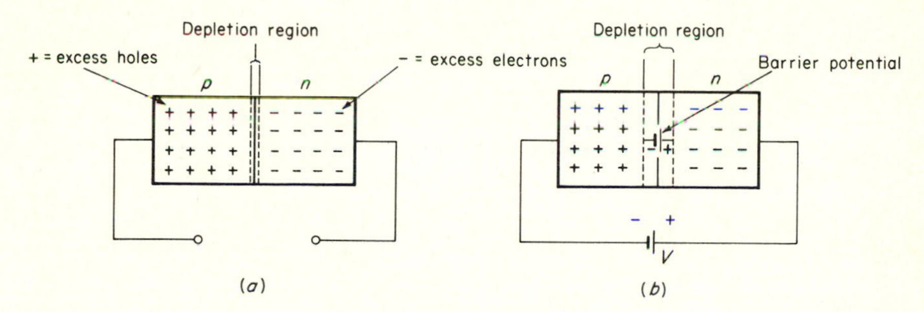

Fig. 1-2 Formation of a depletion region: (*a*) *pn* junction with no bias; (*b*) *pn* junction with bias.

There is a tendency for current carriers to diffuse from a region of high concentration to one of low concentration. Since there is a high concentration of electrons in the *n* region and a low concentration of electrons in the *p* region, electrons tend to diffuse from the *n* region to the *p* region. Similarly, holes tend to diffuse from the *p* region (high concentration of holes) to the *n* region (low concentration of holes). However, the *n* and *p* regions are both electrically neutral; so when electrons leave the *n* region, they "uncover" immobile positive ions, thereby leaving a net positive charge on the *n* side of the *pn* junction. Similarly, a net negative charge is created on the *p* side of the *pn* junction. Thus an internal "battery" or *barrier potential* is set up, and its polarity is such as to preclude further crossing of the junction by charge carriers. Therefore, diffusion of charge carriers is not complete, and instead there is a rearrangement of the holes and electrons within the crystal structure, which produces a narrow region *depleted* of charge carriers. This region is called a *depletion region*.

With no external bias applied as in Fig. 1-2*a*, the depletion region is very thin, and for purposes of explanation of JFET operation we can consider it negligibly thin. However, an externally applied reverse-biased potential (Fig. 1-2*b*) has the effect of *widening* the depletion region. This is because the + side of the external battery attracts electrons and "pulls" them away from the *n* side of the junction. Similarly, the − side of the battery "pulls" holes away from the *p* side of the junction. Thus the barrier potential or "height" is increased.

The significance of depletion regions as far as JFETs are concerned is that it is possible to electrically change the resistance of a piece of semiconductor material by changing the concentration of charge carriers with an externally applied electric field. This is true even though there is no *pn* junction in the normal conduction path of a JFET, which is *along* the depletion region, not *across* the junction as in a bipolar transistor.

Conduction in JFETs is always by majority charge carriers only, which is why they are called *unipolar* transistors. This is in contrast to bipolar transistors, in which both majority and minority charge carriers are involved in current flow. Since there are two kinds of current carrier in semiconductors, electrons and holes, it is natural that there should be two polarities of JFET. Those employing electron conduction are called *n* types, and those employing hole conduction are called *p* types; *n* and *p* types are similar in their modes of operation except for reversal of all polarities.

A JEFT may be visualized (Fig. 1-3) as a conductive bar or channel, usually of silicon, at each end of which are ohmic contacts called the *source* and *drain.* (An ohmic contact is one not involving a *pn* junction.) A third ohmic contact is made to the material on either side of the channel, as shown in Fig. 1-3. When appropriately biased, the gate voltage influences the resistance between the source and the drain and hence the flow of current between these two terminals.

Figure 1-3 represents an *n*-channel device. The starting material is a piece of *p*-type silicon (A) into which is diffused an *n*-type region (B). A *p* region (C) is then diffused into the *n* region, leaving a thin channel of *n* material. Ohmic contacts for the source and drain are made by metallizations at each end of the *n* channel. This *n* channel is the conductive part of the device. With no voltages applied, the impedance between the source and drain will be an ohmic resistance, the resistance depending on the dimensions of the *n* channel. In fact, the resistance can be calculated as for any other resistor from the well-known formula

$$R = \frac{\rho L}{WT} \tag{1-1}$$

where ρ = resistivity of the material
L = length of the channel
W = channel width
T = channel thickness
L, W, and T are indicated on Fig. 1-3.

The third ohmic contact, the gate, is connected to both *p* regions. Notice that the gate and the channel form two *pn* junctions, and that the channel thickness t is the distance between the two transistion regions from *p* to *n* material. Figure 1-4*a* is a two-dimensional representation of a cross section of the device shown in Fig. 1-3.

Application of a reverse gate-to-channel bias will produce a depletion layer at each transition region, as described earlier with the help of Fig. 1-2. The shape

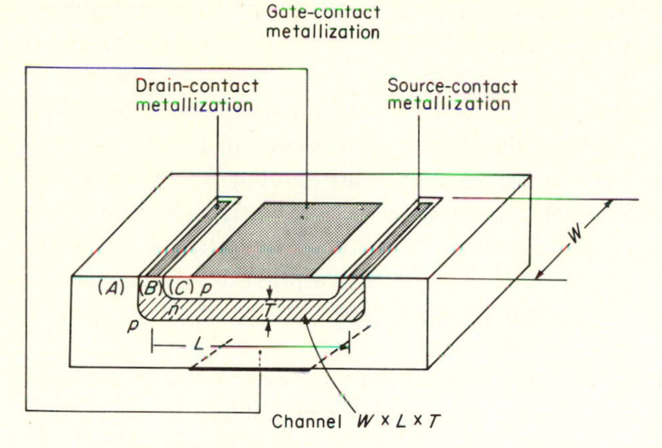

Fig. 1-3 JFET n-channel device showing connections to p *and* n regions.

and extent of the depletion regions and their effect depend on how much bias is applied and where. In Fig. 1-4 we shall assume that the drain-to-source voltage V_{DS} is positive, while the gate-to-source voltage V_{GS} is negative. These are appropriate bias polarities for an n-channel JFET. The same principles apply to a p-channel JFET, but with all polarities reversed.

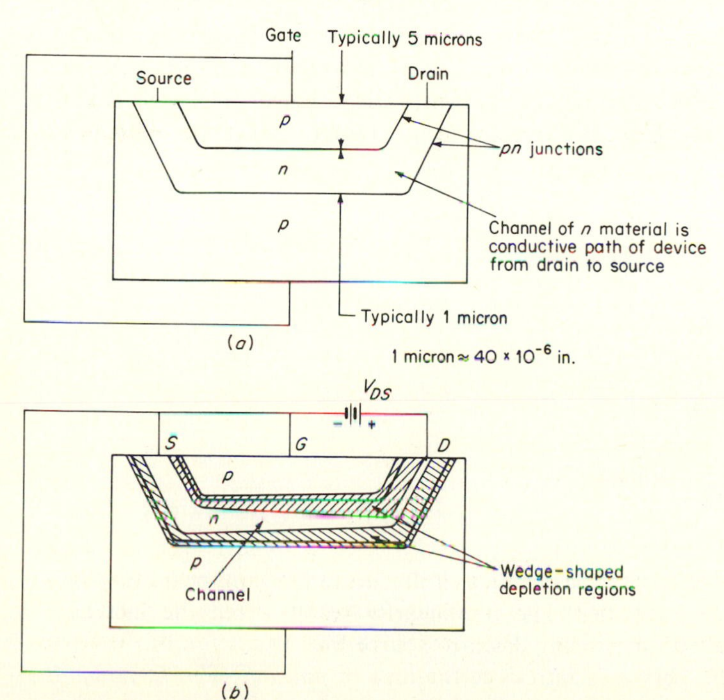

Fig. 1-4 Cross section of an n-channel JFET showing effect of bias on depletion regions: (a) no bias; (b) with bias.

We shall show that various combinations of gate-to-source and drain-to-source potentials can be used to create depletion layers, which control the flow of current from source to drain. To simplify the explanation, we shall first consider the special and important cases in which both source and gate terminals are grounded, so that $V_{GS} = 0$. Since the channel is of *n*-type material, the drain is biased positively with respect to the source by an amount V_{DS}. This set of conditions is illustrated in Fig. 1-4*b*, in which the source, gate, and drain terminals are labeled *S*, *G*, and *D*, respectively. There now exists a situation in which an ohmic resistance, the channel, has along it a potential difference, the drain-to-source bias. Therefore, a potential gradient exists along the channel, increasing from zero at the left-hand or source end to V_{DS} at the right-hand or drain end of the channel. Notice that both *pn* junctions are reverse-biased and that the amount of bias increases along the channel from source to drain. This results in the creation of a depletion region whose thickness increases along the channel from source to drain. The depletion regions therefore have the wedge-shaped appearance illustrated in Fig. 1-4*b*.

The depletion regions are regions in which the number of charge carriers have been substantially reduced. Another way of stating this is to say that the resistivity of the *n*-type channel has been increased. Hence the resistance from source to drain increases as V_{DS} increases and consequently the channel thickness decreases. A plot of V_{DS} versus the drain-to-source current I_D will not therefore be a straight line as it would be for an ordinary resistor, which has a constant resistance. We can refer to the drain-to-source current as simply I_D rather than I_{DS} because the drain current equals the source current. This must be so because current entering the drain can leave only via the source, since the gate current is zero.

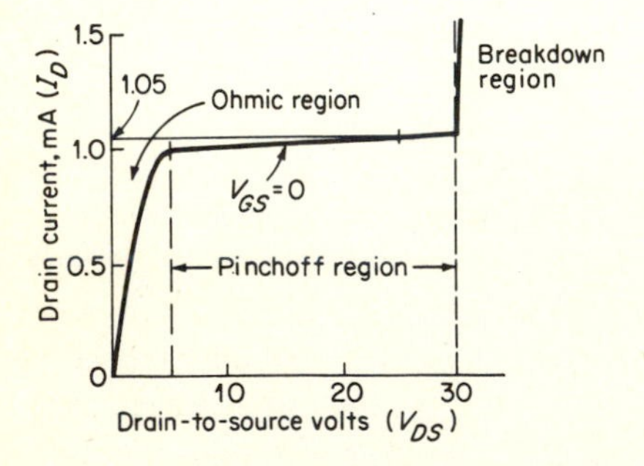

Fig. 1-5 JFET output characteristic for $V_{GS} = 0$.

Figure 1-5 is a plot of I_D versus V_{DS}; it illustrates the nonlinear relationship between these two quantities. The nonlinearity results from the increase of channel resistance with increasing drain-to-source bias V_{DS}.

Figure 1-5 also serves to introduce the idea of *pinchoff*. The I_D versus V_{DS} plot is divided into three areas, an *ohmic region*, a *pinchoff region*, and a *breakdown region*. The region we have discussed so far is the *ohmic region*; it occurs where the applied reverse bias is insufficient to make the depletion layers wide enough to meet. This is the situation illustrated in Fig. 1-4*b*. When sufficient bias is applied to make the depletion layers touch, the channel thickness becomes essentially zero, and the path from drain to source is said to be *pinched off*. It might appear at first sight that the channel current I_D would under these conditions become zero, but this is not the case. As the channel width tries to go to zero, the resistance of the channel increases and the current will tend to decrease. This will cause a drop in the voltage gradient along the channel, which reduces the depletion width, reduces the resistance, and permits more current to flow. The net result is that with sufficient bias an equilibrium condition is realized in which the current reaches a maximum value. After this point, *an increase in* V_{DS} *does not cause a significant increase in* I_D. Figure 1-5 illustrates how the I_D versus V_{DS} curve levels off in the pinch-off region.

Since in the pinchoff region the I_D versus V_{DS} curve is almost horizontal, we can consider the ac or *incremental* resistance to be very large. It would be infinite if the curve were perfectly horizontal. Thus, for *changes* in V_{DS} in the pinchoff region, the channel resistance may be thought of as very large. The dc or *static* resistance of the channel is relatively small and is found by dividing V_{DS} by I_D.

For example, for the curve shown in Fig. 1-5, which is for the special case where $V_{GS} = 0$, the dc channel resistance at $V_{DS} = 15$ V is

$$R_{dc} = \frac{V_{DS}}{I_D} \approx \frac{15}{1} = 15 \text{ k}\Omega \tag{1-2}$$

The ac resistance, which is commonly called r_{ds}, at the same point is

$$r_{ds} = \frac{\Delta V_{DS}}{\Delta I_D} \approx \frac{25-5}{1.05-1} = \frac{20}{0.05} = 400 \text{ k}\Omega \tag{1-3}$$

[For a more precise definition of r_{ds}, see Eq. (1-10)].

In addition to the ohmic and pinchoff regions we have just discussed, there is a third region, which is also illustrated in Fig. 1-5. It is called the *breakdown region*. It occurs when the diode formed by the *pn* junction at the gate-channel interface breaks down, resulting in a large increase in I_D for a small increase in V_{DS}. While in the pinchoff region, the JFET approximates a constant-current device. In the breakdown region, the JFET approximates a constant-voltage device, similar to a zener diode. As with a zener diode, the breakdown is not destructive, and no damage will occur to the JFET if the power dissipation is limited.

So far we have discussed the behavior of the JFET with $V_{GS} = 0$. If V_{GS} is allowed to vary as well as V_{DS} and I_D, a family of curves similar to the one plotted in Fig. 1-5 is obtained. They are commonly referred to as the JFET *drain* or *output* characteristics, and a typical set for an *n*-channel device is shown in Fig. 1-6.

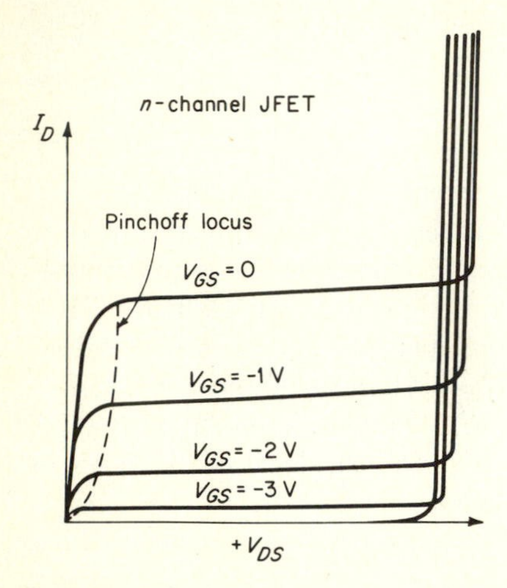

Fig. 1-6 Family of *n*-channel output characteristics for various values of V_{GS}.

The uppermost curve is for the special case where $V_{GS} = 0$, which we have previously discussed. The other curves are for other values of V_{GS}, the polarities of which are such as to reverse bias the gate-channel *pn* junction. For an *n*-channel device, this means that the gate is biased negatively with respect to the source, as in Fig. 1-1*a*.

Notice that the width of the channel produced by the depletion layers depends on combinations of V_{DS} and V_{GS}. A result of this is that the conditions under which the channel becomes pinched off is dependent on *both* values of V_{DS} and V_{GS}, since it is the gate-to-drain voltage that primarily determines the depletion-layer width.

The effect of having both V_{DS} and V_{GS} present simultaneously can be restated another way. Since the extent of the depletion region is determined by the total voltage across the *pn* junction, the effect of increasing V_{GS} is to cause pinchoff to occur at a lower value of V_{DS}. This results in a lower level of pinchoff current, since I_D is less for less V_{DS}. Increasing V_{GS} therefore reduces the value of drain current at pinchoff. If V_{GS} is made sufficiently large, pinchoff can occur at nearly zero drain current. Thus the JFET can be "turned off," that is, I_D made zero, by sufficient reverse bias (V_{GS}). Notice the difference between turning off a JFET, which means zero I_D, and pinching off the channel, which means constant I_D. A turned-off JFET is pinched off, but a pinched-off JFET is not necessarily turned off.

In the extreme case pinchoff can occur due to V_{GS} alone, without any assistance from the potential gradient created by V_{DS}. A sufficiently large value of V_{GS} will cause pinchoff for any value of V_{DS} less than breakdown, including zero. We can therefore think of two extreme situations. One is where pinchoff occurs due to V_{DS}, with $V_{GS} = 0$. The other is where pinchoff occurs due to V_{GS}, with $V_{DS} = 0$. Between these two extreme cases there are an infinite number of other combinations in which neither V_{DS} nor V_{GS} is zero but in which *combinations* of these two quantities produce pinchoff. In Fig. 1-6 the dotted line is the locus of points for which the sum of V_{DS} and V_{GS} is sufficient to pinch off the channel. Calling the pinchoff voltage V_P, we can write

$$V_P = |V_{GS}| + |V_{DS}| \tag{1-4}$$

for this locus.

Equation (1-4) shows that V_P can theoretically be obtained from the output characteristic curves by a number of methods, of which two are the most simple. We can read off V_{DS} from the locus at the point where $V_{GS} = 0$, or we can read off V_{GS} from the locus at the point where $V_{DS} = 0$. Setting either V_{DS} or V_{GS} equal to zero in Eq. (1-4) will clearly mean that the one not equal to zero must be equal to V_P. In practice both methods are very inaccurate, and other means are used to determine V_P.

1-3 JFET PARAMETERS

1-3-1 NOTATION

JFETs are characterized by a number of parameters, such as in Table 1-1, which can be either measured or deduced from the characteristic curves. Certain notational conventions have been adopted in industry, and the notation introduced in this section is widely used.

TABLE 1-1 Summary of JFET parameters and definitions

Parameter	Definition
V_P	Pinch-off voltage. V_{GS} for an arbitrarily low level of I_D, for example 1 μA. Specified for a particular value of V_{DS}.
V_{GSoff}	$= V_P$
g_{fs}	Transconductance. The ratio of the change in I_D to the change in V_{GS} producing it for a specified value of V_{DS}. That is, $g_{fs} = \dfrac{\Delta I_D}{\Delta V_{GS}} \quad V_{DS} = \text{const}$
g'_{fs}	g_{fs} at $V_{GS} = 0$ (that is, at $I_D = I_{DSS}$), for a specified value of V_{DS}.
g_m	g_{fs}
I_{DSS}	I_D with $V_{GS} = 0$ for a specified V_{DS}
I_{GSS}	I_G with $V_{DS} = 0$ for a specified V_{GS}
g_{os}	Output conductance or slope of output characteristic
r_{ds}	Output resistance or $1/g_{os}$
BV_{GSS}	Gate-to-source breakdown voltage with drain shorted to source
BV_{DGO}	Gate-to-drain breakdown voltage with source open
C_{GS}	Gate-to-source capacitance including header capacitance
C_{GD}	Gate-to-drain capacitance including header capacitance
C_{DS}	Drain-to-source capacitance including header capacitance (mainly header capacitance)
C_{iss}	Common-source input capacitance, drain shorted to source $= C_{GS} + C_{GD}$
C_{is}	Common-source input capacitance, drain not shorted to source
C_{rss}	Reverse-transfer capacitance, gate shorted to source $= C_{GD}$
C_{rs}	$= C_{rss}$
C_{oss}	Common-source output capacitance, gate shorted to source $= C_{rss} + C_{DS}$
C_{os}	Common-source output capacitance, gate not shorted to source

JFET parameters are specified with the aid of a terminology in which single-, double-, and triple-subscript notation is employed. For dc parameters such as leakage current, pinchoff voltage, and breakdown voltage, capital letters are usually employed. For ac parameters, such as transconductance, lowercase letters are usual. However, manufacturers have not standaradized on a notation which will distinguish between dc and ac parameters. For example, in Table 1-2, Y_{fs} is used for forward transadmittance, which is an ac quantity even though a capital Y is used. Another manufacturer uses g_{fs} for the same quantity. Where single- and double-subscript notation is used, the meaning is usually obvious. Two examples have already been introduced, namely, V_P, the pinchoff voltage, and V_{DS}, the drain-to-source voltage. The case in which confusion is possible is where triple-subscript notation is used. This is due to the two meanings assigned to the letter S. The rules for understanding the use of the letter S in JFET terminology are as follows:

1. As the first or second subscript, S is an abbreviation for the word "source." Further, the word "source" is used in the sense that it is a JFET terminal, as are the gate and drain. None of the other meanings of the word "source" apply here. As a first or second subscript, S means that the source terminal is being used as a node point.
2. As the third subscript, the letter S is an abbreviation for the word "shorted." It indicates that any terminals not designated by the first two subscripts are shorted to the terminal designated by the second subscript.

For example, consider the parameter I_{DSS}. I has its usual meaning of current. D refers to the drain, and the first S, which is the second subscript, refers to the source terminal. The second S, which is the third subscript, means that the gate (the only terminal not designated by the first two subscripts) is shorted to the source (the terminal designated by the second subscript). Thus I_{DSS} is the drain-to-source current with the gate shorted to the source. When I_{DSS} is defined, the drain-to-source voltage is also specified. Similarly, I_{GSS} is the gate-to-source current with the drain shorted to the source.

The letter O also is used as the third letter in triple-subscript notation. It is an abbreviation of the word "open" and means that the terminal not represented by the first two subscripts is open. For example, V_{DGO} is the drain-to-gate voltage with the source open—that is, not connected to anything.

When X is used as the third letter in triple-subscript notation, it refers to the indicated parameter under specified conditions. Thus, I_{DSX} would be the drain-to-source current for specified values of V_{DS} and V_{GS}.

There is often an interaction among JFET parameters. The magnitude of some parameter will depend on the magnitude of other parameters or on the conditions under which the measurement is made. Thus, to give a value for I_{DSS} is not completely valid unless other information is also given. We shall now discuss the definition of the more important parameters in detail and, where applicable, examine their relationship to JFET characteristic curves.

TABLE 1-2 Electrical characteristics of the 2N5163 *n*-channel JFET (*Courtesy Fairchild Semiconductor*)

Electrical Characteristics (25°C Free Air Temperature unless otherwise noted)

Symbol	Characteristics	Min.	Typ.	Max.	Units	Test Conditions	
Y_{fs}	**Forward Transadmittance** (f = 1.0 kHz)	**2,000**	**6,000**	**9,000**	**μmhos**	V_{DS} = 15 V	V_{GS} = 0
$Re(Y_{fs})$	Forward Transconductance (f = 1.0 MHz)	1,800	5,500		μmhos	V_{DS} = 15 V	V_{GS} = 0
e_n	**Equivalent Input Noise Voltage (f = 1.0 kHz, BW = 150 Hz)**		**12**	**50**	**nV/√Hz**	V_{DS} = 15 V	I_D = 1.0 mA
NF	Noise Figure (f = 1.0 kHz, R_G = 150 kΩ, BW = 150 Hz)			3.0	dB	V_{DS} = 15 V	I_D = 1.0 mA
NF	**Noise Figure (f = 1.0 kHz, R_G = 1.0 MΩ, BW = 150 Hz)**		**<0.1**		**dB**	V_{DS} = 15 V	I_D = 1.0 mA
$r_{ds(on)}$	**Drain "On" Resistance** (f = 1.0 kHz)		**125**	**500**	**ohms**	V_{GS} = 0	I_D = 0
I_{DSS}	Drain Current	1.0	14	40	mA	V_{DS} = 15 V	V_{GS} = 0
$V_{GS(off)}$	**Gate to Source Cutoff Voltage**	**−0.4**	**−3.7**	**−8.0**	**Volts**	V_{DS} = 15 V	I_D = 1.0 μA
V_{GS}	Gate to Source Voltage		−3.5	−7.5	Volts	V_{DS} = 15 V	I_D = 100 μA
I_{GSS}	**Gate Reverse Current**		**0.1**	**10**	**nA**	V_{GS} = −15 V	V_{DS} = 0
I_{GSS}(85°C)	Gate Reverse Current		0.03	0.6	μa	V_{GS} = −15 V	V_{DS} = 0
C_{rss}	Reverse Transfer Capacitance (f = 1.0 MHz)		1.3	3.0	pF	V_{DS} = 15 V	V_{GS} = 0
C_{iss}	Input Capacitance (f = 1.0 MHz)		8.7	12	pF	V_{DS} = 15 V	V_{GS} = 0
Y_{os}	Output Admittance (f = 1.0 kHz)		60	200	μmhos	V_{DS} = 15 V	V_{GS} = 0
BV_{GSS}	Gate to Source Breakdown Voltage	−25			Volts	V_{DS} = 0	I_G = 10 μA

1-3-2 DEFINITIONS

In this section we shall refer to the two characteristic curves for the 2N5163, an *n*-channel JFET, which appear in Fig. 1-7. Figure 1-7*a* is the output characteristic which is a plot of I_D versus V_{DS}. Figure 1-7*b* is the transfer characteristic, which is a plot of I_D versus V_{GS}. Several transfer curves are given in Fig. 1-7*b* since they will not be the same for different 2N5163s. The curves in Fig. 1-7*a* are all for the same typical transistor.

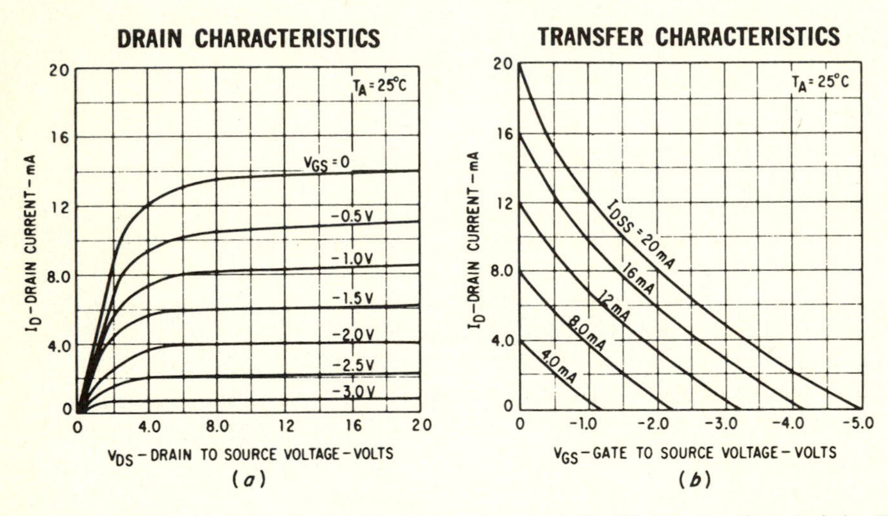

Fig. 1-7 2N5163 characteristics: (*a*) Output characteristic; (*b*) transfer characteristic. *Fairchild Semiconductor*.

V_P, **pinchoff voltage (also called V_{GSoff}).** This can theoretically be obtained from either of the two characteristics given in Fig. 1-7. On the output characteristic, Fig. 1-7*a*, V_P is that value of V_{DS} at which the knee of the $V_{GS} = 0$ curve occurs. Also, it is that value of V_{GS} at which I_D becomes practically zero. On the transfer characteristic, Fig. 1-7*b*, V_P is that value of V_{GS} at which I_D becomes essentially zero. A glance at either characteristic will show that by these definitions V_P is a very indeterminate voltage. In order to be more specific, V_P is usually defined as that value of V_{GS} required to reduce I_D to some low but definite level at some particular drain-to-source voltage. For the 2N5163, V_P is defined as the gate-to-source voltage required to reduce I_D to 1 μA at $V_{DS} = 15$ V. Under these conditions V_P is typically -3.7 V, although it can range from -0.4 to -8.0 V.

I_{DSS}, **drain current for zero gate voltage.** As can be seen from the output characteristics in Fig. 1-7*a*, I_{DSS} is a function of V_{DS}. To give a meaningful value for I_{DSS} therefore involves specifying a value for V_{DS} also. If V_{DS} is equal to or greater than V_P, then I_{DSS} becomes equal to the pinchoff current for $V_{GS} = 0$. From Fig. 1-7*a* it can be seen that for the 2N5163, I_{DSS} is about 14 mA at $V_{DS} = 15$ V. This agrees with the typical value given by the manufacturer for this transistor in Table 1-2.

g_{fs}, **common-source forward transconductance (or transadmittance y_{fs}).** Transconductance is used here in the same sense in which it is used for other active elements, such as bipolar transistors and vacuum tubes. The notation here is similar to that for bipolar transistors. Just as h_{fe} is the forward (f) current gain in the common-emitter (e) configuration, so g_{fs} is the forward (f) transconductance in the common-source (s) configuration. It is a measure of the effect of an input-voltage change, in this case ΔV_{GS}, on the output current, in this case ΔI_D. It is defined for particular values of V_{DS} and V_{GS} and usually at some specified frequency, in the following manner:

$$g_{fs} = \frac{\Delta I_D}{\Delta V_{GS}}\bigg|_{(V_{DS},\ \text{freq. const.})} \tag{1-5}$$

Notice that the units are amperes per volt or mhos, which are the same as for *conductance*. This, coupled with the fact that g_{fs} describes how the device affects the *transfer* of signal from input to output, is why g_{fs} is called the transconductance. Because of the magnitude of the numbers involved, micromhos are more commonly used than mhos, where 1 mho $= 10^6$ μmho. Thus, in Table 1-2, instead of calling the g_{fs} of the 2N5163 JFET 0.006 mho, the manufacturer uses 6000 μmho. g_{fs} can be deduced from the output characteristics by drawing a vertical line at the value of V_{DS} for which g_{fs} is to be defined. This satisfies the requirement in Eq. (1-5) that V_{DS} be constant. To find the g_{fs} at a particular current, note the change in I_D produced by a change in V_{GS} in the region of the I_D in question. From Fig. 1-7*a* at $V_{DS} = 15$ V and $V_{GS} = 0$, changing V_{GS} from 0 to -0.5 V changes I_D by $14 - 11 = 3$ mA. So we have, from Eq. (1-5),

$$g_{fs} = \frac{\Delta I_D}{\Delta V_{GS}} = \frac{(3)(10^{-3})}{0.5} = (6)(10^{-3}) \text{ mho}$$
$$= 6000\ \mu\text{mho} \tag{1-6}$$

This agrees with the value for g_{fs} given by the manufacturer in Table 1-2.

Figure 1-7*a* shows that in the pinchoff region g_{fs} is not very dependent on V_{DS}, since the spacing of the output characteristics does not change much as V_{DS} changes. The same set of curves indicates that g_{fs} is a strong function of I_D, since the characteristics "squeeze" together as I_D becomes smaller. The relationship between g_{fs}, V_{DS}, and I_D is discussed later in this section.

BV_{GSS}, **gate-to-source breakdown voltage with drain shorted to source.** Since the JFET has as its input a *pn* junction, it exhibits the phenomenon of breakdown. As mentioned in Sec. 1-2, breakdown occurs in the region beyond pinchoff, as in Fig. 1-5, and begins at the point where the output characteristic changes from being roughly parallel to the V_{DS} axis to being roughly parallel to the I_D axis. In the pinchoff region before breakdown V_{DS} has little effect on I_D, but in the breakdown region it has a very considerable effect.

Breakdown in a JFET is more complicated than in a simple *pn* junction because of the voltage gradient from drain to source. Figure 1-8, which represents an *n*-channel device, illustrates this point. The drain-to-source path, or channel, is represented by a resistor that, when the device is biased with appropriate voltages, will exhibit a voltage gradient. For purposes of discussion, typical voltages have been placed on Fig. 1-8. It must be stressed that these potentials are internal to the device and cannot be externally measured. The *pn* junction has been represented by a number of diodes, all reverse-biased and having common anodes connected to the gate. The cathodes are connected to taps on the drain-to-source channel resistance. In reality the gate-to-channel path is a single, distributed *pn* junction rather than a large number of individual diodes. The gate is at -5 V as shown. From Fig. 1-8 it can be seen that the diode nearest the drain has 15 V of reverse bias across it, while the diode nearest the source is subjected to only 5 V.

In measuring BV_{GSS}, the drain is shorted to the source so that there is no voltage gradient along the channel, and all diodes, or rather all parts of the distributed *pn* junction, are equally biased. This method has the advantage that the diode with the lowest breakdown voltage is the one used to rate the device. In applying BV_{GSS} to circuit design, it must be remembered that it is the sum of the gate and drain voltages, or V_{GD}, that can cause breakdown. This is why, in Fig. 1-6, the breakdown curves are spread. The breakdown drain voltage for $V_{GS} = 0$ is greater than that for V_{GS} more negative than zero. With some negative V_{GS} present, less V_{DS} is required to break down the junction.

I_{GSS}**, gate-to-source current with drain shorted to source.** Since the gate and channel of a JFET form a *pn* junction, which is usually reverse-biased for amplifier operation, an input leakage current will exist. As with any *pn* junction, this

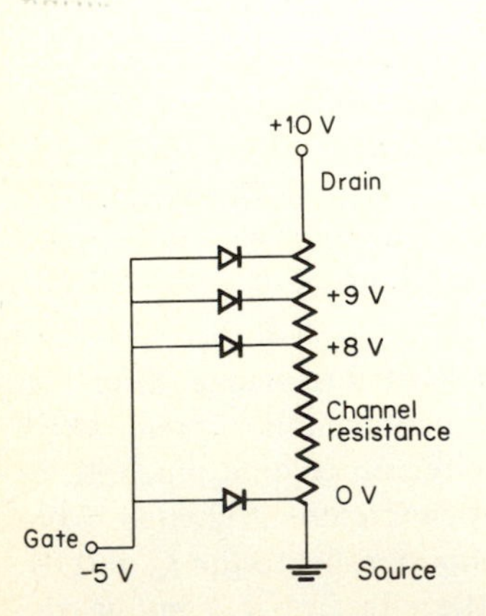

Fig. 1-8 Gate-to-channel breakdown mechanism.

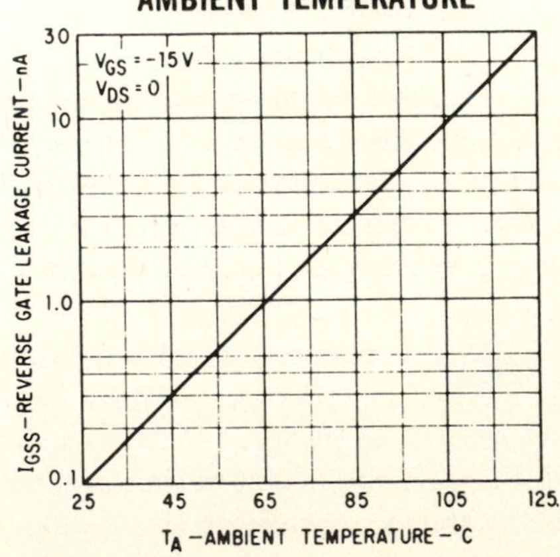

Fig. 1-9 Gate-leakage-current variation with temperature. *Fairchild Semiconductor.*

leakage current will be voltage- and temperature-sensitive. A curve of I_{GSS} versus temperature is shown in Fig. 1-9. On the semilogarithmic plot shown, the characteristic is a straight line. This results from the usual exponential increase of leakage current with temperature, typical of any reverse-biased *pn* junction. It should be realized that the curve is for a *reverse*-biased *pn* junction. If the gate-to-channel junction becomes *forward*-biased, the input resistance will drop to a low value, and heavy current flow could result.

C_{iss}, **input capacitance with the drain shorted to the source.** In a JFET, as with any other amplifying device, the total input capacitance has several components. For a JFET they are, referring to Fig. 1-10,

C_{GS} Gate-to-source capacitance of the JFET

C_{GD} Gate-to-drain capacitance of the JFET

C_{RS} All residual capacitances from gate to source, including header capacitance

C_{RD} All residual capacitances from gate to drain, including header capacitance

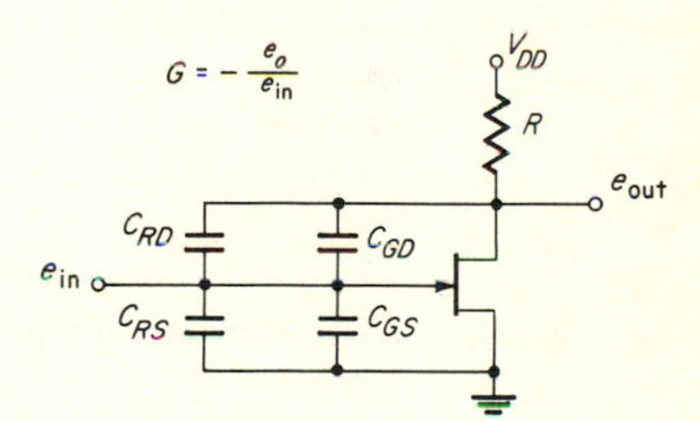

Fig. 1-10 JFET input capacitances.

C_{GS} and C_{RS} are in parallel, as are C_{GD} and C_{RD}. C_{GS} and C_{GD} are voltage-sensitive, since they are the depletion capacitances of a *pn* junction. C_{RS} and C_{RD} are fixed, although in any practical circuit they are always increased by wiring strays.

C_{iss} must be applied to a circuit-design problem with considerable care, because in many cases the JFET is not operated with the drain shorted to the source, from either an ac or dc point of view. In Sec. 1-5-2 it is shown that in a common-source voltage amplifier the total input capacitance C_{is} is greater than C_{iss}, owing to the Miller effect. It can be shown for the circuit in Fig. 1-10 that C_{is} will be, for an amplifier gain of G,

$$C_{is} = (1 - G)(C_{GD} + C_{RD}) + (C_{GS} + C_{RS}) \tag{1-7}$$

For example, suppose in the circuit of Fig. 1-10 that the voltage gain is -10. If $C_{GD} + C_{RD} = 5$ pF and $C_{GS} + C_{RS} = 5$ pF, the total input capacitance will be

$$C_{is} = [1 - (-10)](5) + 5 = 60 \text{ pF} \tag{1-8}$$

For the same transistor, C_{iss} would be the sum of the capacitances specified connected in parallel, since the drain is shorted to the source by the definition of C_{iss}. This situation is illustrated in Fig. 1-11, and we have

$$C_{iss} = 5 + 5 = 10 \text{ pF} \tag{1-9}$$

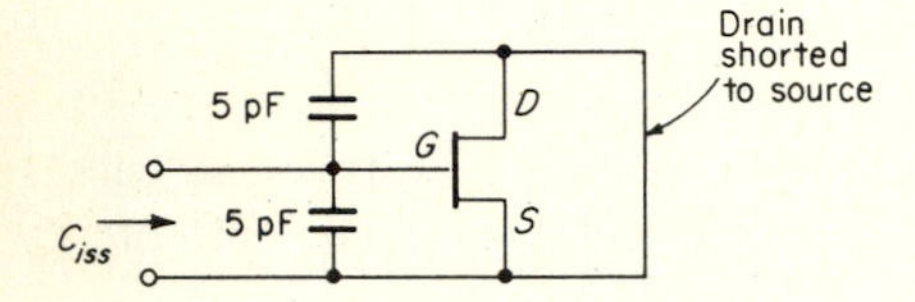

Fig. 1-11 JFET connections for C_{iss}.

JFETs are commonly used for their high input impedance, which decreases as input capacitance increases. Comparing Eqs. (1-8) and (1-9), we can see that the input impedance will be less than C_{iss} alone would indicate. Furthermore, when C_{iss} alone is specified there is no way of knowing how much of it is due to C_{GS} and how much to C_{GD}. As Eqs. (1-7) and (1-8) indicate, C_{GD} and C_{RD} are the ones that predominantly affect the input capacitance where there is appreciable voltage gain.

$r_{ds} = 1/g_{os}$, **the output resistance.** This parameter has already been referred to in Eq. (1-3). A more precise definition takes account of the fact that V_{GS} is held constant. That is,

$$r_{ds} = \frac{\Delta V_{DS}}{\Delta I_{DS(V_{GS}=\text{const})}} \tag{1-10}$$

r_{ds} and $1/g_{os}$ are different ways of expressing the same thing. r_{ds} is the output resistance, and g_{os} is the output conductance. In the pinchoff region where the JFET acts as a constant-current generator, for V_{GS} = const, r_{ds} is large and g_{os} is consequently small.

Table 1-1 is a summary of the JFET parameters covered in this section. In addition, the definitions of some other commonly used parameters are given.

Table 1-2 is from a data sheet for the 2N5163 JFET, the output and transfer characteristics for which appear in Fig. 1-7. For leakage currents, the typical and maximum values only are given since the effects of this parameter are usually undesirable and one must assume the worst case. For the breakdown voltage BV_{GSS}, the minimum value is given since although some JFETs may break down at a higher voltage, the minimum must be assumed to avoid individual component selection. For parameters that affect the circuit from the point of view of performance (for example, gain) a typical design center value is given—I_{DSS}, V_{GSoff}, Y_{fs}, and C_{iss} are specified in this manner. Often a maximum and minimum value are given for this type of parameter, so that the designer can account for parameter spread in his or her design.

1-3-3 PARAMETER INTERRELATIONSHIPS

Work by Shockley as early as 1952 showed that there is a fixed relationship among six JFET parameters. The parameters so related are as follows:

- V_P Pinchoff voltage
- I_D Drain current at some value of $V_{GS} \neq 0$
- I_{DSS} Drain current at $V_{GS} = 0$
- V_{GS} Gate-to-source voltage
- g_{fs} Transconductance at $I_D \neq I_{DSS}$ and $V_{GS} \neq 0$
- g'_{fs} Transconductance at $I_D = I_{DSS}$ and $V_{GS} = 0$

Shockley demonstrated that

$$I_D = I_{DSS}\left(1 - \frac{V_{GS}}{V_P}\right)^2 \tag{1-11}$$

Equation (1-11) is true for diffused junction silicon JFETs and may be regarded as being accurate to a few percent. By differentiating Eq. (1-11) with respect to V_{GS}, it can be shown that

$$g_{fs} = \frac{\Delta I_D}{\Delta V_{GS}} = \frac{2I_{DSS}}{V_P}\left(1 - \frac{V_{GS}}{V_P}\right) \tag{1-12}$$

Since, by definition, $g'_{fs} = g_{fs}$ when $V_{GS} = 0$, it follows from Eq. (1-12) that

$$g'_{fs} = \frac{2I_{DSS}}{V_P} \tag{1-13}$$

For example, for a particular JFET, $I_{DSS} = 1$ mA and $V_P = 4$ V. What drain current will flow for a gate-to-source reverse bias of 2 V? What will be the transconductance under these conditions, and what is g'_{fs}?

From Eq. (1-11) the drain current will be

$$I_D = (10^{-3})\left(1 - \frac{2}{4}\right)^2 = \frac{10^{-3}}{4}\text{ A} = 0.25\text{ mA}$$

From Eq. (1-12) the transconductance will be

$$g_{fs} = \frac{(2)(10^{-3})}{4}\left(1 - \frac{2}{4}\right) = \frac{10^{-3}}{4} = (0.25)(10^{-3})\text{ mho} = 250\ \mu\text{mho}$$

From Eq. (1-13) the transconductance at $V_{GS} = 0$ will be

$$g'_{fs} = \frac{(2)(10^{-3})}{4} = (0.5)(10^{-3}) \text{ mho} = 500 \ \mu\text{mho}$$

Equations (1-11) to (1-13) assume that V_{DS} is constant as the other parameters vary. Even if this is not true, errors introduced due to variations of V_{DS} in the pinchoff region will not produce significant errors for high r_{ds} JFETs, since I_D in the pinchoff region is then relatively unaffected by changes in V_{DS}. (See Fig. 1-7*a*.)

Transconductance and output resistance are related by an expression that will be familiar to those who have worked with vacuum tubes, namely,

$$\mu = g_m r_{ds} \tag{1-14}$$

where μ is the amplification factor and $g_{fs} = g_m$.

Also, since $r_{ds} = 1/g_{os}$, Eq. (1-14) may be written

$$\mu = \frac{g_m}{g_{os}} \tag{1-15}$$

1-4 BIASING

Like the bipolar transistor, vacuum tube, or any other active element, it is unfortunately necessary to bias a JFET to some operating condition before it can be usefully employed as an amplifier. There are many ways of biasing a JFET, and each method has its own advantages and disadvantages. Sometimes those having the advantage of simplicity have other serious disadvantages. For example, consider the circuit shown in Fig. 1-12*a*, which shows a possible biasing arrangement for the 2N5163. The characteristics for this transistor were given in Fig. 1-7, and a list of its parameters was given in Table 1-2.

Being an *n*-channel device, it requires a positive supply voltage, so let us choose $V_{DD} = +10$ *V*. Since the gate and source are both grounded, the gate-to-source diode is off, no current flows through R_G, and $V_{GS} = 0$. With $V_{GS} = 0$ the drain current will be I_{DSS}, and if we wish to make $V_{DS} = +5$ *V*, R_D must drop 5 V. To find R_D we must divide 5 V by I_{DSS}. Referring to Table 1-2, we see that I_{DSS} for the 2N5163 can range from 1 to 40 mA. Thus R_D can range in value from 5 kΩ to 125 Ω, depending on the particular 2N5163 selected. The circuit has two other disadvantages. First, the drain current can be only I_{DSS}, and it may not be desirable to bias at this level. Second, there is no degeneration in the circuit to stabilize it against operating-point drift due to temperature.

The circuit shown in Fig. 1-12*b* is similar to that in Fig. 1-12*a* and has similar disadvantages. Its advantage is that the gate is now biased negatively, so that V_{GS} is not zero and I_D is not I_{DSS}. Selection of any desired operating point, consistent with the specifications of the 2N5163, is therefore possible. Once again, though, this circuit requires individual component selection and is temperature-sensitive. Also it requires both a positive and a negative supply voltage.

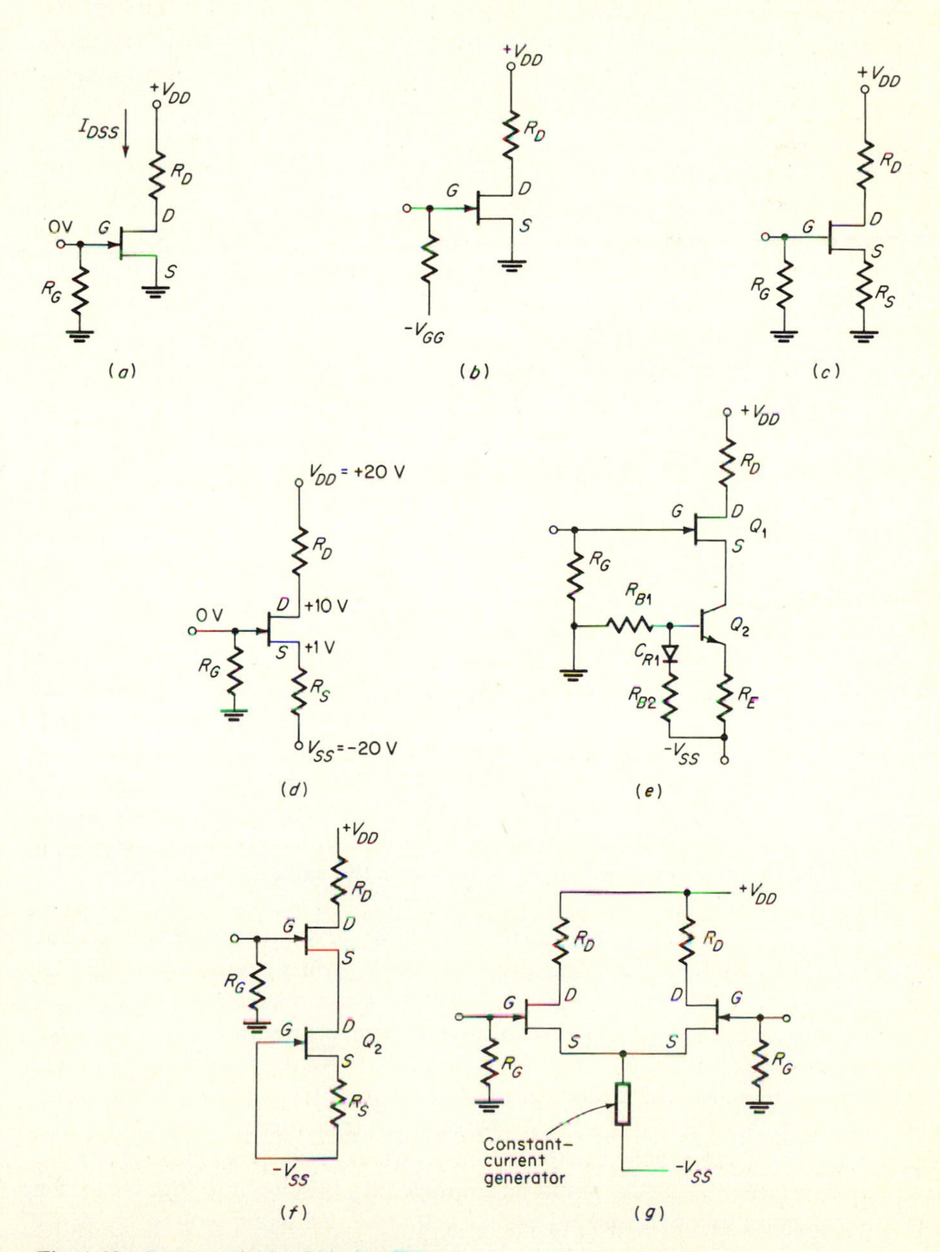

Fig. 1-12 Seven methods of biasing JFETs.

Circuit (*c*) of Fig. 1-12 is an improvement over circuits (*a*) and (*b*) in that it uses negative feedback. Using load-line techniques, R_D and R_S can be chosen to give an operating point that is less dependent on the individual JFET than in (*a*) and (*b*). The source resistor R_S provides degeneration in the same way that an emitter resistor provides feedback for a bipolar transistor circuit. It stablizes the operating point and the gain. R_S can be bypassed with a capacitor if desired, so that R_S retains its bias-stabilizing effect, but does not reduce the gain. Circuit (*c*) shares an advantage over (*b*) with (*a*); namely, only one polarity of power-supply voltage is required.

The circuit in Fig. 1-12*d* shows a biasing technique that is satisfactory for many purposes, especially if relatively high supply voltages are available. For example, suppose it is desired to make V_D with respect to ground $+10$ V, and to operate the JFET with a V_{GS} of -1 V. Assume ± 20-V supplies are available. These voltages, along with their correct polarities, appear in Fig. 1-12*d*. Notice that a V_{GS} of -1 V means that the gate is 1 V negative with respect to the source. Since the gate is at 0 V, the source must be at $+1$ V. This means that V_{DS} is $+9$ V. From the output characteristic for the 2N5163, which is given in Fig. 1-7*a*, I_D is ≈ 8 mA for $V_{GS} = -1$ V and $V_{DS} = +9$ V. The voltage drop across R_D is $20 - 10 = 10$ V so that

$$R_D = \frac{10 \text{ V}}{8 \text{ mA}} = 1.25 \text{ k}\Omega$$

The drop across R_S is 21 V so that

$$R_S = \frac{21 \text{ V}}{8 \text{ mA}} = 2.63 \text{ k}\Omega$$

$I_D = 8$ mA for $V_{GS} = -1$ V is a *design center* condition. Suppose now we plug a different 2N5163 into the circuit of Fig. 1-12*d* and find that V_{GS} is -2 V with this new transistor. How will this 100 percent difference in V_{GS} affect the $+10$ V level we have chosen for V_D if we keep the same resistance values as before?

With $V_{GS} = -2$ V, the drop across R_S will be 22 V (instead of 21), so that I_D will be $22/2.63 = 8.4$ mA. The drop across R_D will therefore be (1.25 kΩ) (8.4 mA) = 10.5 V. V_D will be $+20 - (+10.5 \text{ V}) = +9.5$ V.

This is a 5 percent change in V_D for a 100 percent change in V_{DS}. It illustrates the ability of the circuit of Fig. 1-12*d* to bias a JFET to a desired operating point, even though there may be a considerable spread in JFET parameters. For practical purposes, biasing to this accuracy is usually satisfactory. For critical circuits requiring very accurate gains and bias points, circuits with large amounts of both ac and dc negative feedback should be employed. In the circuit in Fig. 1-12*d*, the relative magnitudes of R_S and R_D are such that to realize any voltage gain, R_S would have to be bypassed. In this situation there would be no ac feedback to stabilize the gain, although the dc degeneration provided by R_S would be retained and would continue to stabilize the dc operating point.

Circuits suitable for high gain accuracy and operating-point stability are discussed in Chap. 4. Simple, single-transistor amplifiers such as that shown in Fig. 1-12*d* should be considered only for applications in which 10 or 20 percent accuracies are required. The source follower is an exception to this rule; it is discussed in Sec. 1-5-1.

The circuit in Fig. 1-12*e* is a variation of that in Fig. 1-12*d*. R_S has been replaced by a constant-current generator in the form of a bipolar transistor Q_2 and its associated temperature-compensated bias network. Action is similar to that for circuit 1-12*d* except that now I_D is forced to be constant at 8 mA, independent of V_{GS}, provided that Q_1 remains biased with the correct polarities. For different 2N5163 transistors, V_D will always be $+10$ V for a constant I_D of 8 mA. Q_1 will automatically adjust itself to provide the correct combination of V_{GS} and V_{DS} to give $I_D = 8$ mA. Again the source may be bypassed to give high ac gain. In an integrated circuit, Q_2 and its associated components would be replaced with a current mirror such as is discussed in Sec. 3-3-2.

The circuit in Fig. 1-12*f* is a variation on that in Fig. 1-12*e* in that a simpler constant-current generator has been incorporated. This biasing method has the disadvantage that the constant current generated will be a strong function of the particular JFET chosen for Q_2, owing to the large spread in JFET parameters. The bias levels for the circuit given in Fig. 1-12*e* will be much more predictable, owing to the smaller variation in V_{BE} for a bipolar transistor for a specified current level, compared with V_{GS} for a JFET.

With matched JFETs, the circuit in Fig. 1-12*g* can provide a very stable, predictable bias, especially if a good, predictable constant-current generator is used. In addition, no bypass capacitors are required to achieve voltage gain, and the circuit has a flat frequency response down to dc. The circuit of Fig. 1-12*g* is a differential amplifier, and an explanation of its properties is given in Chap. 3.

Matched JFET pairs are often used on the inputs of operational amplifiers (see Chap. 4) to provide the high input impedance, low input current, and low noise desirable in such amplifiers. Earlier difficulties in fabricating JFETs and bipolar transistors on the same chip have now been overcome. For example, the National Semiconductor LF100 is an operational amplifier with a JFET input and bipolar circuitry on the same chip, in which the JFET input devices are matched to produce voltage offsets of only 1 mV with a temperature coefficient of 5 μV/°C.

1-5 SMALL-SIGNAL JFET AMPLIFIERS

1-5-1 SINGLE-STAGE AMPLIFIERS

The JFET, like its predecessor the bipolar transistor, can be operated as an amplifier with any of its three terminals common to the input and output. The three modes of operation are illustrated in Fig. 1-13.

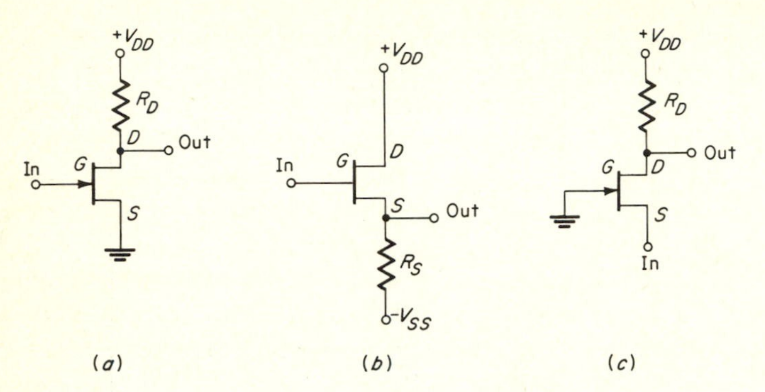

Fig. 1-13 Small-signal JFET amplifiers: (*a*) common source; (*b*) common drain or source follower; (*c*) common gate.

Several means are available by which these basic circuits may be analyzed. The selection of the most suitable means is part of the art of circuit design. The selection process involves eliminating from the statement of the problem those factors that will have a negligible effect. For example, in dc or low-frequency amplifiers it is common to ignore the effects of circuit capacitances. Also, certain impedances are often large compared with others, and this results in the further elimination of components in the circuit to be analyzed.

To illustrate this point, we shall find the gain of the common-source amplifier shown in Fig. 1-13*a* by two methods and compare the results numerically. First, we shall use the equivalent circuit for the JFET that is introduced in Fig. 1-14.

The gate, source, and drain terminals are interconnected by a network consisting of resistors, capacitors, and a current generator. For many types of analysis, the circuit shown can lead to considerable complexity, even though it is itself a very much simplified model of a real JFET. For example, no noise generators are shown, capacitors are assumed to be lumped and fixed, and the effects of nonlinearities are all ignored. Components may be identified as follows: r_{in} is the resistance of the reverse-biased gate-to-source diode. C_{GS} is the gate-to-source capacitance, which, with C_{GD} and the circuit gain, will determine the input capacitance of the amplifier. On a data sheet the JFET output resistance r_{ds} is

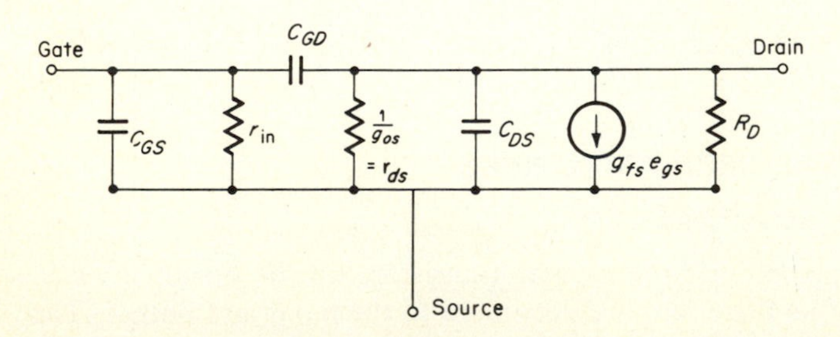

Fig. 1-14 JFET equivalent circuit.

frequently expressed as the reciprocal of the output conductance, that is, as $1/g_{os}$. In either case it can be obtained from the slope of the output characteristic, such as that in Fig. 1-7*a*. C_{DS} is the drain-to-source capacitance and is small enough to be considered nonexistent for all practical purposes. R_D is the external load resistor. Notice that in the small-signal or ac equivalent circuit, R_D is connected between the drain and the supply voltage V_{DD}, as in Fig. 1-13*a*. Since V_{DD} is assumed to be ideal, that is, a zero-impedance supply, and since an ideal battery is an ac short circuit, V_{DD} is an ac ground. The source terminal is both an ac and a dc ground, so in the ac circuit V_{DD} and ground are the same. Finally, there is an output current generator, which generates a current equal to the transconductance times the gate-to-source voltage or $g_{fs}e_{gs}$.

Let us now find the voltage gain of the circuit in Fig. 1-13*a*, making the following assumptions. The transistor is a 2N5163 with the gate biased to ground. V_{DD} is $+15$ V, and V_{DS} is $+10$ V. The voltage applied to the gate is from a zero-impedance generator, and the operating frequency is sufficiently low so that capacity effects can be neglected.

First we must find R_D. From the output characteristic in Fig. 1-7*a*, we find that with $V_{DS} = +10$ V and $V_{GS} = 0$, $I_D = 14$ mA. Since the voltage drop across R_D is 5 V,

$$R_D = \frac{5\text{ V}}{14\text{ mA}} \approx 360\ \Omega$$

Redrawing the equivalent circuit to retain only the essentials, we obtain the circuit in Fig. 1-15. In this circuit we can equate the current from the current generator to the sum of those through $1/g_{os}$ and R_D:

$$g_{fs}e_{gs} = i_1 + i_2 = \frac{-e_{\text{out}}}{1/g_{os}} + \frac{-e_{\text{out}}}{R_D} \tag{1-16}$$

Equation (1-16) can be solved for e_{out}/e_{gs}, giving

$$\text{Gain} = \frac{e_{\text{out}}}{e_{gs}} = \frac{-g_{fs}R_D}{1 + g_{os}R_D} \tag{1-17}$$

From Table 1-2, g_{fs} is 6000 μmho at $V_{DS} = +15$ V and $V_{GS} = 0$. We are operating at $V_{DS} = +10$ V and $V_{GS} = 0$, so we must estimate a value to use for g_{fs} in

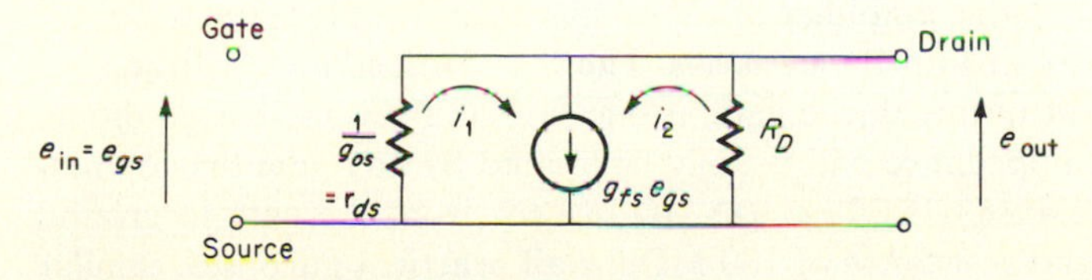

Fig. 1-15 JFET simplified equivalent circuit.

our example. In Eq. (1-12), which is an expression for g_{fs}, notice that V_{DS} does not appear. This implies that g_{fs} is independent of V_{DS}, and to a large extent this is true. Evidence of this can be seen in the output characteristic in Fig. 1-7*a*. Recalling that g_{fs} is the ratio of the change in I_D to the change in V_{GS} producing it, notice that in the region where $V_{GS} = 0$, the same change in V_{GS} produces about the same change in I_D at both $V_{DS} = +15$ V and $V_{DS} = +10$ V. We shall therefore assume a value for g_{fs} of 6000 μmho.

g_{os} is given by the manufacturer as $Y_{os} = 60$ μmho. We have already found R_D to be 360 Ω, and so, from Eq. (1-17), the gain is

$$\frac{e_{\text{out}}}{e_{gs}} = \frac{-(6000)(10^{-6})(360)}{1 + (60)(10^{-6})(360)} = -2.11$$

We can obtain approximately the same answer by an even more simplified approach. The transconductance g_{fs} of a JFET is the ratio of the change in output or drain current to the change in input or gate-to-source voltage producing it, with V_{DS} remaining constant. g_{fs} is 6000 μmho in this case, and let the input change by 1 V. V_{DS} will not be constant, but since we are in the pinchoff region, its effect will be small. So we can write, from the definition of g_{fs}:

$$\Delta I_D = g_{fs}\Delta V_{GS} = (6000)(10^{-6})(1) = 6 \text{ mA} \tag{1-18}$$

Since we are operating in the pinchoff region, the JFET acts like a constant-current generator, which means that the load resistor will have a negligible effect on the drain current. The output voltage change is therefore

$$-(\Delta I_D)(R_D) = -(6000)(10^{-6})(360) = -2.16 \text{ V} \tag{1-19}$$

Since the input is 1 V and the output is −2.16 V, the gain must be −2.16. Notice that what we have done in this second calculation is to neglect the $g_{os}R_D$ term in Eq. (1-17). It turns out that this term is frequently small compared with unity and can therefore often be omitted from Eq. (1-17).

The above example illustrates that with some experience and careful reading of a data sheet, quick calculations can be made that are usually accurate enough for most purposes. If a very accurate value for gain is required, the circuit in Fig. 1-13*a* should not be used anyway because of the large variation found in JFET parameters from transistor to transistor.

For the common-source amplifier of Fig. 1-13*a*, some simple deductions can be made about the circuit impedance levels. The input impedance will be very large because the input terminal, the gate, is one side of a reverse-biased diode. In practice the input impedance will usually be limited by any gate-biasing network used. For the 2N5163 JFET, a 100-MΩ resistor from the gate to ground would result in an input resistance of 100 MΩ for all practical purposes. Similar conclusions are drawn for the circuit output impedance. R_D is in parallel with the JFET output impedance which, since it is the impedance of a current generator, is high. For one particular set of conditions g_{os} was 60 μmho, so that $r_{ds} = 1/g_{os} = 16.7$ kΩ. Since 16.7 kΩ is large compared with $R_D = 360\ \Omega$, we can say that the circuit output resistance is about 360 Ω.

The common drain or source-follower circuit of Fig. 1-13*b* can be analyzed by similar means. Again the input impedance will be large, being limited in practice by the value of any gate bias resistor. An expression for the circuit gain can also be obtained by observing that the source follower is a feedback version of the common-source amplifier of Fig. 1-13*a*. In the latter circuit the input is applied between gate and ground, which is also connected to the source. There is therefore no feedback. In the source-follower circuit of Fig. 1-13*b*, the input is again applied from the gate to ground. Notice that V_{SS} is an ac ground, and that the voltage across R_S is applied in series with the gate-to-source voltage, so that

$$\Delta V_G = \Delta V_{GS} + \Delta V_{R_S} \qquad (1\text{-}20)$$

This means that 100 percent of the output voltage (the voltage across R_S) is fed back in series with the input. The expression for the gain of an amplifier with feedback is

$$A_F = \frac{A_o}{1 + \beta A_o} \qquad (1\text{-}21)$$

where A_F = gain with feedback
A_o = gain with no feedback = $g_{fs} R_S$
β = fraction of the output fed back = 1

Thus, by substitution into Eq. (1-21), the gain of the source follower is

$$A_F = \frac{g_{fs} R_S}{1 + g_{fs} R_S} \qquad (1\text{-}22)$$

The same answer can be obtained by using standard circuit analysis on the equivalent circuit in Fig. 1-14.

A close approximation to the circuit output resistance R_o can be obtained as follows: R_o is the resistance from the source or output terminal to ground and is R_S in parallel with the JFET output resistance. The shunting effect of R_S must clearly be included in R_o, since it is connected from the output to V_{SS}, an ac ground. To simplify the explanation, we shall assume for the moment that R_S is very large so that it has a negligible effect. The little effect it does have can be accounted for, as we shall show later. Now ground the gate, and let the source change by 1 V. This is a change in V_{GS} of 1 V, which will produce a channel-current change of $g_{fs}V_{GS}$ by the definition of transconductance [Eq. (1-5)]. Since R_S is large, virtually all this current will flow into the output so that the output-current change is $(g_{fs})(1) = g_{fs}$. The output impedance is the output-voltage change divided by the output-current change, or $1/g_{fs}$. This is the output impedance ignoring R_S. To account for R_S, we must add R_S in parallel, which gives

$$R_o = \frac{R_S(1/g_{fs})}{R_S + 1/g_{fs}} = \frac{R_S}{1 + g_{fs}R_S} \tag{1-23}$$

To obtain some idea of the magnitude of the gain and output impedance of a source follower, let us assume that the JFET is a 2N5163 operated at $V_{GS} = 0$, $I_D = 14$ mA, and $g_{fs} = 6000$ μmho, with $V_{DD} = +15$ V and $V_{SS} = -15$ V. The source resistor is

$$R_S = \frac{V_S - V_{SS}}{I_D} = \frac{0 - (-15)}{(14)(10^{-3})} = 1.07\ \text{k}\Omega$$

From Eq. (1-22) the gain is

$$A_F = \frac{g_{fs}R_S}{1 + g_{fs}R_S} = +0.87$$

From Eq. (1-23) the output impedance is

$$R_o = \frac{R_S}{1 + g_{fs}R_S} = \frac{1070}{1 + (6000)(10^{-6})(1070)} = 144\ \Omega$$

Notice that the gain is fairly close to unity, as one would expect from a follower type of circuit. The output impedance is fairly low and results from the use of a high-transconductance JFET. The output impedance is in fact approximately equal to the reciprocal of the transconductance as indicated earlier, if the shunting effect of R_S is neglected. In this case $1/g_{fs}$ is $1/(6000)(10^{-6})$ or 167 Ω, which is close to the 143 Ω obtained accounting for R_S.

The gain, input impedance, and output impedance for the common-gate amplifier (Fig. 1-16) are most easily obtained from the formulas already derived for the other two configurations. For example, with the gate grounded and 1 V applied to the source, which is the input terminal for the grounded-gate circuit, the drain current will change by g_{fs}. This change will flow through R_D, producing a voltage change of $g_{fs}R_D$. The circuit gain is therefore $g_{fs}R_D$, which is

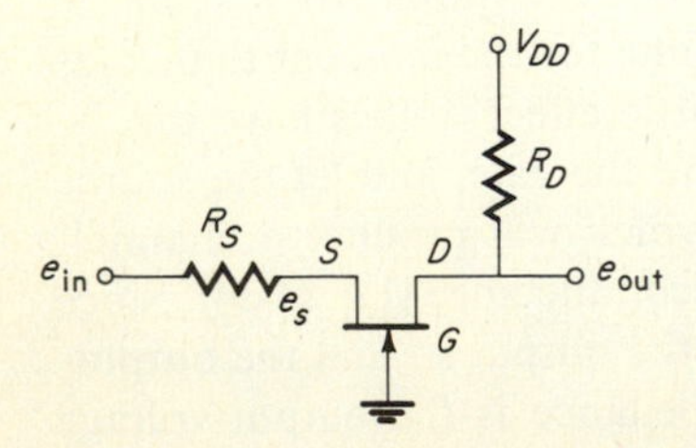

Fig. 1-16 Common-gate amplifier.

the same in magnitude as for the common-source amplifier in Fig. 1-13*a*. However, the gain is positive for the common-gate amplifier, meaning input and output in phase, whereas it was negative, meaning input and output 180° out of phase, for the common-source circuit. The opposite phase arises because +1-V change applied to the gate with the source grounded will cause an increase in drain current, for an *n*-channel device (see Fig. 1-7*a*), whereas a +1-V change applied to the source with the gate grounded will cause a decrease in drain current.

The input impedance seen looking into the source is the same as the output impedance of the source follower would be with R_S absent, or $1/g_{fs}$. This means that the common-gate amplifier has a much lower input impedance than either the common-source amplifier or the source follower. The impedance of the generator driving the amplifier will be much more significant, therefore, when the common-gate amplifier is used. For example, find the gain of the circuit in Fig. 1-16. Assume $g_{fs} = 6000\ \mu$mho, $R_D = 360\ \Omega$, and $R_S = 167\ \Omega$.

First find e_s in terms of e_{in}. The impedance looking into the source, the impedance seen by e_s, is the input impedance of the common-gate amplifier, or

$$\frac{1}{g_{fs}} = \frac{1}{(6000)(10^{-6})} = 167\ \Omega$$

R_S and $1/g_{fs}$ form a potential divider, and since they are both 167 Ω in this example,

$$e_s = \frac{e_{in}}{2} \tag{1-24}$$

The gain of a JFET from source to drain is $g_{fs} R_D$ so that

$$\frac{e_{out}}{e_s} = g_{fs} R_D \tag{1-25}$$

Eliminating e_s from Eqs. (1-24) and (1-25) gives the overall circuit gain:

$$\frac{e_{out}}{e_{in}} = \frac{g_{fs} R_D}{2} = \frac{(6000)(10^{-6})(360)}{2} = +1.08$$

If the common-source amplifier in Fig. 1-13*a* had been used with a 100-MΩ input-gate bias resistor, the effect of the 167-Ω generator impedance would have been negligible. The gain would have been −2.16 as in Eq. (1-19.)

By reasoning similar to that for the common-source amplifier, the circuit output impedance is approximately equal to R_D.

Finally it is interesting to note that the common-gate amplifier has a current gain of approximately unity, whereas the gain for the other two configurations approaches infinity at low frequencies.

1-5-2 EFFECT OF CIRCUIT CAPACITANCES ON THE DIFFERENT CONFIGURATIONS

If a signal generator is driving a capacitor with one side of the capacitor grounded, the effective capacitance is simply the value of the capacitor. This situation is illustrated in Fig. 1-17*a*.

If the capacitor is not grounded but is returned to a terminal at which there exists a voltage related to the signal voltage, the effective capacitance will no longer be C. This situation is illustrated in Fig. 1-17*b*.

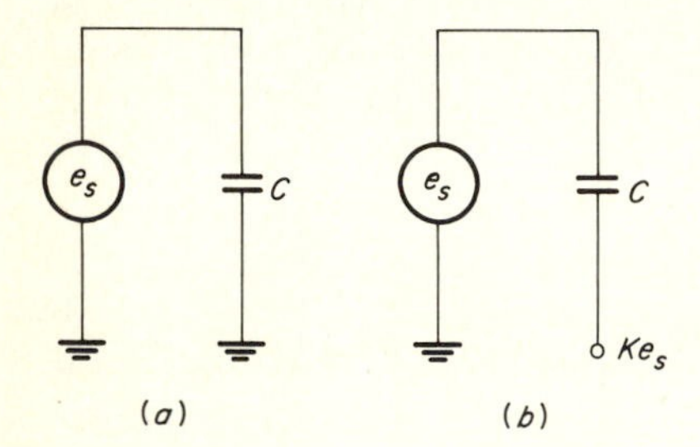

Fig. 1-17 Capacity multiplication.

In circuit (*a*) the current delivered by the generator is e_s/X_C. In circuit (*b*) the voltage across capacitor C is $e_s - Ke_s = (1 - K)e_s$. The current through C is therefore $(1 - K)e_s/X_C$ by Ohm's law. This is $1 - K$ times the current flowing in circuit (*a*) for the same e_s. It is the same current that would be obtained with a grounded capacitor of value $(1 - K)C$. The effective capacitance for circuit (*b*) is therefore

$$C_{\text{eff}} = (1 - K)C \tag{1-26}$$

In a JFET amplifier some or all of the interelectrode capacitances are operated with varying voltages on both sides. Figure 1-18 illustrates the position of the gate-to-source, gate-to-drain, and drain-to-source capacitances in the three amplifier configurations. C_{DS} is usually negligible.

In the common-source circuit of Fig. 1-18, the effective input capacitance, seen looking into the gate, has two major components. The first is C_{GS}, which is grounded so that its effective value is not modified in any way. The second component is related to C_{GD}, which is returned to the drain. The drain potential varies in response to a gate signal by an amount dependent on the gain of the amplifier. K in Eq. (1-26) is therefore the gain of the amplifier, which in this case is a negative number. The common-source input capacitance is therefore

$$C_{CS} = C_{GS} + (1 - K)C_{GD} \tag{1-27}$$

If $C_{GS} = C_{GD} = 5$ pF and $K = -10$,

$$C_{CS} = 5 + [1 - (-10)]5 = 60 \text{ pF}$$

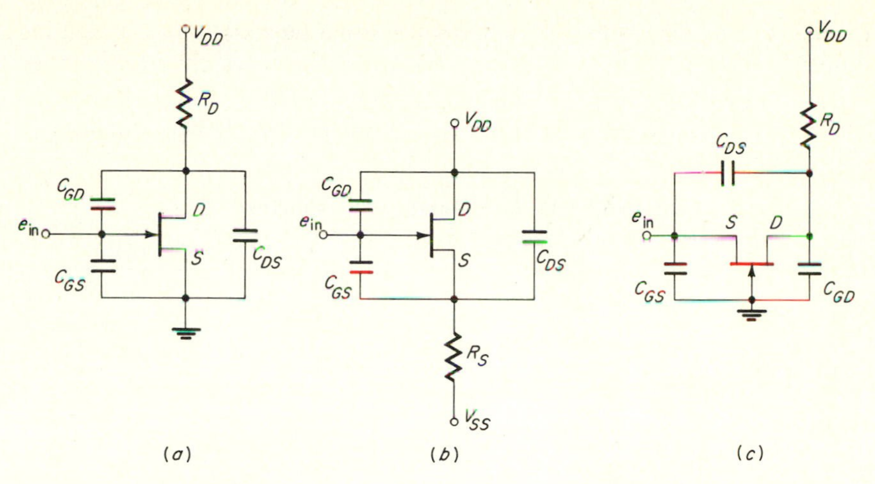

Fig. 1-18 Circuit capacitances in the three amplifier configurations: (*a*) common source; (*b*) common drain or source follower; (*c*) common gate.

C_{DS} is very small compared with C_{GD} and C_{GS} and has negligible effect on the circuit input capacitances. It has therefore been neglected.

For the common-drain or source-follower circuit in Fig. 1-18*b*, C_{GD} is ac grounded, and it is C_{GS} which is modified in value. C_{GS} is connected to the source so that K in Eq. (1-26) is the gain of the source follower. This is a positive number, usually close to unity. The effective circuit input capacitance is therefore

$$C_{CD} = C_{GD} + (1 - K)C_{GS} \tag{1-28}$$

If $C_{GS} = C_{GD} = 5$ pF and $K = +0.95$,

$$C_{CD} = 5 + [1 - (+0.95)]5 = 5.25 \text{ pF} \tag{1-29}$$

Again C_{DS} has a negligible effect.

The common-gate amplifier is more complicated to analyze than the other two configurations. The input terminal, the source, has a relatively low impedance to ground, the input resistance being about $1/g_{fs}$. In the two cases considered previously, the input terminal was the gate, resulting in a very large input resistance. The common-gate amplifier is the least used circuit, although one application is given in Sec. 1-5-4, in which the *cascode* amplifier is described. As a first approximation the input capacitance of the grounded-gate amplifier is C_{GS}, since C_{DS} is very much smaller than C_{GS}.

Which of the three configurations is superior depends entirely on the application. The source follower has the highest input impedance and the lowest output impedance. It is therefore useful as a unity-gain impedance transformer when the driving impedance is high. For applications requiring voltage gain, either the common-source or the common-gate amplifier could be used; but if a high input impedance is required as well, only the common-source circuit would be suitable.

A collection of circuits and formulas for single-stage JFET amplifiers is given in Table 1-3.

TABLE 1-3 Circuits and formulas for single-stage JFET amplifiers

	CS	CD	CG
Circuit	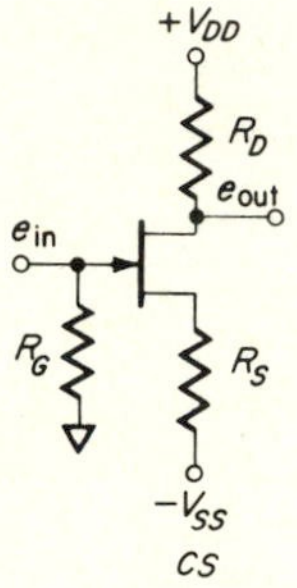 CS	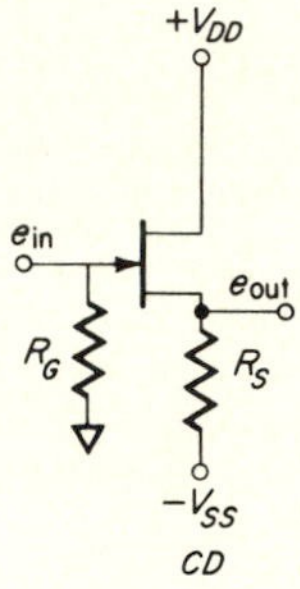 CD	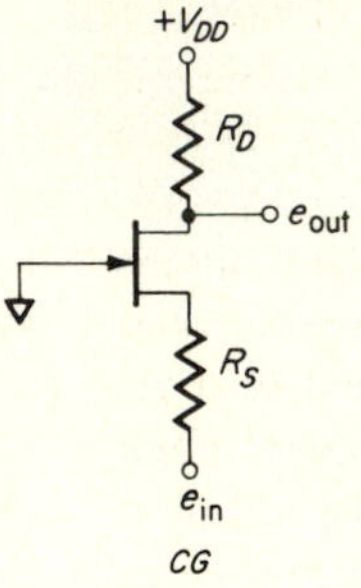 CG
Voltage gain	$\dfrac{-g_{fs}R_D{}^*}{1 + g_{fs}Rs}$	$\dfrac{+g_{fs}R_S}{1 + g_{fs}R_S}$	$\dfrac{+g_{fs}R_D{}^\dagger}{1 + g_{fs}R_S}$
Current gain	∞	∞	Unity
Input resistance	R_G	R_G	$Rs + \dfrac{1}{g_{fs}}$
Output resistance	R_D	$\dfrac{Rs}{1 + g_{fs}R_S}$	R_D

* This reduces to $-g_{fs}R_D$ if $R_S = 0$ or is bypassed.
† This reduces to $+g_{fs}R_D$ if $R_S = 0$ or is bypassed.

1-5-3 JFETs AS FM-RECEIVER FRONT ENDS

At low and medium frequencies the high input impedance of the JFET is a desirable and much used feature from the point of view of circuit design. At high frequencies, however, the input impedance is no longer high because the reactive component is dominant. For example, a JFET circuit might have an input resistance of 10 MΩ and an input capacitance of 10 pF. At 100 Hz, 10 pF has a reactance of about 160 MΩ, so that the input impedance is determined mainly by the 10-MΩ resistive component. At 100 MHz, 10 pF has a reactance of about 160 Ω, so that at this frequency the capacitance would primarily determine the input impedance. Even though the capacity can be tuned out by a parallel inductor to raise the input impedance, the impedance can never be as high at high frequencies as it is at dc and low frequencies, because of Q limitations.

Although JFETs do not have any particular advantage at high frequencies over other devices as far as input impedance is concerned, they do nevertheless possess another advantage. It results from the fact that a square-law relationship exists between the drain current and the gate-to-source voltage. [See Sec. 1-3-3 and Eq. (1-11).] This relationship holds true to a few percent. In contrast, vacuum tubes have a three-halves ($\frac{3}{2}$) power characteristic, and bipolar transsistors have a diode-type characteristic.

The effect of applying a signal of some frequency to the input of a device having a square-law characteristic (such as a JFET) is to produce an output with three components. These are a dc component, an ac component at the same frequency as the input, and an ac component at twice the input frequency. If a tuned RF amplifier is used, the dc and double-frequency components can be eliminated, and only signals having the same frequency as the input appear in the output.

A vacuum tube or bipolar transistor will have in its output, in response to a single-frequency input, a multitude of frequencies because it is not a square-law device. Many of these frequencies are sufficiently close to the input frequency so that they cannot be tuned out by the receiver's tuned circuits, resulting in cross-modulation distortion. This becomes important when an FM receiver is tuned to a very weak station (10 μV/m at the antenna, for example), when the adjacent channel is very strong (1 V/m, for example). If the high-level input is distorted by the introduction of harmonics, some of these harmonic frequencies will mix with the weak signals to which the receiver is tuned. The tuner will, of course, greatly attenuate the strong signal because the receiver is tuned to the weak signal. But the strong signal is 100,000 times stronger than the weak signal, so some of it is bound to get through. The effect in the receiver audio section is audible distortion. In a JEFT-tuned RF amplifier the dc and second harmonic components are sufficiently far away in frequency not to influence an adjacent channel. The JFET circuit therefore places fewer demands on the filtering ability of the tuned elements in the amplifier. This is a primary reason why JFETs are now so popular as front ends for FM receivers of high quality.

Compared with bipolar transistors at RF, the JFET has similar noise performance but has lower conversion gain as a mixer.

1-5-4 JFET-BIPOLAR COMBINATIONS

JFETs and bipolars each have their own peculiar advantages, and circuits are possible which use these devices in combination to achieve better performance than can be obtained with either device alone. The JFET has, in the common-source and common-drain configurations, a very high current gain, which for many applications can be assumed to be infinite. The bipolar transistor has a much larger voltage gain than the JFET for a given supply voltage. This results from the high transconductance of the bipolar transistor which approaches 40,000 μmho with about 1 mA of collector current. From Eq. (1-13) the transconductance of a JFET is $2I_{DSS}/V_P$ with $V_{GS} = 0$. If I_{DSS} is 1 mA and V_P is 2 V, the transconductance will be 1000 μmho. Since the voltage gain for either a bipolar transistor or a JFET is roughly $g_m R_L$, the bipolar would have 40 times the voltage gain of the JFET.

Since there are three bipolar transistor configurations and three JFET configurations, a large number of ways of interconnecting pairs exists. However, since the JFET is almost invariably used as the input device, we can immediately eliminate half the possibilities. Also, input stages using the common-gate configuration are almost never used in JFET-bipolar pair circuits because of their low input impedance. This leaves us with two possible input stages, each of which can be combined with a bipolar transistor in three ways. In this section we shall limit the discussion to the most useful JFET-bipolar combinations.

Figure 1-19 is the basic circuit diagram of a common-source–common-base amplifier, also known as a *cascode* circuit. A simplified analysis of this circuit will serve to demonstrate its main advantages, namely high voltage gain and low input and feedback capacitances. These conflict in a single-stage amplifier because of the Miller effect. [See Eq. (1-27).]

The voltage gain of the first stage is approximately the transconductance times the load impedance. [See the paragraph following Eq. (1-19).] For all practical purposes the load impedance is zero, because it is the emitter of a bipolar transistor with its base gounded. (The impedance at this point is typically 20 to 30 Ω.) The voltage gain of the JFET is therefore essentially zero. Referring to Eq. (1-27), which is

$$C_{\text{in}} = C_{GS} + (1 - K)C_{GD}$$

notice that K is zero. If $C_{GS} = C_{GD} = 5$ pF,

$$C_{\text{in}} = 10 \text{ pF}$$

If Q_1 has a transconductance of 1000 μmho, a 1-V signal on the gate will produce a drain-current change of $g_{fs}v_{gs}$ or 1 mA. Q_2 is a common-base amplifier, which has a current gain of approximately unity, so the collector-current change of Q_2 will also be 1 mA. If R_L is 10 kΩ, the output-voltage change will be 10 V so that the voltage gain is 10. If the bipolar transistor and R_L were replaced by a 10-kΩ resistor alone, the voltage gain would still be 10. However, the input capacitance would be, from Eq. (1-27),

$$C_{\text{in}} = 5 + [1 - (-10)]\,5 = 60 \text{ pF}$$

This is considerably higher than for the cascode circuit.

For some RF applications it is often important to know the value of effective feedback capacitance from output to input, since this affects whether to apply neutralization, or how much to use for purposes of stability. In the cascode circuit this capacity is that from collector to gate, C_{CG} in Fig. 1-19. Since there is no direct connection from collector to gate, C_{CG} is very low. C_{CG} can be shown to be approximately equal to C_{GD} divided by the voltage gain of the bipolar transistor. With $C_{GD} = 5$ pF and a bipolar gain of $g_m R_L = (40{,}000)(10^{-6})(10^4) = 400$, C_{CG} would be about $\frac{5}{400}$ pF. With such a low value the output would have virtually no effect on the input; the cascode circuit is therefore said to be *unilaterized*, and no neutralization is required.

Table 1-4 gives circuits and formulas for the more important JFET-bipolar pairs. These are ac circuits, and for purposes of clarity and simplicity, dc biasing requirements have not necessarily been shown.

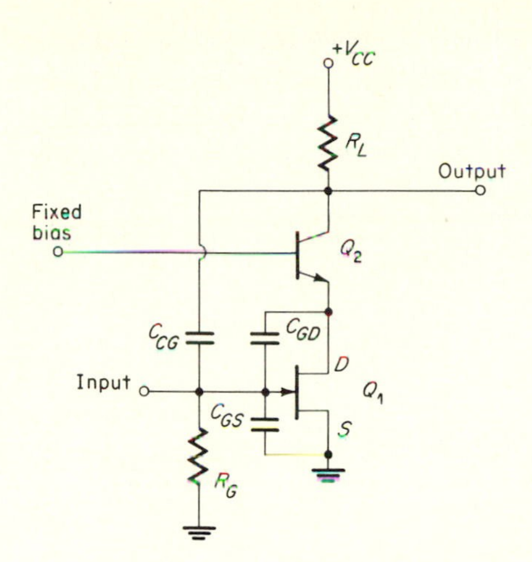

Fig. 1-19 A JFET/bipolar cascode circuit.

Circuits (*a*) and (*b*) in Table 1-4 are both versions of the cascode circuit. In circuit (*a*) all the drain current from the JFET flows into the emitter of the bipolar transistor because there is nowhere else for it to go. In circuit (*b*) virtually all the JFET ac drain current will flow into the bipolar transistor emitter if R is much larger than the emitter impedance. Thus the performances of the two circuits are similar although (*b*) does have two advantages over (*a*). For a given supply voltage a larger output swing is possible with circuit (*b*), and the quiescent drain and collector currents of the two transistors do not have to be equal as they do in circuit (*a*). This permits optimum biasing of each device, which may require different currents. Circuits (*a*) and (*b*) are "unfolded" and "folded" versions of basically the same circuit. Similar variations exist for circuits (*c*) through (*f*), but to avoid repetition they are not shown.

In all the circuits in Table 1-4 the JFET is used as the input stage to take advantage of its high input impedance. Thus for all six circuits, the input impedance is essentially the gate bias resistor shunted by the circuit input capacitance.

The output impedance for circuits (*a*) through (*e*) is approximately R_L, although for greater accuracy other effects such as the bipolar transistor output resistance and capacitance must be taken into account. For circuit (*f*) the output impedance is R_L/h_{fe} by emitter-follower action.

The voltage gains of the cascode circuits in (*a*) and (*b*) have already been discussed. We shall now briefly review how the voltage gains of the remaining four circuits in Table 1-4 may be derived.

In circuit (*c*) a 1-V input change will produce a source-current change of g_{fs} from the definition of transconductance. Since the bipolar transistor is operated grounded-base, the impedance at its emitter will be much less than R, if R has a reasonable value such as 1 kΩ or more. Notice that since the emitter impedance is nearly zero, the JFET is operated grounded-source. Thus, practically all the source-current change will flow into the emitter, and $\Delta I_S = \Delta I_E$. For a grounded-base bipolar transistor the collector-to-emitter current gain is about unity, and so $\Delta I_C = \Delta I_E = \Delta I_S = g_{fs}$. The output voltage $\Delta I_C R_L$ therefore equals $g_{fs} R_L$, and this divided by the 1-V input gives the voltage gain of $g_{fs} R_L$. Neither the common-drain nor the common-base amplifier has a phase inversion, so

TABLE 1-4 Circuits and formulas for JFET/bipolar combinations

Circuit	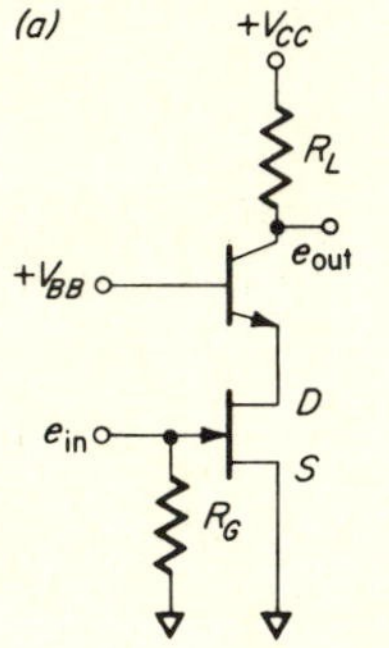 (a) Cascode CS–CB (a)	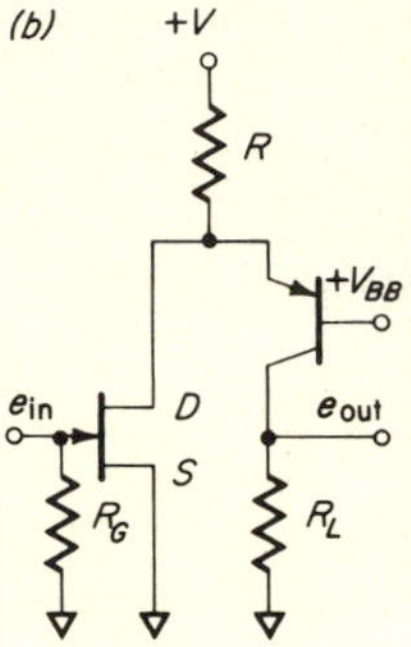 (b) Folded cascode CS–CB (b)	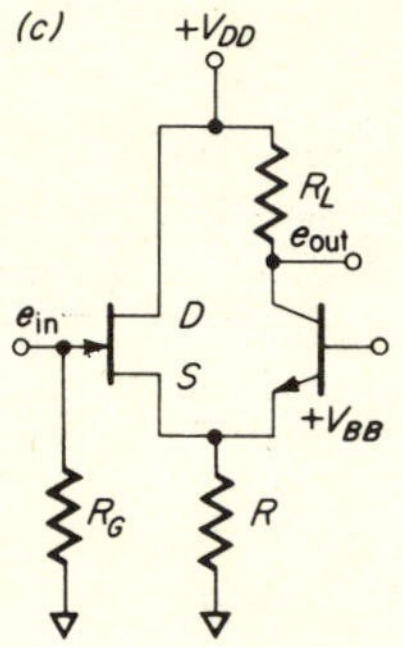 (c) CD–CB (c)
Input resistance	R_G	R_G	R_G
Output resistance	R_L	R_L	R_L
Voltage gain	$-g_{fs}R_L$	$-g_{fs}R_L$	$+g_{fs}R_L$
Circuit	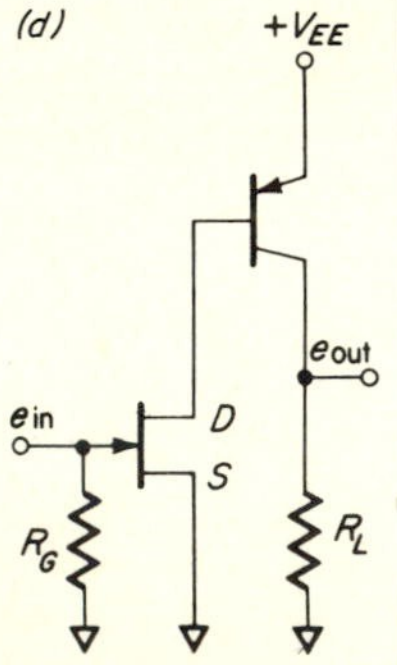 (d) CS–CE (d)	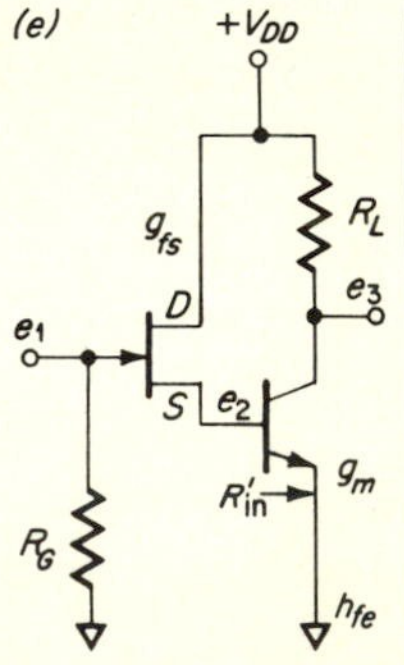 (e) CD–CE (e)	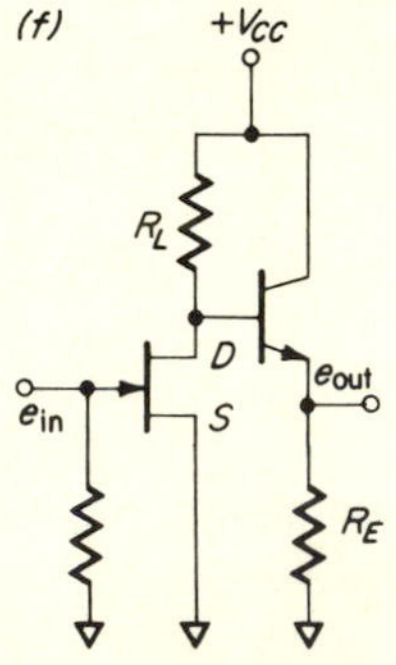 (f) CS–CC (f)
Input resistance	R_G	R_G	R_G
Output resistance	R_L	R_L	R_L/h_{fe}
Voltage gain	$+h_{fe}g_{fs}R_L$	$\dfrac{-h_{fe}g_{fs}R_L}{1 + g_{fs}h_{fe}/g_m}$	$-g_{fs}R_L$

neither does the combination. Hence, the gain is positive, in contrast to that for circuits (*a*) and (*b*).

In circuit (*d*) a 1-V input change again produces a drain-current change of g_{fs}. This becomes the base current of the bipolar transistor, the collector current of which is h_{fe} times the base current, or $h_{fe} g_{fs}$. The output voltage is the collector-current change times the collector load or $h_{fe} g_{fs} R_L$, and this is also the gain since the input is 1 V. Both common-source and common-emitter amplifiers invert, so the net result is that the overall circuit does not invert. The gain is therefore positive.

Circuit (*e*) is slightly more complicated to analyze than the others. First, let g_{fs} and g_m be the transconductances of the JFET and the bipolar transistor, respectively. For analysis we shall need to know the input impedance R_{in} of the bipolar stage. Input impedance is the change in input voltage divided by the change in input current it produces. It may be found for a device characterized by a current gain and a transconductance, as is a bipolar transistor, in the following manner:

$$h_{fe} = \frac{\Delta I_{\text{out}}}{\Delta I_{\text{in}}} \qquad g_m = \frac{\Delta I_{\text{out}}}{\Delta E_{\text{in}}}$$

Therefore,

$$\frac{h_{fe}}{g_m} = \frac{\Delta E_{\text{in}}}{\Delta I_{\text{in}}} = R_{\text{in}}$$

In passing, we may note that this is the way in which the predominant part of the input resistance of the hybrid-π equivalent circuit for a bipolar transistor is specified. (See Appendix.)

R'_{in} is the source impedance for the JFET, which is a source follower. From Table 1-3 its gain is

$$\frac{e_2}{e_1} = \frac{g_{fs} R'_{\text{in}}}{1 + g_{fs} R'_{\text{in}}} = \frac{g_{fs} h_{fe}/g_m}{1 + (g_{fs} h_{fe}/g_m)}$$

The voltage gain of the bipolar stage is $e_3/e_2 = -g_m R_L$. Thus the circuit gain e_3/e_1 may be found as follows:

$$\frac{e_3}{e_1} = \frac{e_3 e_2}{e_2 e_1} = \frac{-g_m R_L g_{fs} h_{fe}/g_m}{1 + (g_{fs} h_{fe}/g_m)} = \frac{-h_{fe} g_{fs} R_L}{1 + (g_{fs} h_{fe}/g_m)}$$

The minus sign indicates that the circuit inverts, which is to be expected since the common-drain circuit does not invert while the common-emitter circuit does.

For circuit (*f*) the JFET stage has a gain of $-g_{fs} R_L$, and the emitter-follower output stage has a gain of about $+1$. The overall circuit gain is therefore $-g_{fs} R_L$.

1-5-5 NOISE CONSIDERATIONS IN USING JFETS

All amplifying devices, including JFETs, add noise to an input signal so that the output is a combination of the signal and the noise. The significance of noise is that it determines the minimum detectable signal, that is, the lower limit of the dynamic range of the amplifier. Drift, which is very-low frequency noise, is important in dc and low-frequency amplifiers. Higher-frequency random noise is important in ac amplifiers.

Figure 1-20 shows the general shape of the noise spectrum for a typical low-noise JFET. It indicates that the noise performance of a JFET is a function of the frequency or band of frequencies over which the device is operated. At fre-

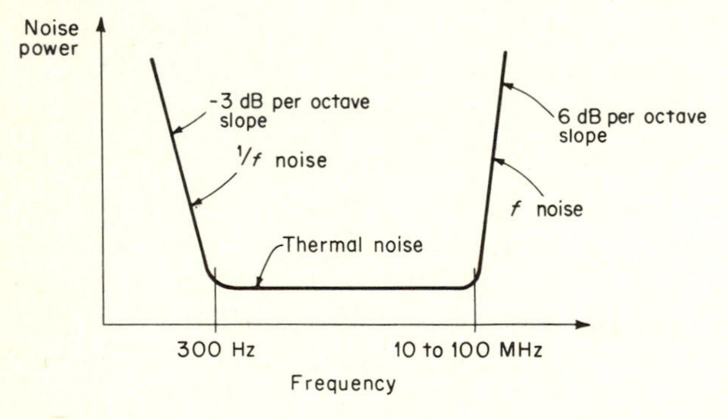

Fig. 1-20 JFET noise spectrum.

quencies below a few hundred hertz, the noise power is approximately inversely proportional to frequency, giving rise to the descriptive term "$1/f$" noise, also called "flicker" noise. The slope of the curve in this region is about -3 dB per octave. At frequencies above about 10 to 100 MHz, the noise increases with frequency at a rate of about 6 dB per octave, giving rise to the descriptive term "f" noise. In between these two extremes is a region where the noise is independent of frequency. In this region the noise is referred to as "thermal," "white," or "Johnson" noise.

JFETs are particularly well suited for low-noise amplification from high source impedances. Their superiority over bipolar transistors in this respect is especially marked at low frequencies since the $1/f$ corner, where the $1/f$ region meets the thermal region, occurs at a much lower frequency for JFETs. For dc amplification from high source impedances, the JFET has the advantage that its input current is that of a reverse-biased diode alone, whereas a bipolar-transistor input current consists of a base bias current in addition to a diode reverse current. As with any amplifier the input stage is usually the most important from the point of view of low-noise design. It is therefore common to find a JFET as the input stage in an application where bipolar transistors comprise the rest of the amplifier because of their peculiar advantages.

Several methods of characterizing the noise performance of amplifying devices have been devised. One is the noise-factor method, along with the related method of noise figure. Noise factor may be defined as

$$\text{Noise factor} = \frac{\text{signal power in/noise power in}}{\text{signal power out/noise power out}}$$

Noise figure is related to noise factor by the expression:

$$\text{Noise figure} = 10 \log \text{(noise factor)}$$

Since the signal and noise work into the same impedances, and since under those circumstances power is proportional to voltage squared, we can write

$$\text{Noise figure} = 20 \log \frac{\text{signal voltage in/noise voltage in}}{\text{signal voltage out/noise voltage out}}$$

Thus the noise factor and noise figure both give an indication of how much the amplifying device degrades a signal as it is passed through the amplifier. *However, these methods have two disadvantages.* Although well suited for characterizing the noise performance of a particular circuit or system for which the bandwidth is fixed, they are not applicable to describing a device in general. Even though the noise factor of a circuit may be known, one cannot determine from this information the noise factor of the device employed in that circuit when it is used in another, different circuit. A second disadvantage is that confusion can arise because a device with a lower noise factor than another under one set of conditions may have a higher noise factor under other conditions. Thus, a JFET operated from a 1-MΩ source over a given bandwidth may have a lower noise figure than a bipolar transistor operated under the same conditions. If the source were then changed to a 600-Ω generator, the bipolar could easily be superior from the point of view of signal-to-noise voltage. This emphasizes the fact that noise factor depends on a particular set of conditions and is not suited to the general characterization of a device. It would be desirable to have a method of specifying the noise performance of a JFET which depended only on the device, but which could be used to determine the equivalent input noise in any given circuit.

Such a method has now gained widespread acceptance. Random noise, like offset and drift, can be represented by a series voltage-noise generator and a parallel current-noise generator, which may or may not be independent of each other. This situation is illustrated in Fig. 1-21.

e_n is a zero-impedance voltage-noise generator, and i_n is an infinite-impedance current-noise generator. e_n and i_n are both proportional to the square root of the bandwidth, so their units are usually given as follows:

e_n Microvolts per square root of cycle $= \mu V/\text{Hz}^{1/2}$

i_n Picoamps per square root of cycle $= p\text{A}/\text{Hz}^{1/2}$

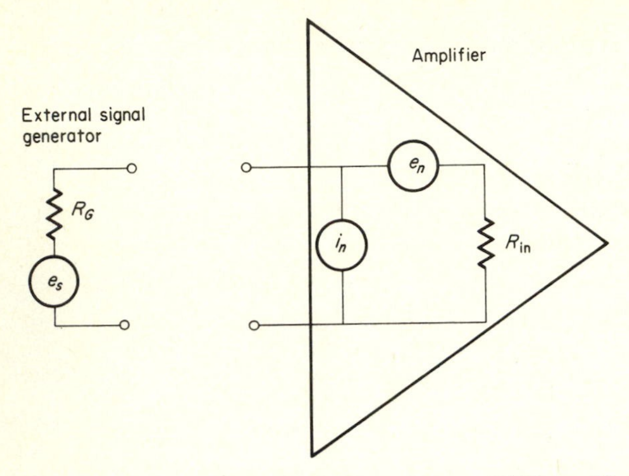

Fig. 1-21 e_n, i_n noise equivalent circuit for a JFET amplifier.

e_n and i_n both vary with frequency and the particular set of bias conditions under which the JFET is operated. An important feature of this method is that they are both independent of external circuit elements. If e_n and i_n are known for a particular device, its noise performance can be calculated for any circuit.

The effect of generator impedance R_G can be seen intuitively from Fig. 1-21. Assuming for the moment that R_{in} is infinite, note that i_n will flow into the external generator impedance and produce an equivalent input noise voltage. If R_G is small, $i_n R_G$ will be small, and e_n may be the dominant source of noise. If R_G is large, $i_n R_G$ will be large and may be the dominant source of noise. If R_{in} is not infinitely large, it can be included by being paralleled with R_G.

Using the e_n, i_n method of noise characterization, it is possible to calculate an equivalent input noise-voltage generator, which can be directly compared with the input signal level. In this way the signal-to-noise ratio for the amplifier can be determined quite directly. The procedure is as follows: First define

e_{eq} Equivalent input voltage-noise generator per root cycle

e_n, i_n Voltage- and current-noise generators, as before

R_G Generator resistance

γ Correlation coefficient between i_n and e_n = approximately 1 for JFETs

k Boltzmann's constant = $(1.372)(10^{-23})$ joules/K

T Absolute temperature in degrees Kelvin = 273 + °C

Then,

$$e_{eq} = \sqrt{(e_n^2 + i_n^2 R_G^2 + 2\gamma e_n i_n R_G + 4kTR_G)(\text{bandwidth})} \tag{1-30}$$

The first term in Eq. (1-30) represents the contribution of the equivalent voltage-noise generator, and the second that of the current-noise generator. The third term represents the effect of correlation between e_n and i_n. The last term is a measure of the thermal noise generated in source resistor R_G. In a more general case R_G would be the equivalent resistance seen by the amplifier input and would be a function of not only R_G, but also the amplifier input resistance and any feedback resistors.

As an example of the use of Eq. (1-30), let us find the signal-to-noise ratio for an amplifier with a Crystalonics 2N6550 JFET input if the signal is 594 μV rms from a 10-MΩ source. The amplifier bandwidth is from 900 to 1100 Hz. $e_n = 1.4\ \text{nV/Hz}^{1/2}$, and $i_n = 0.01\ \text{pA/Hz}^{1/2}$. Assume a correlation coefficient of unity between the voltage- and current-noise generators and that all the amplifier noise is contributed by the JFET input stage. The temperature is 25°C, and the noise power spectrum is flat from 900 to 1100 Hz.

First calculate the value of each of the four terms in Eq. (1-30) separately.

$$e_n^2 \times \text{bandwidth} = [(1.4)(10^{-9})]^2(200) = 3.92 \times 10^{-16}\ \text{V}^2$$

$$i_n^2 R_G^2 \times \text{bandwidth} = (10^{-14})^2(10^7)^2(200) = 2 \times 10^{-12}\ \text{V}^2$$

$$2\gamma e_n i_n R_G \times \text{bandwidth} = (2)(1)(1.4)(10^{-9})(10^{-14})(10^7)(200) = 5.6 \times 10^{-14}\ \text{V}^2$$

$$4kT R_G \times \text{bandwidth} = (4)(1.372)(10^{-23})(298)(10^7)(200) = 33.2 \times 10^{-12}\ \text{V}^2$$

These numbers give

$$e_{eq}^2 = 3.526 \times 10^{-11}\ \text{V}^2$$

$$e_{eq} = 5.94\ \mu\text{V}$$

Since the signal is 594 μV rms, the signal-to-noise (S-N) ratio is

$$\text{S-N ratio} = \frac{594}{5.94} = 100:1$$

In general e_n and i_n vary with frequency, and it is necessary to consult manufacturers' data sheets to obtain an approximate average over the bandwidth of interest.

1-6 JFET SWITCHING CIRCUITS

The perfect switch would have zero ON resistance, infinite OFF resistance, and zero offset voltage, would respond in zero time to the switching signal, and would switch without fail indefinitely. A mechanical switch approximates closely the ON and OFF resistance and zero offset requirements. It is inferior to semiconductor switches in speed and reliability. The bipolar transistor has low ON resistance, 20 Ω being typical, and an OFF resistance of a few megohms. It is fast and reliable, but has an offset voltage due to the *pn* junctions that exist in the path from collector to emitter. A more ideal switch is the JFET.

We have seen (Sec. 1-2) that the path between the drain and source of a JFET is resistive, unlike the path between the collector and emitter of a bipolar transistor, which contains a *pn* junction. By appropriate gate biasing, the resistance between the drain and source of a JFET can be made very low (a few ohms) or very high (thousands of megohms). This means that a JFET can be used as a switch, and there are two biasing conditions to be considered for JFET switching operation, namely, the bias required to turn the switch ON and that required to turn it OFF.

Figure 1-22*a* shows the equivalent circuit for a JFET operated as a switch, and Fig. 1-22*b* through *e* indicate the voltages required to operate the 2N5163 *n*-channel JFET (Fig. 1-7 and Table 1-2) as a switch under various conditions. In Fig. 1-22*a*, *R* is the resistance with the switch open, and *r* is the resistance with it closed. (Since $r \ll R$, the paralleling effects of R upon r can be neglected.) Figures 1-22*b* and *c* show the voltages required for the switch to be closed. From the discussion in Sec. 1-2, we can see that the condition for minimum resistance in a JFET channel is that the gate-to-source voltage V_{GS} be zero. In Fig. 1-22*b* the source is at 0 V, so the gate must be at 0 V also for the switch to be closed. Similarly, in Fig. 1-22*c* the gate must be at +3 V to obtain $V_{GS} = 0$ for the switch to be closed because the source is at +3 V also.

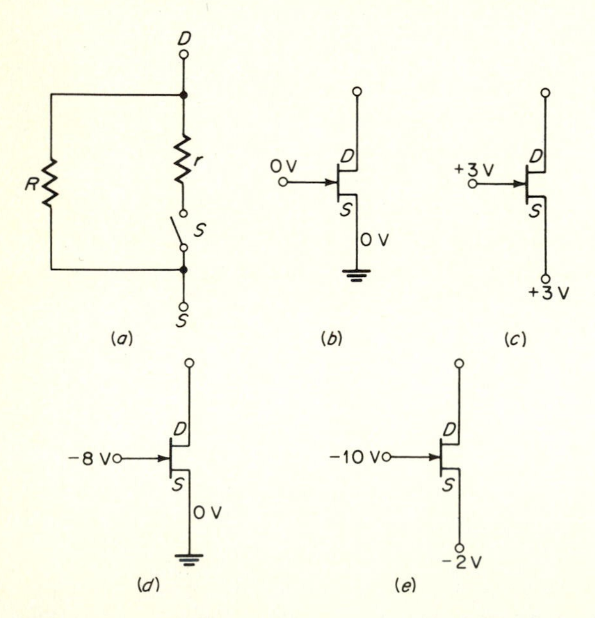

Fig. 1-22 (*a*) JFET switch equivalent circuit; (*b*) and (*c*) bias condition for JFET turned ON; (*d*) and (*e*) bias conditions for JFET turned OFF.

Figure 1-22*d* and *e* show conditions under which the switch is open. From Fig. 1-7*a* we can see that a V_{GS} of more than −3 V is required to turn the JFET completely OFF so that no current flows. This is also the condition that causes the JFET to act as an open switch. How much more than −3 V is necessary for complete turn-OFF can be found from Table 1-2, where the gate-to-source cutoff voltage $V_{GS_{off}}$ is specified as being from −0.4 to −8 V, with −3.7 V being typical. To allow for the large variation in $V_{GS_{off}}$ from JFET to JFET, we should use −8 V, which will turn OFF any 2N5163.

In Fig. 1-22*d* the source is at 0 V, so the gate should be at −8 V to ensure turn-OFF. In Fig. 1-22*e* the gate is at −10 V to ensure turn-OFF, since the source is at −2 V.

JFETs intended for use as switches are commonly supplied in IC packages along with the driver circuitry necessary to turn the JFET ON or OFF. Such IC packages have several convenience features. The JFET geometry used is designed to maximize the OFF resistance and minimize the ON resistance. The drive circuitry is arranged to minimize the feedthrough of voltage spikes from the gate switching voltage via interelectrode capacitances to the source and drain. There is usually more than one JFET switch in an IC package, 2 to 16 being common. The IC switch is often powered from ±15-V supplies, so that if the JFETs have V_{GSoff} of −5 V, the IC can be rated for ±10-V analog signals. Regardless of the internal switching voltages required, the driver circuitry is usually arranged so that 0 and +5 V are the inputs required to open and close the switch. These levels make the switch compatible with other digital systems in which 5-V swings are used. How this is accomplished is shown in Fig. 1-23*a*.

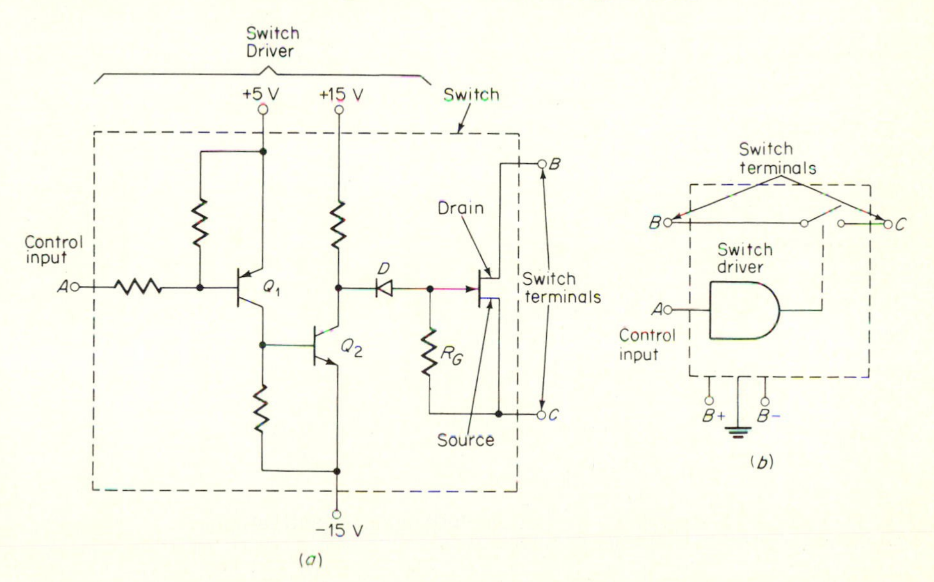

Fig. 1-23 (*a*) JFET switch with driver circuitry; (*b*) equivalent circuit.

A is the switch driver input terminal, and *B* and *C* are the switch terminals, the voltages on them being limited by rating to the range ±10 V. With 0 V on *A*, Q_1 and Q_2 are OFF and Q_2 collector is at +15 V, which turns OFF diode *D*. Since *D* is OFF, there is no current flow in gate resistor R_G and the gate and source will be at the same voltage, which is the condition for a JFET to be ON. With *A* at +5, V, Q_1 and Q_2 are ON and the collector of Q_2 and the anode of *D* are both at −15 V. Because of the polarity of *D*, the gate is also near −15 V, which will turn the JFET OFF, provided that *B* and *C* are both within the range ±10 V.

In practice the driver circuitry is more sophisticated than indicated in Fig. 1-23*a* to overcome feedthrough from *A* and *B* to *C*, and speed and temperature problems. From a user's point of view, however, the IC still appears as a switch to which one can apply ±10-V signals and which is operated by 0 and +5 V. An equivalent circuit for a single-pole single-throw JFET switch is shown in Fig. 1-23*b*. With extra JFETs, multipole and double-pole double-throw switches are possible.

Figure 1-24 shows several applications of JFET switch-drivers. The circuit in Fig. 1-24*a* is a sample-and-hold circuit consisting of two operational amplifiers (see Chap. 4), a capacitor, and a JFET switch/driver. The voltage at *B* from

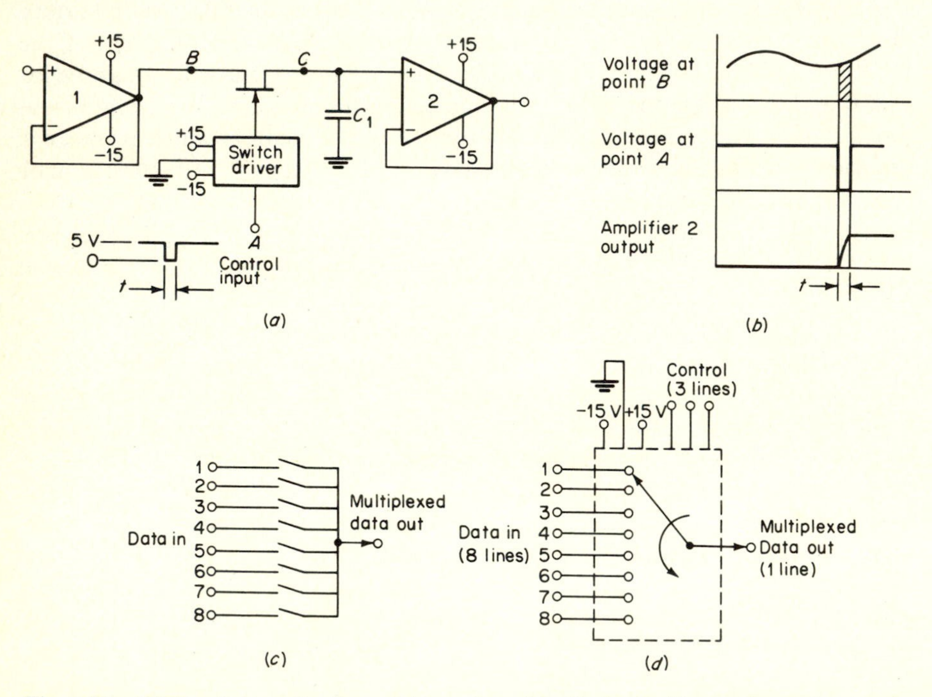

Fig. 1-24 JFET switch applications: (*a*) sample-and-hold circuit; (*b*) waveforms for sample-and-hold circuit; (*c*) and (*d*) multiplexer circuits.

amplifier 1 is changing as shown in Fig. 1-24*b*, and it is desired to sample and hold its value over some time interval *t*. Normally, input *A* is held at +5 V, which keeps the switch open. During time *t* the voltage on *A* is changed to 0 V, which connects *B* to *C* and charges capacitor C_1 to the same voltage as exists on point *B*. When the voltage on *A* rises to +5 V again, the switch opens and the charge on C_1 cannot leak away owing to the high-resistance paths presented by the open JFET switch and the high input impedance of amplifier 2. C_1 therefore remains charged to the voltage present at *B* during time *t*, even though *B* changes in voltage later. The sampled voltage is therefore held on C_1, and it may be read from the output of unity-gain amplifier 2 without discharging C_1.

Figure 1-24*c* and *d* indicate how analog multiplexing may be accomplished with JFET switches, which are shown in the diagrams as ordinary switches to simplify the explanation. In multiplexing, signals from a number of sources are sampled sequentially, and the sampled data are placed on a single line. This is useful, for example, in situations where data must be sent over long distances, in which case it is cheaper to use one wire rather than many, or in radio transmission, to conserve bandwidth by using one channel instead of many. In Fig. 1-24*c* there are eight sources of data and eight JFET switches, which would close one at a time in a fixed sequence. Figure 1-24*d* is an equivalent circuit for Fig. 1-24*c*, in which the eight switches have been replaced by a single rotating or commutating switch. In reality, there would still be eight JFETs (at least) on the chip.

With a view toward reducing the number of lines in a system, most JFET switches have on-chip decoding. Although there are eight JFET switches to control in Fig. 1-24*d*, there are three (not eight) control inputs. Since three binary bits can represent eight different numbers (Chap. 5), the signals on the control lines are decoded internally (on-chip decoding) to operate the appropriate switch. For example, a control input of 000 might close the switch in position 1; 001 would close the switch in position 2; and so on up to 111, which would close the switch in position 8.

JFETs are also used as choppers in dc amplifiers. A serious limitation of such amplifiers is drift. One method of overcoming the problem is to convert the dc signal to be amplified to an ac signal so that gain can be achieved with an ac amplifier. This is called "chopping." Since the dc output level of an ac amplifier is not susceptible to drift caused by input-level shifts, thermal effects are eliminated or, in practice, substantially reduced. After ac amplification, the signal is demodulated to reproduce the original dc or low-frequency waveform. Three simple mechanical choppers and their JFET equivalents are shown in Fig. 1-25.

Circuit (*a*) is used more frequently than (*b*) because it is always easier to switch a JFET which has its source lead grounded. Circuit (*c*) gives more perfect switching action than (*a*) or (*b*), as is indicated by the following simple analysis.

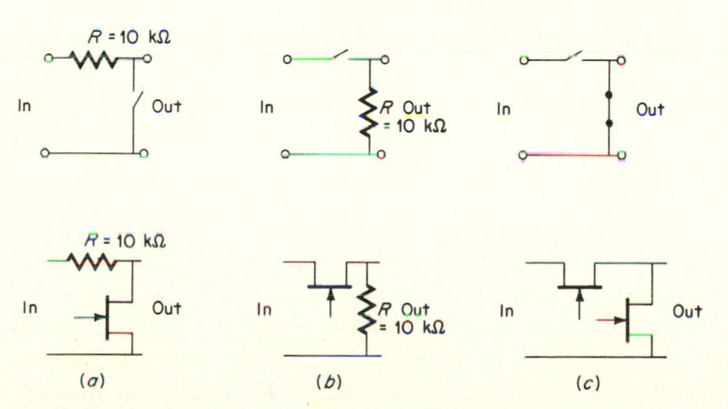

Fig. 1-25 JFET switching configurations: (*a*) shunt; (*b*) series; (*c*) series shunt.

If $R_{on} = 100\ \Omega$ and $R_{off} = 10\ \mathrm{M}\Omega$, consider the output voltages in (*a*) and (*c*) for 1-V inputs:

Circuit (*a*), JFET ON: By potential divider action,

$$e_{out} = \frac{10^2}{10^4 + 10^2} = 0.01\ \mathrm{V}$$

Circuit (*a*), JFET OFF: By potential divider action,

$$e_{out} = \frac{10^7}{10^7 + 10^4} = 0.999\ \mathrm{V}$$

Circuit (*c*), with one JFET ON and with the other OFF as shown: By potential divider action,

$$e_{out} = \frac{10^2}{10^7 + 10^2} = 0.00001\ \mathrm{V}$$

Circuit (*c*), with one JFET OFF and with the other ON as shown: By potential divider action,

$$e_{out} = \frac{10^7}{10^7 + 10^2} = 0.99999\ \mathrm{V}$$

In the ON condition the ideal switch would have $e_{out} = 1$ V, and when OFF would have $e_{out} = 0$ V. The calculations show by how much circuit (*c*) more closely approaches this ideal.

Although the JFET is a more perfect switch than the bipolar transistor from the point of view of voltage offsets, such offsets *do* occur in JFETs; however, their magnitude is much less than for bipolars. Feedthrough of the gate voltage through C_{GD} produces transient spikes at the chopping node, which in turn represent a voltage offset. Also, when the JFET is turned off, the usual leakage current associated with a reverse-biased *pn* junction exists. This gate leakage current flows through circuit resistances and produces voltage offsets. The total effect of these two offsets produces an equivalent input voltage offset of typically less than 1 μV.

1-7 JFETS AS VOLTAGE-CONTROLLED RESISTORS

It was indicated in the previous section that it is possible to change the ohmic resistance between the drain and source of a JFET from a few ohms in the ON condition to many megohms in the OFF condition. These are extremes that are achieved by applying gate-to-source voltages of zero and greater than the pinchoff voltage, respectively. By applying gate voltages between these limits, the drain-to-source resistance can be varied between the lowest and the highest value.

Figure 1-5 showed a plot of the output characteristic for a JFET with $V_{GS} = 0$. The characteristic was divided into three regions named ohmic, pinch-off, and breakdown. When used as a voltage-controlled resistor JFETs are operated in the *ohmic* region and, further, operation is limited to that portion of the characteristic for which V_{DS} is low—typically less than 100 mV. With this limitation the drain-to-source resistance is linear and bidirectional. That is to say, I_D is proportional to V_{DS}, and V_{DS} can be positive or negative regardless of whether a *p*- or *n*-channel device is being used. Figure 1-26 shows the voltage-controlled resistance characteristic for a typical 2N5613. These curves are merely a repeat plot of the output characteristic (Fig. 1-7*a*) for this transistor, expanded near V_{DS} and $I_D = 0$. The slope of the curves $\Delta I_D/\Delta V_{DS}$ represents the drain-to-source conductance. Notice that the conductance is linear, provided that V_{DS} is less than about ± 0.1 V. The effect of changing V_{GS} is to rotate the curves, which always pass through the origin, $V_{DS} = I_D = 0$. This in turn changes

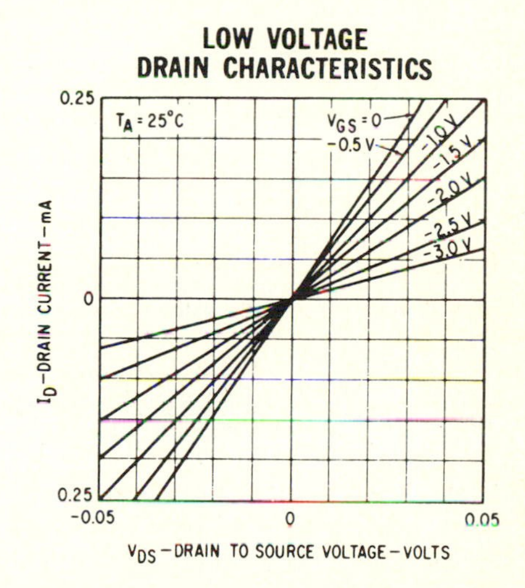

Fig. 1-26 Voltage-controlled resistance characteristic for the 2N5163 JFET. *Fairchild Semiconductor.*

the drain-to-source resistance since the slope of the line is changed. Thus a voltage V_{GS} is controlling a resistance $\Delta V_{DS}/\Delta I_D$. In practice it is not possible to stably vary the resistance of the channel over the full range from hundreds of ohms to many megohms. Instead, a change of about three orders of magnitude in resistance can be achieved with a gate control voltage of a few volts.

The drain-to-source resistance is temperature-sensitive and behaves similarly to drain current in this respect.

Figure 1-27 is a collection of several circuits in which JFETs are used as voltage-controlled resistors. Figure 1-27*a* is a simple potential divider. Figure 1-27*b* is an amplifier in which the gain may be varied by varying the amount of negative feedback. (See Chap. 4.) In Fig. 1-27*c* the amount of phase shift is controlled by varying the JFET resistance. Figure 1-27*d* is a voltage-controlled phase-shift oscillator using an amplifier and three elements similar to that used in Fig. 1-27*c*. Such a circuit will oscillate at the frequency at which there is 360° phase shift around the loop. By varying the three R_{DS} values electrically, the frequency at which zero phase shift, and hence oscillation, occurs can be varied.

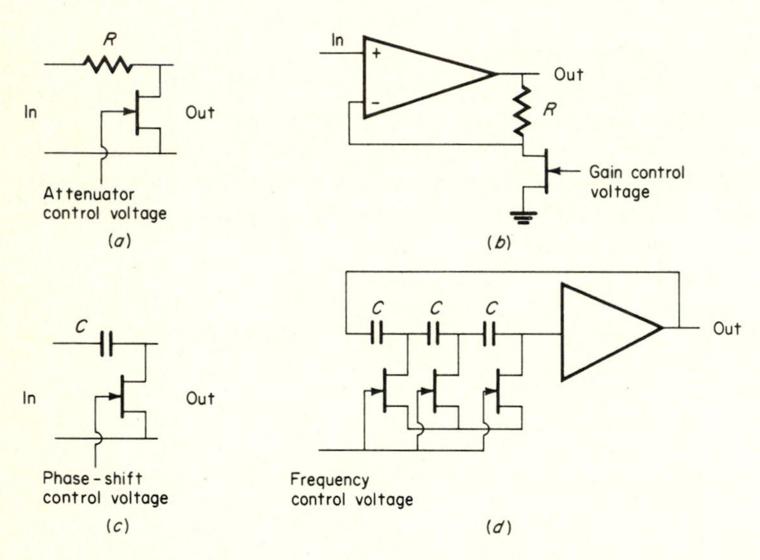

Fig. 1-27 Typical applications of JFETs as voltage-controlled resistors: (*a*) voltage-controlled attenuator; (*b*) voltage-controlled variable-gain amplifier; (*c*) voltage-controlled phase shifter; (*d*) voltage-controlled oscillator.

1-8 COMPARISON OF *n*- AND *p*-CHANNEL JFETS

The current carriers in an *n*-channel JFET are electrons, while those in a *p*-channel device are holes. Many of the differences that exist between *p*- and *n*-channel JFETs are attributable to the differences between the mobilities of holes and electrons. Electrons are superior to holes by almost two to one in this respect. For otherwise identical devices, the transconductance of an *n*-channel JFET will therefore be greater than that of a *p* channel. Since the interelectrode capacitances are the same, the figure of merit for an *n*-channel device, g_{fs}/C, will be better than that for a *p*-channel device. g_{fs}/C is a figure of merit because it is a measure of the amount of current available to charge circuit capacitances and hence affects frequency response. Also, since $R_{DS} \approx 1/g_{fs}$ for a turned-ON JFET, lower drain-to-source resistances can be achieved with *n*-channel JFETs, assuming other parameters are equal.

Finally, it turns out that *n*-channel JFETs exhibit lower equivalent input noise than do their *p*-channel counterparts.

In practice other factors, such as available power-supply polarities, often determine the selection of a *p*- or an *n*-channel device. However, for critical circuit applications the slight superiority of the *n*-channel device should be noted.

PROBLEMS

1-1 Draw the symbols for a *p*- and an *n*-channel JFET. If the source is grounded and the drain and gate voltages are 10 and 2 V, respectively, indicate on your drawings the polarities of these voltages for both devices, if they are to be biased as small-signal amplifiers.

1-2 How does the biasing of a JFET and a bipolar transistor account for the high input impedance of one and the low input impedance of the other? Give typical ranges of values for both devices.

1-3 Describe briefly how a depletion region occurs in a *pn* junction.

1-4 Why are JFETs called unipolar transistors? Why does a JFET have no offset voltage in the drain-to-source path, while a bipolar transistor does in the collector-to-emitter path?

1-5 Show by means of a sketch what is meant by the " "channel" in a JFET. What is meant by "pinchoff"? Indicate the shape of the depletion region at pinchoff.

1-6 Sketch the output characteristics for a typical JFET, and explain why the three regions, ohmic, pinchoff, and breakdown, occur.

1-7 For the output characteristic given in Fig. 1-7*a* find the dc and ac channel resistances at $V_{DS} = +10$ V and $V_{GS} = -1$ V. Assume the characteristic is a straight line from $V_{DS} = +6$ V to $V_{DS} = +16$ V.

1-8 Explain what the following JFET parameters mean.

(*a*) V_P (*b*) I_{DSS}
(*c*) g_{fs} (*d*) g_m
(*e*) BV_{GDS} (*f*) I_{GSS}
(*g*) C_{iss} (*h*) g_{os}

1-9 Give two methods for obtaining the value of the pinchoff voltage from a JFET output characteristic, and explain the limitations of the two methods.

1-10 A JFET has $I_{DSS} = 2$ mA and $V_P = -5$ V. Find

(*a*) I_D at $V_{GS} = -5$ V
(*b*) I_D at $V_{GS} = 0$
(*c*) I_D at $V_{GS} = -2.5$ *V*

1-11 Find the transconductance of the JFET in Prob. 1-10 for the same three values of V_{GS}.

1-12 Explain why the circuit in Fig. Prob. 1-12*a* is inferior to that in Fig. Prob. 1-12*b* from the point of view of bias-point repeatability with different transistors. Using

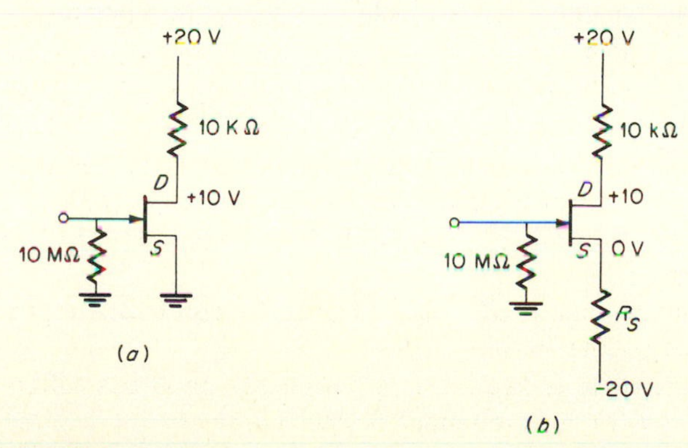

Fig. Prob. 1-12

the voltages indicated in Fig. Prob. 1-12*b*, find the value of R_S. Also find I_{DSS} and the JFET transconductance if $V_P = -2$ V.

1-13 Find the gain and input and output resistance for the circuit of Fig. Prob. 1-12b under the following conditions if $V_P = -2$ V:

(*a*) Input to gate. Output from drain. Source bypassed to signal frequencies.

(*b*) Input to gate. Output from source.

14-1 Calculate the input capacitance of common-source and common-drain amplifiers having the following characteristics: $C_{GD} = C_{GS} = 5$ pF. Common-source gain $= -10$, and common-drain gain $= +0.95$.

1-15 Sketch the cascode configuration. Give two advantages of this circuit over the common-source amplifier.

1-16 Why are JFETs commonly used in high-quality radio receivers?

1-17 Sketch the general shape of the noise spectrum for a typical JFET. Indicate the typical frequencies at which significant changes in noise performance occur. Name the three regions separated by break points. What are the slopes of the curves in the three regions?

1-18 What is the equivalent noise input voltage for a JFET with an equivalent voltage-noise generator of 0.1 $\mu V/Hz^{1/2}$ and an equivalent current-noise generator of 0.05 $pA/Hz^{1/2}$? The resistance of the signal source is 50 MΩ and the voltage- and current-noise generators have a correlation coefficient of unity. The circuit bandwidth is 1,000 Hz, and the circuit operates at 25°C.

1-19 In what ways does a JFET approach the ideal switch? In what ways does it not? In what ways are JFETs superior to bipolars as switches? How do C_{DG} and reverse leaking current affect the performance of a JFET switch?

1-20 An *n*-channel JFET operated as a switch has $V_P = -5$ V. If the source is at $+2$ V, what gate voltage is required to

(*a*) Turn the device ON so that its resistance is a minimum?

(*b*) Turn the device OFF?

1-21 A JFET switch IC, similar to that in Fig. 1-24*d*, is designed to multiplex 16 channels of data to a single line. How many control inputs are required?

1-22 Discuss the use of JFETs as voltage-controlled resistors. Give two examples.

1-23 Why are *n*-channel JFETs generally superior to *p*-channel devices?

1-24 Calculate the gain of a common-drain–common-emitter amplifier with a 10-kΩ collector load, as shown in Table 1-4*e*. The JFET has a transductance of 1000 μmho. The bipolar transistor has a current gain of 50 and a transconductance of 35,000 μmho.

1-25 Deduce and explain how a curve tracer may be used to find the zero-temperature-coefficient bias point for a JFET.

chapter 2

Metal Oxide Semiconductor Field Effect Transistors

2-1 INTRODUCTION

The metal oxide semiconductor field effect transistor (MOSFET) has a number of similarities to the junction field effect transistor (JFET) discussed in Chap. 1. Like the JFET, the MOSFET is a low-power semiconductor device that combines the high input impedance of the vacuum tube with the low power requirements of the bipolar transistor. Both MOS and junction field effect transistors have a drain, source, and gate, and in both there is a conducting channel, the resistance of which is varied by means of a potential applied to the gate. There are also *p*- and *n*-channel MOS transistors, and there is an important class of integrated circuits called *complementary MOS*, or CMOS, which uses both (Sec. 5-3-10). In contrast to bipolar transistors, in which the controlling effect is a current (the base current), both JFET and MOS devices have only very small input currents, which are due to leakage effects. The controlling *effect* is an electric *field* produced by an input voltage, which is why JFETS and MOS transistors are called *field effect* transistors.

The difference between JFETs and MOSFETs is that whereas the gate-to-source path in a JFET is a reverse-biased *pn* junction, in the MOSFET it is not. In a MOSFET a thin layer of insulating material is placed over the channel before deposition of the gate electrode. Because the layer is so thin, typically less than 1 μ, the field produced by the gate potential still penetrates and influences significantly the conductivity of the channel. Even though the insulating layer is very thin, its resistance is made high by using silicon dioxide as the insulating material. Since a MOSFET has an insulator in series with its gate, the gate current is even lower than for a JFET, and the input resistance is much higher. Typical values of gate current and input resistance for a MOSFET are 10^{-14} A and 10^{14} Ω.

Figure 2-1 compares the symbols for an *n*-channel MOSFET, an *n*-channel JFET, and a vacuum triode. More is said about MOSFET symbols in Sec. 2-2-2, after the various types of MOSFET have been discussed. The symbol for a *p*-channel MOSFET is similar to that for an *n* channel, except that the arrow is reversed. Notice that in Fig. 2-1*a* the gate connection is isolated from the rest of the MOSFET. This symbolizes the capacitative or insulated connection of the gate to the *n* channel; it contrasts with the JFET shown in Fig. 2-1*b*, in which

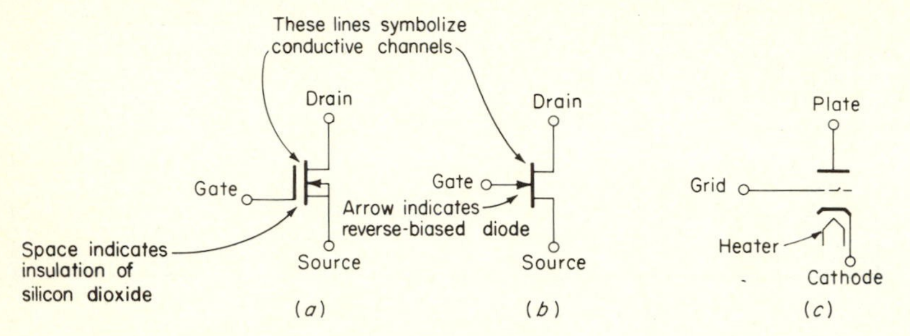

Fig. 2-1 MOSFET, JFET, and vacuum-triode symbols: (*a*) depletion-type *n*-channel MOSFET; (*b*) *n*-channel JFET; (*c*) vacuum triode.

the gate is connected to the channel via a reverse-biased diode, represented by an arrow. The significance of the arrow in Fig. 2-1*a* will become apparent when the construction and theory of operation of the MOSFET are discussed.

2-2 THEORY OF OPERATION OF THE MOSFET

2-2-1 MOSFET FABRICATION AND OPERATION

In a JFET the polarity of the gate potential must normally be such as to maintain the input *pn* junction reverse-biased. Since there is an insulator in the input circuit of a MOSFET, the gate potential is not restricted in polarity. There are therefore two possible modes for MOSFET operation, which are called "enhancement" and "depletion" modes. The construction and typical characteristics of the two modes are given in Fig. 2-2. One of the structures, that in Fig. 2-2*d*, represents a device which can operate in either mode. Two sets of characteristics are therefore given for this structure.

Figure 2-2*a* represents a cross section of an *n*-channel enhancement MOSFET. The starting material is a *p*-type substrate or wafer upon which is grown an insulating layer of silicon dioxide. The oxide layer is coated with a photosensitive material in a dark room and then exposed through a very-high-resolution mask. Areas not exposed can be removed by a suitable solvent rinse, and regions of unprotected oxide are etched away to expose the *p*-type substrate. An *n*-type diffusion is then performed by placing the wafer in a high-temperature oven which contains an *n*-type impurity, such as boron gas. The boron diffuses into the *p*-type substrate only where the oxide has been removed by the etching process. Thus, two islands of *n*-type material are formed, as shown in Fig. 2-2*a*. These islands form the source and drain of the MOSFET. At this point there are frequently other steps which involve reoxidization and further etching to thin the oxide over the channel between the two islands where the gate will be formed. Description of these steps will be omitted here, since they are not essential to an understanding of MOSFET operation, although they are necessary to produce a practical device. Final steps involve metallization to

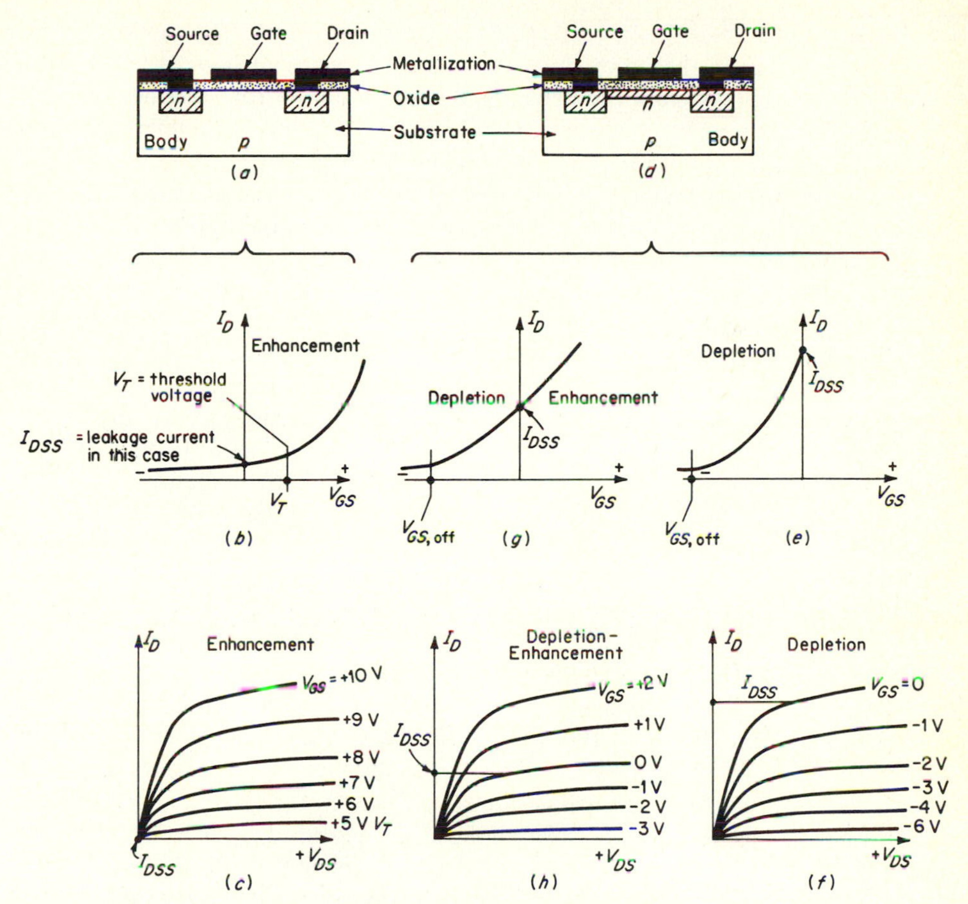

Fig. 2-2 Construction and characteristics of *n*-channel enhancement- and depletion-mode MOSFETs: (*a*) structure of an *n*-channel enhancement-mode MOSFET; (*b*) forward transfer characteristic of an *n*-channel enhancement-mode MOSFET; (*c*) output characteristics for an *n*-channel enhancement-mode MOSFET; (*d*) structure of an *n*-channel enhancement-depletion-mode MOSFET; (*e*) forward transfer characteristic of an *n*-channel depletion-mode MOSFET (can also be used in enhancement mode); (*f*) output characteristics for an *n*-channel depletion-mode MOSFET (can also be used in enhancement mode); (*g*) forward transfer characteristic of an *n*-channel enhancement-depletion MOSFET; (*h*) output characteristics for an *n*-channel enhancement-depletion MOSFET.

contact the source and drain regions as well as the oxide layer comprising the gate. It should be understood that MOSFETs, like all semiconductors, are not made one at a time. Instead, many transistors are made simultaneously on a thin wafer of silicon which is a few inches in diameter. After testing on the wafer, acceptable and unacceptable transistors are appropriately marked, and the wafer is sliced to separate the individual devices. These are then individually packaged in containers which are easier to handle than the minute transistors themselves.

Integrated circuits and their fabrication are discussed in Chap. 3, but it is worth noting at this point that the fabrication of a MOSFET is, in many ways, easier than for integrated circuits and many other semiconductor devices. There are fewer steps in the processing, no epitaxial layer (see Chap. 3) is required, and there is only one diffusion compared with up to four for other devices.

The enhancement-mode device illustrated in Fig. 2-2*a* is also referred to as a "normally off" MOSFET. This is because with zero gate bias voltage the source and drain contacts are separated by two *pn* junctions connected back to back. Thus, no drain current will flow, even with potential applied from drain to source (assuming the potential is less than that required to break down the reverse-biased junction). However, suppose bias voltages are applied to the drain and gate, both biases being positive with respect to the source. Positive charges on the metallized gate induce corresponding negative charges in the *p* material on the other side of the oxide, just as in a capacitor. The positive charges may also be thought of as repelling positive charges in the *p* material, leaving free electrons. Thus, the density of holes in the *p*-channel material will be reduced, and the electron density will be enhanced. With enough positive bias the electron enhancement will be sufficient to *convert the p region under the gate to an n channel.* In this way the two *n* islands, the source and drain, become joined by an *n* region, so that a conductive channel exists, rather than a path including reverse-biased *pn* junctions. Since the drain is biased with respect to the source, current will flow along the channel.

Figure 2-2*b* and *c* are the forward transfer and output characteristics, respectively, for an enhancement-mode MOSFET. Figure 2-2*b* is a plot of the amount of drain current I_D that will flow as V_{GS} is varied. The plot is made with V_{DS} constant at some arbitrary level, typically 5 or 10 V. Notice that with $V_{GS} = 0$ or negative, I_D is small and relatively constant. This drain current, with $V_{GS} = 0$, is referred to as I_{DSS} as with the JFET described in Chap. 1. Since it is the current which flows with reverse-biased *pn* junctions in the drain-to-source path, it is a leakage current, typically a fraction of a nanoamp at room temperature. If Fig. 2-2*b* were drawn to scale, I_{DSS} would be virtually touching the V_{GS} axis. As V_{GS} is made more positive, channel enhancement occurs as described above, and I_D increases.

Figure 2-2*c* is a plot of the drain current that will flow as V_{DS} is varied, for a number of values of V_{GS}. Once again the $V_{GS} = 0$ curve almost touches the V_{DS} axis, since I_D is almost zero with zero V_{GS} for an enhancement-mode MOSFET. As V_{GS} is made more positive, I_D increases. The shapes of the curves are similar to output characteristics for other devices such as JFETs and vacuum-tube pentodes. As V_{DS} increases, I_D increases until the "knee" of the curve is reached. I_D then remains almost constant, even as V_{DS} is increased, indicating a high dynamic output impedance.

Figure 2-2*d* represents a depletion-mode device, which is also referred to as a "normally on" MOSFET. As we shall shortly see, it is possible to operate any depletion-mode MOSFET as an enhancement-mode device also.

Depletion-mode devices are called "normally on" because with zero gate bias voltage the drain and source islands, which are of *n*-type material, are connected by an *n* channel. Notice that the *n* channel is not diffused in this case, but is the result of trapped charges in the oxide. Thus, with potential applied from drain to source, and with $V_{GS} = 0$, current will flow along the resistive channel, which has no *pn* junctions in the current path. Suppose that a negative bias is applied to the gate. Negative charges on the metallized gate induce corresponding positive charges in the *n* material on the other side of the oxide. The negative charges may also be thought of as repelling negative charges in the *n* channel, thereby reducing its conductivity. Thus the density of electrons in the *n* channel will be depleted, and the density of holes will be enhanced. With enough negative bias the electron depletion will be sufficient to convert the *n* material in the channel to *p*-type material. In this way, the two *n* islands, the source and drain, become separated by a *p* region. Two *pn* junctions therefore exist in the signal path, one of which will be reverse-biased by V_{DS}. Any I_D flow will be a leakage current, and the MOSFET is essentially OFF.

Figure 2-2*e* and *f* are the forward transfer and output characteristics, respectively, for a depletion-mode MOSFET. Notice the similarity in shape to the characteristics for an enhancement-mode device (Fig. 2-2*b* and *c*). For the transfer characteristic in Fig. 2-2*e*, the difference is that the whole curve is shifted into the negative V_{GS} region. With $V_{GS} = 0$, $I_D = I_{DSS}$ and is not a low-level leakage current as it is for the enhancement MOSFET. As V_{GS} is made more negative, I_D decreases until it becomes the leakage current of the reverse-biased *pn* junction.

Figure 2-2*f* is a plot of the drain current that will flow as V_{DS} is varied, for a number of values of V_{GS}. Unlike Fig. 2-2*c*, the $V_{GS} = 0$ curve does not almost touch the V_{DS} axis. Instead, it is the curve at which maximum current flows. As V_{GS} is made more negative, I_D reduces until it becomes a low-level leakage current.

The structure shown in Fig. 2-2*d*, in addition to being operated as a depletion-mode device, can be operated as an enhancement device. With negative bias, the device operates as described for the depletion-mode MOSFET. However, positive bias can be applied also, and this enhances the *n* region which connects the source and drain *n* islands. This enhancement increases the conductivity of the channel, resulting in more drain current for a given V_{DS}. The resulting characteristics are shown in Fig. 2-2*g* and *h*.

The JFET discussed in Chap. 1 is also a depletion-mode or "normally on" device. JFETs are not used in the enhancement mode (except under small-signal conditions with $V_{GS} = 0$) because this would involve forward biasing the gate-to-source input *pn* diode. This could result in excessive gate-current flow and would lower the input impedance or even damage the device, since there is no insulator in series with the gate as there is in a MOSFET. Figure 2-3 summarizes all possible types of FETs.

It should be noticed that a MOSFET *does have response down to dc*, even though the gate is isolated from the substrate by an insulator, thereby forming a capacitor.

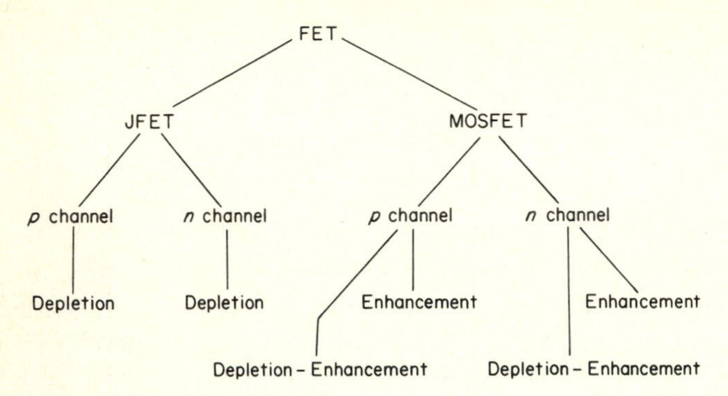

Fig. 2-3 Types of available FETs.

2-2-2 MOSFET SYMBOLS

Figure 2-4 shows the standard symbols for the various types of MOSFET. Typical bias voltages have been included for each device operated as a small-signal amplifier. More is said about biasing in Sec. 2-4.

In all four symbols in Fig. 2-4, the gate is shown separated from the rest of the transistor, indicating that the gate metallization is electrically isolated from the channel by a silicon dioxide layer. As for a JFET, an *n*-channel device is indicated by an arrow pointing into the transistor, and a *p*-channel by an arrow pointing out. Enhancement-mode MOSFETs are shown with a broken vertical line joining the source and drain, indicating a "normally off" or zero-current device with zero gate-to-source voltage. Depletion-mode MOSFETs

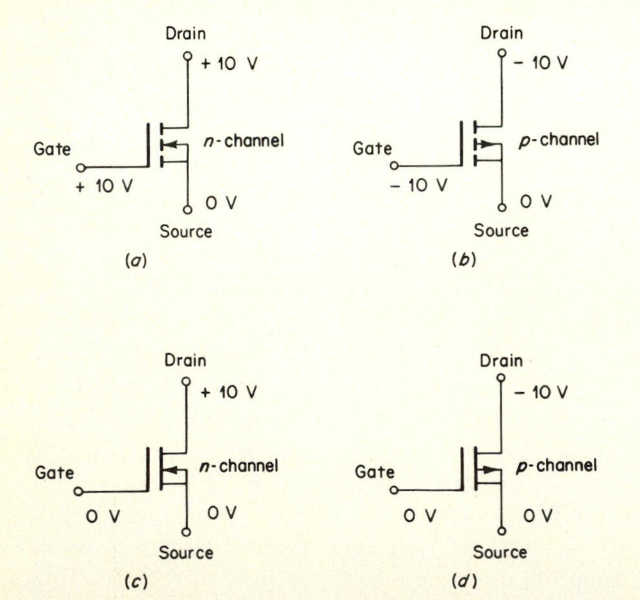

Fig. 2-4 MOSFET symbols: (a) *n*-channel enhancement type; (*b*) *p*-channel enhancement type; (*c*) *n*-channel depletion type; (*d*) *p*-channel depletion type.

are indicated by an unbroken line between source and drain as are JFETs, which are depletion devices, indicating the existence of a conducting channel with zero gate-to-source voltage.

In all four symbols in Fig. 2-4, the substrate is indicated by the arrow and is shown connected to the source. Occasionally this is done internally by the manufacturer, but MOSFETs are also supplied as four-lead devices, that is, with the substrate not connected to the source but brought out as a separate lead. In this case the MOSFET symbol is modified accordingly.

2-2-3 GATE BREAKDOWN

MOSFETs have one limitation which must be understood to avoid device failure in installation and use. The silicon dioxide layer in series with the gate is sufficiently thin so that, if too large a voltage is applied from gate to channel, the oxide ruptures owing to excessive electrostatic field stress. The effect is similar to breakdown in a capacitor due to overvoltage stress. One hundred volts applied to a silicon dioxide layer 1000 angstroms thick produces a stress of 10^6 V/cm, and 100 V can easily be applied to the gate by electrostatic discharge from handling on a dry day, or from a soldering iron with poor insulation from the line.

Manufacturers now supply some MOSFETs with a protective zener diode built in from gate to source. The diode is designed to break down nondestructively at about 50 V, which is less than the voltage which produces destructive breakdown of the oxide layer. Unfortunately this technique degrades the input of the MOSFET in several ways. The input leakage current is greater and the input resistance is less, because there is a diode across the input as with a JFET. Leakage currents of 10^{-10} A and input resistances of 10^{11} Ω are typical for diode gate-protected MOSFETs. In addition, the input capacitance is somewhat larger but not usually significantly so.

Some circuit requirements are such that a MOSFET with no gate protection must be used. One example is an electrometer, which is a type of amplifier used for measuring very low currents. (See Sec. 2-5-2.) An electrometer design might call for an input transistor with an input leakage current of less than 10^{-14} A, in which case a MOSFET without a shunt diode would be the only device suitable. Where the gate must be left unprotected, there are precautions which can be taken to avoid damage to the MOSFET. For example, any soldering iron used should have its tip grounded, and the gate should be grounded or connected to any other lead on the device before it is touched.

2-3 MOSFET PARAMETERS

Figure 2-5*a* is an equivalent circuit for a MOSFET which uses lumped linear circuit elements and is suitable for small-signal, linear amplifier analysis. Although the circuit is similar to that for the JFET shown in Fig. 1-14, there are some significant differences, mainly in the typical values of the parameters. For example, R_{GS}, the resistance of the path from the gate to the source, is much higher than the corresponding input resistance of a JFET. This is because a MOSFET has a high-quality insulator—a layer of silicon dioxide—in the gate-to-source path, while a JFET input consists of a reverse-biased *pn* junction. Such a junction has a high resistance, about 10^{10} or 10^{11} Ω, but the input resistance of a MOSFET is even higher, typically 10^{14} to 10^{15} Ω. For similar reasons the input, or gate-leakage, current of a JFET is higher than that for a MOSFET, being typically 10^{-10} A for the junction device compared with 10^{-14} A for a typical MOSFET. Also, the JFET leakage current is temperature-sensitive in the same way as for any semiconductor diode, increasing exponentially with increasing temperature. A MOSFET gate-leakage current will change somewhat with temperature, but the change is not exponential and is due to surface leakage effects which are much less dependent on temperature. The leakage resistance from gate to drain, R_{GD}, is similar in value to R_{GS} and has similar properties.

r_d is the incremental or dynamic output resistance of the MOSFET. If the MOSFET is biased so that its operating point is in the region where the output characteristic is nearly parallel to the V_{DS} axis, r_d will be high. In this region $r_d = 100$ kΩ is typical. This parameter is often described as an output conductance in the same manner as for a JFET. All the remarks about r_d made in Sec. 1-3-2 pertain also to the MOSFET.

C_{GS}, C_{DS}, and C_{GD} are the interelectrode capacitances, and C_{DS} is usually insignificant. It is common for C_{GD} in a MOSFET to be lower than for a JFET, often a fraction of a picofarad. Since this is the reverse transfer capacitance (also called C_{rss}—see Table 1-1), it is that capacitor which affects feedback from output to input. Its low value is useful in RF circuit design. Also, since C_{GD} is the Miller capacitor (see Sec. 1-5-2), a low value is desirable to maintain high input impedance in voltage-amplifier applications.

The terminal marked *B* (for "body") represents a connection to the substrate. By referring to Fig. 2-2, it can be seen that for an *n*-channel MOSFET both source and drain are connected to the substrate via *pn* junctions designated D_1 and D_2 in Fig. 2-5. The body terminal is sometimes connected to the source, so that D_1 is shorted out. D_2 is reverse-biased for the following reason. A junction diode is reverse-biased if the *n* material is biased positively with respect to the *p* material. (See Fig. 1-2.) D_1 and D_2 are shown in the direction they would occur for an *n*-channel device such as in Fig. 2-2. For an *n*-channel MOSFET the drain is biased positively with respect to the source. If the body is connected to the source, the drain will be positive with respect to the body. Since the drain is connected to an *n* region and the body or substrate to a *p* region, the diode thus formed is reverse-biased.

MOSFETS are also available as four-terminal devices, that is, with the substrate available for external connection. To maintain a reverse bias on both diodes (for an *n*-channel device), the body may be connected to a voltage supply, the potential of which is always more negative than the drain or source when

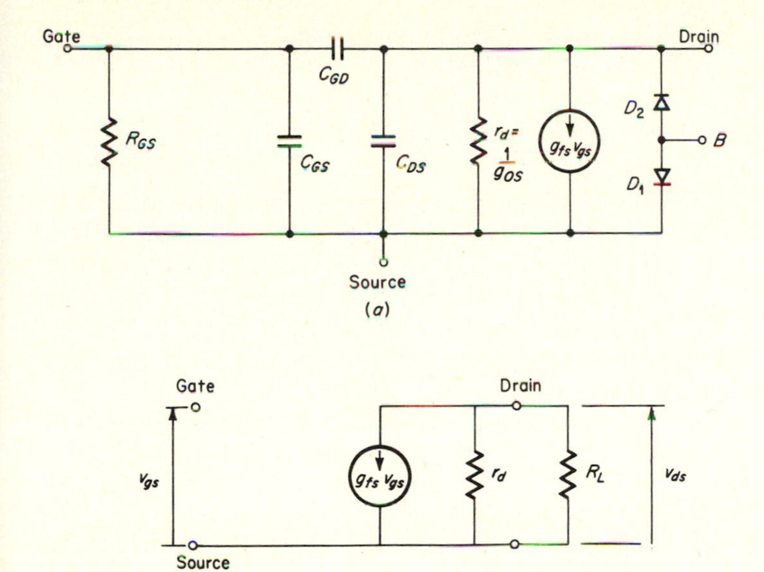

Fig. 2-5 (*a*) MOSFET equivalent circuit; (*b*) MOSFET equivalent circuit simplified.

the MOSFET is in use. Similar remarks apply to *p*-channel MOSFETs except that the polarities of all potentials are reversed.

The active element in the equivalent circuit of Fig. 2-5*a* is a current generator $g_{fs}v_{gs}$. The voltage gain of a MOSFET amplifier is found in the same way as for a JFET. All the formulas in Tables 1-3 and 1-4 apply to MOSFET, and MOSFETs are frequently combined with bipolar transistors in the same manner as are the JFETs in Table 1-4.

For example, a MOSFET with $g_{fs} = 2$ mA/V is used in the common-source configuration shown in Table 1-3. R_D is 5 kΩ, R_S is 1 kΩ, and R_G is 100 MΩ. Find the voltage gain and the input and output resistances. What would the voltage gain be if R_S were completely bypassed at the signal frequency?

Notice that the units of the forward transconductance g_{fs} are milliamps per volt, in contrast to the μmos used in Chap. 1. In practice both units are used for either JFETs or MOSFETs. Milliamps per volt is simply a larger unit of forward transconductance than μmhos since 1 mho $= 10^3$ mA/V $= 10^6$ μmho.

From Table 1-3 the voltage gain G of a common-source JFET (or MOSFET) amplifier is

$$G = \frac{-g_{fs}R_D}{1 + g_{fs}R_S} = \frac{-(2)(10^{-3})(5)(10^3)}{1 + (2)(10^{-3})(1)(10^3)}$$

$$= \frac{-10}{1 + 2} = -3.33$$

From Table 1-3 the input resistance is 100 MΩ, and the output resistance is 5 kΩ. With R_S bypassed, the gain G_B is

$$G_B = -g_{fs}R_D = -(2)(10^{-3})(5)(10^3) = -10$$

Figure 2-5*b* is a simplified low-frequency equivalent circuit for the MOSFET. It is similar to the simplified JFET equivalent circuit in Fig. 1-15. For a given input voltage v_{gs}, the output voltage v_{ds} may be found by calculating the current in the external resistor R_L. The current comes from the generator $g_{fs}v_{gs}$ and divides between r_d (the MOSFET output resistance) and R_L. If the MOSFET approximates a constant-current generator so that $r_d \gg R_L$, r_d can be neglected. See, for example, Eqs. (1-16) to (1-18).

Enhancement-mode MOSFETs have a gate-to-source voltage level which is called the *threshold voltage* V_T. No current will flow in the device until $V_{GS} = V_T$.

2-4 BIASING

As with the JFET, bipolar transistor, and vacuum tube, it is unfortunately necessary to apply dc bias voltages to a MOSFET before it can be put to use as an amplifier. Information about how to bias a MOSFET can be obtained from the output characteristics, as with any device. Using load lines, or related techniques, it is possible to design a bias arrangement for a MOSFET which will result in a circuit having suitable drain, gate, and source voltages and drain currents.

MOSFETs do, however, have some unique properties which permit interesting biasing circuits to be used. The primary feature of such circuits is the simplicity of the biasing, and this arises from the use of enhancement-mode devices. Such MOSFETs have the property that they can be biased to an operating point at which the gate and drain voltages have the same magnitude and polarity. In vacuum tubes, bipolar transistors, and depletion FETs (both junction and MOS), the plate, collector, or drain must be at a different potential to the grid, base, or gate, respectively, for small-signal operation.

Figure 2-6*a* shows a three-stage, directly coupled amplifier employing *n*-channel enhancement-mode MOSFETs which has response down to dc. Note the absence of interstage *RC* coupling networks, since all drains and gates are at the same potential of +10 V. The circuit input is biased to +10 V also, which may or may not be a disadvantage, depending on the type of signal source used.

Suppose all three MOSFETs in Fig. 2-6*a* have $g_{fs} = 2000$ μmho, and that the reactance of input capacitor *C* is negligibly small at the signal frequencies being amplified. What will the output be with (*a*) $e_{in} = 2$ mV, and (*b*) $e_{in} = 20$ mV, both peak to peak?

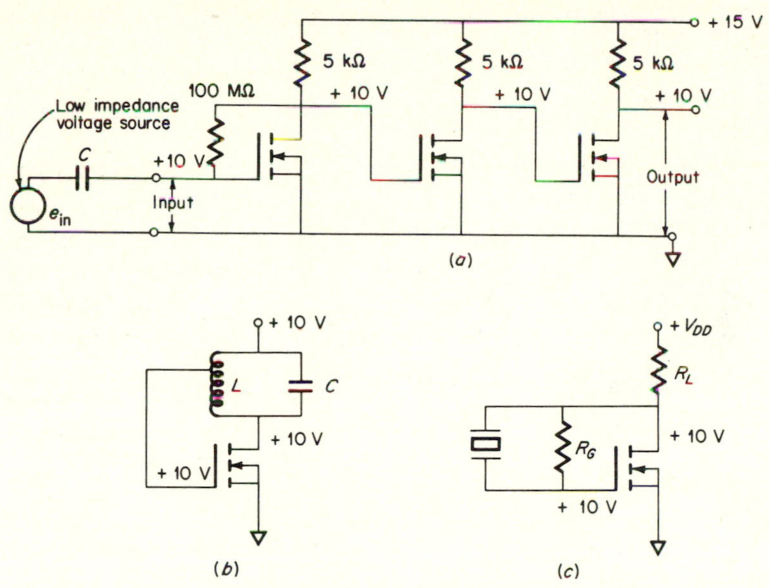

Fig. 2-6 Circuits taking advantage of unique MOSFET biasing characteristics: (*a*) three-stage directed-coupled amplifier; (*b*) Hartley oscillator; (*c*) crystal oscillator.

We already found the gain per stage to be -10 in Sec. 2-3. The MOSFET in that example had $R_D = 5$ kΩ and $g_{fs} = 2$ mA/V $= 2000$ μmho, and having R_S bypassed is equivalent to having $R_S = 0$ at signal frequencies. The 100-MΩ input-bias resistor provides no feedback, owing to the low impedance of the signal source, and its shunting effect on the first-stage load resistor of 5 kΩ is negligible. The overall gain G_T is therefore:

$$G_T = (-10)(-10)(-10) = -1000$$

In case (*a*), e_{in} is 2 mV so the output is (2 mV) (1000) = 2 V peak to peak. In case (*b*), the output will try to swing (20 mV)(1000) = 20 V peak to peak, but due to power-supply limitations can swing only 15 V peak to peak. In fact, the output will swing from the output quiescent point of +10 V to +15 V in the positive direction and to 0 V in the negative direction, so that the output sine wave will be clipped going positive and unclipped going negative.

Figure 2-6*b* shows an interesting version of a Hartley oscillator. An oscillator must contain a frequency-selective network, an amplifying device, and a positive feedback path. Any other components such as biasing resistors and decoupling capacitors are extraneous to the basic circuit and may be thought of as necessary evils. In Fig. 2-6*b* there are no such extraneous parts, and this is possible only because of the unique biasing characteristics of the enhancement-mode MOSFET. If *L* is a high-quality inductor, its resistance will be negligible and there will be no dc drop across it. Thus the drain and gate will be at the same dc potential, which is a valid bias condition for an enhancement-mode MOSFET. The connection between the gate and the tap on the inductor provides the feedback path necessary. The *LC* circuit gives frequency selectivity, and the MOSFET provides the required gain. No other components are needed.

Figure 2-6*c* shows the basic circuit for a crystal oscillator of the Pierce type. In this circuit a MOSFET has the advtange that a very large value for R_G can be tolerated since the gate leakage current is so small. Since R_G shunts the crystal, a large value is desirable to maintain high circuit Q and hence good frequency stability. The fact that the MOSFET has such a large input impedance helps in this respect also. In an actual MOSFET Pierce oscillator, certain other components would be required to make the circuit practical. However, the circuit shown in Fig. 2-6*c* serves to illustrate the two advantages that MOSFETs have in crystal-oscillator circuits.

Theoretically, the substrate or fourth electrode of a MOSFET could be used as a control electrode. In practice it turns out to be of little value in this respect. Instead the substrate electrode is often connected to the source terminal, especially in amplifier applications. If for some reason it cannot be tied to the source, the substrate is biased to a potential which assures that the substrate-to-channel diode is never forward-biased. In Fig. 2-4*a* and *c* it can be seen, from the direction of the substrate diode, that the substrate electrode should always be held more negative than the source potential for an *n*-channel MOSFET. For the *p*-channel devices shown in Fig. 2-4*b* and *d*, a bias voltage more positive than the source is required.

2-5 CIRCUITS TAKING ADVANTAGE OF SPECIAL MOSFET CHARACTERISTICS

2-5-1 MOSFETS AS LOAD RESISTORS

In integrated circuits it is often desirable to use one MOSFET as the load resistance for another, as illustrated in Fig. 2-7*a*. In the inverting amplifier shown, Q_1 is the amplifying MOSFET and Q_2 acts as its load resistor. The configuration shown in Fig. 2-7 is useful because of the large difference in physical size between an integrated MOSFET and an integrated resistor. A typical MOSFET occupies an area of about 1 mil by 1 mil. The area occupied by a 10-kΩ resistor can be estimated as follows. The sheet resistivity of silicon doped for use in MOSFET integrated circuitry is about 200 Ω per square [see. Sec. 3-3-1 and Eq. (3-2)]. To make a 10-kΩ resistor would therefore require $10/0.2 = 50$ squares. Assume the minimum practical linewidth due to resolution is 0.5 mil, although 0.1 mil is possible. The resistor would then have to be 25 mils long by 0.5 mil wide to obtain 50 squares of 0.5×0.5 mil. For a 100-kΩ resistor, an area 250×0.5 mil would be required. Thus, resistors of even reasonable value are considerably larger than a typical MOSFET. Larger values in less area could be obtained using differently doped silicon, but this is impractical because doping levels are set by the MOSFET requirements.

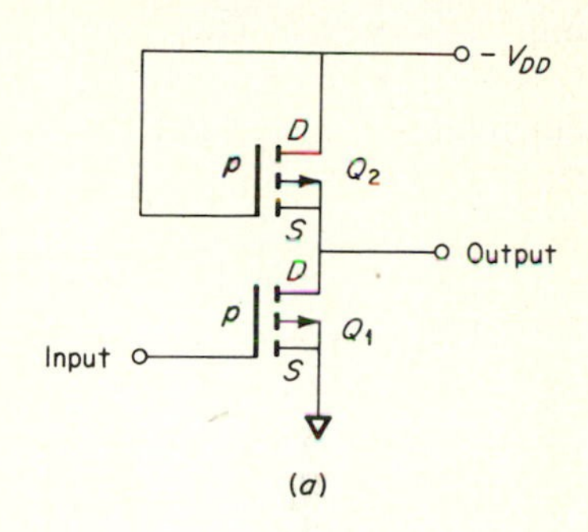

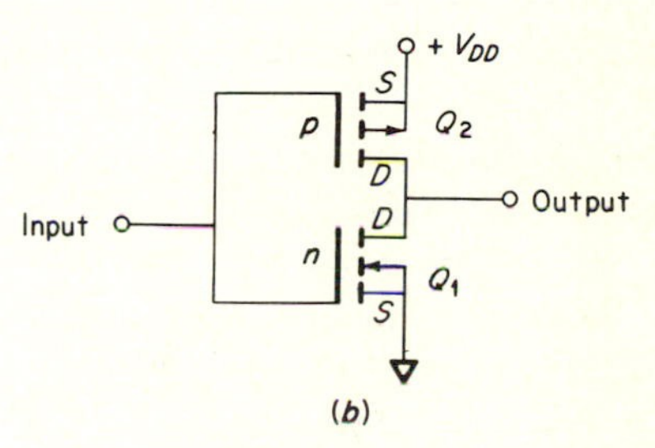

Fig. 2-7 Use of one MOSFET as the load resistor of another: (*a*) inverting amplifier using two *p*-channel enhancement-mode MOSFETs (*b*) complementary MOS (CMOS) inverter using one *p*-channel and one *n*-channel MOSFET.

These considerations become especially important in digital circuits in which it is desirable to have many MOSFET circuits in a small area to achieve a high bit density and hence low cost per bit. Fortunately the resistor can be replaced with another MOSFET, as in Fig. 2-7*a*.

In Fig. 2-7*a*, Q_1 is the amplifier, and Q_2 the load. Since Q_2 is an enhancement-mode device (as in Q_1), it requires a certain voltage from gate to source before it will conduct, the voltage at which conduction begins being called the threshold voltage V_T. For the *p*-channel device shown, there will be a conductive path between the drain and source of Q_2 if the gate-to-source potential of Q_2 exceeds the threshold voltage. No conductive path will exist if the gate-to-source potential is less than the threshold voltage.

Another method of eliminating resistors in MOSFET digital circuits is shown in Fig. 2-7*b*. Here, complementary enhancement-mode MOSFETs are required, resulting in a CMOS circuit. The lower transistor Q_1 is an *n*-channel device, and the upper transistor Q_2 is a *p*-channel. When the input is near ground, Q_1 is OFF because its V_{GS} is below the threshold voltage V_T. Q_2 is ON because its gate is sufficiently negative with respect to its source, which is the requirement for a *p*-channel enhancement MOSFET to be ON. Since Q_2 is ON, its V_{DS} is small, and the output goes to a high or positive level.

When the input is near $+V_{DD}$, Q_1 is ON and Q_2 is OFF so that the output is near ground.

The CMOS circuit has the advantage that power is consumed only during the switching phase. With the input statically high or low, the power dissipation is essentially zero. This is not true for the circuit in Fig. 2-7*a*, in which power is consumed when Q_1 and Q_2 are ON. However, both circuits share the same advantage, namely, that the output can be coupled directly to the gate of the following stage. This follows from the use of enhancement-mode devices as discussed in Sec. 2-4.

Since the input level must exceed the threshold voltage, enhancement-mode MOSFETs have good noise immunity, which is important in digital applications. A circuit with high noise immunity can have large noise spikes on its input without the output level changing.

Both complementary (CMOS) and noncomplementary (PMOS and NMOS) circuits are used extensively in digital systems. Their application in this area is expanded upon in Chap. 5.

2-5-2 ELECTROMETERS

An electrometer is a type of amplifier used for the measurement of very low currents. Figure 2-8*a* shows a configuration using an operational amplifier such as is discussed in Chap. 4. I_{in} is the current to be measured, and it divides at the inverting or negative input terminal of the amplifier into two components, I_a and I_f. I_a is that part of the input current which flows into the amplifier, and

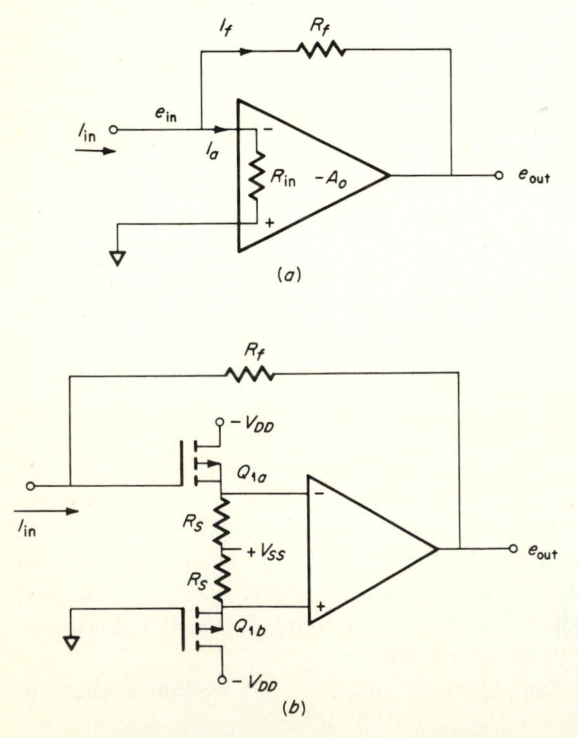

Fig. 2-8 Electrometers: (*a*) basic circuit; (*b*) practical circuit using a dual MOSFET.

it has several components. If there is some output voltage e_{out}, it must result from an input voltage e_{out}/A_o, where $-A_o$ is the voltage gain of the operational amplifier and is large. Also, due to offsets in the amplifier, there will be some offset voltage at the input even when e_{out} is zero or when I_{in} is zero. Thus there will be a voltage from the inverting terminal to ground due to the finite amplifier gain and to offsets. Part of I_a results from the current produced by this input voltage driving a current into the amplifier input resistance R_{in}. There is yet another component of I_a, which arises from the use of active circuit elements such as bipolar transistors, JFETs, or MOSFETs as input devices. All these devices have input leakage currents, input bias currents, or both.

In an ideal amplifier I_a would be zero as indicated by the following:

$$I_{in} = I_a + I_f \tag{2-1}$$

$$I_f = \frac{e_{in} - e_{out}}{R_f} \tag{2-2}$$

$$e_{out} = -A_o e_{in} \tag{2-3}$$

Equations (2-1) through (2-3) can be solved to give

$$e_{out} = -(I_{in} - I_a)R_f \frac{A_o}{1 + A_o} \tag{2-4}$$

If A_o is very large and $I_a \ll I_{in}$, Eq. (2-4) can be simplified to

$$e_{out} = -I_{in} R_f \tag{2-5}$$

e_{out} is therefore proportional to I_{in}, *if* $I_a \ll I_{in}$. If I_a is not $\ll I_{in}$ and is a random, temperature-sensitive drift current, e_{out} will not be an accurate indication of I_{in} since I_a subtracts from I_{in} as shown in Eq. (2-4).

If it is desired to measure very low current levels of, say, 10^{-13} or 10^{-14} A, I_a must be even lower, say 10^{-15} A. Using a MOSFET as the front end of an operational amplifier has two advantages. Because its input resistance is so high, any input offset voltage will drive a negligible current into the MOSFET input resistance. For example, a 10-mV input offset on the gate of a MOSFET having 10^{15} Ω input resistance would produce a current of 10^{-17} A. Secondly, the gate leakage current of a MOSFET can be as low as 10^{-15} A, which is desirable as explained above. This combination of high input resistance and low input current are the two major reasons for the use of MOSFETs in electrometer amplifiers. Other reasons include freedom from microphonics, low power requirements, and small size.

Figure 2-8*b* shows how a dual MOSFET matched pair might be used in conjunction with an operational amplifier to form an electrometer circuit.

In a *hybrid* design the MOSFET stage would be on one chip, and the operational amplifier on another. In a *monolithic* design (see Chap. 3) both the MOSFET stage and the operational amplifier would be on the same chip. Since MOS devices and bipolar transistors can now be fabricated on the same chip, the operational amplifier could be either all MOSFET, all bipolar transistor or a combination of both. The dual MOSFET is not used in this case primarily to obtain matched and hence canceling temperature coefficients, as with the matched bipolar transistor inputs used in dc amplifiers. This is not necessary, since the coefficients can be adjusted to zero or at least close to zero by appropriate selection of the drain current. A matched dual MOSFET is used in this case to simplify the biasing of the operational amplifier which follows. The gain of the operational amplifier is so high, typically 10^5, that virtually no voltage is required across its input terminals to produce any output level up to a few volts. This means that the potential of the negative (or inverting input) must nearly equal that on the positive (or noninverting input) of the operational amplifier. The use of dual matched MOSFETs as source followers simplifies the realization of this requirement. Both gates are at or near 0 V, and since equal source resistors R_S and matched MOSFETs are used, the source potentials are virtually equal to each other.

Resistor R_S is chosen to fix the source current (and hence drain current) to the level required for V_{GS} to have a zero temperature coefficient.

A simplified design procedure for an electrometer circuit with a MOSFET front end, such as shown in Fig. 2-8*b*, might be as follows. It is desired to measure currents from 10^{-13} to 10^{-11} A. Full-scale output voltage is to be -5 V. The matched MOSFETs to be used require $V_{DS} = -20$ V, $V_{GS} = -5$ V, and $I_D = 1$ mA for zero-temperature-coefficient biasing.

First choose supply voltages of $V_{DD} = -15$ V and V_{SS} of $+15$ V. Since the gates are at 0 V, a V_{GS} of -5 V will result in a V_{DS} of -20 V as desired. To set I_D at 1 mA, notice that the drop across each R_S is 10 V so that $R_S = 10\ \text{k}\Omega$. The selection of the operational amplifier is not particularly critical in this case since its relatively large input currents are isolated by the MOSFETs from the low input currents being measured. However, it must be capable of linear, differential amplification with both inputs at -5 V, since they are connected to the sources of Q_{1a} and Q_{1b}.

R_F can be chosen using Eq. (2-5). With an input current of 10^{-11} A and an output voltage of -5 V,

$$R_F = \frac{-e_{\text{out}}}{I_{\text{in}}} = -\frac{-5}{10^{-11}} = 5 \times 10^{11}\ \Omega \tag{2-6}$$

With an input current of 10^{-13} A and $R_F = 5 \times 10^{11}\ \Omega$,

$$e_{\text{out}} = -(10^{-13})(5 \times 10^{11}) = -5 \times 10^{-2} = -0.05\ \text{V} \tag{2-7}$$

Thus as I_{in} varies from 10^{-13} to 10^{-11} A, e_{out} will vary from -0.05 to -5 V.

2-5-3 AMPLIFIERS FOR SIGNALS FROM HIGH SOURCE IMPEDANCES

Certain types of input transducers, such as piezoelectric or capacity transducers, are very-high-impedance devices, especially at low frequencies. Consider, for example, the capacity transducer circuit shown in Fig. 2-9. Q_1 is an enhancement-mode *p*-channel MOSFET used, in this circuit, as an impedance-transforming source follower. This means that it has a voltage gain of slightly less than unity. It has a very high input resistance and a low output resistance, which enables it to couple a signal voltage from a high source impedance to a relatively low-impedance (say 100 kΩ) amplifier without undue attenuation. In this sense the source follower acts as an impedance transformer. R_G is a gate bias resistor, and C is the capacity transducer. A capacity transducer is one in which a capacitance is varied in sympathy with some external influence being measured, for example, pressure. If the pressure is varying at, say, 1 Hz, the transducer reactance will be about 1600 MΩ or 1.6×10^9 Ω. To prevent undue attenuation and phase shift, both R_G and the input resistance R_{in} of Q_1 must be large compared with 1600 MΩ. If 100 times larger provides sufficient accuracy, R_G and R_{in} in parallel must be equal to or greater than 1.6×10^{11} Ω. MOSFETs are available with input resistances of 10^{14} or 10^{15} Ω so that for all practical purposes R_G must equal 1.6×10^{11} Ω.

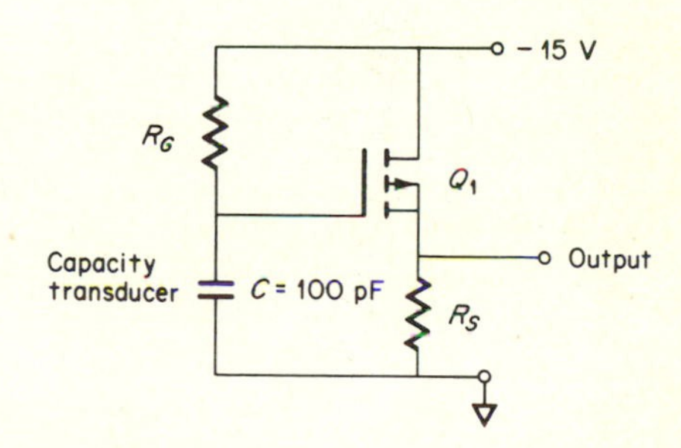

Fig. 2-9 MOSFET source follower used as a capacity transducer preamplifier.

It is possible to use such a large value resistor in the circuit of Fig. 2-9 because the gate leakage current of the MOSFET is so low, typically 10^{-14} A. Such a current level would result in a voltage drop across R_G of only $1.6 \times 10^{11} \times 10^{-14} = 1.6 \times 10^{-3}$ V or 1.6 mV. If a JFET had been used, the drop would have been considerably more. The gate leakage current of a JFET might be 10^{-11} A, and the drop across R_G would then be 1.6 V. This 1.6 V would vary drastically with temperature, since a JFET gate current results from a reverse-biased *pn* junction, in contrast to that of a MOSFET. In a capacity-transducer circuit such as that in Fig. 2-9, the sensitivity, or "volts out for a given pressure change in," for example, depends on the bias voltage across the capacity transducer. With a JFET the bias would vary appreciably with temperature, whereas with a MOSFET the effect would be negligible.

The MOSFET has an additional advantage over the JFET in this application, since for source impedances of about 100 MΩ or more the MOSFET is a lower-noise device.

2-5-4 MOSFETS AS OUTPUT DEVICES

It is often important for the voltage of the output stage of a multistage amplifier (such as the operational amplifiers discussed in Chap. 4) to be able to swing close to the plus and minus power-supply potentials. For example, if ± 15-V power supplies are used, an amplifier capable of delivering ± 14.99 V is more versatile than one which can supply only ± 14 V. In this respect output stages made from MOSFETs are superior to those built with bipolar transistors.

Figure 2-10*a* shows the outline of a multistage amplifier with a complementary-bipolar-transistor output stage. Owing to the construction of a bipolar transistor, the output terminal is always connected to the supply rails via a *pn* junction in this circuit. Figure 2-10*b* shows a *pn*-junction (diode) current/voltage characteristic, and it can be seen that any significant current flow, such as I', produces a voltage drop. This voltage drop is called an *offset*, and it prevents the output terminal of Q_1 or Q_2 in Fig. 2-10*a* from getting to within 0.6 V of either of the plus or minus supply rails when any significant current is drawn

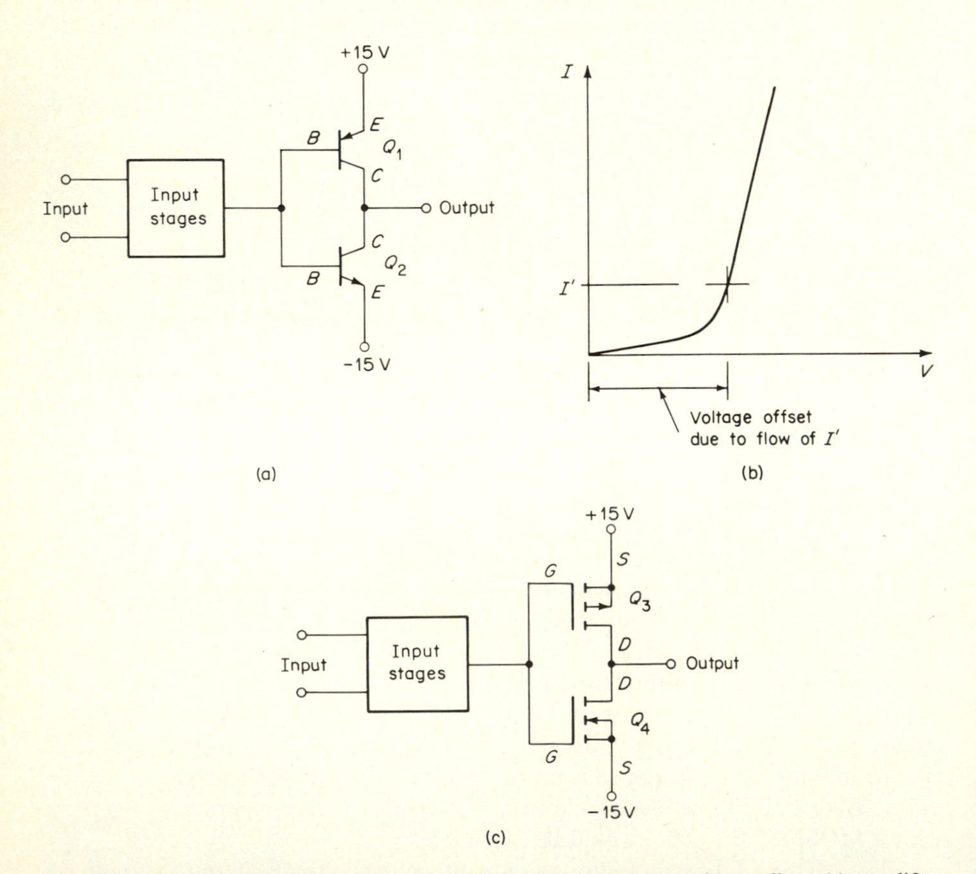

Fig. 2-10 (*a*) Amplifier with bipolar output stage; (*b*) *pn*-junction voltage offset; (*c*) amplifier with MOSFET output stage.

from the amplifier. This is true even if the series resistances in the transistor are low, since it is the *pn* junction that produces the offset. Since the amplifier in Fig. 2-10*c* uses MOSFETs in the output stage, there are no *pn* junctions in the paths between the output and either supply rail. Instead there is a purely resistive path or channel. Since the resistance of the channel can be made very low, current flow through it will not produce any appreciable voltage drop, and the output voltage can approach the supply-rail voltages to within a few millivolts. The RCA 3130 is an example of an operational amplifier with a complementary MOS (CMOS) output stage. This amplifier is a good example of an IC with different devices on a single chip, since it has PMOS, NMOS, bipolar transistors, a zener diode, regular signal diodes, and resistors all on one chip. (See Sec. 3-4-3.)

2-5-5 MOSFETS AS CHOPPERS

MOSFETs can be used as choppers in much the same way as JFETs, as described in Sec. 1-6. Like the JFET, there are no *pn* junctions between source and drain when the device is ON, so there are no junction offset voltages. As with a JFET, the OFF resistance is very high.

MOSFETs have certain advantages over JFETs in chopper circuits. The gate leakage current of a JFET, although low, increases exponentially with temperature. When measuring microvolt signals the leakage current can give rise to dc offset errors, since it must flow through the device channel. The gate leakage current of a MOSFET is much lower at room temperature than that of a JFET and does not increase as rapidly with temperature. For applications involving high-temperature environments, MOSFETs are superior in this respect to JFETs.

Another advantage that MOSFETs have over JFETs in switching applications results from the simpler gate circuitry required by a MOSFET switch. In a MOSFET the gate is connected to the channel via an insulator, so that no protective diode is necessary. To turn a MOSFET ON, it is necessary only to apply a large enough gate signal of the appropriate polarity. To turn a MOSFET OFF, a large enough signal of the opposite polarity is all that is required. Reasonable overvoltages (less than those that would damage the device) have no effect on the ON and OFF properties of a MOSFET. Protective diodes are usually incorporated to protect against overvoltages, however.

2-5-6 THE MOSFET AT HIGH FREQUENCIES

MOSFETs have properties which make them highly suitable as RF amplifiers, oscillators, and mixers at both VHF and UHF. Many of the reasons outlined in Chap. 1 for the use of JFETs at high frequencies apply also to MOSFETs. The transfer characteristic (I_D versus V_{GS}) closely approximates that of a square-law device. This reduces cross modulation, which is the transfer of information from an unwanted carrier to the carrier frequency to which a receiver is tuned. Also, MOSFETs can be made with very low values of feedback capacitance.

2-5-7 MOSFETS AT VERY LOW TEMPERATURES

It is sometimes necessary to operate amplifiers at temperatures as low as $-269°C$, which is only about 4°C from absolute zero, the lowest temperature possible. An example of such an application occurs in infrared (IR) measurements. Any heat in the IR detector or its surroundings results in their radiating IR which mixes with the IR being measured; the two signals cannot be distinguished (Sec. 7-1). Such detector assemblies are therefore cooled with liquid helium whose temperature is about 4°C above absolute zero. Since the IR detectors often have high impedance, it is advantangeous to mount a high-input-impedance, low-output-impedance preamplifier close to the detector. This means that the preamplifier is also cooled to the low temperature.

A MOSFET source follower makes an ideal impedance transformer (high impedance in, low impedance out), and it has been found that many MOSFETs will operate satisfactorily at the required low temperature.

2-6 MOSFETS IN INTEGRATED CIRCUITS

Both JFETs and MOSFETs are suitable for various types of integrated circuits. In practice, however, most FET integrated circuits use MOSFETs, and most of these are digital.

MOSFETs are preferred over JFETs in digital integrated circuits for several reasons. JFETs are always operated in the depletion mode, so that a *p*-channel device requires a negative supply for the drain and a positive gate turn-OFF voltage. An *n*-channel JFET requires a positive drain supply and a negative gate turn-OFF voltage. In both cases, either dual-polarity power supplies must be used or relatively elaborate biasing schemes must be employed to achieve operation with a single power supply. Enhancement-mode MOSFETs, on the other hand, can be turned ON and OFF with a single-polarity supply. Biasing problems are further simplified because various logic elements can be direct-coupled. Even resistors are not required in the coupling network, because the gate and drain of enhancement-mode devices can be designed to operate at similar potentials. Resistors are also not required as drain loads, as explained in Sec. 2-5-1. The elimination of resistors from MOSFET digital circuits means that many more circuit elements can be fabricated on a standard-size chip, since an average resistor is so much larger than a MOSFET. This in turn makes possible the fabrication of very complex circuitry on a single chip if MOSFETs are used, and it results in lower cost.

Since a given circuit function can be made in such a small area using MOSFETs, the number of unacceptable circuits on a wafer due to wafer imperfections is reduced. This is illustrated in Fig. 2-11. Figure 2-11*a* shows a standard-sized wafer upon which have been fabricated four integrated circuits. The shaded portion represents an area of the wafer containing imperfections which spoil circuits fabricated on this portion of the wafer. Such areas are often present on

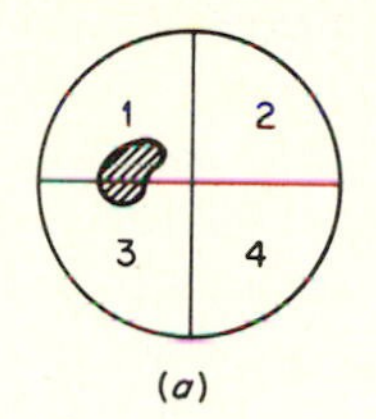

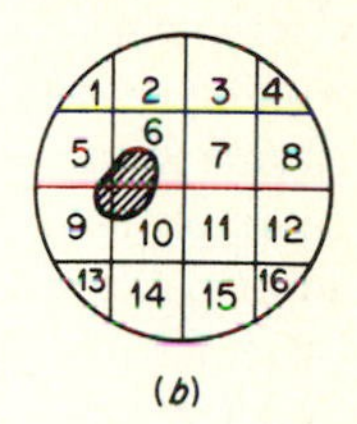

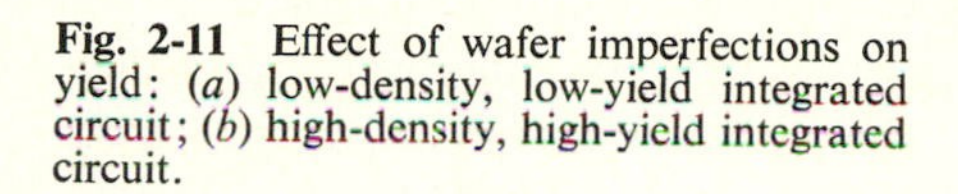
Fig. 2-11 Effect of wafer imperfections on yield: (*a*) low-density, low-yield integrated circuit; (*b*) high-density, high-yield integrated circuit.

wafers and may be due to crystal imperfections or faulty processing. Only two of the four circuits on the wafer shown in Fig. 2-11*a* are "good." The yield, which is the ratio of "good" circuits to the maximum possible number of circuits (see Chap. 3), is therefore 50 percent. In contrast, the yield for the wafer shown in Fig. 2-11*b* is 75 percent, since only four circuits are spoiled, and this is a direct result of making smaller-area circuits. If the shaded, or imperfect, area of the wafer occupies 10 percent of the total available area, the yield obtained for infinitely small circuits would be 90 percent. This is, of course, an unattainable limit, but we have just demonstrated that a 75 percent yield is possible with only 16 circuits. The small size of MOSFET circuits produces high yields, which is a principal factor in the low cost of MOSFET circuitry.

The high input impedance of MOSEFTs means that high *fan-outs* are possible. Fan-out is a measure of the number of digital circuits that can be connected to the output of the digital circuit whose fan-out is being specified. Theoretically the fan-out for MOSFETs should be extremely large because of their high input impedance. In practice, very high fan-outs are possible only at dc because of limitations due to excessive load capacity if the fan-out is too high.

Digital integrated circuits are among the most important applications of MOSFETs. They are discussed further in Chap. 5, as is a type of bipolar-transistor logic called I^2L, which has advantages similar to MOS logic, since load resistors are eliminated in I^2L also.

2-7 MOSFET, JFET, AND BIPOLAR-TRANSISTOR COMPARISON

The selection of a device for a particular circuit depends, naturally, on the application. Very often there is considerable overlap in the suitability of MOSFETs, JFETs, and bipolars for a particular application. For example, all three can be used quite successfully as switches. Sometimes an external factor will cause a device to be selected which might not theoretically be the ideal. There are ways, however, in which each device is superior to the other two, and the most distinct differences result from the increasing input impedances as one considers bipolars, JFETs, and MOSFETs.

Consider, for example, circuits in which drift and low-frequency noise are the most important factors. The selection of the most suitable device depends on the source resistance. For source resistances of a few thousand ohms or less, the bipolar transistor will have the lowest noise and drift. With resistances from roughly 10 kΩ to about 10 or 100 MΩ, the JFET is best. Above 100 MΩ the MOSFET gives the best drift and noise performance. Thus, a dc amplifier for a thermocouple (<1 Ω) sensor would probably be made from bipolar transistors. For operational amplifiers (see Chap. 4), JFETs are very suitable because they have adequately low voltage offset and drift, very low current offset and drift, and high input impedance. Operational amplifiers with JFET front ends are more universal in application than those made with bipolars, because a wider range of values of external feedback elements can be used. For electrometer circuits, where exceptionally low currents are to be measured, MOSFETs are most nearly ideal.

The superiority of the devices mentioned for each of the applications above can be explained by a means which is very helpful in selecting the best device for a particular job. Any transistor, be it a bipolar, JFET, or MOSFET, can be represented by an equivalent circuit consisting of an ideal device with certain nonideal characteristics shown separately as in Fig. 2-12. The ideal transistor is

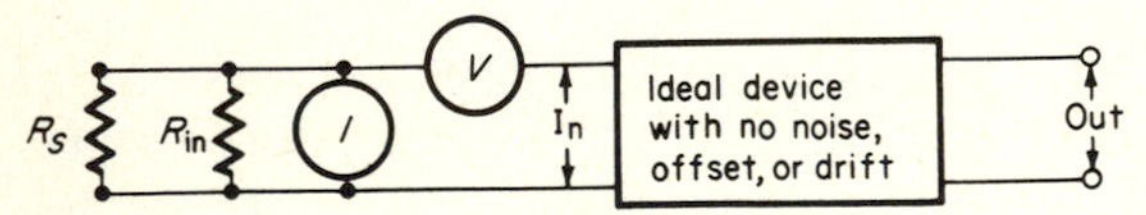

Fig. 2-12 Ideal device (MOSFET, JFET, or bipolar) with nonideal characteristics shown externally.

represented by the reactangular block. V and I are voltage and current generators and can be noise, drift (which is very-low-frequency noise), or offsets. R_{in} is the input impedance of the transistor, and R_s is the impedance of the signal source.

Current I will flow through the parallel combination of R_s and R_{in} and produce a voltage drop. Thus, at the input of the device, there will be two voltage generators, one due to V, and one due to I, R_s, and R_{in}. Their sum will be

$$V + I\frac{R_{\text{in}} R_s}{R_{\text{in}} + R_s}$$

and will represent the total equivalent input noise, drift, or offset. Correlation between V and I is assumed zero to simplify the explanation.

It turns out that V is lowest for bipolars and highest for MOSFETs, with JFETs in between but nearer to bipolars. This applies to noise, drift, and offsets. I is highest for bipolars and lowest for MOSFETs, with JFETs once more in between. Again, this applies to noise, drift, and offsets.

If R_s is low in value, $(I)(R_s \| R_{\text{in}})$ will be small and V will be the dominant term. Since bipolar transistors have the lowest V, they are best for low-source-impedance circuits.

If R_s is very large, R_{in} must be large to prevent undue signal attenuation, and $(I)(R_s \| R_{\text{in}})$ will be the dominant term. Since MOSFETs have the lowest I, they are best for extremely-high-source-impedance circuits.

For intermediate values of R_s, JFETs are best since their V is lower than that of a MOSFET and their I lower than that of a bipolar transistor.

Device selection is also made on many other grounds. MOSFETs and JFETs make very good switches and gates, since they are equivalent to low ohmic resistors when ON. Bipolar transistors, however, are faster. MOSFETs and JFETs are good for RF applications because they are square-law devices, although at RF their noise is comparable to that of bipolars. MOSFETs are very small and can be both active and passive (resistors) in digital circuits. This small size, coupled with the relatively few processing steps required in their fabrication, makes MOSFETs suitable for highly complex digital arrays.

Bipolar transistors have a higher transconductance (and therefore higher voltage gain) in general than do FETs. Bipolars are therefore still the "workhorses" of many circuits, even where FETs are used for their special characteristics. For example, in the electrometer amplifier shown in Fig. 2-8*b*, MOSFETs are used in the input stage, but the rest of the amplifier could use bipolars, probably in integrated-circuit form.

PROBLEMS

2-1 Describe the construction of an enhancement- and a depletion-mode MOSFET, both p channel. Say which is referred to as normally ON and which as normally OFF, and why.

2-2 Draw the symbols for enhancement and depletion-mode MOSFETs, and explain the reason for:

(*a*) The gap separating the gate from the rest of the symbol

(*b*) The solid line in depletion MOSFETs and the broken line in enhancement types

(*c*) The direction of the substrate-to-channel diode arrow

2-3 Explain why the output and transfer characteristics of enhancement- and depletion-mode MOSFETs have the shape, voltage, and current polarities they do.

2-4 Discuss "gate breakdown" in MOSFETs. What are the advantages and disadvantages of zener-diode gate protection?

2-5 Design a source follower using a p-channel enhancement-mode MOSFET. Supply voltages are ± 20 V, and the gate is biased at 0 V with $I_D = 3$ mA and $V_{GS} = +7$ V. $R_{in} = 100$ MΩ.

2-6 For the circuit of Prob. 2-5, find the gain and output resistance. The MOSFET transconductance is 2000 μmho.

2-7 Discuss the biasing of a MOSFET substrate.

2-8 Enhancement-mode MOSFETs have certain advantages over all other amplifying devices in their biasing. What are these advantages? Illustrate them with a circuit.

2-9 Why are MOSFETs especially suited to digital integrated-circuit applications?

2-10 What is an electrometer, and why are MOSFETs particularly useful in electrometer circuits?

2-11 Design a feedback electrometer circuit such as in Fig. 2-8*b* to measure currents from 10^{-12} to 10^{-10} A with a 10-V full-scale output.

2-12 Give three reasons why a MOSFET would be used as the front end of an amplifier intended for use with a 100-pF piezoelectric transducer at low frequencies.

2-13 Compare MOSFETs with JEFTs and bipolar transistors, giving the advantages and disadvantages of each.

2-14 Explain the logic behind the standard MOSFET symbols shown in Fig. 2-4.

2-15 Why are MOSFETs available in both enhancement and depletion modes while JFETs operate almost invariably in the depletion mode? Describe a situation in which a JFET is used in the enhancement mode.

2-16 An *n*-channel enhancement-mode MOSFET is to be used as a common-source amplifier as in Fig. Prob. 2-16. $V_{GS} = +5$ V, and $g_{fs} = 10{,}000$ μmho. C bypasses R_S completely at frequencies of interest. Find:

(*a*) R_S and R_D

(*b*) V_{DS}

(*c*) The voltage gain from gate to drain

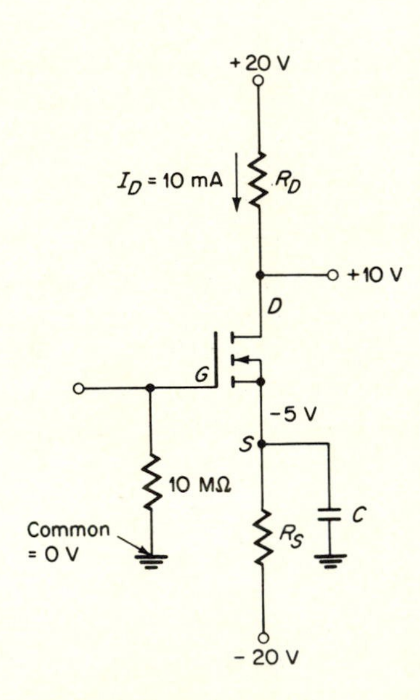

Fig. Prob. 2-16

2-17 If a 1-kΩ resistor is inserted in series with C in Fig. Prob. 2-16, find:

(*a*) The voltage gain from gate to drain if the reactance of C is negligible compared with 1 kΩ

(*b*) V_{DS}

part 2 Integrated Circuits

Integrated circuits can be divided into two general categories, monolithic and hybrid, according to how they are made. The word *monolithic* is derived from two Greek words, *monos* meaning single and *lithos* meaning stone, and therefore literally means "of a single stone." As the name implies, a monolithic device is one in which a complete circuit, including active and passive elements and all interconnections (except those going to external leads), are formed upon and within a single piece of silicon crystalline material. *Hybrid* integrated circuits are made by a variety of techniques, but generally involve mounting separate parts on an insulating substrate, with interconnections made by wires or a metallization pattern.

Thus, monolithic integrated circuits are those fabricated upon and *within* a semiconducting substrate, which in fact acts as the circuit "chassis." The chassis for hybrid integrated circuits, on the other hand, is an insulator which is added to, not itself modified.

Figure 3-1*a* and *b* show the appearance of circuits made by the two methods. Figure 3-1*a* is a monolithic microprocessor, while Fig. 3-1*b* is a circuit fabricated using hybrid techniques. More is said about monolithic and hybrid construction later in this chapter,

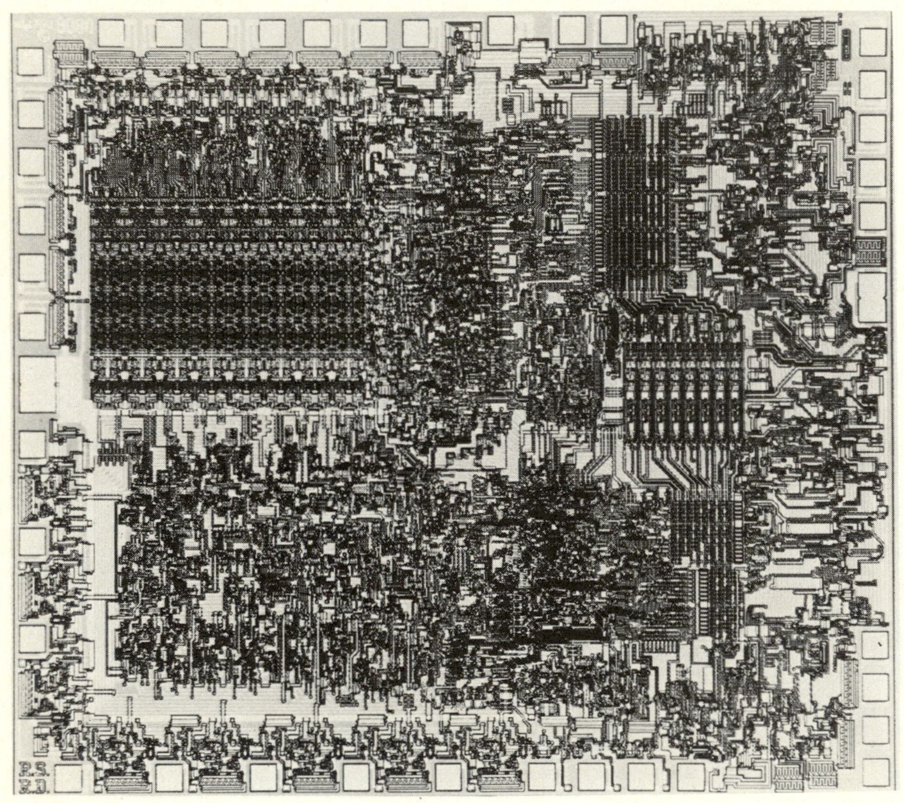

Fig. 3-1 (*above*) Monolithic integrated circuit, the Intel 8008 microprocessor. *Intel Corp.* (*below*) Hybrid integrated circuit. *Hybrid Systems Corp.*

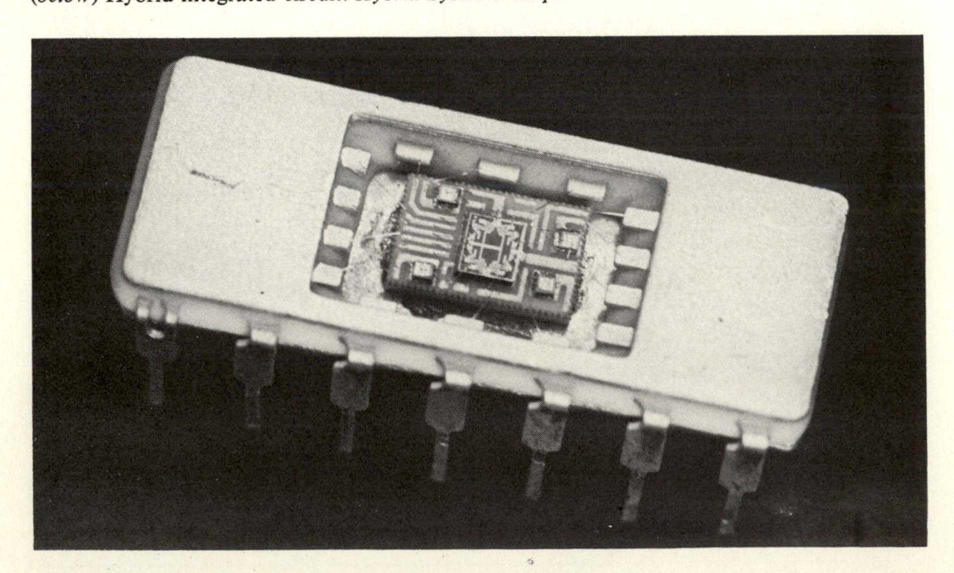

but the photographs are included here to give the reader some idea of the appearance of typical integrated circuits.

The small size of integrated circuits is advantageous in applications such as spacecraft electronic systems, where weight and volume are critical. It turns out, however, that small size is not the primary reason for the use of integrated circuits. They have many other advantages over discrete-component circuits apart from size, and a major one is low cost.

Integrated circuits, and especially monolithic integrated circuits, are made by a sequence of production steps in which each step is shared by hundreds or even thousands of other identical circuits. Thus fabrication is by a means which is conducive to high-volume production, which of course means low cost. The reduction in handmade or wired interconnections helps reduce cost, and also is a principal factor in the high reliability of integrated circuits. Connections, whether soldered, welded, or of the connector variety, have long been known to be the weak link in the struggle to achieve high reliability. The increase in reliability in integrated circuits, due to the decrease in the number of handmade interconnections, has another more subtle but far-reaching effect. The degree of complexity it is theoretically possible to achieve in an electronic system is limited to a large extent by the number of interconnections in that system. This is because if the number of interconnections is large enough, some of them, and hence the system, will be unreliable. By using integrated circuits, systems of increased complexity with fewer interconnections and greater reliability are possible. This in turn means that systems of greater sophistication are more possible than before, now that integrated circuits are available.

Integrated circuits can be classified according to their complexity. An operational amplifier or a *JK* flip-flop would simply be called an integrated circuit. The term *large-scale integration* or LSI is applied to integrated circuits of much greater complexity than these two basic devices. LSI therefore refers to an oversized integrated circuit, which is to an integrated circuit what the integrated circuit is to the individual transistor. MSI or *medium-scale integration* is an intermediate step between integrated circuits, such as operational amplifiers or *JK* flip-flops, and LSI. Both LSI and MSI are discussed in more detail in Chap. 5, but their significance is that they are building blocks at the system level.

A further factor which reduces cost is the use of multiapplication integrated circuits. Examples of such circuits are the operational amplifier and the microprocessor. These are discussed in Chaps. 4 and 5, respectively. Although one is a linear circuit and the other digital, they have the property in common that a wide variety of different circuit functions can be achieved with each. They are in fact circuit building blocks, and they eliminate the necessity for a complete detailed redesign with each new circuit requirement. Since the same basic circuit can be used in many applications, production is high volume and costs are reduced.

Sometimes low power is given as a further advantage of integrated circuits, but there is nothing about integrated circuits that makes them particularly low power. In fact, if low power were the major consideration, discrete-component circuits would usually be better.

This part on integrated circuits has three chapters. Chaper 3 is on linear integrated circuits (LICs) and covers fabrication, circuit design, and specific LICs. Chapter 4 covers operational amplifiers, which, when integrated, are also LICs. However, operational amplifiers are so widely used that they are important enough to warrant a separate chapter. Operational amplifiers are also made in discrete-component form, but are supplied in a package that makes them, nevertheless, essentially a single device. Chapter 5 covers digital integrated circuits.

chapter 3

Linear Integrated Circuits

3-1 INTRODUCTION

Since this chapter discusses *linear* integrated circuits, it is worth explaining what is meant here by linear. Electronic circuits can be broadly divided into two areas, namely, analog and digital. In digital circuits the active elements are used primarily as switches, since they are either ON or OFF, but are not operated in between. In analog circuits, various input levels produce corresponding levels at the outputs of the active elements. Frequently there is a linear relationship between the input and the output, and integrated circuits operating in this mode are called linear integrated circuits. "Analog" and "linear" tend to have somewhat overlapping meanings. One difference is that "analog" often refers to computing circuits or signal-conditioning equipment in telemetry systems. Also, analog circuits can be nonlinear, as for example in the generation of square-law functions, but they are still not ON or OFF, as in a switch. The best-known example of a linear integrated circuit is the operational amplifier, which will be discussed in Chap. 4. In this chapter we shall be considering the other types of linear integrated circuit.

A discussion of the means by which integrated circuits are fabricated involves the use of a number of words having specific meanings when applied to such circuits. To facilitate the description of integrated-circuit fabrication, important definitions are tabulated in Table 3-1.

3-2 INTEGRATED-CIRCUIT FABRICATION

Although integrated-circuit fabrication takes many forms, it is convenient, for purposes of discussion, to divide the fabrication methods into two general categories, *hybrid* and *monolithic*. A *monolithic* device is one in which a complete circuit, including active and passive elements and all interconnections (except those going to external leads), are formed upon and *within* a single piece of crystalline silicone material. *Hybrid* integrated circuits are made by a variety of techniques, but generally involve mounting separate parts on an insulating substrate with interconnections made by wires or a metallization pattern. Thus, monolithic integrated circuits are those fabricated upon and within a semiconducting substrate which acts as the "chassis" for the circuit. The chassis for

TABLE 3-1 Integrated circuit definitions

Bonding	The joining together of two materials. For example, the attachment of wires to an integrated circuit or the mounting of an integrated circuit to a substrate.
Bonding pads	Areas of metallization on the integrated circuit die, which permit connection to the die of fine wires or circuit elements.
Cermet	A mixture of ceramic and metal powders from which film resistors are made.
Chip	A part of a wafer of silicon upon which an integrated circuit or a component such as a transistor is fabricated.
Dice	Plural of die.
Die	Same as chip.
Die bonding	The attachment of an integrated circuit chip to a substrate or header.
Epitaxial growth	The process in which the single crystal structure of a silicon substrate is extended and modified at its surface by the addition of dopants and more silicon. The latter are passed over the slice of silicon in an epitaxial furnace, in which the slice is heated by radio-frequency energy. Dopant and silicon atoms are located and held in the crystal lattice by means of interatomic forces, thereby modifying the electrical characteristics of the slice. The essential feature of epitaxial growth is that layers of semiconductor material are deposited on the substrate with the same crystal orientation as exists in the substrate. This enables very accurate control of the composition and thickness of the deposited layers. Since the deposited layers are used to create circuit elements, this accuracy is of great value to the circuit designer.
Diffusion	The introduction of controlled but small quantities of material into a crystal structure. Quantities, purity, and conditions are accurately controlled and temperatures are high. The effect of diffusion is to modify the electrical characteristics of the crystal structure.
Dopant	An impurity, introduced under highly controlled conditions, in very small but accurately known quantities into a silicon slice. Dopants modify the electrical characteristics of the silicon material by creating *p* or *n* regions and hence *pn* junctions.
Etching	Applied to integrated circuits this means the removal of surface material from a chip by chemical means. Etching is selective since photoresist materials prevent removal of material from protected regions. The selectivity permits windows to be cut, through which subsequent diffusions can be performed.
Film integrated circuit	A circuit in which the elements are formed by various processes upon an insulating substrate. A thin-film circuit employs films, vacuum-deposited by sputtering, evaporation, or chemical vapor deposition. Thin-film thicknesses range from 10 to 10^4 angstroms. In a thick-film circuit, films are obtained by printing patterns on an insulating substrate and curing them at high temperatures. Screen printing is the most commonly used technique. Thick films range from 10^2 to 10^6 angstroms thick.

TABLE 3-1 (Cont.)

Flip-chip	A type of unpackaged chip in which connection to the substrate combines two operations. The chip is both bonded to the substrate and its terminals are connected in the same operation. This is in contrast to "unflipped" chips, which are first mounted face up and then have their connections made to the substrate by means of fine wires.
Hybrid integrated circuit	A circuit in which the components are mounted on a ceramic substrate and interconnected by wire bonds or a metallization pattern. A hybrid circuit may contain one or more monolithic integrated circuits and may use a combination of film and monolithic techniques.
Ion implantation	A precise method of doping. Ions of the appropriate dopant (such as boron or phosphorus) are accelerated to high enough energies that they can penetrate an IC wafer target. The energy controls the depth of penetration and since the amount of energy can be accurately controlled the doping can be very precise. Areas not to be doped are masked with oxide or aluminum.
Isolation	A technique for electrically separating circuit elements. In dielectric isolation components are isolated by means of insulating layers. In diode isolation components are isolated by means of reverse-biased *pn* junctions.
Metalization	The step during which the various circuit components are interconnected. Aluminum for example is evaporated in a vacuum chamber over the entire wafer and contacts components through holes cut in an insulating oxide layer. The desired pattern is left after photo-etching.
Monolithic integrated circuit	A circuit in which all components are formed upon or within a single piece of silicon crystalline material.
Planar process	A technique in which the various semiconductor junctions are brought to a common plane surface. The significance of the technique is that it permits the formation of a junction beneath a protective layer of insulating material, usually silicon dioxide. (See Fig. 3-2.) Thus, many of the problems arising from devices with exposed junctions are avoided.
Substrate	An insulating support upon or in which an integrated circuit is fabricated or mounted.
Thermal compression bonding	The bonding of wires to metal pads on an integrated circuit chip by means of a combination of heat and pressure, which causes plastic flow of the two materials. No solder or melting is used.
Yield	The ratio of the number of acceptable integrated circuits produced in a production run to the total number that were attempted to be produced.

hybrid integrated circuits, on the other hand, is an insulator which is added to, not modified itself.

Hybrid-integrated-circuit fabrication itself has several variations. In thick-film hybrid integrated circuits, an interconnection pattern and circuit resistors are deposited on an insulating base or substrate, which is usually a ceramic material. Transistors, diodes, and capacitors are then bonded by various means to the interconnect pattern to complete the circuit. Thus, the construction of thick film circuits has similarities to the construction of printed-circuit-board assemblies. However, in thick-film circuitry there is another step prior to the addition of external components—namely, the curing and firing of the resistors by baking at high temperatures. Notice that on a printed-circuit board only an interconnect pattern is printed, whereas in a thick-film circuit both the interconnect pattern and circuit resistors are bonded onto the insulating substrate. Another difference is that the resulting thick-film circuit is physically very much smaller than a corresponding printed-circuit version of the same circuit.

In thin-film hybrid integrated circuits, screen printing is not used. Instead, films are deposited by sputtering, evaporation, or chemical vapor deposition through a mask. Resistors, capacitors, and even active elements such as transistors can be deposited in this manner, although thin-film transistors are still experimental. As the names imply, thick films are usually thicker than thin films, but not always. (See Table 3-1.) The essential difference between the two methods is not in the relative thickness, but rather in the manner in which the films are deposited.

Multichip hybrid integrated circuits are, as the name indicates, made by connecting several chips together. The chips may be resistors, capacitors, transistors, or even fairly complicated monolithic integrated circuits. The chips are mounted on an insulating substrate and interconnected by wire bonding.

Monolithic integrated circuits are made by a sequence of steps in which a complete circuit, including active and passive elements, are formed in a single piece of silicon crystalline material. Monolithic circuits represent an extreme example of the differences in construction that exist between integrated and discrete-component circuits. For this reason their fabrication will be covered in more detail than that of hybrid integrated circuits.

Figure 3-2*a* shows a circuit which contains four of the components most commonly used in integrated circuits—namely, a capacitor, a resistor, a diode, and an *npn* transistor. We shall use this circuit to illustrate the basic principles of monolithic circuit fabrication. It should be understood, however, that there are many refinements and variations on the methods outlined here, and that these are necessary to produce practical, economical circuits.

Figure 3-2*b* shows the four basic materials that will go to make up the final circuit. Many other materials are involved in the actual processing. These include dopants to change the intrinsic (or electrically neutral) silicon into *p* or *n* material, an oxidizer to produce silicon dioxide, etchants to remove various layers, and photosensitive materials. A reminder of the meaning of dimensions commonly used in integrated-circuit work is included in Fig. 3-2*b*. These are important because the cross sections shown in Fig. 3-2 are not drawn to scale. If they were, the oxide layers and diffusions would be too thin to distinguish.

First we shall describe the fabrication steps involved, paying particular attention to the *npn* transistor. We shall then examine the structure we have obtained to see how the diode, resistor, and capacitor were made in the process.

The starting material is a bar of *p*-type silicon, a few inches long and 2 to 3 in [50.8 to 76.2 mm] in diameter. The bar is actually a single crystal of silicon to which a *p*-type dopant, often boron, was added during its growth. The bar is sliced with a diamond cutter into thin wafers roughly 0.01 in [0.254 mm] thick. After lapping and etching to produce a smooth surface, the wafers are ready for the sequence of steps that will result in the monolithic circuits desired.

The circuit shown in Fig. 3-2*a* will occupy only a small portion of the area available on a single wafer, and many such circuits can be made simultaneously on one wafer. Therefore, although we shall be describing the manufacture of one circuit, it should be remembered that many are being made at the same time. This high-volume production is one feature of monolithic circuits which results in their low cost.

In the next step an *n*-type layer is epitaxially grown upon the *p*-type substrate. In epitaxial growth (see Table 3-1), the *n*-type layer is formed upon the substrate by placing the wafers in a high-temperature reaction furnace (or epitaxial furnace) into which appropriate gases are introduced. By chemical reaction, the grown *n* layer assumes the same crystal orientation as the *p* substrate, so that the resulting *pn* wafer is still a single crystal. Notice that the epitaxial process results in physical growth of the crystal by perhaps 10 percent, so that the wafer actually becomes thicker. This is in contrast to the subsequent diffusion steps in which material diffuses *into* the water but does not thicken it.

The *n*-type layer is then covered with a thin coating of silicon dioxide, a glasslike material that is similar to quartz. As we shall see, this oxide performs several functions. As an insulator it isolates the final interconnect metallization pattern from circuit elements and can be used as a capacitor dielectric. As a chemically inert material it seals any *pn* junctions, thereby protecting them from contamination; it also forms a mask through which diffusions can be selectively made.

The oxide layer is grown by placing the wafer in an oxidizing atmosphere at high temperature. After this step, the resulting structure is as shown in Fig. 3-2*c*.

"Windows," which are openings in the oxide layer, are now cut so that the first of three diffusions, the isolation diffusion, can be made through the windows. The windows are cut by the following sequence of steps. First the wafer is covered with a photosensitive material and exposed to light through a mask. Special high-resolution masks are required since linewidths of as little as 0.0001 in [0.0025 mm] are commonly used. The photosensitive material not exposed is dissolved away, and the oxide not protected by the exposed photosensitive material is etched away, forming windows as shown in Fig. 3-2*d*. The wafer is now ready for the isolation diffusion.

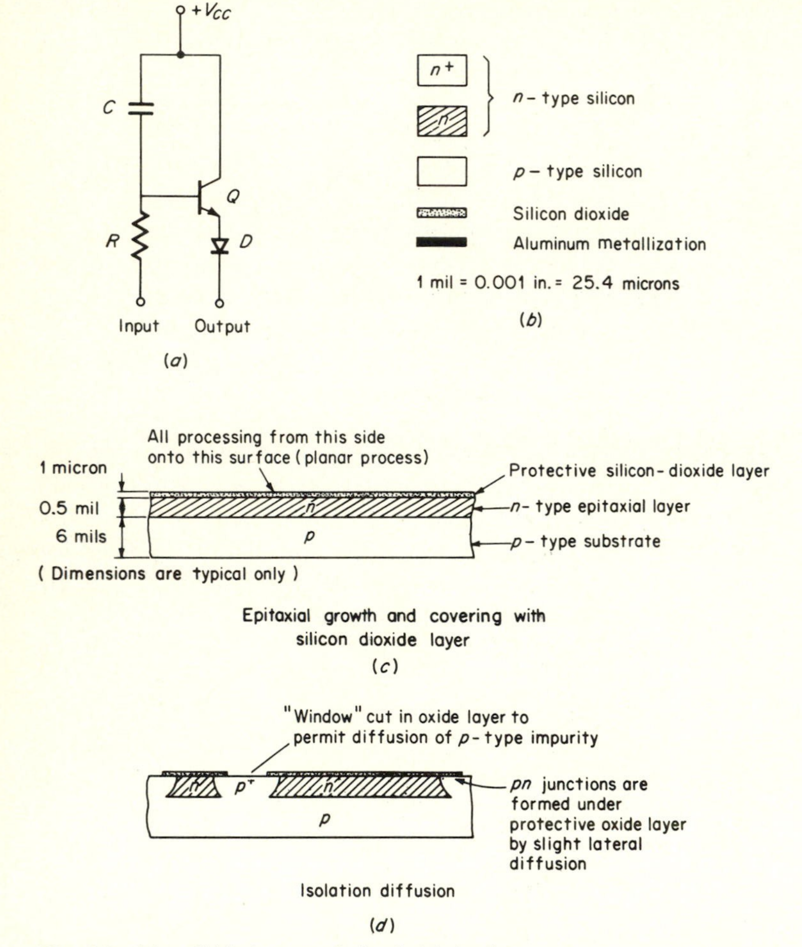

Fig. 3-2 Monolithic integrated-circuit fabrication.

The isolation diffusion is performed by placing the wafer in a furnace having an atmosphere which contains a p-type dopant, often boron in the form of one of its compounds. The p-type dopant diffuses through the epitaxial n-type layer into the original p-type substrate, thereby *isolating n* regions or *pools*, as shown in Fig. 3-2*d*.

Notice that the p regions under the windows in Fig. 3-2*d* are marked p^+. This indicates that the p doping is highly concentrated, which is necessary to overcome the original n-type epitaxial doping. Notice also that the two pools of n material are isolated electrically by the two back-to-back (*np-pn*) diodes formed by the isolation diffusion.

Figure 3-2*e* shows the structure of the wafer after the next diffusion, the base diffusion. This follows the reoxidization of the plane surface through which all processing is taking place. Windows are cut as before, so that dif-

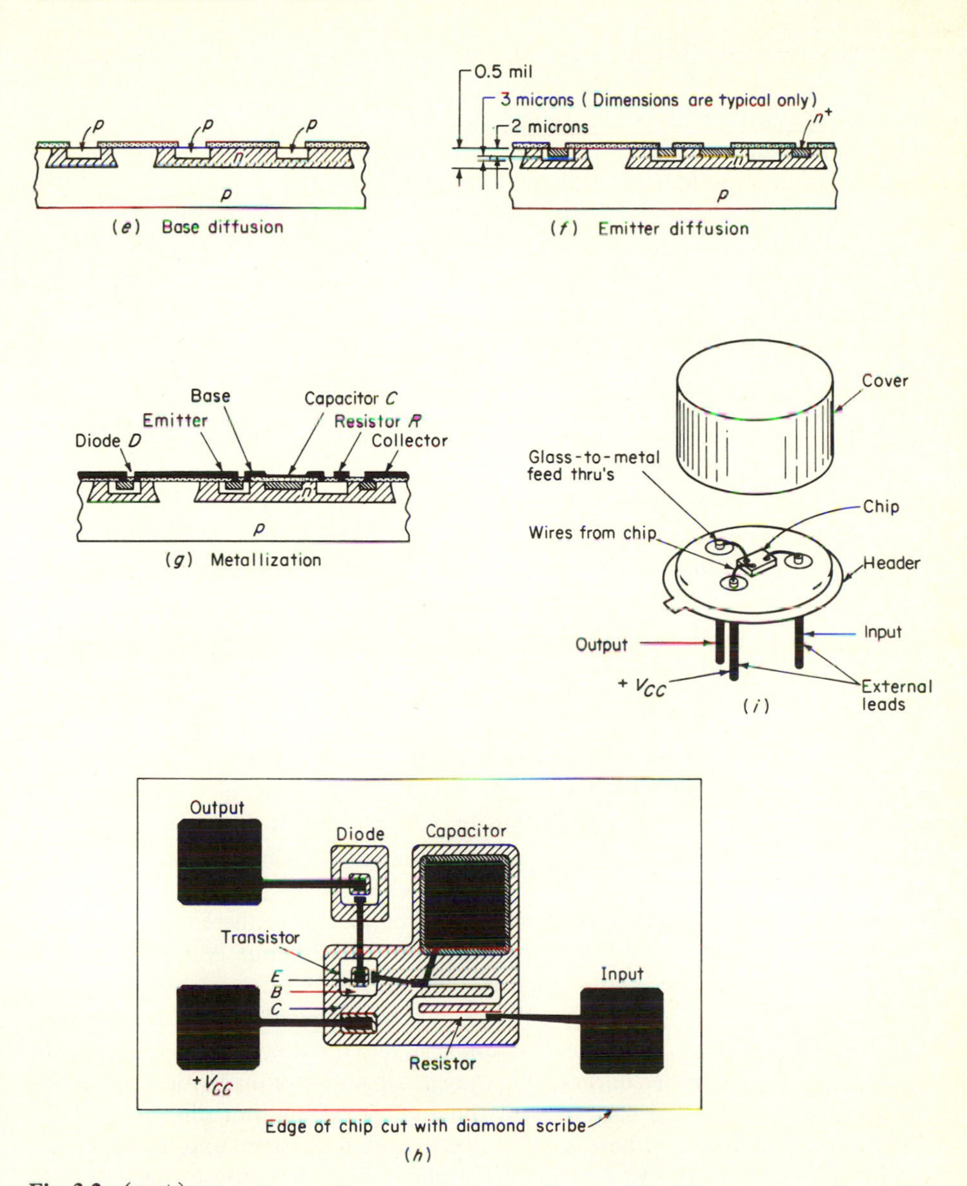

Fig. 3-2 (cont.)

fusion will take place in selected areas only. The dopant is again *p* type (boron) since this is the base diffusion, and an *npn* transistor is being made. Although both the isolation and base diffusions use *p*-type dopant, a different diffusion is required. This is because the isolation diffusion goes through the epitaxial *n* layer to the *p* substrate and uses heavy doping concentration. The base diffusion penetrates only a small distance into the epitaxial layer and is not as concentrated.

The emitter diffusion which follows results in the structure shown in Fig. 3-2*f*. Again, the wafer is reoxidized and windows are cut in the new oxide. An n^+-type impurity is diffused through the windows to form the emitter layer of the *npn* transistor and to ensure an ohmic collector contact when the aluminum metallization is performed. Aluminum is a *p*-type dopant in silicon, so if the aluminum collector contact were connected directly to the *n*-collector region, there would be the possibility of a rectifying (*pn* junction) contact existing. When n^+ material is used to contact the aluminum metallization, no *pn* junction is formed because the aluminum *p* dopant is not concentrated enough to overcome or "invert" the n^+ doping. Notice that where aluminum contacts *p* material, as in the case of the resistor in Fig. 3-2*g*, there is no problem because two *p* materials do not form a *pn* junction.

The final steps are reoxidization of the wafer, window cutting, and metallization. The purpose of metallization is to provide interconnections or "wiring" between components. Metal is evaporated over the whole slice by placing it in a vacuum chamber containing a metal evaporator, in which aluminum (the metallization material) is boiled from a hot filament. Since windows have been cut through the oxide where it is desired to contact various components, the metal deposits on these components as well as over the rest of the wafer. In a further sequence of photosensitizing, masking, and etching, metal is selectively removed to leave the desired interconnect pattern. A cross section of the completed circuit is shown in Fig. 3-2*g*.

Figure 3-2*g* is a two-dimensional representation of a structure which actually exists in three dimensions. Figure 3-2*h* is a top view of the chip from the side from which processing took place, showing how the circuit might be arranged in practice. Notice the relatively large area taken up by the three metallization contacts which will be connected to the external leads in Fig. 3-2*i*. The large size is necessary for ease of alignment, since wires are connected to these pads manually.

Referring again to Fig. 3-2*g*, let us see how the four components were formed and interconnected to give the circuit in Fig. 3-2*a*.

The diode was formed from the *pn* junction created by the base (*p*) and emitter (n^+) diffusion. The metallization contacts the diode in two places. The cathode contact (emitter diffusion) is the circuit output connection. The anode contact (base diffusion) goes over to the emitter (n^+ emitter diffusion) of the transistor. The transistor base is connected to both the resistor and the capacitor. A capacitor consists of two conductors isolated by a dielectric or insulator. As shown in Fig. 3-2*g*, the metallization pattern is one conductor or capacitor plate and the dielectric is a thinned area of silicon dioxide. The other plate is an n^+ layer formed during the emitter diffusion. Although strictly speaking a semiconductor, emitter-diffused material has a high enough conductivity to make it suitable for use as a capacitor plate in this application. Notice that the capacitor plate formed by the emitter diffusion contacts the *n* pool, which forms the collector of the transistor. Since this *n* pool is connected to the V_{CC} metallization contact, both collector and capacitor are connected to $+V_{CC}$ as desired. Connection to $+V_{CC}$ is through the *n*-collector material via an n^+ emitter diffusion. The latter provides a low-resistance ohmic contact to the aluminum and, since it is n^+ material, no junctions are formed. The resistor is simply a bar of *p*-type silicon, again with no *pn* junctions.

In Fig. 3-2*g* the various diffusions are shown separated by sharp boundaries. In practice the boundaries are much less distinct although, if ion implantation had been used (Table 3-1), sharper boundaries would have resulted.

Up to this point in the processing we have produced a wafer containing many identical circuits. The wafer is now tested by an automatic device that makes contact with the metallization contacts through miniature probes. Bad circuits are marked with a colored dye so that they can be separated out when the wafer is sliced. Slicing involves the separation of the individual dice, each of which contains a circuit, from the wafer. It is done by scratching the wafer between circuits with a diamond scribe and mechanically breaking the wafer into uniform, rectangular dice. This process is the same, although on a smaller scale, as that used to cut panes of glass.

To make the circuit a practical device, that is, one that can be handled without special microtools, it is mounted on a header into which pins or wires have been previously attached. Wires from bonding pads on the integrated circuit are attached to the header pins, and the package is sealed. For high-quality applications such as space or military, a hermetic seal, such as that used in the conventional TO-5 transistor, would be used. For low-cost, or commercial, applications the integrated circuit might be sealed with a special epoxy. Figure 3-2*i* shows the appearance of a chip mounted in a TO-5 size can. Notice the small size of the chip relative to the size of the final package.

A monolithic circuit made by the techniques described above can be thought of as a four-layer device. The thickest layer is the *p*-type substrate. Of the remaining layers, three are used to make *npn* transistors. Diodes can be made by using *pn* junctions or by appropriate connection of transistors. Resistors are made using single layers of *p*- or *n*-type silicon, or by deposition of films which, if compatible with monolithic technology, are called *compatible film resistors*. Capacitors can be made by using *pn*-junction capacitance or by compatible techniques in which the silicon dioxide layer is used as a dielectric. Compatible resistors and capacitors may require extra processing steps as, for example, in Fig. 3-2*g*, where the oxide under the capacitor has been thinned to provide higher capacitance.

Examination of Fig. 3-2*g* reveals the existence of a feature of integrated circuits which must be borne in mind by the designer, namely, the presence of parasitics. These are unwanted but inevitable components or connections. For example, the *pn* isolation junctions are not perfect insulators and have resistance, capacity, and leakage currents, all voltage- and temperature-dependent. There are also several transistors present, apart from the one required in Fig. 3-2*a*. For example, the *p* substrate, the *n* isolation diffusion, and the *p* base diffusion form what is called a *substrate pnp* transistor.

3-3 CIRCUIT DESIGN FOR LINEAR INTEGRATED CIRCUITS

3-3-1 GENERAL CONSIDERATIONS

The advent of integrated circuits has produced a trend whereby the engineer tends to design more from a system viewpoint, rather than being involved in detailed circuit design. This is because an integrated circuit is a subsystem, which can be considered to be a device characterized by appropriate parameters that define its performance. In the ideal case the engineer merely has to interconnect these integrated circuits or subsystems to achieve a desired objective. However, in practice it is important to have an understanding of basically how the integrated circuit works, in order to avoid pitfalls and to fully take advantage of the capabilities of these circuits. This fact is recognized by manufacturers who almost invariably provide a schematic diagram on the data sheet for a linear integrated circuit.

Integrated-circuit design differs from the design of discrete-component circuits in a number of ways. The following remarks deal primarily with monolithic integrated circuits, since they are an extreme example of the differences between integrated and discrete-component circuits.

An important factor in the design of a monolithic integrated circuit is the fact that the number of types of components available is more limited than in discrete-component circuits. Coils, transformers, and very-large-value resistors and capacitors are impossible or impractical to integrate and cannot be used. The components that *are* available will have much poorer tolerances than can be achieved with discrete-component circuits, because of the complexity and number of steps involved in making monolithic components and the fact that component selection is impossible. The diffusion processes used in the fabrication of monolithic integrated circuits are sensitive to temperature variations of a few degrees out of temperatures that exceed 1000°F, and this can produce errors in the absolute value of a component. Other influences include inevitable imperfections, however small, in the monitoring and stabilizing equipment, nonuniform removal of photoresists, and variations in the properties of all materials used in the processes.

Fortunately, although it is difficult to closely control the *absolute value* of components in monolithic integrated circuits from chip to chip or even over a whole single chip, components which are located *closely* together on a chip tend to have not only the same values but also the same temperature coefficients. This is because the source providing the material to be diffused can be located sufficiently far from the substrate onto which the diffusion is taking place that any variations over a given area will be very gradual.

This ability to fabricate *matched* pairs of components, especially resistors and transistors, is frequently employed to overcome the disadvantage of not being able to fabricate components with close tolerances on their *absolute* values.

A glance at most monolithic designs will reveal one particularly striking difference compared with discrete circuits, namely, the higher proportion of active to passive elements. It used to be that, since active elements were more expensive and less reliable than passive elements, the latter were used wherever possible. Neither of these factors applies to integrated circuits, and other reasons dictate the use of as many active and as few passive elements as possible. Capacitors and resistors of average values are larger than transistors and diodes, which means that if they are used in profusion, fewer circuits can be made on a chip of a given size. Thus, active elements are preferred over passive ones because the resulting circuit may be physically smaller, which means more circuits per wafer and greater economy.

There are many other factors which influence the design of monolithic integrated circuits. Capacitors may be voltage-dependent and resistors non-linear. Even transistors are somewhat different compared with discrete types, since their collector contacts are made from the top and not through the substrate. Isolation between components is not perfect, although dielectric isolation, which reduces stray capacity and leakage currents, helps.

Table 3-2 gives a list of the values and temperature coefficients for components used in integrated circuits. Most of the terms used are self-explanatory, but the term *sheet resistivity* which is measured in *ohms per square* may require some explanation. These expressions are commonly used in integrated-circuit terminology since the resistance of a conductive sheet is a function only of its resistivity and thickness for a square of the material, independent of the size

TABLE 3-2 Typical integrated circuit components

Monolithic		
Transistors	Emitter-base breakdown	6 V typically
	Collector-base breakdown	40–100 V typically
	Bandwidth	500 MHz
	Current gain	50 typically*
Diodes	Use transistor structures	
Resistors	Emitter-diffused sheet resistivity	2.5 Ω/square at 100 ppm/°C used for up to 1 kΩ
	Base-diffused sheet resistivity	200 Ω/square at 1,000 ppm/°C used for up to 30 kΩ
	Pinch type	Up to 0.5 mΩ
Capacitors	Reversed-biased junction type	0.2 pF/mil², used for up to 400 pF
Thin Films		
Transistors	Still experimental	
Resistors	Nichrome sheet resistivity	500 Ω/square at 100 ppm/°C used for up to 30 KΩ
Capacitors	Silicon dioxide dielectric	0.3 pF/mil², used for up to 750 pF
Thick Film		
Transistors	Discrete type chips used	
Resistors	Up to 100 MΩ available	
Capacitors	Up to 3 μF available	

* Super-beta transistors can be fabricated with current gains in excess of 5000.

of the square. This arises as follows. Figure 3-3 shows a uniformly doped conductive sheet with thickness t, length, l, width w, and resistivity ρ. The resistance between surfaces A and B is, by definition,

$$R = \frac{\text{(resistivity)(length)}}{\text{area}} = \frac{\rho l}{wt} \tag{3.1}$$

If the sheet is a square, $l = w$ and

$$R = \frac{\rho}{t} \tag{3-2}$$

The units of ρ are ohm-meters, and those of t meters, so that R is in ohms. Thus a square sheet of the material will have a resistance which depends only on the resistivity and thickness of the sheet and is independent of the size of the square.

For example, if an area 0.008 by 0.004 in [0.203 by 0.102 mm] is available to make a resistor in an integrated circuit, and the material has a sheet resistance of 100 Ω per square, how would one make a resistor with approximately 800 Ω resistance? A gap is left as shown in Fig. 3-4. By inspection the pattern can be

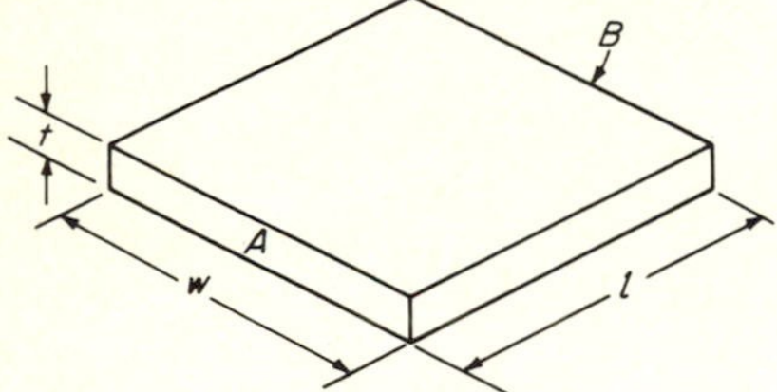

Fig. 3-3 Definition of ohms per square.

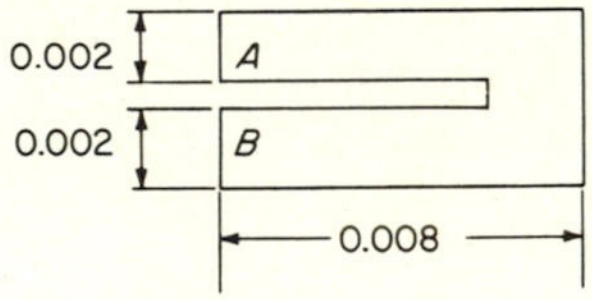

Fig. 3-4 Example of use of sheet resistance.

divided into eight squares, each 0.002 by 0.002 in [0.051 by 0.051 mm], and each will have a resistance of 100 Ω. Consequently the resistance between contacts placed on areas A and B will be approximately 800 Ω. In practice corrections must be applied to account for corner and other effects, but the principles are the same.

Most circuits can be designed without resorting to high-value resistors, but occasionally, especially in designs where low power dissipation is important, high-value resistors are required. So-called "pinch resistors" are frequently used in this application, and such a resistor is actually an ordinary diffused-base resistor in which the effective cross-sectional area has been reduced by diffusing an emitter on top of it. This results in an increase in resistance due to two effects. One of course is the reduction in cross-sectional area. A less obvious effect is the increase in resistivity due to the decreasing concentration of dopant away from the surface. Pinch resistors are not truly passive, tend to be very nonlinear, and have a low breakdown voltage and a large temperature coefficient.

However, by using them in such applications as operational amplifiers, where their function is to produce large open-loop gains which are reduced by negative feedback to produce stable, smaller gains, the effect of these characteristics is minimized.

3-3-2 CURRENT SOURCES

Figure 3-5*a* is a circuit diagram of one type of constant-current generator commonly employed using discrete components. The bias stabilization technique is well known and will not be elaborated upon further here, except to note that temperature stability is achieved through the use of emitter-degenerative resistor R_3 and diode CR_1. The voltage across CR_1 tracks the base-emitter voltage of Q_1, thereby maintaining an approximately constant voltage across R_3, which means that the emitter current of Q_1 is constant. This in turn means that the collector current is held essentially constant as temperature changes.

In circuit (*b*) of Fig. 3-5, two resistors have been eliminated and CR_1 has been replaced by a diode-connected transistor Q_1. The bases of both transistors are at the same potential, since they are connected together. If the transistors are located very close together on the same chip, their properties will be very similar and their collector currents will tend to be equal, regardless of temperature, because they have identical base-to-emitter voltages. The collector current of Q_1 will be determined by the value of R_4 and the supply voltage, which jointly determine the base current if Q_1 and Q_2 have reasonable current gains. Q_1 collector current is consequently largely independent of temperature and therefore so is the collector current of Q_2. This circuit is often called a *current mirror*. Figure 3-7*c* is another version of a current mirror using *n*-channel enhancement-mode MOSFETs.

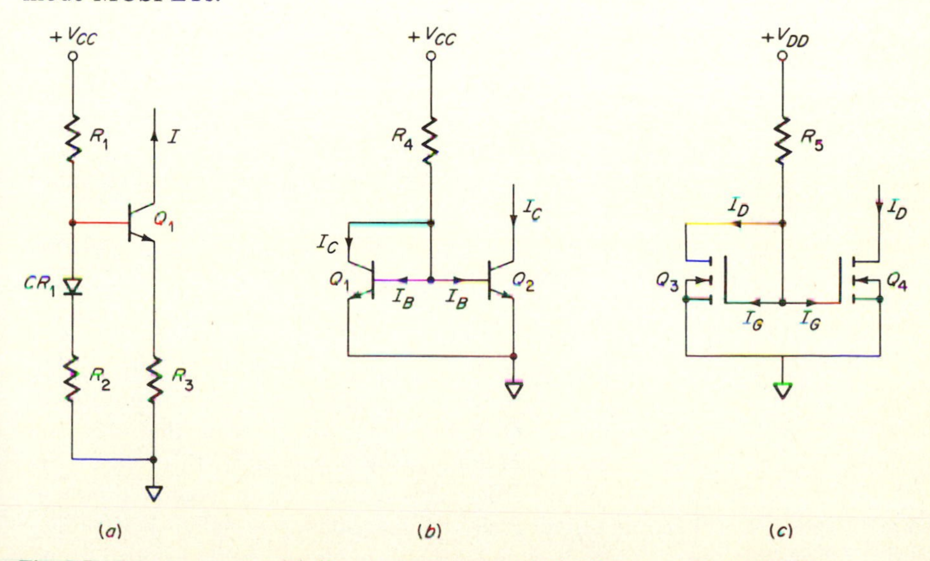

Fig. 3-5 Current sources: (*a*) discrete component; (*b*) IC bipolar; (*c*) IC MOSFET.

If, in Fig. 3-5b, $R_4 = 144\ k\Omega$, $V_{CC} = 15$ V, $V_{BE} = 0.6$ V, and $h_{FE} = 50$, we can find I_C as follows: The current in R_4 is the voltage drop across it (15 − 0.6 = 14.4 V) divided by its value (144 kΩ) or (14.4 V)/(144 kΩ) = 0.1 mA. Due to the matched characteristics of Q_1 and Q_2, which are transistors located physically adjacent to each other on an IC chip, their collector currents I_C will be equal and so will their base currents I_B. The current in R_4 is equal to the sum of the collector and base currents of Q_1 and the base current of Q_2, or,

$$I_C + I_B + I_B = 0.1 \text{ mA} \tag{3-3}$$

or

$$I_C + 2I_B = 0.1 \text{ mA} \tag{3-4}$$

Also, from bipolar-transistor current-gain considerations we have,

$$I_B = \frac{I_C}{h_{FE}} = \frac{I_C}{50} \tag{3-5}$$

Equations (3-4) and (3-5) give

$$I_C + \frac{2I_C}{50} = 0.1 \text{ mA} \tag{3-6}$$

$$1.04I_C = 0.1 \text{ mA}$$

$$I_C = 0.0962 \text{ mA}$$

3-3-3 DIFFERENTIAL PAIRS

In its simplest form a differential pair consists of two emitter-coupled transistors, and for good differential action the common emitters are usually fed by a constant-current generator. The circuit diagram for a basic differential amplifier is given in Fig. 3-6.

Q_1 and Q_2 are the matched pair with equal collector loads R_1 and R_2. Q_3 and Q_4 comprise a current source of the type shown in Fig. 3-5b.

Circuit operation is as follows. Suppose a signal of +1 mV is applied to Q_1 base, and a signal of −1 mV is applied to Q_2 base. Owing to the symmetry of the circuit, the potential of the common emitters of Q_1 and Q_2 will not change. Each transistor therefore acts as though its emitter were grounded, and the impedance represented by Q_3 is bypassed, which means that it has no effect on gain. Unlike bypassing achieved using a capacitor, this type of bypassing works down to dc. This is a desirable feature for an integrated circuit, since large-value capacitors cannot easily be integrated, especially where monolithic fabrication is employed. If circuit values are chosen so that both Q_1 and

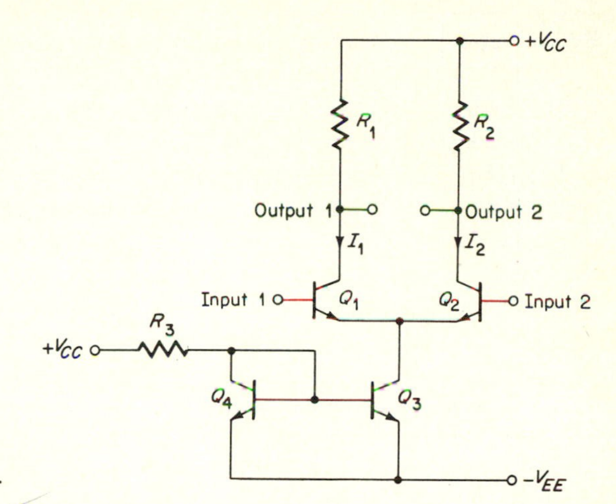

Fig. 3-6 Differential pair.

Q_2 have grounded-emitter voltage gains of, for example, 100, Q_1 collector will change by -100 mV, and Q_2 collector will change by $+100$ mV. If the output is taken off differentially, that is, from between the two collectors, it will be a 200-mV signal. Since the input was a differential signal of 2 mV, the differential-to-differential gain is 100. If a single-ended output is desired, the gain will be only 50, regardless of which collector is chosen for the single-ended output. Suppose now that one input, say Q_2 base, is grounded, and $+2$ mV is applied to the base of Q_1. Owing to circuit symmetry, the common emitters will now change by $+1$ mV. The base-emitter voltage of Q_1 will therefore change by $+1$ mV, and that of Q_2 by -1 mV. If each transistor provides a gain of 100, Q_1 collector will see a change of -100 mV, and that of Q_2 $+ 100$ mV. If the output is taken off differentially, the gain is once again 100, this time in response to a single-ended input. If the output is taken off single-ended, the gain will be only 50 regardless of which collector is used. However, the gain can be plus or minus 50, depending on which collector is used.

The above discussion may be summarized by saying that for differential outputs the circuit gain is the same as if one of the transistors was used alone, single-ended in to single-ended out. This is true whether the input to the differential amplifier is single-ended or differential. For single-ended outputs the circuit gain is one-half that obtained with a single-transistor circuit.

The common-mode rejection of the circuit depends on how large an impedance the collector of Q_3 presents to the common emitters of Q_1 and Q_2. Call this impedance R_{cm}, and suppose for example that

$$R_{cm} = 100R_1 = 100R_2$$

By definition, a common-mode signal is the same signal applied to both inputs simultaneously, so assume that $+1$ mV is applied to both Q_1 and Q_2 bases.

By emitter-follower action the common emitters will also experience a change of $+1$ mV, and the collector current of Q_3 will change by $1/R_{cm}$ mA if R_{cm} is in ohms. By symmetry, this current change will divide equally into R_1 and R_2, via Q_1 and Q_2, respectively. Thus the current change in R_1 equals that in R_2 and is $1/2R_{cm}$. The potentials of both collectors will change by the product of the current change and the load resistance or $R_1/2R_{cm}$, which will be $\frac{1}{200}$ mV. If the output is taken off differentially it will be zero, since the direction and magnitude of both collector-voltage changes are the same. The common-mode gain will therefore be zero, and the common-mode rejection, which is the ratio of signal gain to common-mode gain, will be $\frac{100}{0}$ or infinity.

If the output is taken off single-ended, it will be equal to $\frac{1}{200}$ mV. The common-mode gain will therefore be $\frac{1}{200}$, and the common-mode rejection will be 50 divided by $\frac{1}{200}$ or 10,000.

It can be seen that for differential outputs the common-mode rejection can theoretically be infinite since it is a matter of how well each side can be matched. In practice slight mismatches and other factors, neglected here as second-order effects, result in a finite common-mode rejection.

Where a single-ended output is desired, as for example in operational amplifiers, the common-mode rejection is determined largely by the relative magnitudes of the collector impedances R_1 and R_2 compared with the impedance presented by the collector of Q_3. It is possible to apply feedback to the base of Q_3 to oppose any collector-current changes in this transistor due to common-mode inputs, and this has the effect of raising the effective impedance at the collector of Q_3.

The effects of temperature on the amplifier can be deduced from common-mode considerations since temperature, in the absence of any gradients, is a common-mode effect, being applied equally to both Q_1 and Q_2. For differential outputs the effects of temperature can be eliminated by perfect matching, which of course can never be achieved in practice. For single-ended outputs the stability of the output depends somewhat on the degree of match between Q_1 and Q_2, but also on the stability of the current generated by Q_3. For dc amplifiers with single-ended outputs (such as operational amplifiers), the bias stability of Q_3 is critical.

The circuit of Fig. 3-6 is capable of separating out-of-phase signals from in-phase (common-mode) signals such as 60-Hz interference. It can do this only over a limited range of input levels because, if the common-mode signal is large enough to swing the common emitters of Q_1 and Q_2 down close to the emitter voltage of Q_3, then Q_3 will saturate. Also, if the common-mode input signal swings sufficiently positive, Q_1 and Q_2 can saturate as their emitter and collector voltages becomes nearly equal. Notice that since the common-mode signal is attenuated, the maximum allowable common-mode signal in a multistage amplifier is limited by the input stage, in contrast to the limitation on the size of a wanted signal, which is the maximum swing permitted by the output stage.

3-4 LIC APPLICATIONS

3-4-1 TEMPERATURE-STABILIZED DC AMPLIFIERS

As indicated in the previous section, monolithic circuits, by virtue of the close component matching and close thermal coupling inherently possible with this fabrication technique, have many of the characteristics desirable in dc amplifiers. The achievable temperature performance is limited, however, by the degree of match attainable in the characteristics of two transistors located adjacently on a monolithic chip. On a production basis the best voltage drift that can be achieved is a few tenths of microvolts per degree Centigrade.

A number of techniques are available to lower this drift limit. They include chopper stabilization, which is complex and expensive and often leaves only one input terminal, individual temperature compensation, which is not suited to volume production, and temperature stabilization. In the latter technique, a monolithic transistor pair is stabilized at a temperature above ambient by an integrated-circuit heater. As the ambient temperature changes, the power fed to the heater is changed to maintain the temperature of the differential pair at a constant level. Since the temperature of the differential pair does not change appreciably, the effective input-voltage drift is considerably reduced.

Figure 3-7 is a schematic of the Fairchild μA726, and Fig. 3-8 is a photomicrograph of the same device, which is a temperature-stabilized dc amplifier. An external resistor connected to the collector of Q_6 develops a control voltage as a result of the temperature-sensing action of Q_6 and Q_7. This temperature-sensing action results from the fact that Q_6 base is held at a constant voltage so that the usual -2.5 mV/°C base-emitter-voltage changes in Q_6 and Q_7 cause the collector current of Q_6 to change. The control voltage is coupled via emitter

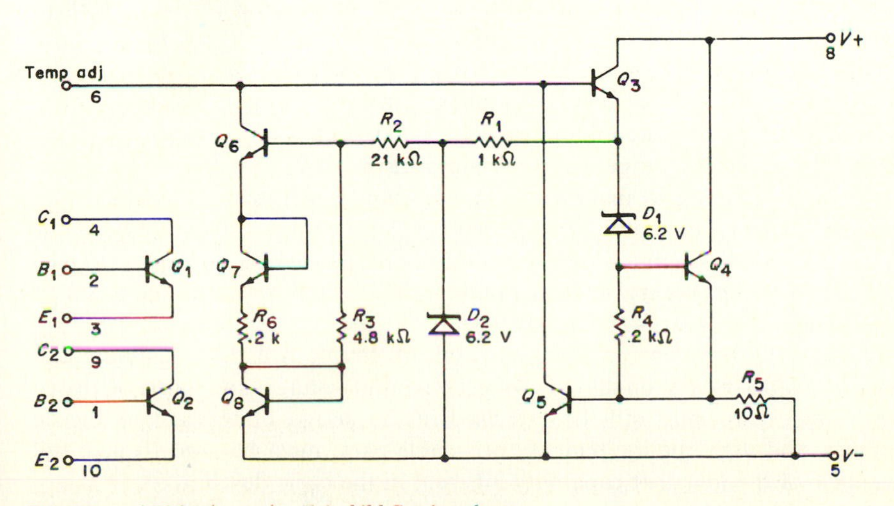

Fig. 3-7 μA726 schematic. *Fairchild Semiconductor.*

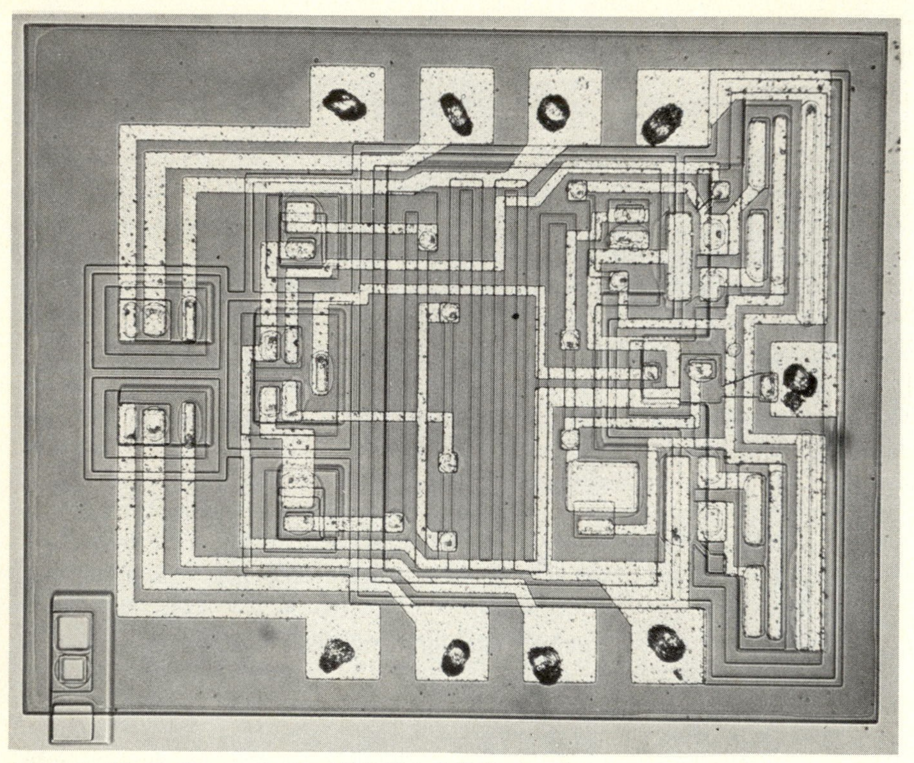

Fig. 3-8 Photomicrograph of μA726. *Fairchild Semiconductor.*

follower Q_3 and level-shifting diode D_1 to the base of Q_4, causing the power dissipated by Q_4 to change in a direction which produces a chip-temperature change in opposition to the original change. Q_4 is located physically at the opposite end of the chip from the differential pair Q_1 and Q_2, so that by the time the heat reaches them thermal gradients are minimized. Transistor Q_5 is a current limiter to prevent unduly large turn on surges. D_2 provides a stable voltage source to bias temperature-sensing transistors Q_6 and Q_7.

Because of the small mass involved, the chip temperature stabilizes within a few seconds. When stabilized, the input offset voltage drift is only about 0.2 μV/°C, and the input offset current drift is reduced to about 10 pA/°C. The control system is capable of maintaining the chip temperature constant to better than ± 2°C from -55 to $+125$°C.

It must be remembered that the resulting circuit is a pair of differential transistors with exceptionally low effective input-voltage and current drifts. Considerable care must still be exercised in the design of the rest of the dc amplifier, and high-quality, temperature-stable parts must be used to take full advantage of the low drift capability inherent in the basic device.

3-4-2 PHASE-LOCKED LOOPS

The basic phase-locked loop (PLL) system is shown in Fig. 3-9*a*. Power connections are not shown (to simplify the explanation), but it should be remembered that a PLL is an integrated circuit that will require power. It is useful to know the meanings of several terms as they specifically apply to PLLs; their definitions are summarized in Table 3-3.

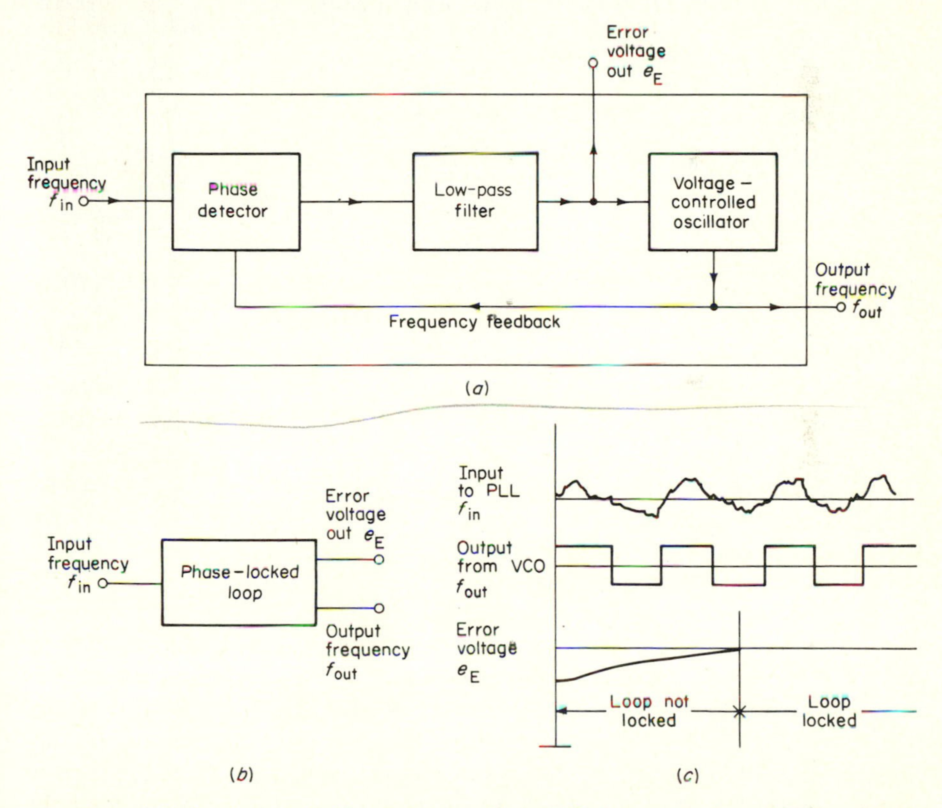

Fig. 3-9 (*a*) Basic phase-locked loop system; (*b*) simplified representation of PLL; (*c*) PLL waveforms.

The PLL's signal inputs and outputs total three. There is an input terminal to which is fed the signal to be processed. There are two outputs, one an *error-signal output* whose *level* e_E is proportional to the phase difference between the input and output signals, f_{in} and f_{out}, and the other a *frequency output* whose *frequency* f_{out} is made to try to follow the input-signal frequency. The three signals are shown in the simplified PLL block diagram in Fig. 3-9*b*.

A PLL system has three main parts: a phase detector which gives an output voltage that is a function of the relative phase of two input signals; a low-pass filter which allows the passage of dc and low-frequency signals while removing

TABLE 3-3 Phase-locked loop terminology

Voltage-controlled oscillator (VCO)	An oscillator whose frequency can be controlled with a dc or low-frequency voltage.
Low-pass filter	A circuit that permits the passage of low frequencies and dc while suppressing higher frequencies. In a PLL the low-pass filter sets the bandwidth.
Input level	The range of signal strengths over which the PLL will operate. Typically, a PLL will lock with signals ranging from 100 μV to 1 V.
Lock range	The range of frequencies over which the PLL input can vary before the lock is lost.
Capture range	The range of frequencies that the PLL is capable of locking onto. Capture range is usually less than lock range.
Frequency range	The range of frequencies over which the PLL can be used. 1Hz to 30 Mhz is common.
Phase detector	That part of a PLL in which the input signal and the VCO output are compared for phase and in which a voltage proportional to the phase difference is generated.

high-frequency components; and a voltage-controlled oscillator or VCO whose output frequency is a function of its input-signal voltage level. The three parts are shown in Fig. 3-9*a*.

These three parts are combined to make the PLL an electronic servo capable of locking onto an input-signal frequency so that its output frequency is the same as that of the input. A significant feature of PLLs is that the output can be a large, clean signal, frequency-locked to a low-level, noisy input signal. One application for a PLL therefore is as a combination filter-limiter in broadcast or telemetry FM receivers, whose function is to take a low-level, noisy signal clean it up, amplify it, and fix or limit its amplitude prior to feeding it to the discriminator. In this sense the PLL functions partly as a filter, although unlike linear, analog filters its output-signal *amplitude* is fixed and, within the specifications of the device, independent of the input-signal amplitude. The fact that the output amplitude is fixed should be borne in mind when a PLL is referred to as a filter.

The basic PLL of Fig. 3-9*a* operates in the following manner. The input-signal frequency f_{in} and the feedback-signal frequency f_{out} are the two inputs to the phase detector. The phase detector is a device that has zero output voltage when its two inputs are in phase, but that has an output when the two inputs are not in phase. If the two phase-detector input signals have different frequencies, phase differences will exist between them, and the phase detector will output a voltage. This voltage will fluctuate about some dc level, and the purpose of the low-pass filter is to remove the higher frequency fluctuations to give a resulting error signal e_E as in Fig. 3-9*c*. The error signal from the low-pass filter controls the frequency of the VCO, and the error-signal polarity is made such as to force the VCO frequency to move in the direction of the input frequency. When the input and VCO frequencies are equal, the error voltage e_E controlling the VCO will be zero, and the VCO frequency remains equal to the input frequency. The PLL is then said to be *locked* as in Fig. 3-9*c*.

In addition to functioning as an FM filter-limiter, the PLL can also be used as an FM discriminator, which is a device whose output-voltage level is linearly related to its input frequency. The input is the frequency-modulated carrier in which information is carried in the form of frequency variations about some center frequency. As the frequency varies, the error voltage varies in sympathy, to try to force the VCO to track the input frequency. The error voltage (Fig. 3-9*a*) is therefore the demodulated carrier, which is the audio in FM broadcasting, for example. A capacitor, external to the chip, but in the low-pass filter circuit, would remove the RF carrier from the error voltage, while allowing it to vary at audio frequencies. Notice that the cleaned-up carrier (the VCO output) is not fed to a separate discriminator external to the PLL chip. All amplification, filtering, limiting, and frequency discriminating are done in the PLL. A difference between such a PLL system and a conventional broadcast receiver system is that in the PLL filtering is done at audio frequencies with a very simple type of filter, while in a conventional receiver much of the filtering is done at high frequencies in the intermediate-frequency (IF) strip. IF strips require initial alignment and can become misaligned with time.

PLLs may also be used as programmable frequency synthesizers in which a variety of precise frequencies can be generated under digital and crystal control. (See Prob. 3-9 and Fig. Prob. 3-9.) They are also used as AM receivers, as frequency-shift-keying modulators and demodulators, in analog light-coupled isolators, as phase modulators and demodulators, in tone decoders and burst generators, and in speech scramblers and metal detectors. These and many other applications are described in Ref. 1.

3-4-3 MOS/BIPOLAR AMPLIFIER

Figure 3-10*a* is the circuit diagram of a linear integrated-circuit operational amplifier. Operational amplifiers are discussed in Chap. 4, but we can discuss the circuit in Fig. 3-10*a* simply in terms of a linear IC amplifier.

The amplifier is the result of processing techniques that allow a variety of different components to be fabricated on the same chip. It contains PMOS, NMOS, and *npn* bipolar transistors, as well as diodes, both signal and zener, and resistors. Circuit operation may be explained with the help of the simplified circuit in Fig. 3-10*b*.

First, ignore biasing and current generators, and trace the gain path through the amplifier. Input transistors Q_6 and Q_7 form a PMOS version of the differential amplifier discussed in Sec. 3-3-3. A single-ended signal is taken from the drain of Q_7 and fed to bipolar transistor Q_{12}, which provides most of the voltage gain as explained in Sec. 2-7. The output from the collector of Q_{12} feeds the complementary MOS (CMOS) output driver stage composed of Q_{10} and Q_{11}.

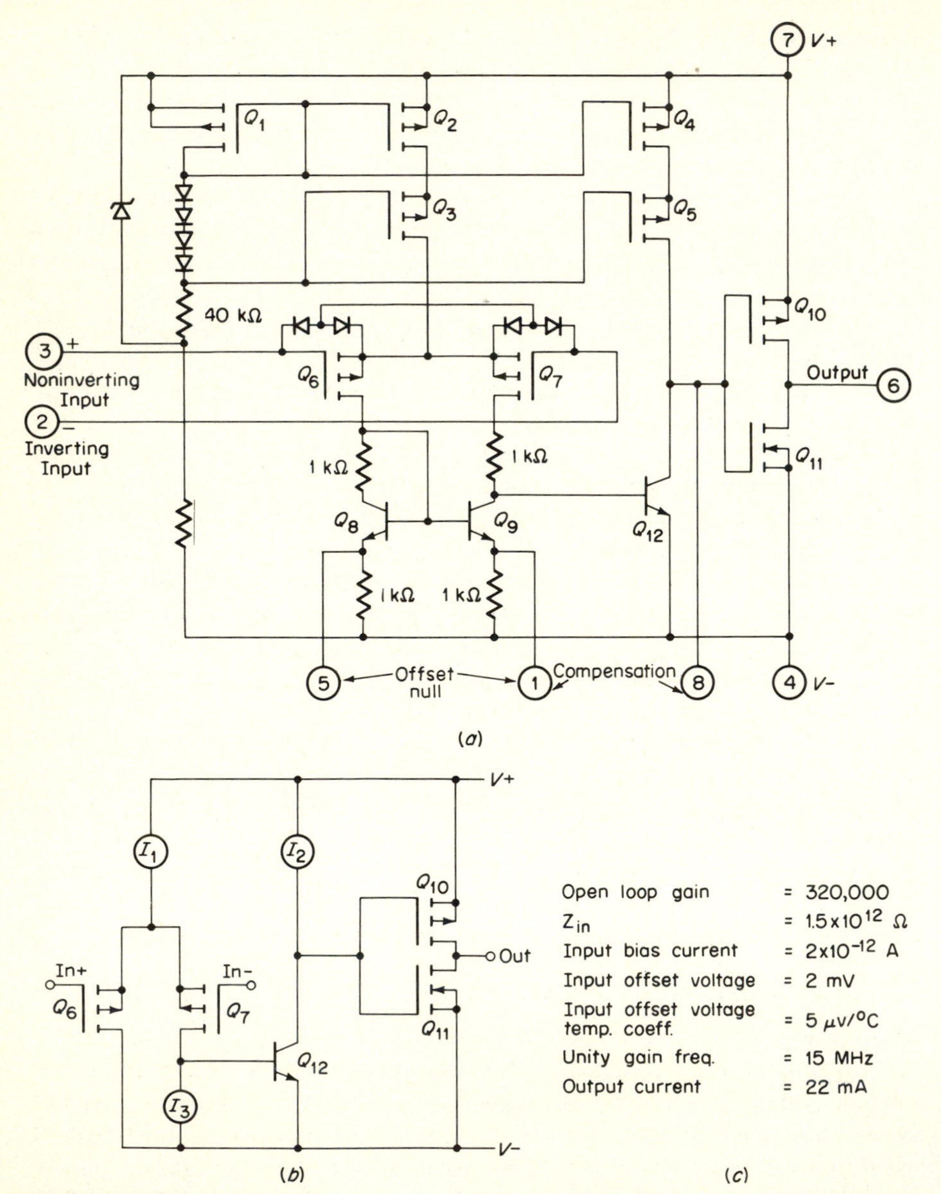

Fig. 3-10 The RCA 3130 amplifier: (*a*) circuit diagram; (*b*) simplified schematic; (*c*) characteristics. *RCA.*

There are several current sources of the type discussed in Sec. 3-3-2. Q_1, Q_2, and Q_3 comprise current source I_1 in Fig. 3-10*b*, and Q_1, Q_4, and Q_5 comprise I_2. Stacking Q_2 and Q_3, and Q_4 and Q_5, increases the impedances of the respective current sources, making them more closely approach the ideal. (An ideal current source has an infinite output impedance, just as an ideal voltage source has a zero output impedance.) Q_8 and Q_9 form another current source, and the 1-kΩ emitter resistors serve to stabilize the current (from bipolar-transistor bias-stability theory) and to raise the current-source output impedance.

The voltage gain of Q_{12} is high because it has a large effective collector impedance, namely the current generator formed by Q_1, Q_4, and Q_5.

The CMOS output stage formed by Q_{10} and Q_{11} has already been discussed in Sec. 2-5-4, where it was pointed out that such an arrangement allows the output to swing to within millivolts of the supply-rail voltages, thereby maximizing the output swing possible with a particular power supply.

Note the use of four diodes on Q_6 and Q_7 to protect them from overvoltage inputs.

An unusual feature of this amplifier is that the input can swing 0.5 V below the negative supply-rail voltage, owing to the bias operation of PMOS transistors Q_6 and Q_7, a feature which is useful in comparators. Other diodes in the circuit are for biasing and temperature compensation.

Notice the high ratio of active (transistors and diodes) to passive (resistors) components in the circuit, typical of integrated circuits.

As explained in Chap. 4, operational amplifiers are intended to be used with feedback from the output to one or both inputs. This provides such desirable features as stable gain and linearity. To achieve these features requires a high gain *before* feedback is applied, and this gain for the RCA3130 is 320,000. Other performance parameters are listed in Fig. 3-10*c*.

Another operational amplifier, the Harris HA2900, also contains MOSFETs plus bipolars on the same monolithic chip. In this circuit, however, the MOSFETs are used as switches to periodically apply correction voltages to the input to cancel drift, an example of an auto-zeroing technique. Components are dielectrically isolated from each other—that is, separated by an insulating film rather than by reverse-biased *pn* junctions as in conventional integrated circuits—and the active elements number over 200.

3-4-4 INTEGRATED-CIRCUIT PRESSURE TRANSDUCERS

Linear integrated circuits are commonly thought of as providing such functions as amplification or other conditioning of signals obtained from various sources, including transducers. A *transducer* is a device that converts one type of signal, for example temperature, pressure, or acceleration, into another type of signal that can be processed electronically, namely, an electric signal.

It is now possible to make many types of transducers using integrated circuit techniques. For example, temperature can be measured using the 2-mV/°C change in base-to-emitter voltage that occurs when a bipolar transistor is heated. A pressure transducer can be fabricated by the method indicated in Fig. 3-11.

The starting point is an *n*-type silicon chip, a portion of which is etched away to provide either a vacuum reference or some reference pressure. The etching also serves to thin the material so as to form a diaphragm, which can

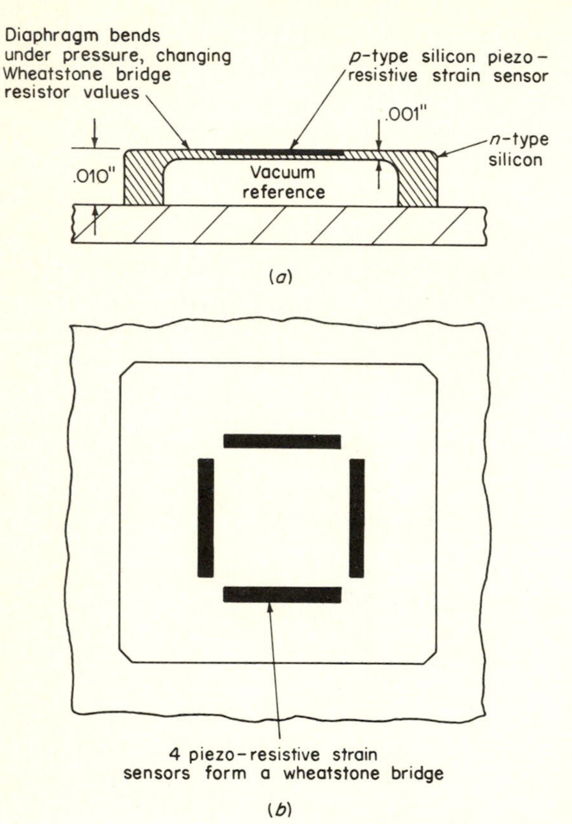

Fig. 3-11 Integrated-circuit pressure transducer: (*a*) cross section; (*b*) plan view.

deflect when pressure is applied. Four piezo-resistive strain gages are diffused into the top of the diaphragm. (A *strain gage* is a device for measuring small changes in dimensions, and a *piezo-resistive* element is a resistor intended to have one of its dimensions changed by some external force, thereby changing its resistance.) Such strain-sensitive resistors can be designed to temperature-compensate each other in pairs, so that minimization of drift due to temperature is possible.

On the same chip might be a temperature sensor, a temperature compensator, a voltage-supply regulator, and some signal processing. All these functions could be performed with integrated circuit elements such as bipolar transistor and diodes.

Reference 2 gives applications of a commercially available series of integrated circuit pressure transducers as a pressure-to-frequency converter, an altimeter, and a digital-readout barometer.

3-4-5 MONOLITHIC RMS-VOLTAGE MEASURING DEVICE

A root-mean-square (rms) voltage is that value which would cause the same power dissipation in a resistive load as an equivalent dc voltage. For example, if 1 V dc is found to dissipate a certain amount of power in a resistive load, it turns out that an ac sine-wave voltage of $2\sqrt{2}$ V, peak to peak, will dissipate the same amount of power. The $2\sqrt{2}$ V peak-to-peak ac voltage is then said to have an *rms value* of 1 V. The rms–to–peak-to-peak conversion factor is therefore $2\sqrt{2}$, so that 115 V rms, for example, is $(115)(2\sqrt{2}) = 325.3$ V peak to peak.

Various means have been devised to measure the rms values of voltages with sine and other waveforms. For sine waves, one way is to full-wave rectify, as in Fig. 3-12*a*, and apply the resulting signal to an averaging meter, which would read 0.636 times the peak of the rectified wave. For a 1 – V rms signal,

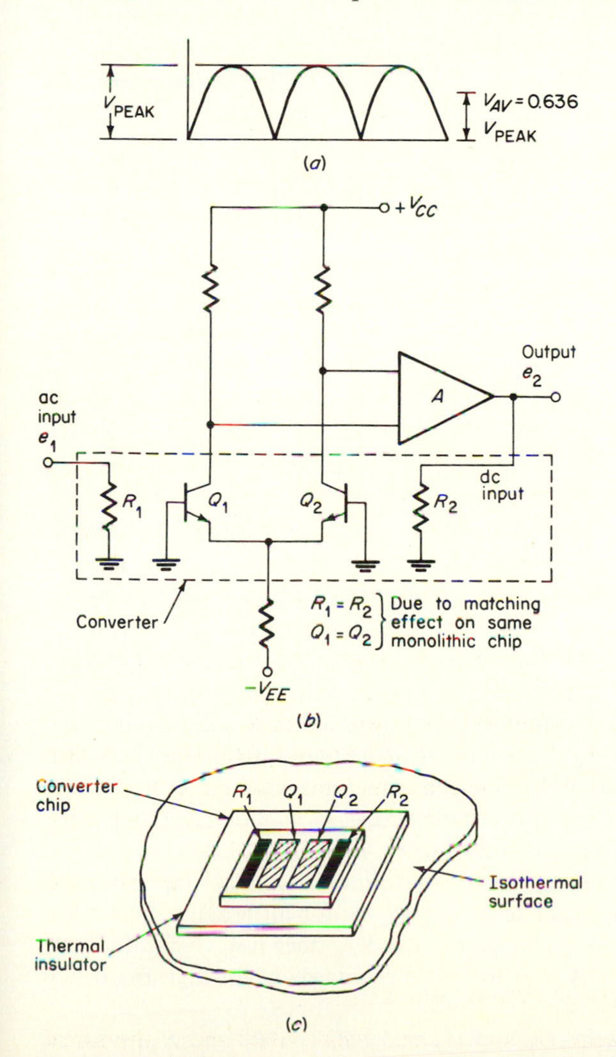

Fig. 3-12 Rms measuring device: (*a*) relationship between peak and average values for a full-wave rectified sine wave; (*b*) block diagram of rms converter circuit; (*c*) converter chip. *Burr-Brown.*

the meter would read $(0.636)(\sqrt{2}) = 0.899$ V. By adding an amplifier with a gain of $1/0.899 = 1.112$, or by recalibrating the meter, the system could be made to measure rms directly.

Such a method has two main disadvantages. First, the calibration would only be good for sine-wave voltages. Other waveforms such as square waves, pulse trains, or random noise would require factors other than the 0.636 used for sine waves. Calibration would be feasible, but only if the shape of the waveform were known exactly, which is impossible for example with noise. Secondly, the system would only be accurate at relatively low frequencies, because of problems in accurately rectifying and averaging high-frequency signals.

Figure 3-12*b* shows an rms measuring circuit that overcomes both these disadvantages. It is the Burr-Brown BB4131. The heart of the device is the monolithic converter shown in Fig. 3-12*c*. It contains two *npn* bipolar transistors, Q_1 and Q_2, and two diffused resistors, R_1 and R_2. The circuit in Fig. 3-12*b* operates as follows.

The voltage e_1, whose rms voltage level is to measured, is applied to the ac input. It can be a sine wave, square wave, pulse train, random noise signal, or any other waveform, even dc, although a simple dc voltmeter could be used equally well to measure the dc directly. The input signal could also be a mixture of ac and dc—that is, an ac level on a dc bias. The measurement is independent of the waveform because the input voltage is used to *heat* resistor R_1. As we recalled earlier, the definition of rms has to do with power dissipation which, for a resistor, results in heat. Because the circuit is integrated, it is physically *small*, and therefore very little power is required to produce a measurable heating effect, especially since amplification of the voltage resulting from the heating effect is employed.

When R_1 is heated by the input signal it heats Q_1. When a transistor is heated, its base-to-emitter voltage V_{BE} changes by about 2 mV/°C, which in this case unbalances the differential amplifier comprised of Q_1 and Q_2 in Fig. 3-12*b*. The unbalance appears as a signal to amplifier A, whose resulting output is fed to R_2 on the monolithic converter chip. This heats R_2, which changes the base-to-emitter voltage of Q_2. The polarity of the feedback is chosen so that Q_1 and Q_2 tend to rebalance, which stabilizes the power fed to R_2. An equilibrium condition is reached in which R_1 and R_2 are heated equally. From the definition of an rms voltage, they must both have the same rms voltage level applied, since they are heated equally. But the voltage across R_2 is a dc level and is the system output. The rms level of e_1 is therefore e_2.

If e_1 is a high-frequency signal, the thermal time constant of input resistor R_1 will average it, so that the signal seen by Q_1 is dc for all practical purposes. The amplification of the resulting change in Q_1's V_{BE} does not, therefore, have to be done at high frequencies. Also, there is no requirement for high-frequency rectification.

The close matching between Q_1 and Q_2 and between R_1 and R_2, both of which are necessary for this application, are a direct consequence of the small size and monolithic construction of the four-component converter chip. The circuit is an example of a design using the inherent matching achievable with IC components mounted close together on a common chip. The circuit does not require accurate *absolute* values, just good *matching*.

The BB4131 has an accuracy of better than 2 percent up to 100 MHz and has a 30-dB dynamic range. At lower frequencies accuracies of 0.05 percent are achieved.

3-4-6 MONOLITHIC VOLTAGE REGULATORS

The regulators discussed in this section are dc-to-dc voltage regulators. They accept as an input an unregulated dc voltage, which may vary with the load current, local line, or battery voltage, and which may have an ac ripple superimposed on the mean dc level. Their output is a regulated dc level, which remains relatively constant as the input voltage or load current changes, and which has a drastically reduced ripple level.

All voltage regulators operate on the same general principles, which are explained with reference to Fig. 3-13*a*. A portion of the output voltage deter-

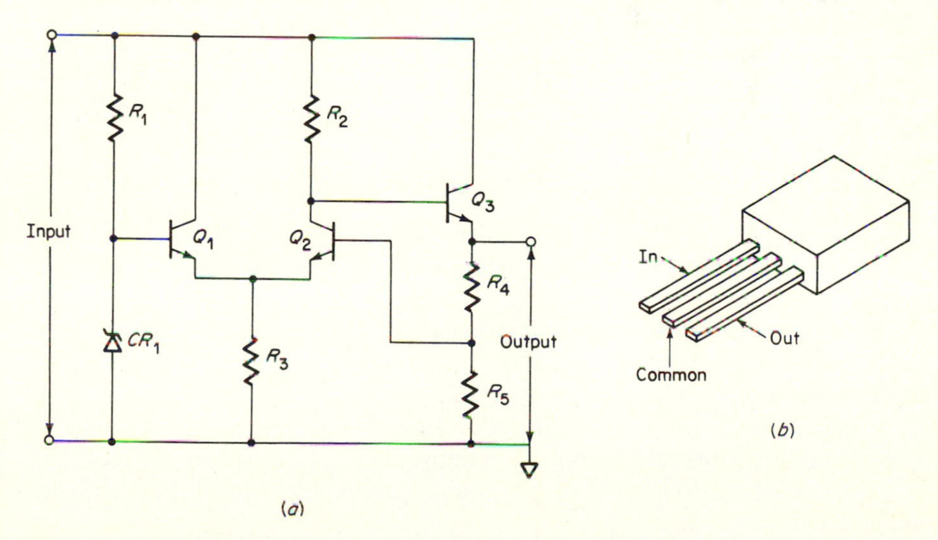

Fig. 3-13 (*a*) Principle of voltage-regulator action; (*b*) typical package.

mined by R_4 and R_5 is fed to one input of a differential amplifier consisting of Q_1, Q_2, R_2, and R_3, the other input of which is tied to a reference voltage supplied by R_1 and reference diode CR_1. Current flow from input to output is via the series-pass transistor Q_3. If for some reason the output tries to go positive, Q_2 base will also go positive and Q_2 collector will go negative. Since Q_3 is an emitter follower, its emitter will try to follow its base, which is tied to Q_2 collector. The result of Q_3 emitter going positive is an opposing change tending to stabilize the output. This will happen regardless of why the output changed, whether it was due to a load change or an input-voltage change. All regulators consist of a sensing circuit, in this case R_4 and R_5, a reference circuit, in this case R_1 and CR_1, a comparison circuit, in this case the differential amplifier Q_1, Q_2, R_2, and R_3, and a controlled element, in this case Q_3.

Another way of thinking about this circuit is to regard Q_3 as an impedance in series with the load so that the two form a simple potential divider. As the voltage across the load tries to change, the impedance of Q_3 is made to change in such a manner as to maintain the load voltage constant. Q_3 is often called a *series-pass transistor.*

Notice that no regulation can take place unless there is a change in the output voltage to initiate a correction, so that the regulation can never be perfect.

How well a regulator regulates depends on a number of factors. The reference must be stable with temperature, time, and variations in the unregulated input. The comparison amplifier must have a high gain so that very small output changes produce a strong corrective action, and it must also be stable since if it drifts the output will drift also. The overall frequency response of the system is important also, since this influences the degree of regulation in response to rapid input-voltage or load-current changes.

Note that the circuit in Fig. 3-13*a* requires a package with only three terminals, an input, an output, and a common; many voltage regulators are now supplied in packages such as that shown in Fig. 3-13*b*.

3-4-7 INTEGRATED-CIRCUIT MULTIPLIERS

A device capable of multiplying two *time-variable voltages*, which a simple digital calculator cannot do, has many applications. Multiplying equal voltages results in a squaring circuit. Multiplying by a reciprocal is equivalent to division. By generating voltages which are proportional to other quantities, other useful mathematical operations can be performed. For example, if voltages proportional to current and phase angle are multiplied by a voltage, then power can be obtained from $VI \cos \theta$. In addition to these fairly obvious applications in performing mathematical operations, multipliers are also useful for frequency doubling, phase detection, electronic gain control, and modulation and demodulation.

Electronic multiplying devices have been in existence for a long time and operate on a variety of principles. However, one type, the so-called *g_m multiplier*, is particularly suited to fabrication by integrated circuits. This is because essential features of the circuit include a differential amplifier, a constant-current generator, and the use of a high ratio of active to passive elements.

Figure 3-14 is a circuit diagram of a basic g_m multiplier. It is called a g_m multiplier because it relies on the fact that, for a bipolar transistor, the transconductance g_m is proportional to the dc value of the emitter current to an accuracy of about 1 or 2 percent. Thus we can write, for transistor Q_2,

$$g_m = K_1 I_E \tag{3-7}$$

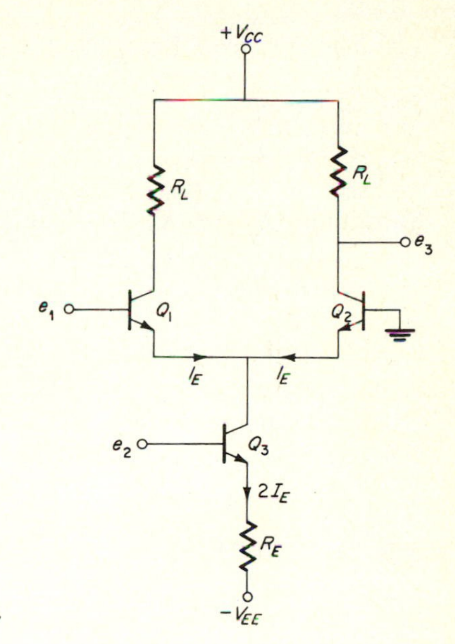

Fig. 3-14 Principle of a g_m multiplier.

The small-signal ac gain from the base of Q_1 to the collector of Q_2 is

$$\frac{e_3}{e_1} = \tfrac{1}{2} g_m R_L \tag{3-8}$$

[Eq. (3-8) is obtained from Sec. 3-3-3, where the single-ended gain of a *differential stage* is shown to be *one-half* the gain of a *single-transistor stage* using the same transistor. The single-stage gain is obtained from Eq. (A-7) and is $+g_m R_L$ because there is no phase inversion between e_1 and e_3.] Eliminating g_m from Eqs. (3-7) and (3-8) gives

$$\frac{e_3}{e_1} = \tfrac{1}{2} K_1 I_E R_L \tag{3-9}$$

The emitter current of Q_3 is $\approx 2I_E$ and by emitter-follower action is *linearly related* to e_2. In fact, by appropriate level shifting of e_2 before it is applied to the base of Q_3, the emitter current of Q_3 can be made *proportional to* e_2. This is not shown in Fig. 3-14 because it is not essential to the point under discussion and would unnecessarily complicate the analysis and the circuit. Assuming that such level shifting *were* done, we could write

$$2I_E = \frac{K_2 e_2}{R_E} \qquad \text{where } K_2 = \text{constant} \tag{3-10}$$

Eliminating I_E from Eqs. (3-9) and (3-10) gives

$$\frac{e_3}{e_1} = \frac{K_1 K_2 R_L e_2}{4R_E} \tag{3-11}$$

Substituting $K_3 = K_1 K_2 R_L/4R_E$ in Eq. (3-11) gives

$$e_3 = K_3 e_1 e_2 \tag{3-12}$$

Equation (3-12) shows that the output voltage e_3 is proportional to the product of the two input voltages e_1 and e_2.

Analog multipliers are typically 1 percent, or at best 0.1 percent, devices. For greater accuracy, digital techniques are used.

3-5 HIGH-FREQUENCY INTEGRATED CIRCUITS

Integrated circuits intended for high-frequency applications cannot be divided into completely separate categories because of the considerable overlap in the functions they perform. (The PLL of Sec. 3-4-2 is an example of a multifunction high-frequency circuit.) It is convenient, nevertheless, to categorize the applications and select integrated circuits which will satisfy the requirements of each application.

3-5-1 VIDEO AMPLIFIERS

Video amplifiers are wide-band amplifiers covering the frequency range from about 10 Hz to about 10 MHz but sometimes to 100 MHz. They do not have very high gains, a typical value being about 10. Such circuits are used as pulse amplifiers in which their wide bandwidth permits faithful reproduction of all the harmonic frequencies in the pulse, thereby reproducing at the output the same waveform as the pulse input. Pulse amplifiers are used in television and wide-band data systems such as telemetry and video tape recording.

Although integrated circuits are frequently more complex in terms of the number of active components than their discrete-component counterparts, integrated video amplifiers are an exception. In days when vacuum tubes were common, video amplifiers tended to be quite complex, with several cascaded stages employing local feedback and having interstage inductive peaking. AC coupling between stages and the use of overall feedback to achieve bandwidth made such amplifiers hard to design without undue compromise. Figure 3-15, which is a circuit diagram of an integrated-circuit video amplifier, the RCA CA3001, illustrates the simplicity of the modern video amplifier. There is still local feedback, supplied by emitter resistors R_5 and R_6, but the circuit consists simply of an emitter-coupled differential amplifier with a constant-current emitter source, and with emitter-follower inputs and outputs. Since the circuit

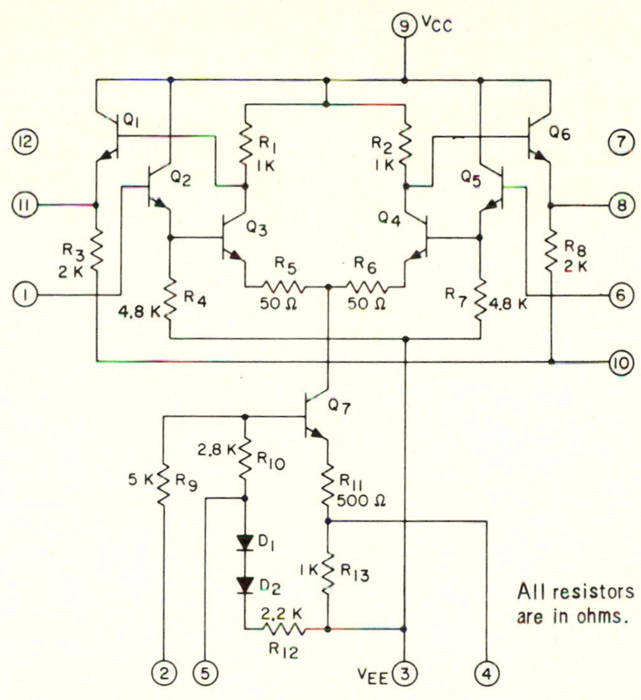

Fig. 3-15 Schematic of CA3001. *RCA.*

is direct-coupled throughout, the low-frequency response is determined by external coupling capacitors, which can be small owing to the high input impedance. The high-frequency response is determined by the *RC* time constant of the load on the differential-pair collectors. The circuit has an inherently wide bandwidth, being flat down to dc and extending to a 3-dB frequency of about 16 MHz. Higher gains can be achieved by merely cascading stages, using *RC* coupling between them. Since the inherent bandwidth is so high, it is not necessary to resort to overall feedback with its attendant stability problems in order to increase the bandwidth.

A connection is brought out from pin 2 on the CA3001 to permit external biasing of the constant-current generator. This changes the current level in, and consequently the transconductance of, the differential pair. Since the stage gain is proportional to the transconductance, pin 2 can be used as an automatic-gain-control input.

3-5-3 INTERMEDIATE-FREQUENCY AMPLIFIERS

Integrated-circuit amplifiers intended for IF use are good examples of the compromises that must be made when it is impractical to construct a complete circuit using integrated techniques. An IF amplifier is really a narrow-band filter with gain, which must be tunable to some preset frequency. The filtering is most economically achieved by the use of tunable transformers, the tuning being accomplished either by varying the transformer inductance or with a trimmer capacitor. It is practical and economical to make the gain element in an IF amplifier by integrated-circuit techniques. Figure 3-16, which is a circuit for the Fairchild μA703, is an example of such a circuit. This versatile device, in addition to IF application, may also be used as an RF amplifier, either limiting or nonlimiting, or as a harmonic mixer. All biasing is internal, and the only external components required are input and output transformers. The circuit consists of a differential amplifier using transistors Q_3 and Q_4, with constant-current generator Q_5 supplying their emitter currents. R_2 and diode-connected transistors Q_1 and Q_2 form the bias string. Input bias for Q_3 is through the secondary of the input transformer, which is connected between pins 3 and 5. The output is single-ended, the primary of the output transformer being connected to pins 1 and 7, thus providing a dc path for the collector current of Q_4.

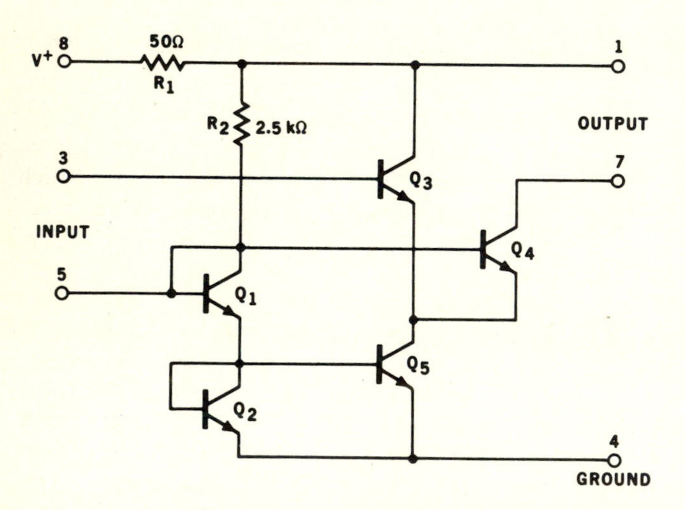

Fig. 3-16 μA703 schematic. *Fairchild Semiconductor.*

3-5-3 RADIO-FREQUENCY AMPLIFIERS

Unlike IF amplifiers, which are designed to operate over a fixed and relatively low range of frequencies, an RF amplifier must be capable of being tuned over the entire band of frequencies to be covered by the receiver. It is even more true of an RF amplifier that it cannot be fully integrated, and a typical integrated-circuit RF amplifier will have many external trimming and tuning components.

Figure 3-17 is a schematic for the RCA 3028A, which is an integrated-circuit RF amplifier. It is similar to the video amplifier described in Sec. 3-5-1 in that it consists of a differential pair with a constant-current emitter source. There are no input and output emitter followers, and the collector loads have been omitted. This is to permit maximum flexibility so that the source and load impedances can be selected to optimize performance. There is no local feedback

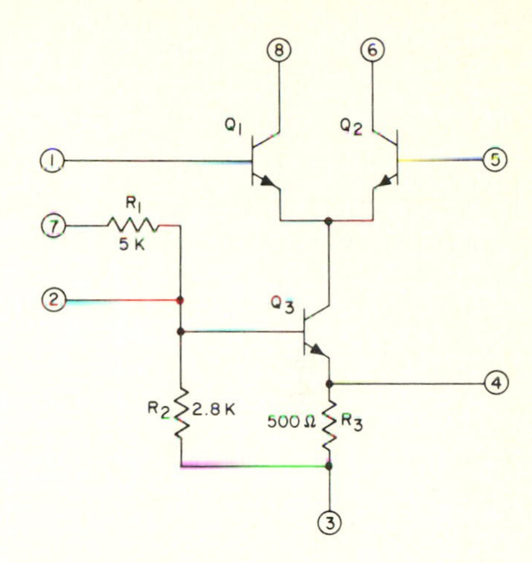

Fig. 3-17 Schematic of CA3028A. *RCA.*

in the form of emitter degeneration as in the circuit described in Sec. 3-5-1. Thus the circuit gain is higher, but the allowable dynamic range of the input signal for linearity is lower, the latter being of little consequence since RF amplifiers are intended to handle low-level antenna signals. Notice the large number of connections brought out from the integrated-circuit package, which is done to allow the circuit to be employed in push-pull, cascode, and a number of other modes.

3-6 COMPUTER-AIDED DESIGN

Although techniques such as breadboarding are available to pretest a circuit design intended for monolithic construction, in general an accurate prediction of the performance of a monolithic design is best obtained by analytical means. Such analysis is performed by using a detailed model which can include parasitic effects and the effects of tolerances and mismatching of assumed matched components. The value of each parameter of an element can be varied without changing other parameters in the circuit. High-frequency and transient effects can be analyzed, and, if the model is an accurate one, the results will be more useful than those obtained from a breadboard, where layout could produce significant differences when compared with the final monolithic circuit.

The complexity of an accurate model, and the large number of variations of interactions that it is desired to investigate, virtually dictate the use of a computer to optimize a design, increase the resultant yields, and hence reduce costs. It must be borne in mind, however, that the computer can follow only simple instructions and lacks the creativity necessary to actually design the circuit. Insofar as circuit design is an art as much as a science, the computer will always be relegated to the role of a tool which the designer can use to enhance intuitive techniques, the inner workings of which can never be spelled out in computer-program form. It should be pointed out, of course, that the word "design" can be taken to have several meanings, especially when applied to integrated circuits. If one is referring to *circuit* design, the role of the computer is usually that of circuit *analysis* rather than circuit *design*. However, the *layout* of integrated circuits is a sufficiently complex procedure so that the word "design" is also applied to this procedure. The complexity is such also that a computer can be an extremely useful aid in this kind of design. The computer is often equipped with an interactive graphic terminal, which consists of a large-screen cathode-ray tube and a light-sensing pen. The designer conducts what is essentially a graphical conversation with the computer. He talks to the computer by pointing to various objects and commands, and the computer responds by displaying the words, symbols, and pictures of the components that will make up the design. For example, the designer might specify a particular type of logic gate. The integrated-circuit layout of this gate will appear on the face of the cathode-ray tube and can be located by the designer where he chooses. Other integrated circuits, diodes, resistors, and transistors can be displayed and located in a like manner, and the computer can design a specified resistor if its value and such processing information as resistivity and junction depth are provided. Using a light pen, it is also possible to draw the interconnection pattern used for the metallization step.

The information stored in the computer at this point is used to create tapes which can drive an automatic drafting machine, which in turn produces typically 100-times-full-size mask masters. These are used to produce actual-size production masks, usually in rectangular arrays of several hundred.

In addition to speed, the computer method has the advantage over the manual method of being more accurate. This is because the computer has a more reliable memory than a human and will not forget some vital, though apparently trivial, operation such as one interconnect out of hundreds. Also, all components used in the layout have accurately known characteristics, since they can be stored in the computer memory in some detail.

PROBLEMS

3-1 For the sheet of monolithic integrated-circuit material shown in Fig. Prob. 3-1, what must be the sheet resistance if the total resistance between A and B is 2400 Ω? Ignore end effects.

3-2 In Fig. 3-5*b*, $V_{CC} = +10$ V, $I_C = 1$ mA, $V_{BE} = 0.6$ V, and $h_{FE} = 100$. Find R_4.

3-3 In Fig. 3-5*c*, $V_{DD} = +15$ V, $V_{GS} = +5$ V, and the gate current I_G of each MOSFET is 10^{-12} A. If $R_5 = 100$ kΩ, find I_D accurate to three significant figures. Assume Q_3 and Q_4 are perfectly matched.

3-4 In Fig. 3-6, $V_{CC} = +15$ V, $V_{EE} = -15$ V, $R_3 = 100$ kΩ, all V_{BE} are 0.6 V, and all $h_{FE} = 100$. If the bases of Q_1 and Q_2 are at 0 V, find R_1 and R_2 to give Q_1 and Q_2 collector voltages of $+10$ V. What will the collector voltages be if Q_1 and Q_2 bases are raised to $+1$ V? Assume Q_3 is an ideal current source.

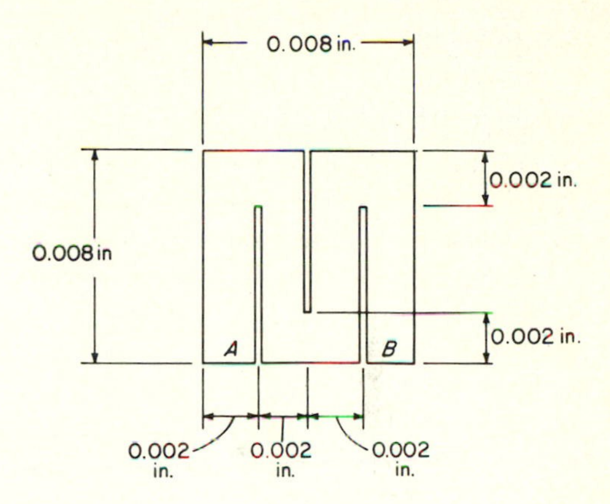

Fig. Prob. 3-1

3-5 Given a pair of complementary transistors and as many resistors as necessary, draw circuits having approximately unity gain and having the following low-frequency properties:

(*a*) Minimum output impedance
(*b*) Minimum offset voltage
(*c*) Minimum voltage drift
(*d*) Minimum input capacitance
(*e*) A gain as close to unity as possible

3-6 Explain how the circuit in Fig. Prob. 3-6 acts as a voltage regulator, and compare its performance with the circuit shown in Fig. 3-13*a*.

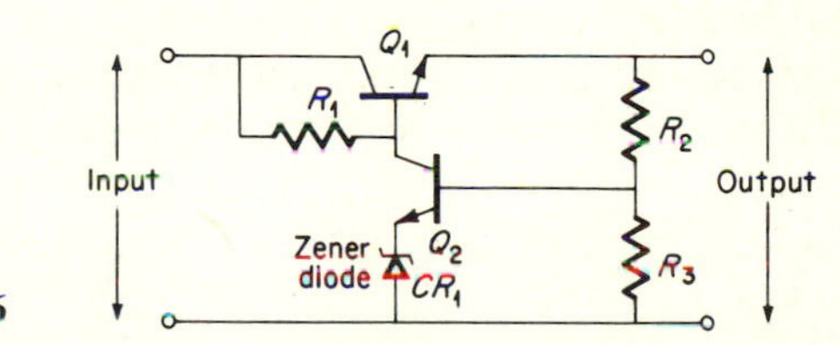

Fig. Prob. 3-6

3-7 For the circuit of Fig. 3-6, discuss the effect of replacing Q_3, Q_4, and R_3, by a single resistor producing the same collector currents, on:

(*a*) The differential output in response to +1 mV on Q_1 base and −1 mV on Q_2 base

(*b*) The differential output in response to +2 mV on Q_1 base with Q_2 base grounded

(*c*) The common-mode rejection for differential outputs and for single-ended outputs

(*d*) The stability of the output with respect to temperature when taken differentially

(*e*) The stability of the output with respect to temperature when taken single-ended

Answer by saying whether there is any difference and, if so, what kind of difference, rather than answering numerically.

3-8 Describe the principle of operation of a temperature-stabilized differential pair suitable for use in a stable dc amplifier.

3-9 In Fig. Prob. 3-9, $\div n_1$ and $\div n_2$ counters have been inserted into a PLL system as shown. If n_1 and n_2 are variable, explain how the circuit can be used to generate frequencies which are multiples of f_0. If $n_1 = 3$, $n_2 = 4$, $f_0 = 30$ kHz, and the loop is locked, what are the frequencies of signals at points 1, 2, and 3 in Fig. Prob. 3-9?

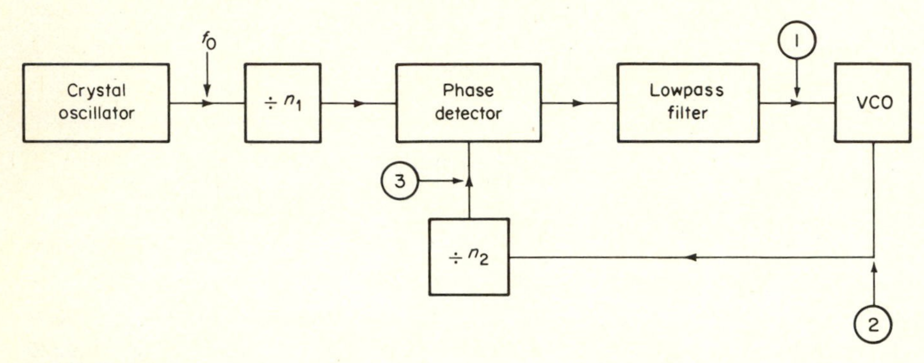

Fig. Prob. 3-9

3-10 In Fig. 3-13*a*, the supply is 15 V, all transistors have $I_C = 1$ mA, and $V_{BE} = 0.6$ V. The h_{FE} is high, so assume $I_B = 0$. CR_1 is a 5-V zener diode requiring 1 mA of bias current. Find R_1, R_2, R_3, R_4, and R_5 for an output voltage of 10 V. What is the total current drawn from the supply?

3-11 Discuss why a video amplifier constructed using the circuit of Fig. 3-15 would have a flatter frequency response in the passband than one constructed from discrete components which employed overall feedback, interstage shunt peaking, and *RC* interstage coupling.

3-12 In progressing up the frequency spectrum from dc to video to IF to RF amplifiers, would one expect to find an increasing or decreasing number of external components used to optimize the performance of a monolithic integrated circuit? Why?

3-13 Outline the differences between the construction of discrete, monolithic, and hybrid circuits. Give an example of when each would be used. In what ways does the design of a circuit made by a monolithic process differ from the design of one to be constructed from discrete components?

3-14 PLL action tends to force the VCO and input frequencies to be the same under steady-state conditions when the loop is locked. Under these conditions the error voltage is zero. Why is there *ever* an error voltage, since PLL action tries to force the error voltage to zero?

REFERENCES

1. Signetics handbook on digital, linear, and MOS applications, pp. 6-1–6-73, 1973.
2. National Semiconductor: *Appl. Note AN-94.* "LX series pressure transducers: design and applications information," by Stephen Calebotta, August 1973.

chapter 4

Operational Amplifiers

4-1 INTRODUCTION

Consider the simple circuit shown in Fig. 4-1*a*. If an input voltage V_1 is applied as shown, the output voltage will be

$$V_2 = V_1 \frac{R_2}{R_1 + R_2} \tag{4-1}$$

The circuit therefore acts as an attenuator, the amount of attenuation depending only on the values of the two resistors in the attenuating network.

Equation (4-1) can be solved to give V_1 as a function of V_2:

$$V_1 = V_2 \left(\frac{R_1 + R_2}{R_2} \right) \tag{4-2}$$

Unfortunately, with V_2 as an input the network in Fig. 4-1*a* will not produce an output voltage V_1 to satisfy Eq. (4-2), since this would involve the creation of energy. However, if a circuit were possible which did satisfy Eq. (4-2), it would be highly desirable. Instead of attenuation there would be gain, and the gain would depend only on the values of resistors, so that it could be very stable, accurate, and predictable.

Figure 4-1*b* shows a circuit which *will* satisfy Eq. (4-2). It does so by the addition of another element to the attenuator of Fig. 4-1*a*. The element is an amplifier, or more precisely an *operational amplifier*, which is represented by the triangle in Fig. 4-1*b*. It takes power from an external source (a power supply, which is not shown) and feeds it into the resistive network R_1 and R_2 so that Eq. (4-2) *is* satisfied. Thus energy is merely transformed from one form to another (not created). How the addition of the operational amplifier makes the circuit in Fig. 4-1*b* satisfy Eq. (4-2) is explained in Sec. 4-3-4.

In practice errors exist because the operational amplifier is not perfect or "ideal," but the errors can be made to be very small fractions of a percent. In fact, in most cases the amplifier is sufficiently ideal that Eq. (4-2) can be applied without any corrections. We shall be analyzing many circuits using the assumption that the operational amplifier employed is ideal. We shall also be analyzing the effects of circuits using nonideal amplifiers.

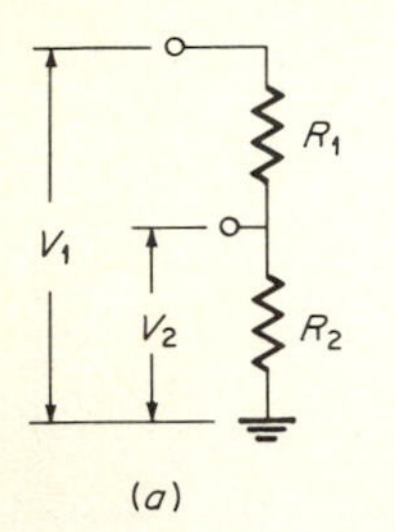

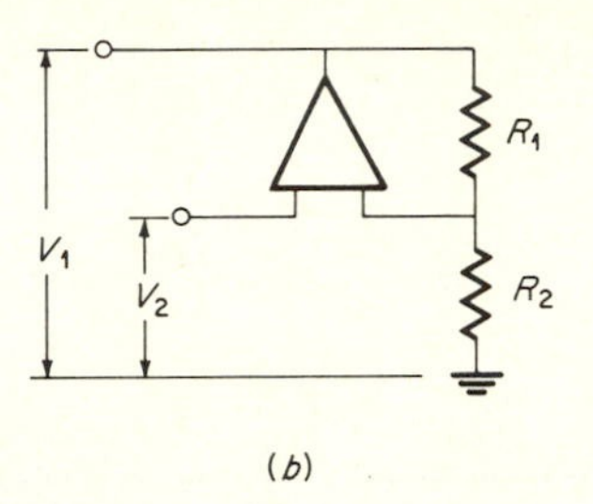

Fig. 4-1 Use of an operational amplifier with an attenuator to achieve precise gain.

The term *operational amplifier* was originally coined to denote a type of amplifier used in analog computers to perform such mathematical operations as summation, subtraction, integration, and differentiation. Its versatility is such, however, that its use has been extended into virtually every phase of electronics. In addition to performing mathematical operations, these amplifiers are now used in signal conditioners, power regulators, active filters, function generators, instrumentation, process control, analog-to-digital and digital-to-analog converters, and many other applications.

The reason for such widespread use is that an operational amplifier is a device that has a very large open-loop gain, but which is operated in a closed-loop condition with a closed-loop gain small compared with the open-loop gain. Such a condition implies the use of large amounts of negative feedback. This in turn means that the characteristics of the resulting circuit depend almost entirely on those of the feedback elements used (such as R_1 and R_2 in Fig. 4-1*b*), not on the characteristics of the transistors, resistors, and capacitors that make up the operational amplifier. Since the feedback elements are usually passive, operational-amplifier circuits can be both very stable and predictable in performance.

The operational amplifier may be considered to be a device because it is but one element in a circuit. Monolithic integrated-circuit operational amplifiers are in fact fabricated on a single chip of silicon. Discrete-component operational amplifiers are made of numbers of transistors, resistors, and capacitors, but are packaged for use as a single circuit element. Between the monolithic and conventional types are hybrid amplifiers. However, all operational amplifiers, whether monolithic, hybrid, or discrete, can be analyzed using certain common basic principles, which we shall employ. The operational amplifier is only one of many linear circuits that can be supplied in integrated form. Other linear integrated circuits are discussed in Chap. 3.

Table 4-1 contains a collection of circuits, all of which employ operational amplifiers. It is given to provide an indication of the breadth of applications of these amplifiers. Each circuit is identified according to the general function it is intended to perform. The transfer function of each circuit is listed along with some information about other relevant properties. The table is intended to give an overall picture of how operational amplifiers are commonly used, and more detailed information on many of the circuits is presented later in this chapter.

Table 4-1 Operational amplifier circuits

Name	*Circuit*	*Gain*	R_{in}	R_{out}	*Other*
Inverting amplifier	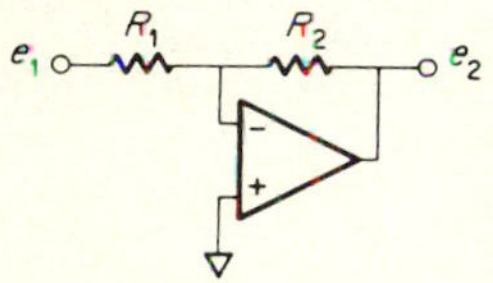 e_1, R_1, R_2, e_2	$\frac{e_2}{e_1} = -\frac{R_2}{R_1}$	R_1	Low	Inverts
Noninverting amplifier	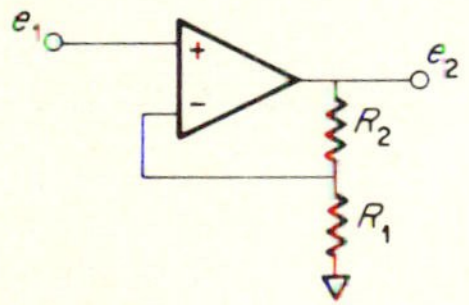 e_1, e_2, R_2, R_1	$\frac{e_2}{e_1} = 1 + \frac{R_2}{R_1}$	High	Low	No inversion
Follower	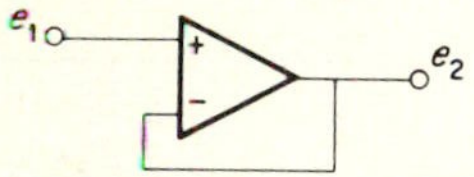 e_1, e_2	$\frac{e_2}{e_1} = 1$	Very high	Low	No inversion
Comparator	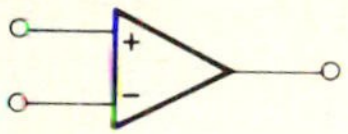	Same as operational amplifier open-loop gain	10 kΩ → 1 MΩ Typical	1 kΩ → 10 kΩ Typical	Used as level detector
Adder	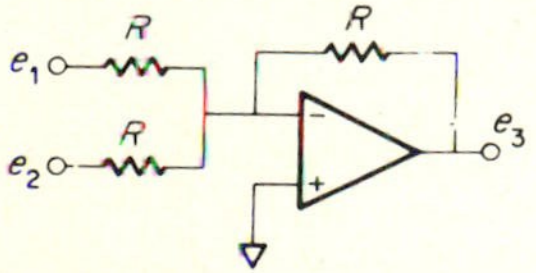 e_1, e_2, R, R, R, e_3	$e_3 = -(e_1 + e_2)$	R	Low	Can sum time-varying input voltages

Table 4-1 (cont.)

Name	*Circuit*	*Gain*	R_{in}	R_{out}	*Other*
Subtracter, differential amplifier, common-mode circuit	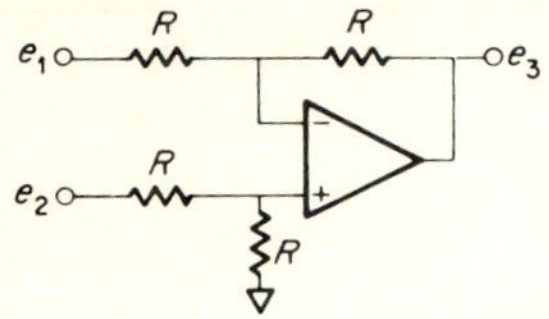	$e_3 = (e_2 - e_1)$		Low	Can subtract time varying input voltages; input impedance is voltage-variable (no virtual ground)
Add/subtract	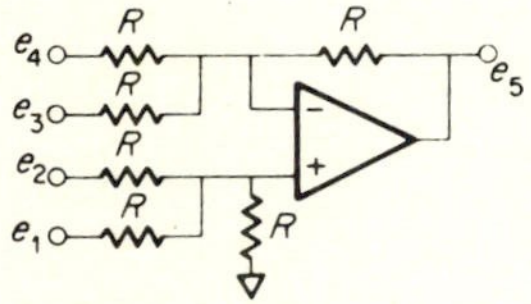	$e_5 = (e_1 + e_2) - (e_3 + e_4)$		Low	Can add or subtract time-varying voltages; input impedance is voltage-variable (no virtual ground)
Integrator	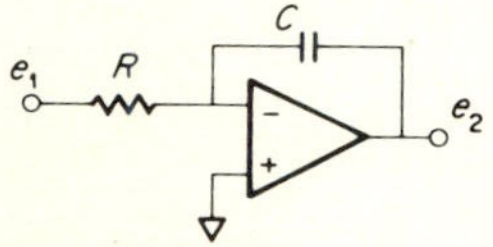	$e_2 = -\frac{1}{RC}\int e_1\,dt$	R	Low	Can integrate a time-varying voltage
Differentiator	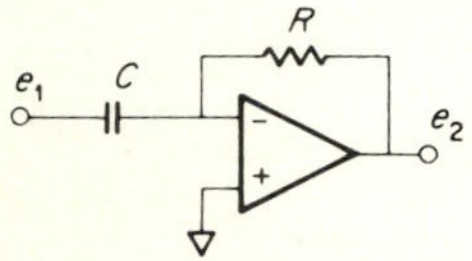	$e_2 = -RC\frac{d}{dt}e_1$	$\frac{1}{j\omega C}$	Low	Can differentiate a time-varying voltage; input impedance varies with frequency
Instrumentation amplifier	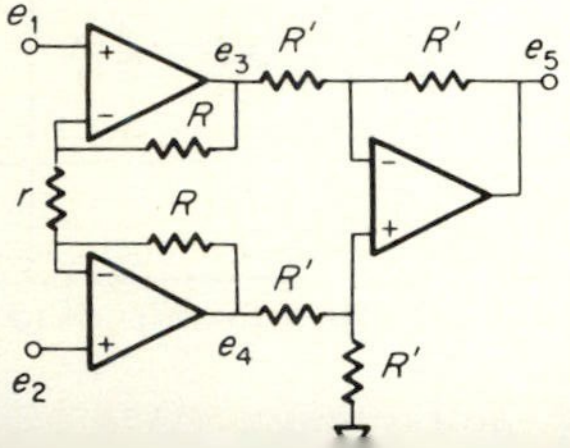	$e_5 = -(1 + \frac{2R}{r})(e_1 - e_2)$	High	Low	Differential to single-ended converter with high common-mode rejection

Table 4-1 (cont.)

Name	*Circuit*	*Gain*	R_{in}	R_{out}	*Other*
Phase shifter	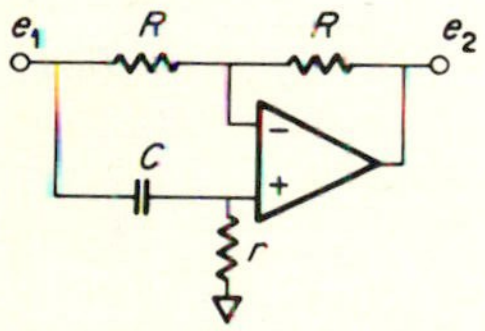	$e_2 = -\left(\frac{1 - pCr}{1 + pCr}\right) e_1$		Low	Circuit gain is always 1, phase $0 \rightarrow 180°$ as r varies
Precision rectifier	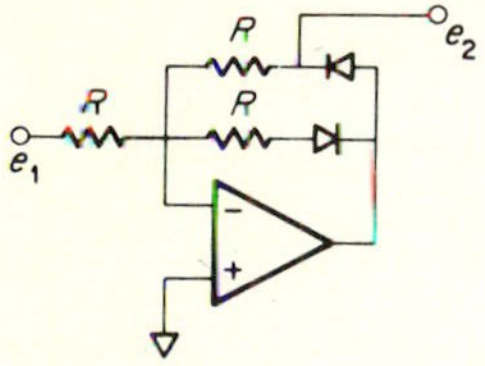	Half-wave rectifies	R	Low	Rectifies below the diode voltage
Current amplifier	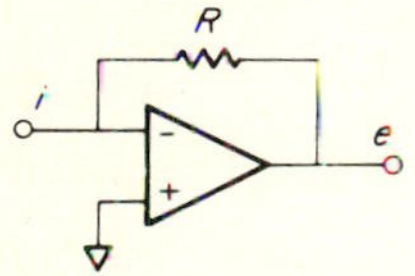	$e = -iR$	0	Low	Also called current-to-voltage converter or transconductance amplifier
Logarithmic amplifier	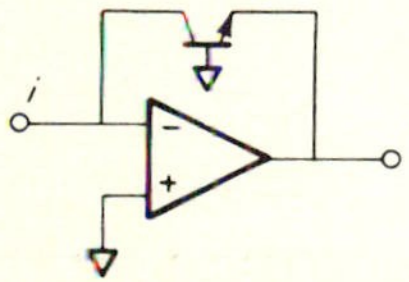	$e = -K \log i$	0	Low	9 decades possible 10^{-3} to 10^{-12} A

Table 4-1 (cont.)

Name	*Circuit*	*Gain*	R_{in}	R_{out}	*Other*
Vector sum	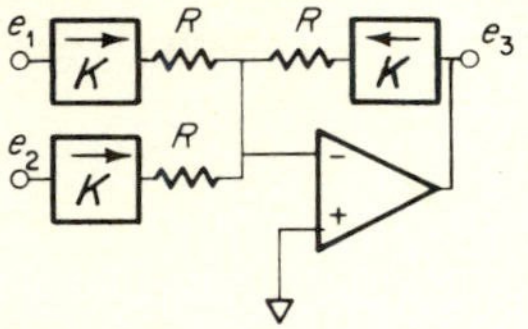	$e_3 = -\sqrt{e_1^2 + e_2^2}$		Low	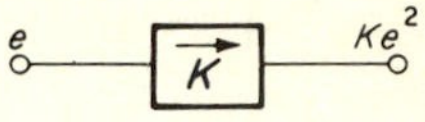 Uses squaring element
High-input-impedance differential amplifier	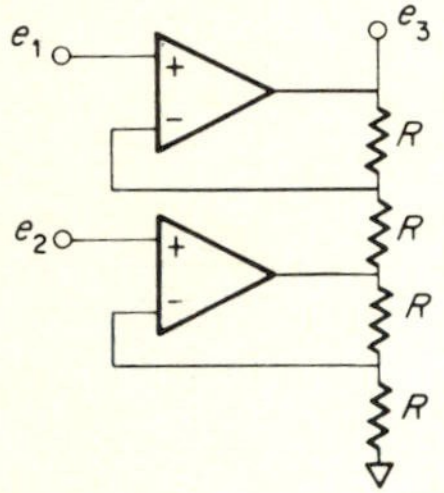	$e_3 = 2(e_1 - e_2)$	Very high	Low	Differential to single-ended converter (see preferred circuit, bottom figure page 118
Gyrator	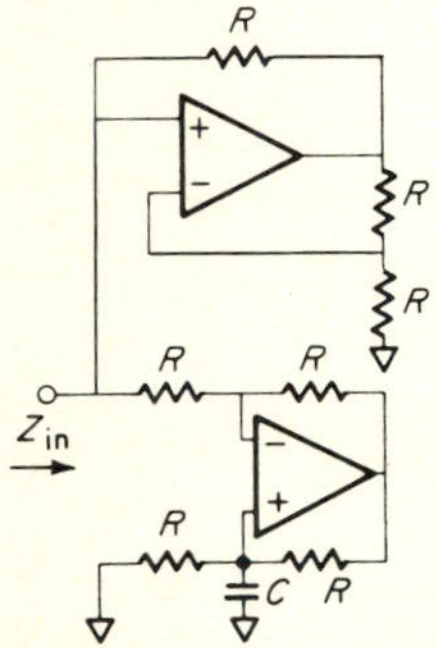		$Z_{in} = j\omega CR^2$	Not applicable, is 2-terminal device	Z_{in} is inductive; $j\omega(L) = j\omega(CR^2)$ so that $L = CR^2$

4-2 THE IDEAL OPERATIONAL AMPLIFIER

In its most common form, an operational amplifier consists of a dc amplifier with a differential input and a single-ended output, as shown in Fig. 4-2. The ideal

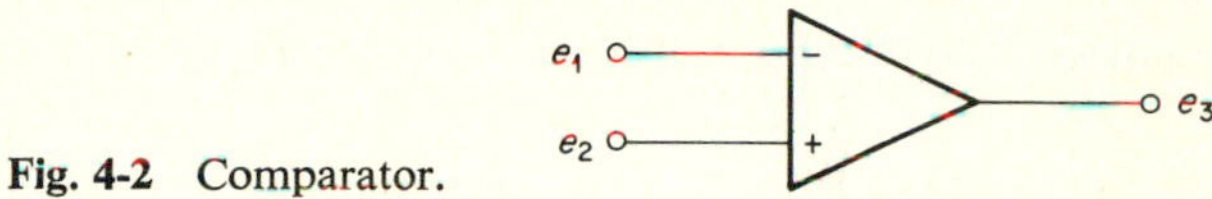

Fig. 4-2 Comparator.

operational amplifier would have the following properties:

1. Infinite gain
2. Infinite bandwidth
3. Infinite input impedance, both between the two input terminals and from each terminal to ground
4. Infinite output-current drive capability
5. Zero output impedance
6. Zero input-voltage offset
7. Zero input current
8. True differential amplification with infinite common-mode rejection
9. The above eight points true at all temperatures

Clearly such a device cannot be built, but amplifiers can be designed so that for a very large number of applications the compromises made do not significantly affect the performance of the resulting circuit, and for many calculations the above properties can be assumed to be true. The effect of nonideal characteristics will be examined in Sec. 4-4.

Some important and highly simplifying consequences arise if the ideal assumptions listed above are used. These are:

1. Any current flowing into either input terminal must leave via another path. It cannot flow into the amplifier, because of the infinite input impedance and the zero current bias.
2. To obtain any output voltage, no potential difference is required between the input terminals, because of the infinite gain.
3. Frequency and loading effects can be ignored, and zero offsets imply zero voltage out for zero voltage in.

We shall now analyze a number of operational-amplifier circuits, often simplifying the analysis by using the above assumptions.

4-3 ANALYSIS OF OPERATIONAL-AMPLIFIER CIRCUITS

4-3-1 THE COMPARATOR

As pointed out earlier, an operational amplifier is used with feedback elements to obtain a desired circuit. However, the first circuit to be introduced has no feedback elements. The comparator circuit shown in Fig. 4-2 is the easiest to analyze since it consists of only one component, the amplifier itself.

The amplifier is represented by a triangle as shown, with two inputs marked + and −. Note that these do not refer to power-supply connections. It is usual, for the sake of simplicity, to omit the power-supply connections which are, of course, always present. The output is taken off the right-hand apex of the triangle. The significance of the positive and negative input notations is as follows: if input voltage e_1 changes with e_2 held constant, e_3 will also change, but with the opposite polarity to e_1. The − terminal is therefore called the *inverting input.* If e_2 changes with e_1 held constant, e_3 will again change, but this time in the same direction as e_2. The + terminal is therefore called the *noninverting input.*

In a comparator, no feedback is employed, so the circuit operates with full gain. With an ideal amplifier when $e_1 = e_2$, $e_3 = 0$. If e_1 is more positive than e_2, then e_3 will be negative; but if e_1 is more negative than e_2, e_3 will be positive. Thus, if e_2 is a reference voltage, the circuit output will indicate whether or not e_1 is more positive than e_2. Comparators are widely used in applications requiring level detection, such as analog-to-digital converters, digital-to-analog converters, time-to-voltage converters, and zero-crossing detectors.

4-3-2 THE VOLTAGE FOLLOWER

Surpassed only by the comparator for simplicity is the voltage follower. Again, no feedback elements are used, but there is a feedback connection from the output to the inverting input as shown in Fig. 4-3. We can find the circuit gain as

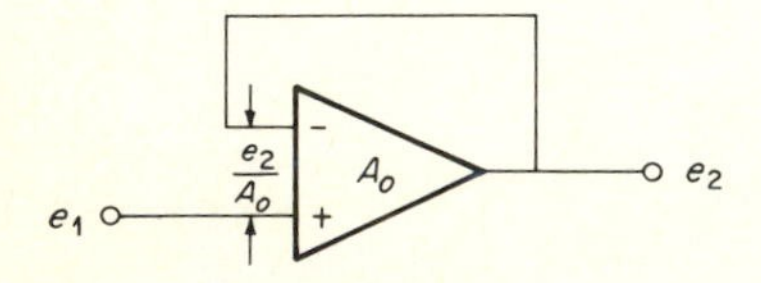

Fig. 4-3 Voltage follower.

follows: Suppose e_2 changes by 1 V. This must result from a voltage change, between the input terminals, of $1/A_o$ V, where A_o is the open-loop gain of the operational amplifier, that is, the gain without feedback. If A_o is infinite, as in the ideal operational amplifier, there will actually be no potential difference between the input terminals. Since the negative input is directly connected to the output, e_1 and e_2 will always be at the same potential, and the gain of the circuit is unity. This circuit also has a very high input impedance, even if a nonideal operational amplifier with finite input resistance is used, which arises as follows: Suppose an impedance Z exists between the input terminals. If the output is 1 V, it must arise from a difference in potential between the inverting and noninverting inputs of $1/A_o$ V. e_1 must therefore be $1 + 1/A_o$ V. The current flowing through Z is the voltage across Z divided by the impedance of Z, or $1/A_oZ$. The effective input impedance of the circuit with feedback is the circuit input voltage divided by the circuit input current, which is

$$\frac{1 + 1/A_o}{1/A_oZ} = Z(1 + A_o) \tag{4-3}$$

Thus the closed-loop input impedance is 1 plus the open-loop gain times the open-loop impedance.

It can also be shown that the voltage follower has a very low output impedance, but this will not be shown at this point because more general methods of obtaining the closed-loop characteristics of operational-amplifier circuits are indicated later on in this chapter. Because of the three properties noted here, namely unity gain, high input impedance, and low output impedance, the voltage follower finds wide application as an impedance transformer to couple a high source impedance to a relatively low load impedance. Notice that it is the *impedance* that is increased, which means that not only is the input resistance increased, but also the input capacitance is decreased since $X_c = 1/2\pi fC$. This reduction in input capacitance can be important for high-frequency applications.

Suppose the voltage follower in Fig. 4-3 is built with an operational amplifier having an open-loop gain of 10^5 and a resistance between the input terminals of 200 kΩ. A 100-kΩ resistor is connected from the + input to ground. What are the gain and the circuit input resistance?

The 100-kΩ does not affect the gain, which is unity since the circuit is a voltage follower. The circuit input resistance will be that due to the amplifier, or $Z(1 + A_o) = (2)(10^5)(1 + 10^5) = 20{,}000.2$ MΩ [from Eq. (4-3)], in parallel with the 100 kΩ input resistor. But 20,000.2 MΩ in parallel with 100 kΩ is 99.9995 kΩ. The circuit input resistance is therefore equal to the value of the 100-kΩ input shunt resistor for all practical purposes. In general, the input resistance of an operational amplifier used as a voltage follower is usually so large that the effective intput resistance is that of any external resistance from the + input to ground.

4-3-3 THE INVERTING AMPLIFIER

The inverting amplifier enables the very useful concept of a *virtual ground* to be introduced. Figure 4-4 shows an inverting amplifier with an input resistor R_1 and a feedback resistor R_2. Again, because of the very large open-loop gain, e is very small and, since the + input is grounded, the − input will also be close to ground potential. In other words, although not actually connected to ground the − input is held *virtually* at ground by negative feedback, regardless of the magnitude of potentials e_1 and e_2. If an input e_1 is applied and the amplifier is

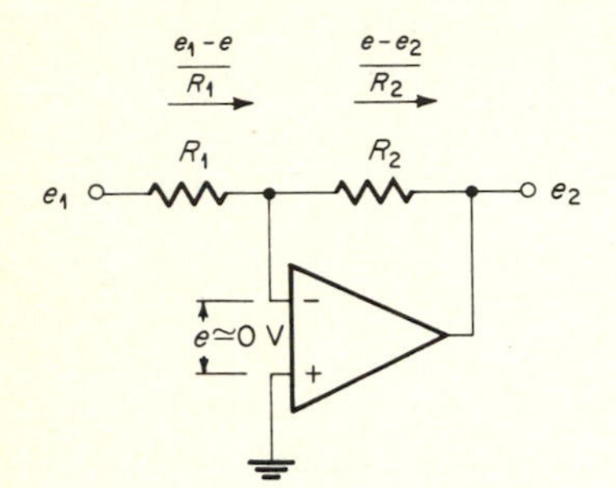

Fig. 4-4 Inverting amplifier.

ideal so that all the input current through R_1 flows into R_2, then equating the currents through these resistors gives

$$\frac{e_1 - e}{R_1} = \frac{e - e_2}{R_2} \tag{4-4}$$

If the open-loop gain is large, e is essentially zero, and Eq. (4-4) becomes

$$\frac{e_2}{e_1} = -\frac{R_2}{R_1} \tag{4-5}$$

The voltage gain is therefore the ratio of the feedback resistance to the input resistance, the significance of the minus sign being that the output is inverted with respect to the input. Note this difference between the inverting amplifier and the follower circuit previously discussed.

The input resistance is the input voltage divided by the input current, or

$$R_{\text{in}} = \frac{e_1}{(e_1 - e)/R_1} \approx R_1 \qquad \text{since } e \text{ is small} \tag{4-6}$$

The output resistance of this circuit, like that of the voltage follower, is made very low by the feedback.

Suppose the inverting amplifier in Fig. 4-4 has $R_1 = 10$ kΩ and $R_2 = 100$ kΩ. If the input is +1 V, find the output voltage, the input resistance, and the input current.

From Eq. (4-5), $e_2 = -(R_2/R_1)e_1 = -(10/1)(1) = -10$ V. From Eq. (4-6), the input resistance is equal to R_1 or 10 kΩ. The input current is the input voltage divided by the input resistance, or (1 V)/(10 kΩ) = 0.1 mA.

4-3-4 THE NONINVERTING AMPLIFIER

The circuit of Fig. 4-4 can be converted to that shown in Fig. 4-5 by grounding R_1 and applying the input signal to the ungrounded positive terminal of the amplifier. The result is a noninverting amplifier, the gain of which can again be

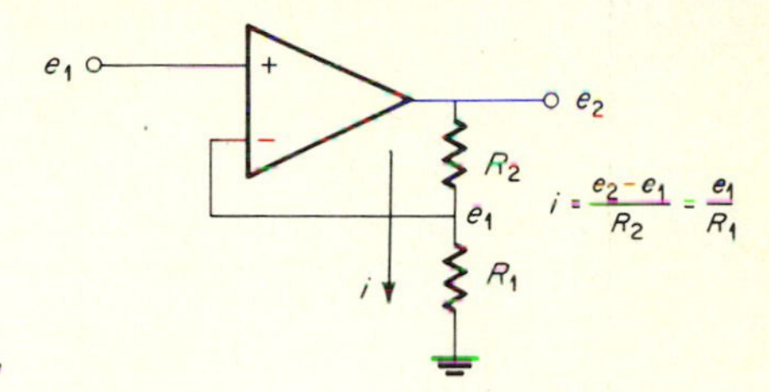

Fig. 4-5 Noninverting amplifier.

found by assuming that the two input terminals are at the same potential and that no current flows into the amplifier. Because of these two assumptions, the currents in R_1 and R_2 are equal and the potential at their junction is e_1 so that

$$\frac{e_2 - e_1}{R_2} = \frac{e_1}{R_1} \tag{4-7}$$

and the gain is

$$\frac{e_2}{e_1} = \frac{R_2 + R_1}{R_1} = 1 + \frac{R_2}{R_1} \tag{4-8}$$

Note that the sign this time is positive and that the gain is 1 plus the resistor ratio. The noninverting amplifier is a more general version of the follower circuit. The circuit in Fig. 4-5 can be converted to that in Fig. 4-3 by making $R_1 = \infty$ and $R_2 = 0$. It has the same general properties as the follower in that it has a high input impedance, a low output impedance, and input and output in phase, but has a gain greater than unity.

Suppose the noninverting amplifier in Fig. 4-5 has $R_2 = 9$ kΩ and $R_1 = 1$ kΩ. Find the output voltage for an input of +1 V.

From Eq. (4-8), the circuit gain is $1 + 9/1 = +10$, so the output voltage is $(+1)(+10) = +10$ V. Notice that the noninverting amplifier gives +10 V out for a +1 V input, while the inverting amplifier of Fig. 4-4 gives −10 V out for a +1 V input.

4-3-5 THE SUMMING AMPLIFIER

Referring to Fig. 4-6 and noting that the current flowing into the virtual ground must equal that flowing out of it, we can write

$$\frac{e_1}{R_1} + \frac{e_2}{R_2} = -\frac{e_3}{R_3} \tag{4-9}$$

or

$$e_3 = -\left(\frac{R_3 e_1}{R_1} + \frac{R_3 e_2}{R_2}\right) \tag{4-10}$$

If $R_1 = R_2 = R$,

$$e_3 = -\frac{R_3}{R}(e_1 + e_2) \tag{4-11}$$

The output voltage e_3 is therefore equal to a constant, $-R_3/R$, times the sum of the input voltages. If $R_1 = R_2 = R_3$, e_3 will be equal to $-(e_1 + e_2)$.

Suppose, for the circuit in Fig. 4-6, $R_1 = 10\ \text{k}\Omega$, $R_2 = 100\ \text{k}\Omega$, and $R_3 = 1\ \text{M}\Omega$. Find the circuit output voltage if $e_1 = +1$ V and $e_2 = -10$ V.

From Eq. (4-10), the output voltage e_3 is

$$e_3 = -\left[\frac{(10^6)(+1)}{10^4} + \frac{(10^6)(-10)}{10^5}\right] = 0\ \text{V}$$

4-3-6 THE SUBTRACTING AMPLIFIER

In the circuit shown in Fig. 4-7 neither of the operational-amplifier input terminals is connected directly to ground, so it cannot be assumed that they are both at ground potential. Therefore, in writing the current-flow equations, a

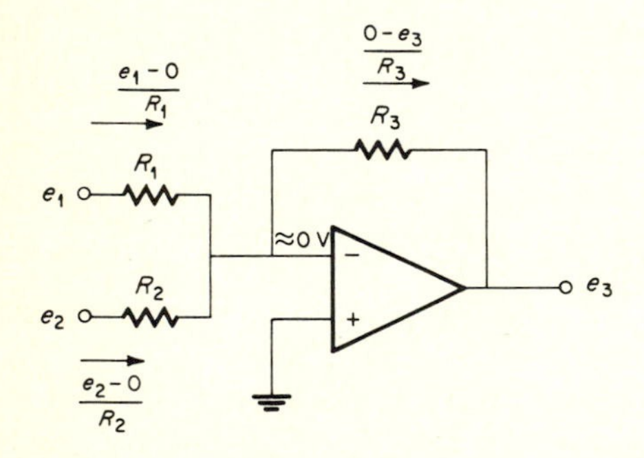

Fig. 4-6 Summing amplifier.

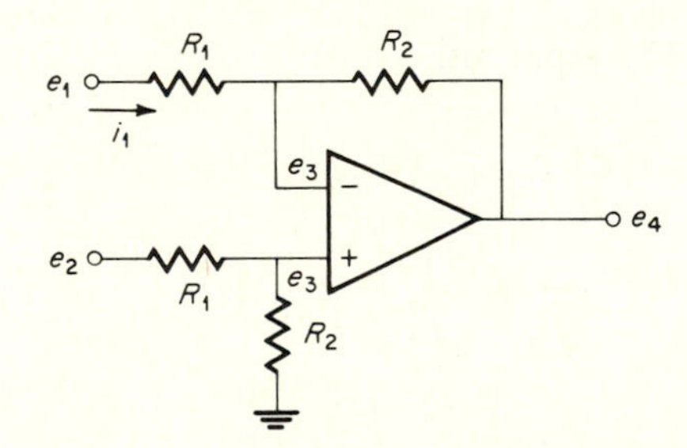

Fig. 4-7 Subtracting amplifier.

value e_3 is assigned to the voltage level of both the inverting and noninverting inputs, measured with respect to ground. That both can be considered to be at the same potential is a consequence of the fact that any value of e_4 can be produced by a negligibly small value of potential difference between the two input terminals. Noting once again that no current flows into the amplifier input terminals, we can write

$$\frac{e_1 - e_3}{R_1} = \frac{e_3 - e_4}{R_2} \tag{4-12}$$

and

$$\frac{e_2 - e_3}{R_1} = \frac{e_3}{R_2} \tag{4-13}$$

We solve Eq. (4-13) for e_3, substitute e_3 into Eq. (4-12), and solve for e_4:

$$e_4 = \frac{R_2}{R_1}(e_2 - e_1) \tag{4-14}$$

If $R_1 = R_2 = R$, e_4 is equal to the difference between e_2 and e_1.

The subtracting amplifier is really a combination of the inverting and noninverting amplifiers discussed in Secs. 4-3-3 and 4-3-4, but its input resistance is quite different from either. Referring to Fig. 4-7, let $R_1 = R_2 = R$. The noninverting input of the circuit, that seen by e_2, then has an input resistance of $2R$, since the path to ground is via two resistors of value R in series. The + terminal of the operational amplifier does not load the circuit, because the amplifier is assumed to be ideal and therefore has an infinite input resistance. The resistance seen by the inverting input, that seen by e_1, is more complicated because it is voltage-variable, being a function of e_2. This can be seen as follows. Since $R_1 = R_2$, the voltage at the + terminal of the operational amplifier will be $e_2/2$. Therefore e_3 in Fig. 4-7 is $e_2/2$. The resistance seen by e_1 is e_1/i_1. i_1 is $(e_1 - e_3)/R_1$ or

$$i_1 = \frac{e_1 - (e_2/2)}{R_1} \tag{4-15}$$

The input resistance is therefore

$$\frac{e_1}{i_1} = \frac{e_1 R_1}{e_1 - (e_2/2)}$$

This expression can be positive or negative depending on the magnitude and sign of e_1 and e_2. If $e_1 = e_2/2$, then the input resistance seen by e_1 is infinite.

For the circuit in Fig. 4-7, let $R_1 = 10\ \text{k}\Omega$ and $R_2 = 100\ \text{k}\Omega$. Find the circuit output voltage and the current i_1 as defined in Fig. 4-7 if (*a*) $e_1 = +3$ V and $e_2 = +1$ V and (*b*) $e_1 = +1$ V and $e_2 = +3$ V.

Case (*a*): From Eq. (4-14) the circuit output voltage e_4 is:

$$e_4 = \frac{(10^5)(1 - 3)}{10^4} = -20\ \text{V}$$

From Eq. (4-15),

$$i_1 = \frac{3 - (1/2)}{10^4} = (2.5)(10^{-4})\ A = 0.25\ \text{mA}$$

Case (*b*): From Eq. (4-14),

$$e_4 = \frac{(10^5)(3 - 1)}{10^4} = 20\ \text{V}$$

From Eq. (4-15),

$$i_1 = \frac{1 - (3/2)}{10^4} = -(0.5)(10^{-4})\ A = -0.05\ \text{mA}$$

In case (*a*) current flows from the source into the R_1 on the operational-amplifier – input, while in case (*b*) current flows from R_1 back into the source.

4-3-7 THE INTEGRATOR

The circuit shown in Fig. 4-8*a* is capable of integrating a time-variable voltage e_1 with respect to time. In practice this means that the output of the circuit is proportional to the *area* under the voltage/time curve for the input voltage over a given period of time. This idea will be clarified with a numerical example after the equation for the integrator is derived. If the voltage/time curve has a simple shape, as in the example that follows, we can deduce the integrator output voltage by calculating the area under the curve. If the shape of the curve is fairly complex, but can still be represented by some mathematical equation, we can use calculus to integrate the equation to obtain the area. If the shape of the curve is so complex that it cannot be represented by any reasonable mathematical expression, it is still possible to perform the integration—that is, find the area—using an integrator such as that shown in Fig. 4-8*a*.

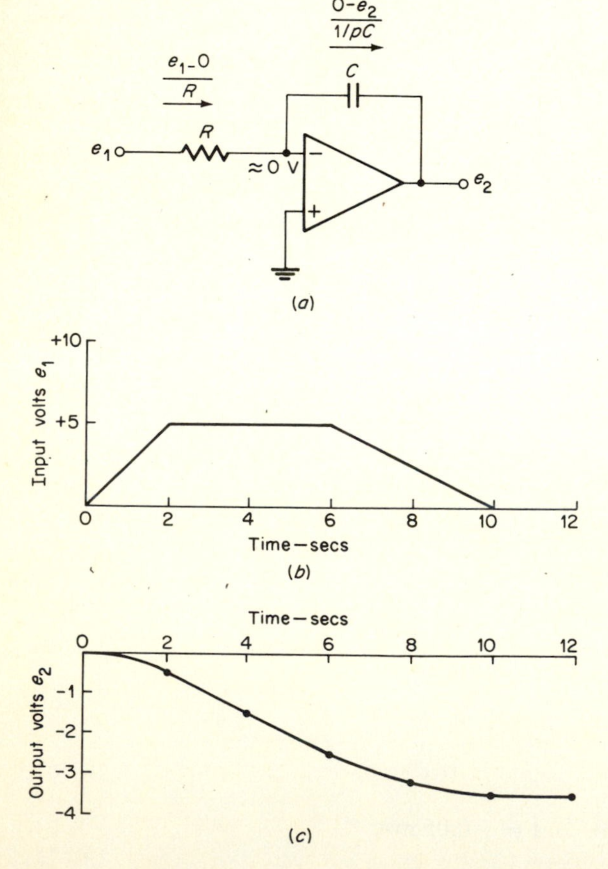

Fig. 4-8 (*a*) Integrator; (*b*) input voltage; (*c*) output voltage.

Although the output voltage of an integrator represents an input area, integrators are mainly used as ramp or sweep generators, as filters, and in simulators to realize mathematical transfer functions which represent the performance of, for example, an airplane under design.

The circuit is most easily analyzed by using operational calculus in which $1/p$ represents integration. Assuming this, we can write the relationship for a capacitor,

$$v = \frac{q}{C} = \frac{\int i\,dt}{C} = \frac{(1/p)i}{C} = \frac{i}{pC}$$

The impedance of the capacitor is therefore $v/i = 1/pC$. Equating currents in Fig. 4-8*a*, we can write

$$\frac{e_1}{R} = -\frac{e_2}{1/pC}$$

so that

$$e_2 = -\frac{1}{pCR} e_1 \tag{4-16}$$

The interpretation of Eq. (4-16) is that e_2 is equal to a constant, $-1/CR$, times the integral of the input e_1 with respect to time.

Suppose the waveform in Fig. 4-8*b* is applied to the input of the integrator in Fig. 4-8*a*. $R = 1$ MΩ, and $C = 10$ μF. Until time $t = 0$ there is a short across the capacitor; it is removed at $t = 0$. Find the circuit output voltages at $t = 0$, 2, 4, 6, 10 and 12 s.

In performing the integration implied by the $1/p$ term in Eq. (4-16), the integrator output voltage is proportional to the *area under the input-voltage–versus–time* curve. This is in contrast to many operational-amplifier circuits in which the output is related in some way to the *actual input voltage*.

Integral calculus is useful when the shape of the input-voltage–versus–time curve is complex. However, in this case, we can find the relevant areas by simple arithmetic, and then substitute into Eq. (4-17) below, which is obtained by rewriting Eq. (4-16) as follows:

$$e_2 = -\frac{1}{CR}\frac{1}{p}e_1 = -\frac{1}{CR}\text{(area under the input-voltage–versus–time curve)}$$

$$= -\frac{1}{CR}A \tag{4-17}$$

$1/CR$ is $1/(10)(10^{-6})(10^6) = 0.1$ s. In addition,

At $t = 0$ s: $A = 0$ volt-second
At $t = 2$ s: $A = 5$ volt-seconds
At $t = 4$ s: $A = 5 + 10 = 15$ volt-seconds
At $t = 6$ s: $A = 5 + 20 = 25$ volt-seconds
At $t = 10$ s: $A = 5 + 20 + 10 = 35$ volt-seconds
At $t = 12$ s: $A = 5 + 20 + 10 = 35$ volt-seconds

Also, at $t = 0$ the voltage across the capacitor is zero because it has just been unshorted. The voltage on both sides of the capacitor is therefore the same, and e_2 must be zero since the other side of the capacitor is at a virtual ground which is 0 V. So we have, from Eq. (4-17),

At $t = 0$: $e_2 = 0$
At $t = 2$: $e_2 = -(0.1)(5) = -0.5$ V
At $t = 4$: $e_2 = -(0.1)(15) = -1.5$ V
At $t = 6$: $e_2 = -(0.1)(25) = -2.5$ V
At $t = 10$: $e_2 = -(0.1)(35) = -3.5$ V
At $t = 12$: $e_2 = -(0.1)(35) = -3.5$ V

Notice that there is no change in output voltage from $t = 10$ s to $t = 12$ s, since the *area* does not change. We can also note that an integrator can integrate or find an area, in real time, while the input change is actually occurring. A mathematical integration, on the other hand, would have to wait until the end of the event so that the equation of the curve could be written, if it could be written at all.

Figure 4-8*c* shows how the output of the integrator changes in response to the input of Fig. 4-8*b*.

4-3-8 THE DIFFERENTIATOR

The circuit realization of a differentiator is similar to that of an integrator, except that the resistor and capacitor are interchanged. As for the case of the integrator, the circuit is most easily analyzed using operational calculus, in which p represents differentiation. For a capacitor we can write

$$v = \frac{q}{C}$$

Hence,

$$\frac{dv}{dt} = \frac{dq/dt}{C} = \frac{i}{C}$$

Since p represents differentiation we can substitute $p = d/dt$, which gives

$$pv = \frac{i}{C}$$

The impedance of the capacitor is v/i, which again is $1/pC$. Referring to Fig. 4-9 and equating currents, we have

$$\frac{e_1}{1/pC} = -\frac{e_2}{R}$$

so that

$$e_2 = -CRpe_1 \tag{4-18}$$

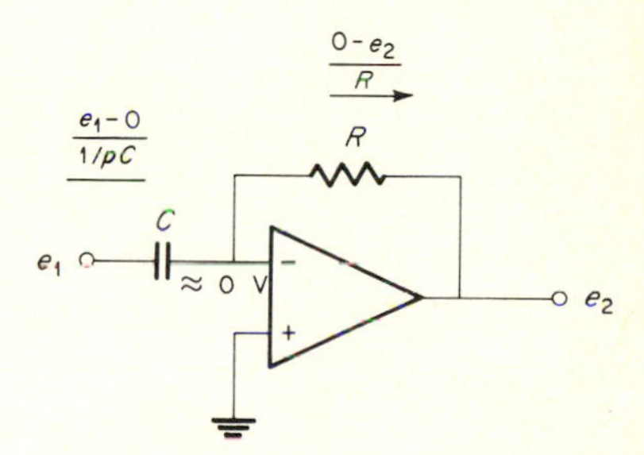

Fig. 4-9 Differentiator.

According to Eq. (4-18), e_2 is equal to a constant, $-CR$, times the differential of the input e_1 with respect to time, which is the *slope* of the e_1 input curve.

Suppose the waveform in Fig. 4-8*b* is applied to the circuit in Fig. 4-9 with $R = 1\ \text{M}\Omega$ and $C = 1\ \mu\text{F}$. At $t = 0$ the circuit output voltage is zero. Find the circuit output voltage at $t = 1$, 4, and 8 s.

In performing the differentiation implied by the p term in Eq. (4-18), the differentiator output voltage is proportional to the *slope* of the input-voltage–versus–time curve. This is in contrast to many operational-amplifier circuits in which the output is related to the *actual input voltage*.

Differential calculus is useful in finding the slope of the input-voltage–versus–time curve when the slope is complex and can be represented by some mathematical expression. In this case, however, we can find the relevant slopes by simple arithmetic, and then substitute them into Eq. (4-19) below, which is obtained by rewriting Eq. (4-18) as follows:

$$\begin{aligned} e_2 &= -CRpe_1 = -(CR)(\text{slope of input-voltage–versus–time curve}) \\ &= -(CR)(S) \end{aligned} \tag{4-19}$$

CR is $(10^{-6})(10^{6}) = 1$ s. In addition,

At $t = 1$ s: $S = 5\ \text{V}/2\ \text{s} = 2.5\ \text{V/s}$
At $t = 4$ s: $S = 0$
At $t = 8$ s: $S = -5\ \text{V}/4\ \text{s} = -1.25\ \text{V/s}$

From Eq. (4-19) we have

At $t = 1$ s: $e_2 = -(1)(2.5) = -2.5$ V
At $t = 4$ s: $e_2 = -(1)(0) = 0$
At $t = 8$ s: $e_2 = -(1)(-1.25) = +1.25$ V

4-3-9 A CONSTANT-AMPLITUDE PHASE SHIFTER

The circuit shown in Fig. 4-10 can be analyzed by once again writing a pair of simultaneous equations describing the current flow in the circuit. We have

$$\frac{e_1 - e_2}{R} = \frac{e_2 - e_3}{R}$$

and

$$\frac{e_1 - e_2}{1/pC} = \frac{e_2}{r} \tag{4-20}$$

Solving these equations for e_3 gives

$$e_3 = -\frac{1 - pCr}{1 + pCr} e_1 \tag{4-21}$$

It can be shown that Eq. (4-21) describes an all-pass transfer function, which means that if e_1 is held constant in magnitude as frequency is changed, e_3 will also remain constant and equal to e_1 in magnitude, while the phase of e_3 with respect to e_1 changes. Notice that this transfer function is independent not only of the operational-amplifier characteristics but also of R. It can be shown also that if r is varied from 0 to ∞ at a particular frequency, e_3 will shift in phase 180° with respect to e_1 while its magnitude remains constant. We can gain an intuitive idea of why this is so by considering the extreme cases, first when $r = 0$ and then when $r = \infty$.

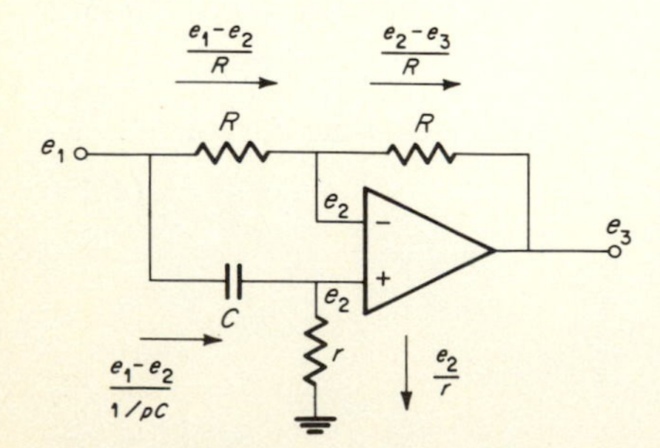

Fig. 4-10 Constant-amplitude phase shifter.

When $r = 0$ the + input of the operational amplifier is grounded, and the circuit becomes the same as that in Fig. 4-4 which, with $R_1 = R_2$, is an inverting amplifier with unity gain. The inversion means that there is a 180° phase shift.

When $r = \infty$, $e_1 = e_2$ at any frequency, because C and r form a potential divider and $1/pC$ is infinitely smaller than r. The input is therefore coupled without loss to the noninverting input. The resistor R between the input of the circuit and the inverting terminal of the operational amplifier is effectively shorted out, since the noninverting and inverting inputs are at essentially the same potential. Since there is no voltage across the input R, no current can flow through the feedback R, so that all of e_3 is fed back to the inverting input. Under these conditions the circuit becomes the same as that in Fig. 4-3, which is a noninverting amplifier with a gain of unity. The noninversion means that the phase shift is zero.

In a practical phase shifter, r would be a variable resistor which could be varied from 0 Ω to some large value, so that the amount of phase shift would vary from 180° to almost zero.

4-3-10 *RC*-ACTIVE FILTERS

Active filters are filters that employ passive elements, usually resistors, and capacitors, in conjunction with active elements, such as operational amplifiers, to obtain characteristics similar to those of *LCR* passive filters. Since *RC*-active circuits contain no inductors, it is possible to integrate them. Also, they can be made to work at very low frequencies (fractions of a hertz) where it is impractical to use inductors, even in discrete circuits.

It is outside the scope of this book to fully discuss the synthesis of *RC*-active filters. Instead we shall confine ourselves to an example which illustrates the role that operational amplifiers have to play in *RC*-active filter design.

Suppose it is desired to realize a second-order, low-pass, Butterworth (maximally flat) *RC*-active filter. The cutoff frequency is to be $10{,}000/2\pi$ Hz, and the gain in the passband is to be unity.

Consider the circuit in Fig. 4-11. It consists of an operational amplifier connected as a voltage follower so that it has a gain of unity. Two resistors and two capacitors are required to make the filter second order. We can write two simultaneous equations for this circuit. The current in R_1 equals the sum of those in R_2 and C_1:

$$\frac{e_1 - e_2}{R_1} = \frac{e_2 - e_3}{1/pC_1} + \frac{e_2 - e_3}{R_2} \tag{4-22}$$

The current in R_2 equals that in C_2:

$$\frac{e_2 - e_3}{R_2} = \frac{e_3}{1/pC_2} \tag{4-23}$$

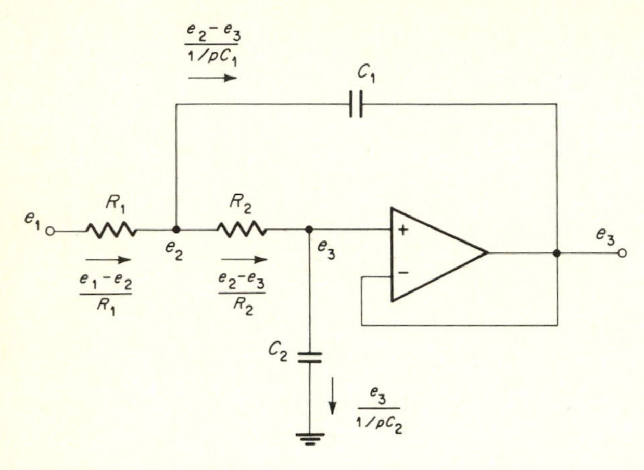

Fig. 4-11 Low-pass filter.

Notice that these equations are written assuming that the same voltage e_3 appears at both the noninverting input and the output of the operational amplifier. This is justified, since the amplifier is connected as a voltage follower.

Equations (4-22) and (4-23) can be solved to give

$$\frac{e_3}{e_1} = \frac{1/R_1 R_2 C_1 C_2}{p^2 + p(C_2 R_2 + C_2 R_1)/R_1 R_2 C_1 C_2 + 1/R_1 R_2 C_1 C_2} \tag{4-24}$$

Now let us substitute $R_1 = R_2 = 1\ \Omega$, $C_1 = \sqrt{2}$ F, and $C_2 = 1/\sqrt{2}$ F into Eq. (4-24). We obtain

$$\frac{e_3}{e_1} = \frac{1}{p^2 + \sqrt{2}\,p + 1} \tag{4-25}$$

Equation (4-25) is the transfer function of a low-pass Butterworth filter with a 3-dB cutoff frequency of $1/2\pi$ Hz, from elementary filter theory. The gain in the passband is unity because at dc, $p = 0$, and the transfer function simplifies to $\frac{1}{1} = 1$. The cutoff frequency ω_c is the square root of the constant term in the denominator, so that $\omega_c = 1$ and $f_c = 1/2\pi$. The filter is maximally flat because the coefficient of the p term in the denominator is $\sqrt{2}$.

However, we do not want a cutoff frequency of $1/2\pi$ Hz; our design calls for a cutoff of $10{,}000/2\pi$ Hz. It can be shown that the form of a transfer function such as that in Eq. (4-25) is not changed by scaling *RC* products in the circuit. For example, we can multiply the *RC* products in Eq. (4-25) by 10^{-4} to scale the circuit frequency from $1/2\pi$ to $10{,}000/2\pi$ Hz as desired. To obtain practical values, let us multiply the *RC* products by 10^{-4} by using

$$\text{Factor for all capacitors} = 10^{-8} \tag{4-26}$$

$$\text{Factor for all resistors} = 10^{+4} \tag{4-27}$$

since 10^{+4} multiplied by 10^{-8} gives an overall factor of 10^{-4} as we want. So we have

$$R_1 = R_2 = 10\ \text{k}\Omega$$

and

$$C_1 = 0.01414\ \mu\text{F}$$
$$C_2 = 0.00707\ \mu\text{F}$$

These new values for R_1, R_2, C_1, and C_2 in the circuit in Fig. 4-11, in conjunction with the unity-gain amplifier, will give the required filter function with a cutoff frequency of $10{,}000/2\pi$ Hz.

The important point to be remembered from this design is that the properties of the operational amplifier permit it to be regarded as a single device in the circuit. In this case the relevant properties are high input impedance so that no current is drawn from the frequency-selective network, low output impedance so that the feedback is entirely through capacitor C_1, precise gain due to follower action, and flat frequency response so that the shape of the final frequency-response curve is determined only by the external frequency-selective network.

4-3-11 A LINEAR RECTIFIER

In the rectifier circuit of Fig. 4-12, $R_1 = R_2 = R_3$, and CR_1 and CR_2 are silicon diodes having very low reverse leakage current, very low forward impedance, and very high reverse impedance. The circuit is linear because the diodes are so placed in the feedback loop that the gain of the operational amplifier is employed to make the diodes turn ON just sufficiently to give a half-wave rectified version of the input, *regardless of how small the input is.* This is in contrast to a conventional rectifier, in which the signal must be greater than the diode drop—about 0.6 V for silicon—before an output can occur, and in which the diode drop continues to change as the input increases, thereby producing a nonlinear output.

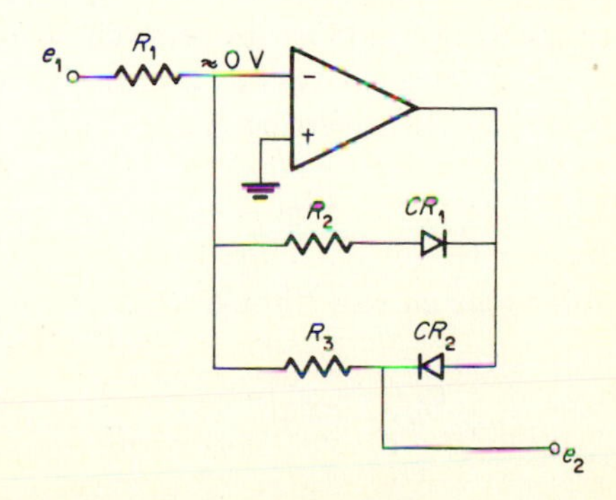

Fig. 4-12 Linear rectifier.

To understand how the circuit works, suppose e_1 is a negative voltage. The operational-amplifier output will be positive, and CR_1 will be reverse-biased, effectively switching R_2 out of the circuit. Current flows through CR_2 and R_3 into the virtual ground and into R_1. Since $R_1 = R_3$ and they have the same current in them, e_2 must equal $-e_1$. The operational amplifier automatically supplies just sufficient voltage to the anode of CR_2 so that this diode is biased to pass the correct amount of current. This is true even though e_1 may be much less than the voltage required to overcome the diode voltage drop. If e_1 now goes positive the operational-amplifier output will go negative, thereby reverse biasing CR_2 and preventing current flow through R_3. Since one side of R_3 is connected to the virtual ground, e_2 must be zero. e_2 will therefore be a positively rectified half-wave signal.

R_2 and CR_1 keep the circuit symmetrical, preventing the operational amplifier from saturating on positive input half-cycles, which avoids recovery-time problems in a practical amplifier and incidently provides a negatively rectified half-wave signal at the anode of CR_1.

4-3-12 INSTRUMENTATION AMPLIFIERS

A general-purpose instrumentation amplifier is one which will work from a variety of sources and drive a variety of output equipment. To be able to do this, it must have:

1. A differential input
2. A single-ended output
3. A high input impedance
4. A simple means for adjusting the gain
5. High common-mode rejection

The circuit in Fig. 4-7 meets requirements 1 and 2 but, in general, fails to meet requirements 3 to 5. Figure 4-13 shows a circuit which meets all five requirements.

Let us first find the output $e_3 - e_4$ of the differential input pair of operational amplifiers A_1 and A_2 in response to an input $e_1 - e_2$ as shown in Fig. 4-13.

First note that the *inverting* inputs of amplifiers A_1 and A_2 *also* have potentials of e_1 and e_2, respectively, on them by operational-amplifier action.

In the ideal case no current flows into any operational amplifier input, so we can write, referring to Fig. 4-13,

$$i_1 = i_2 = i_3 = i \tag{4-28}$$

Also, the current through resistor r is

$$i_2 = \frac{e_2 - e_1}{r} = i \tag{4-29}$$

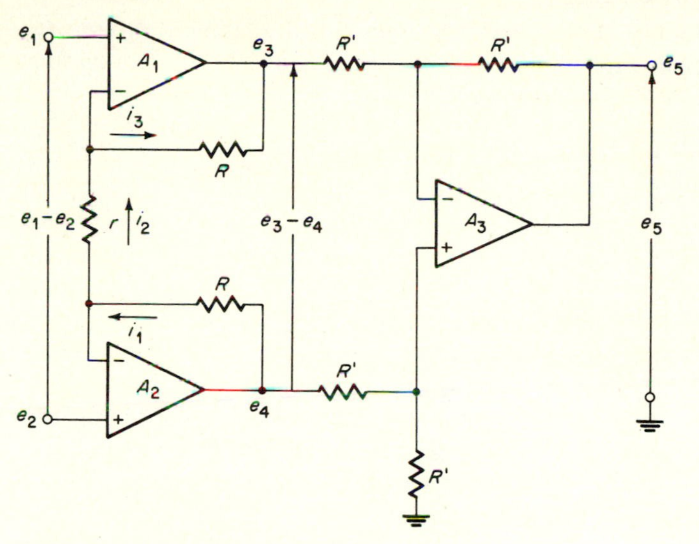

Fig. 4-13 Instrumentation amplifier.

The voltage drop across the three resistors R, r, and R equals $e_3 - e_4$, so that we can write

$$e_3 - e_4 = -(i_1 R + i_2 r + i_3 R) \tag{4-30}$$

From Eqs. (4-28) and (4-30) we can obtain

$$e_3 - e_4 = -i(r + 2R) \tag{4-31}$$

From Eqs. (4-29) and (4-31) we can write

$$e_3 - e_4 = -\frac{e_2 - e_1}{r}(r + 2R) \tag{4-32}$$

Rewriting Eq. (4-32) gives

$$e_3 - e_4 = -(e_2 - e_1)\frac{r + 2R}{r} \tag{4-33}$$

Simplifying Eq. (4-33) gives

$$e_3 - e_4 = -(e_2 - e_1)\left(1 + \frac{2R}{r}\right) \tag{4-34}$$

This is the output of the input pair of operational amplifiers A_1 and A_2. If $e_3 - e_4$ is now applied to a circuit such as that in Fig. 4-7, but with all four resistors equal to R' as in Fig. 4-13, we will have

$$e_5 = -(e_3 - e_4) \tag{4-35}$$

and, from Eq. (4-34),

$$e_5 = -\left(1 + \frac{2R}{r}\right)(e_1 - e_2) \tag{4-36}$$

Now let us see how the circuit of Fig. 4-13 meets the five requirements of an instrumentation amplifier listed at the beginning of this section. Requirements 1 and 2 are met because the output is single-ended and the input is differential. Requirement 3 is met because there are no resistors on either input terminal to shunt the inherently high input impedance of the + inputs of the operational amplifiers. Requirement 4 is met because, unlike the circuit of Fig. 4-7, only a single resistor r is needed to adjust the gain, as can be seen from Eq. (4-36).

Finally it turns out that the common-mode-rejection (CMR) ratio of the circuit in Fig. 4-13 can be more easily adjusted to have a high value than that in Fig. 4-7. CMR is the ratio of the gain for in-phase signals to the gain for out-of-phase signals. It should be high for a good instrumentation amplifier. Noise pickup, such as 60 Hz interference, appears as an in-phase signal, since the input leads are physically close and pick up similar noise. With a high CMR the interference is not amplified as much as is an input difference signal. CMR ratios of 100 dB (10^5) are practical for the circuit in Fig. 4-13. A key factor in the achievement of a high CMR ratio for the circuit in Fig. 4-13 is that the common-mode performance of the output stage containing A_3 is improved by the gain of the preceding differential input stage consisting of A_1 and A_2.

4-4 NONIDEAL OPERATIONAL AMPLIFIERS

Modern operational amplifiers are very good approximations to the ideal amplifier so far discussed, and in most cases circuits can be designed assuming that the device is in fact ideal. However, there are many occasions when a knowledge of the effects of nonideal characteristics is important. These effects are now considered.

4-4-1 EFFECT OF FINITE OPEN-LOOP GAIN

Consider the inverting amplifier shown in Fig. 4-14. Assume that the operational amplifier is ideal except for finite open-loop gain A_o. Because A_o is finite, a voltage e_2 is required at what we have previously considered as a virtual ground point, in order to produce output voltage e_3, and

$$e_2 = -\frac{e_3}{A_o} \tag{4-37}$$

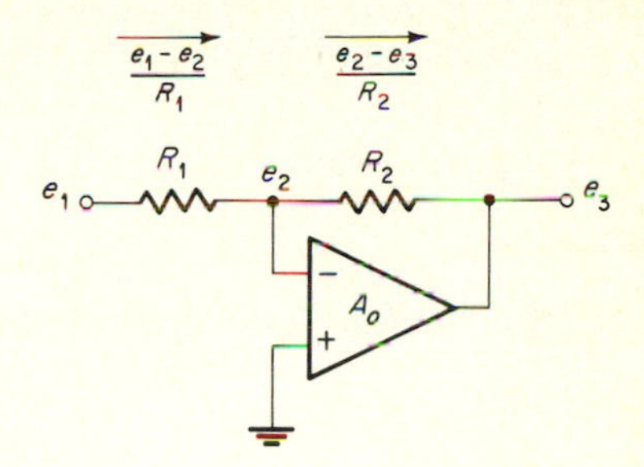

Fig. 4-14 Inverting amplifier with finite open-loop gain.

Equating the currents in R_1 and R_2 gives

$$\frac{e_1 - e_2}{R_1} = \frac{e_2 - e_3}{R_2} \tag{4-38}$$

Solving Eqs. (4-37) and (4-38) for e_3/e_1, we obtain

$$\frac{e_3}{e_1} = -\frac{R_2}{R_1}\frac{1}{1 + (1/A_o)(1 + R_2/R_1)} \tag{4-39}$$

Equation (4-39) is an expresson for the gain of the amplifier in Fig. 4-14. More precisely, it is the *closed-loop* gain of the system, the gain with feedback, which will be referred to from now on as A_F to avoid confusion with the open-loop gain A_o, the gain with no feedback.

Analysis of other operational-amplifier configurations, to be performed in the next few pages, will reveal that equations of the form of Eq. (4-39) occur very frequently in such analyses. In particular we shall find that $1/A_o$ is often multiplied by a term such as that in parenthesis in Eq. (4-39), which we now define as $1/\beta$. In this case,

$$\frac{1}{\beta} = 1 + \frac{R_2}{R_1} \tag{4-40}$$

and the expression $A_o\beta$ is called the loop gain. The loop gain is a significant and recurring quantity in the analysis of operational-amplifier circuits, and it is convenient at this point to gather together the three gains we have defined so far. They are:

A_o Open-loop gain or gain with no feedback

A_F Closed-loop gain or gain with feedback

$A_o\beta$ Loop gain

The relationship between these three quantities will be examined further as we proceed.

Equation (4-39) can now be written

$$A_F = -\frac{R_2}{R_1}\frac{1}{1 + 1/A_o\beta} \tag{4-41}$$

Expanding Eq. (4-41) by the binomial theorem gives

$$A_F = \frac{-R_2}{R_1}\left[1 - \frac{1}{A_o\beta} + \left(\frac{1}{A_o\beta}\right)^2 \cdots\right] \tag{4-42}$$

$$= \frac{-R_2}{R_1}\left(1 - \frac{1}{A_o\beta}\right) \qquad \text{if } A_o\beta \gg 1 \tag{4-43}$$

Now consider the case of a noninverting amplifier, again assuming that the operational amplifier is ideal except for finite open-loop gain A_o. Referring to the circuit in Fig. 4-15, we have

$$e_2 = \frac{e_3}{A_o} \tag{4-44}$$

and, by equating currents through R_1 and R_2,

$$\frac{e_3 - (e_1 - e_2)}{R_2} = \frac{e_1 - e_2}{R_1} \tag{4-45}$$

Solving these equations for e_3/e_1 gives us the closed-loop gain

$$A_F = \frac{1 + R_2/R_1}{1 + (1/A_o)(1 + R_2/R_1)} \tag{4-46}$$

Defining the coefficient of $1/A_o$ once again as $1/\beta$ and assuming that $A_o\beta \gg 1$, as is usually the case, we obtain

$$A_F = \left(1 + \frac{R_2}{R_1}\right)\left(1 - \frac{1}{A_o\beta}\right) \tag{4-47}$$

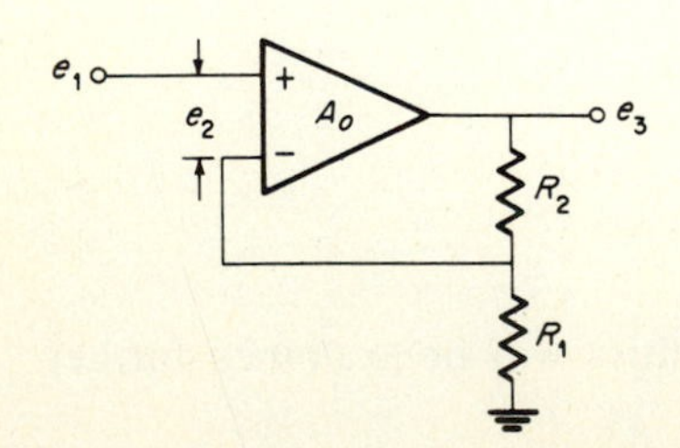

Fig. 4-15 Noninverting amplifier with finite open-loop gain.

Note the similarity between Eqs. (4-43) and (4-47). In each case the closed-loop gain, assuming finite open-loop gain, is equal to that assuming infinite open-loop gain as found in Sec. 4-2, multiplied by the quantity $1 - 1/A_o\beta$.

Now consider the effect of variations in open-loop gain on the closed-loop gain of the amplifier. For an inverting amplifier with $A_o = 10^4$ and $R_2/R_1 = 10$, the gain will be, from Eqs. (4-40) and (4-43),

$$A_F = -10\left(1 - \frac{1}{10^4/11}\right) = -9.989 \qquad (4\text{-}48)$$

Suppose now that the open-loop gain doubles, giving $A_o = (2)(10^4)$. Now

$$A_F = -10\left[1 - \frac{1}{(2)(10^4)/11}\right] = -9.995 \qquad (4\text{-}49)$$

Equations (4-48) and (4-49) illustrate that a very large change in open-loop gain has a very small effect on the closed-loop gain. This gain stabilization is one of a number of the important consequences of having a large open-loop gain. Notice also that the amount by which A_F with finite A_o differs from the ideal, where infinite A_o is assumed, is a function of the loop gain $A_o\beta$, so that the *loop gain is the term* that reveals *how much error* has been introduced in deviating from the ideal.

A further very useful conclusion can be drawn for the case when the closed-loop gain is about 5 or more. In this case it can be shown that

$$\text{Loop gain} \approx \frac{\text{open-loop gain}}{\text{closed-loop gain}} \qquad (4\text{-}50)$$

Since the loop gain is an indicator of the errors introduced by practical amplifiers, and since the closed- and open-loop gains are known at the beginning of a design, Eq. (4-50) is a useful rule of thumb for estimating the final quality of a circuit being designed.

If for the case illustrated by Eq. (4-48), a gain of exactly 10 were required, it could be achieved by adjusting R_1 or R_2. Thus a finite open-loop gain need not introduce an initial error at all, although a large value of loop gain is desirable to stabilize the amplifier against changes in open-loop gain.

4-4-2 EFFECT OF FINITE BANDWIDTH

It is a characteristic of an amplifying system in which feedback is applied that the possibility of oscillation exists. Further, as frequency is increased, the characteristics of all operational amplifiers deviate increasingly from the ideal. These two limitations are the primary factors to be considered when the effects of finite bandwidth are analyzed.

Both problems are conveniently studied by means of Bode plots, which are named after H. N. Bode, who was the first to use this method of analyzing the effects of frequency response on feedback-amplifier design. A Bode plot is a graph showing the frequency response of an amplifier. The absolute value of voltage gain is plotted in decibels on the vertical scale, and frequency is plotted logarithmically on the horizontal scale. For this application decibels are defined as

$$\text{Decibels} = 20 \log_{10} (\text{gain}) \tag{4-51}$$

From this definition it can be seen that 0 dB = unity gain, 20 dB = a gain of 10, 40 dB = a gain of 100, 60 dB = a gain of 1000, and so on.

Any high-gain amplifier will consist of a number of stages, each with its own gain, frequency response, and phase shift. A Bode plot is a graph of the total gain versus frequency. The point at which the frequency response of a particular stage becomes significant is shown as a "break point" on the Bode plot. To understand the significance of these break points, consider the simple lag network shown in Fig. 4-16*a* and its frequency and phase plots. For this circuit,

$$\frac{e_{\text{out}}}{e_{\text{in}}} = \frac{1/j\omega C}{R + 1/j\omega C} = \frac{1}{1 + j\omega CR} \tag{4-52}$$

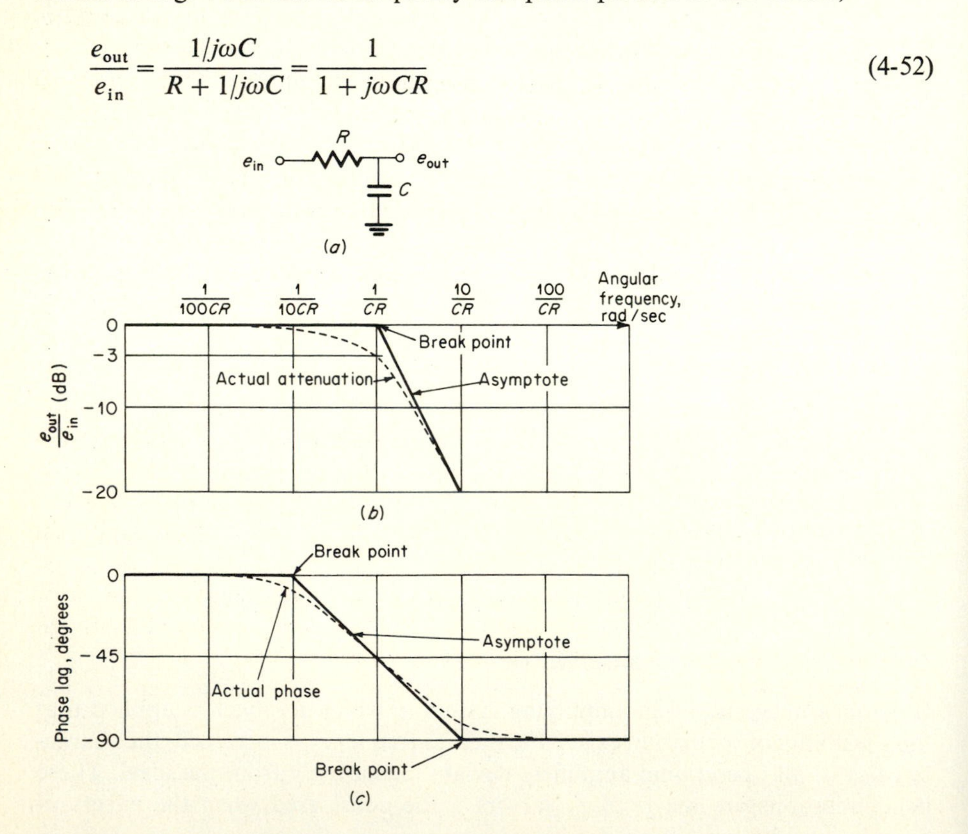

Fig. 4-16 Amplitude and phase plots for an *RC* network.

Analysis of this expression will show that there are three frequencies at which significant attenuations and phase shifts occur. They are

At $\omega = 1/10CR$: Attenuation is almost 0 dB, and phase shift is almost 0°
At $\omega = 1/CR$: Attenuation is 3 dB, and phase shift is $-45°$
At $\omega = 10/CR$: Attenuation is almost 20 dB, and phase shift is almost $-90°$

In Fig. 4-16*b* and *c* the actual attenuation and phase plots are shown along with the asymptotic approximations used in operational-amplifier analysis. It can be seen that the errors introduced by using these asymptotes amount to only a few decibels or a few degrees. The frequency and phase plots reveal two things. First, a simple *RC* lag network can be assumed to have a frequency-response plot consisting of two straight lines, one with zero slope representing no attenuation and one with a slope of 20 dB per decade of frequency (or 6 dB per octave, which is the same slope). The intersection of these two lines is called a *break point* and occurs at $\omega = 1/CR$. Second, a simple *RC* lag network can be assumed to have a phase plot which consists of three straight lines, one representing zero phase shift, one representing $-90°$ phase shift, and a third representing a changing phase shift, the rate of change with frequency being $-45°$ per decade.

Operational amplifiers are almost invariably used in a negative-feedback mode, which means that part of the output is fed back 180° out of phase with the input. If in the forward path between the input and the output, that is, in the operational amplifier, another 180° of phase shift accumulate, the feedback will be 360° out of phase with the input and will therefore be positive. If sufficient gain exists at the frequency of 360° phase shift, oscillations will occur.

Due to combinations of load resistors and transistor capacities, for example, in operational amplifiers, such phase shifts will occur and will accumulate. The problem of designing a stable operational amplifier is one of ensuring that 180° of amplifier phase shift does not occur until after the loop gain has dropped below unity. More important to the user of operational amplifiers is the fact that many operational amplifiers require the use of external stabilizing networks, which are different for different operational-amplifier configurations. To optimize a particular circuit, the right network must be chosen, and this is why a knowledge of Bode plots is important not only to the designer but also to the user of operational amplifiers.

The phase plot for an amplifier is not usually drawn because as attenuation and phase shift accumulate, the phase shift becomes 90°, 180°, etc., as the attenuation slope becomes 20 dB per decade, 40 dB per decade, etc.

The requirement for a stable operational amplifier may be expressed in another way. As was pointed out by H. N. Bode, a closed-loop amplifier will be stable if the open-loop gain slope is less than 40 dB per decade (or 12 dB per octave) in the region where the open-loop and closed-loop gain curves intersect. Notice that 40 dB per decade and 180° phase shift can be obtained from two *RC* sections in cascade. (See Fig. 4-16.) Practical operational amplifiers are designed to have attenuation slopes of less than 40 dB per decade, often 20 dB per decade. However, if stability can be achieved, a slope as close to 40 dB per decade as possible is desirable for reasons that will soon become apparent.

Figure 4-17 is a Bode plot for a typical operational amplifier. The amplifier has a low-frequency open-loop gain of 100 dB or 100,000, with a 3-dB point near 10 Hz. These figures are typical for many operational amplifiers, although it is often surprising to discover that the break point occurs at such a low frequency, especially when one is planning to use the amplifier, at say, 1000 Hz. Provided the ramifications are understood, the use of an amplifier with 10-Hz break point at 1000 Hz is quite feasible and common.

Figure 4-17 shows how deceptive is the high value for dc gain found in most operational amplifiers. The main reason for high dc gain, which for some amplifiers exceeds a billion, is that adequate loop gains can be achieved at frequencies

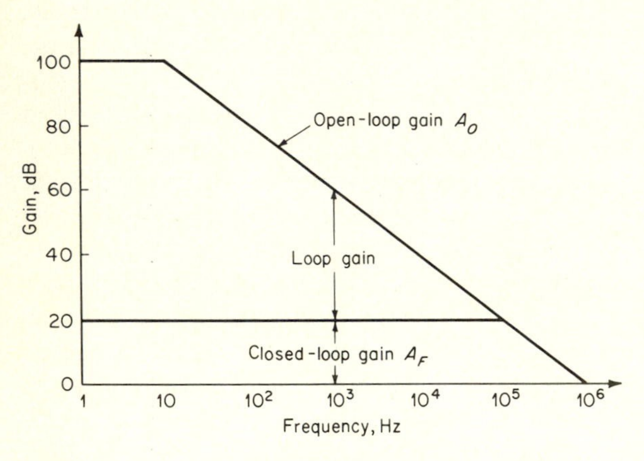

Fig. 4-17 Bode plot for a typical operational amplifier.

of a few hundred hertz. The significant factor is loop gain, which can be obtained by simple subtraction from the curve in Fig. 4-17, since decibels are logarithmic and dividing the open-loop gain by the closed-loop gain to obtain loop gain becomes a process of subtraction. At 1000 Hz, for example, and with a closed-loop gain of 20 dB as shown, the loop gain is 40 dB compared with 80 dB at 1 Hz. We have previously indicated that it is the loop gain that is the dominant factor in determining errors and in evaluating the improvement in circuit performance due to various degrees of feedback. To quickly determine the loop gain at various frequencies, the idea of gain-bandwidth product is useful. Since the open-loop gain falls to 0 dB (=1) at 10^6 Hz, the gain-bandwidth product for this amplifier is 10^6. To obtain the open-loop gain at other frequencies, the gain-bandwidth product is divided by the frequency. For example, at 1000 Hz, A_o is $10^6/10^3 = 10^3 = 60$ dB, and with a closed-loop gain of 20 dB the remaining loop gain is 40 dB.

How negative feedback improves frequency response can be seen clearly from Fig. 4-17. First define the frequency response of an operational amplifier as that range of frequencies from dc to the 3-dB point. Then, the frequency response of the operational amplifier of Fig. 4-17 operated open-loop is 10 Hz, while that for one operated closed-loop is 10^5 Hz. It can be seen that the feedback does not really extend the response of the basic amplifier; rather it reduces the gain of the amplifier to a constant level at frequencies where the loop gain is high.

Although harder to design, a stable operational amplifier with an attenuation slope of 40 dB per decade is superior to one having a slope of 20 dB per decade, as can be seen from Fig. 4-18. Three attenuation slopes are shown. One is for the amplifier before frequency-response-shaping networks are added. Notice that its attenuation slope exceeds 40 dB per decade while it still has more than unity gain. Such an amplifier would be very likely to oscillate when used in a feedback configuration. Networks may be added to modify the attenuation slope, and two cases are shown: one is for a 20-dB-per-decade, slope, and the other is for a slope of 40 dB per decade. Consider the performance of a closed-loop amplifier with an open-loop gain of 20 dB at a frequency ω_n, as shown in Fig. 4-18. It is clear that the loop gain available with an attenuation slope of 40 dB per decade is much greater than for the case with only 20 dB per decade. Although amplifiers with 40-dB slopes are not common, they are frequently designed with slopes in excess of 20 dB.

The transient response of operational amplifiers is intimately tied up with bandwidth and attenuation slope, but the relationship is not a simple one. Limitations exist which will not permit the amplifier's output to change as rapidly as the bandwidth specification might indicate. There are a number of ways of expressing this fact on a data sheet, one of the most common being to give the *slewing rate*. This is the maximum rate at which the output can change; it is usually given in volts per microsecond.

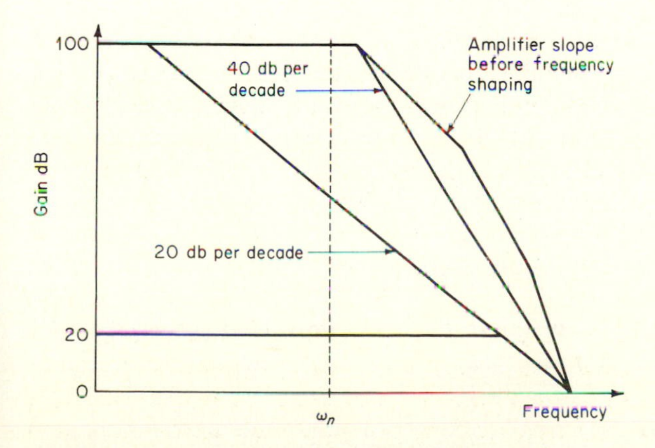

Fig. 4-18 Effect of attenuation slope on loop gain.

An understanding of the effect of bandwidth limitations is more important to the user of monolithic integrated-circuit operational amplifiers than types made from discrete components. The effects of stray capacity, which cause unwanted phase shifts, are harder to control in monolithic circuits than other types. Owing to the nature of monolithic construction, isolation between components is not as complete, and active components such as pinch resistors and diodes are used where linear, more ideal components are used in discrete-type circuits. However, as the technology progresses, means to overcome these disadvantages are being found.

A discussion of the effects of finite bandwidth on the performance of operational amplifiers would not be complete unless a common error in applying Bode plots is pointed out. The error arises because it is frequently forgotten that, in that part of the Bode plot where the gain is falling off at 6 dB per octave, the *gain is a vector quantity*. The result of failing to notice this is that, when the exact closed-loop amplification formula is used to compute gain, sizable errors can result. The problem is best explained with the aid of a numerical example.

Figure 4-19 shows an LM101 monolithic integrated-circuit operational amplifier. Since the input and feedback resistors are equal, the gain is nominally

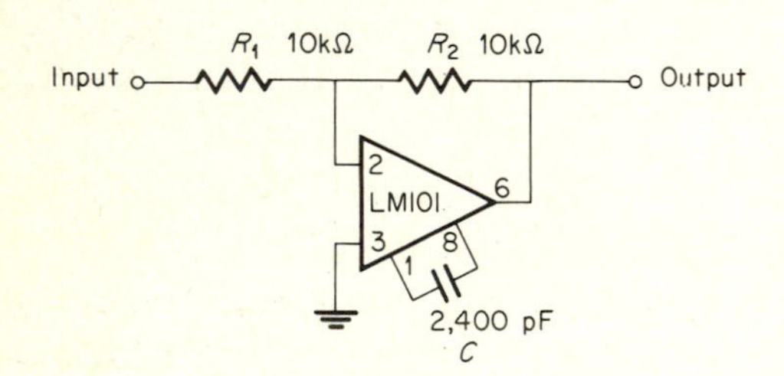

Fig. 4-19 Inverting amplifier with low open-loop gain.

unity. Numbers 1, 2, 3, 6, and 8 in the figure refer to pin numbers on the amplifier. The 2400-pF capacitor connected between pins 1 and 8 is a much larger capacitor than would normally be used with the LM101. The capacitor is used to prevent oscillations and usually has a value of 30 pF or less. A large value was chosen for this example to reduce the open-loop gain A_o to a low value at a reasonably low frequency, so that the results obtained could be checked by measurement. The following data are needed for calculations on the circuit:

A_o (open-loop gain) at 1000 Hz with a C of 2400 pF = 10
Gain roll-off slope = 6 dB per octave
Phase angle θ associated with 6-dB slope = $-90°$

Ignoring the fact that A_o is a vector quantity, we obtain the closed-loop gain at 1000 Hz from Eq. (4-39) as

$$A_F = -\frac{R_2/R_1}{1 + (1/A_o)(1 + R_2/R_1)} = \frac{-1}{1 + \frac{1}{10}(1+1)} = -0.833 \tag{4-53}$$

Equation (4-39) is correct, but in applying it we must take note of the fact that A_o is a vector quantity; that is,

$$A_o = |A_o|(\cos\theta + j\sin\theta)$$

where $|A_o|$ is the magnitude of A_o and $j = \sqrt{-1}$. For $\theta = -90°$,

$$A_o = |A_o|(0 - j \cdot 1) = -j|A_o|$$

Thus

$$\frac{1}{A_o} = \frac{1}{-j|A_o|} = \frac{+j}{|A_o|}$$

By substitution into Eq. (4-39),

$$A_F = -\frac{1}{1 + (j/10)(1+1)} = -\frac{1}{1 + 0.2j}$$

$$|A_F| = -0.981 \qquad (4\text{-}54)$$

The exact gain as calculated in Eq. (4-53) was about 17 percent lower than the gain of unity that would be obtained with an ideal operational amplifier. However, Eq. (4-53) did not account for the fact that A_o is a vector quantity. The exact gain as calculated in Eq. (4-54) was only about 2 percent lower than unity and is the correct answer that is confirmed by measurement.

It should be stressed that usually a C of much less than 2400 pF is used, so that a much higher open-loop gain A_o is obtained at 1000 Hz. Under these circumstances a gain of unity accurate to a small fraction of 1 percent would be realized.

4-4-3 EFFECT OF FINITE INPUT RESISTANCE

A practical operational amplifier will not have the infinite input impedance we have so far assumed. Not only will there be some impedance between the two input terminals, but between each input and ground also. Figure 4-20 illustrates where these impedances occur. R_{in} is the impedance between the inverting and noninverting terminals, and R_c is the common-mode impedance from both inputs to ground. One effect of these impedances is to reduce the loop gain, with the consequent reduction in circuit performance that this implies. For example, consider the inverting amplifier shown in Fig. 4-21. Notice that one of the common-mode resistors is shorted out in this configuration, so that the impedance from the inverting input terminal to ground is R_{in} in parallel with $2R_c$. Operational amplifiers are often designed so that $R_c \gg R_{\text{in}}$, so we can assume that R_{in} is the impedance to ground in this case.

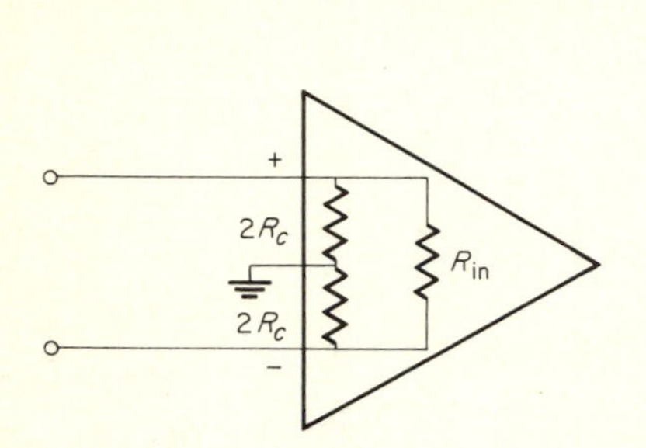

Fig. 4-20 Differential and common-mode input impedances of an operational amplifier.

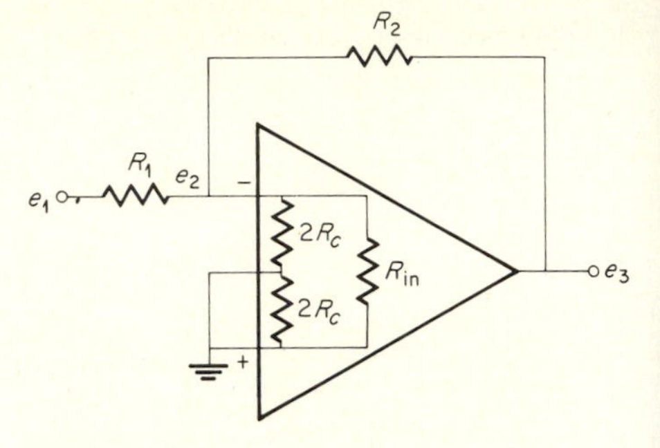

Fig. 4-21 Inverting amplifier, showing operational-amplifier input impedances.

The equations for this circuit are very similar to those written for the circuit in Fig. 4-14, the difference being that we now assume that some current flows into the inverting input of the amplifier. So we have

$$\frac{e_1 - e_2}{R_1} = \frac{e_2}{R_{in}} + \frac{e_2 - e_3}{R_2} \tag{4-55}$$

and

$$e_2 = \frac{-e_3}{A_o} \tag{4-56}$$

Solving Eqs. (4-55) and (4-56) gives

$$A_F = \frac{-e_3}{e_1} = \frac{-R_2}{R_1}\left[\frac{1}{1 + (1/A_o)(1 + R_2/R_1 + R_2/R_{in})}\right] \tag{4-57}$$

As in Eq. (4-40) define $1/\beta$ as the coefficient of $1/A_o$ so that

$$\frac{1}{\beta} = 1 + \frac{R_2}{R_1} + \frac{R_2}{R_{in}} \tag{4-58}$$

If $R_{in} = \infty$, $1/\beta$ becomes $1 + R_2/R_1$ as in Eq. (4-40), which was obtained assuming that $R_{in} = \infty$. The effect of finite R_{in} is therefore to increase $1/\beta$, and the loop gain is reduced since it equals $A_o\beta$.

We can gain some idea of the magnitude of this effect by considering a numerical example. Let $A_o = 10^4$, $R_1 = 100\ \text{k}\Omega$, $R_2 = 1\ \text{M}\Omega$, and $R_{in} = 1\ \text{M}\Omega$. These values give

$$\frac{1}{\beta} = 1 + 10 + 1 = 12 \tag{4-59}$$

The loop gain is therefore

$$A_o\beta = \frac{10^4}{12}$$

and

$$A_F = -10\left(1 - \frac{1}{10^4/12}\right) = -9.988 \tag{4-60}$$

The value for A_F for the same values, except with $R_{\text{in}} = \infty$, was -9.989 [see Eq. (4-48)], so that in this case the effect is negligible. It should be borne in mind that as frequency increases and if R_{in} is too small, a combination of values could exist whereby a more significant error might arise, because $A_o\beta$ decreases as frequency increases.

Equation (4-56) shows that e_2 is very small for large values of A_o, which means that the input resistance is still R_1 for all practical purposes.

The noninverting amplifier may be analyzed by similar methods. Unlike the case of the inverting amplifier, the input impedance of a noninverting amplifier is a very strong function of the operational-amplifier input impedance. If $R_c = \infty$, it can be shown (see Prob. 4-21) that the circuit input impedance is $R_{\text{in}}(1 + A_o\beta)$, demonstrating that once again the loop gain is the significant factor. Theoretically, fantastically high input impedances can be achieved, but the effects of leakage and R_c usually limit practical values to a few hundred meg ›hms for the more common operational amplifiers.

The loop gain of the noninverting amplifier is reduced by the existence of a finite R_{in} in a manner similar to that for the noninverting case. (See Prob. 4-22.)

4-4-4 EFFECT OF NONZERO OUTPUT RESISTANCE

Consider the circuit shown in Fig. 4-22. For this circuit we can write

$$\frac{e_1 - e_2}{R_1} = \frac{e_2 - e_3}{R_2} = \frac{e_3 - e_4}{R_o} \tag{4-61}$$

and

$$e_2 = \frac{-e_4}{A_o} \tag{4-62}$$

The open-loop gain A_o of the amplifier amplifies e_2 to give e_4. However, e_4 is not available at the outut of the operational amplifier, owing to the presence of R_o, the output resistance. The usable open-loop gain will be less than A_o as we can confirm by solving Eqs. (4-61) and (4-62) for the closed-loop gain e_3/e_1.

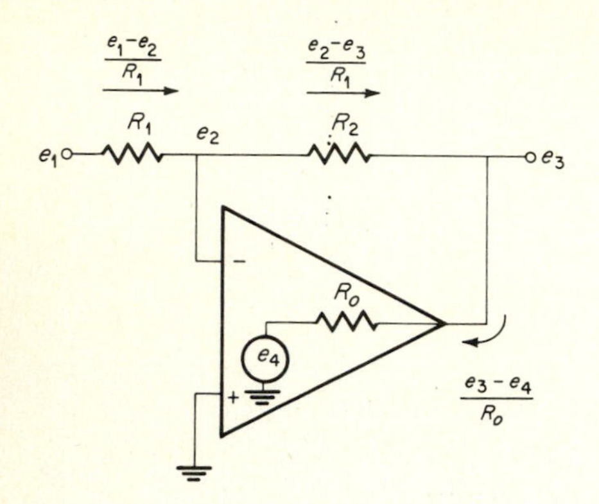

Fig. 4-22 Inverting amplifier with nonzero output impedance.

To simplify the procedure, first obtain an expression for e_4 in terms of e_3, namely,

$$e_4 = e_3 \frac{R_2 + R_o}{R_2 - R_o/A_o} \tag{4-63}$$

For all practical purposes $R_2 \gg R_o/A_o$ so that Eq. (4-63) simplifies to

$$e_4 = e_3 \left(1 + \frac{R_o}{R_2}\right) \tag{4-64}$$

Eliminating e_4 from Eq. (4-61) and using Eqs. (4-62) and (4-64), we obtain

$$\frac{e_3}{e_1} = -\frac{R_2}{R_1} \frac{1}{1 + (1 + R_o/R_2)(1 + R_2/R_1)(1/A_o)} \tag{4-65}$$

As in Eq. (4-40), we define $1/\beta$ as the coefficient of $1/A_o$ so that

$$\frac{1}{\beta} = \left(1 + \frac{R_o}{R_2}\right)\left(1 + \frac{R_2}{R_1}\right) \tag{4-66}$$

If $R_o = 0$, $1/\beta$ becomes $1 + R_2/R_1$, as in Eq. (4-40), which was obtained assuming that $R_o = 0$. The effect of nonzero R_o is therefore to increase $1/\beta$, and the loop gain is reduced since it is equal to $A_o \beta$.

We can obtain some idea of the magnitude of this effect by considering a numerical example. Let $A_o = 10^4$, $R_1 = 10\ \text{k}\Omega$, $R_2 = 100\ \text{k}\Omega$, and $R_o = 1\ \text{k}\Omega$. These values give

$$\frac{1}{\beta} = (1 + \tfrac{1}{100})(1 + 10) = 11.11 \tag{4-67}$$

The loop gain is therefore

$$A_o\beta = \frac{10^4}{11.11} \tag{4-68}$$

and

$$A_F = \frac{e_3}{e_1} = -10\left(1 - \frac{1}{10^4/11.11}\right) = -9.98889 \tag{4-69}$$

A_F for the same values, except for $R_o = 0$, was -9.989 [see Eq. (4-48)], so that in this case the effect is negligible. Comparison of the results of Eqs. (4-69), (4-60), and (4-48) illustrates a characteristic quality of operational amplifiers, namely that the closed-loop performance is relatively unaffected by changes in internal parameters. It should be borne in mind that A_o decreases as frequency increases. Thus, if R_o is too large, a combination of values could result in a larger error owing to a too small value for $A_o\beta$.

It can be shown (see Prob. 4-23) that to a very close approximation the output impedance of an operational amplifier with feedback is equal to the open-loop output impedance divided by the loop gain. At frequencies where the loop gain is high, the closed-loop output impedance is very low, often less than 1 Ω. However, at high frequencies the closed-loop output impedance can become appreciable, owing to the reduction in loop gain that occurs with frequency. Because of this some operational amplifiers have a tendency to oscillate when their loads are coupled by long cables, or when they are loaded by other types of capacitative load. The high output impedance at high frequency and the capacitative load form a phase-shift network, and, as shown previously, if sufficient gain exists at these frequencies, oscillations can occur. Figure 4-23 illustrates this effect. A commonly used method of overcoming this problem is to connect a resistor between the output of the amplifier and the capacitative load. (See Prob. 4-18.) Then the feedback applied back to the input via R_2 in Fig. 4-23 is not phase-shifted as much because of the isolating effect of the additional resistor.

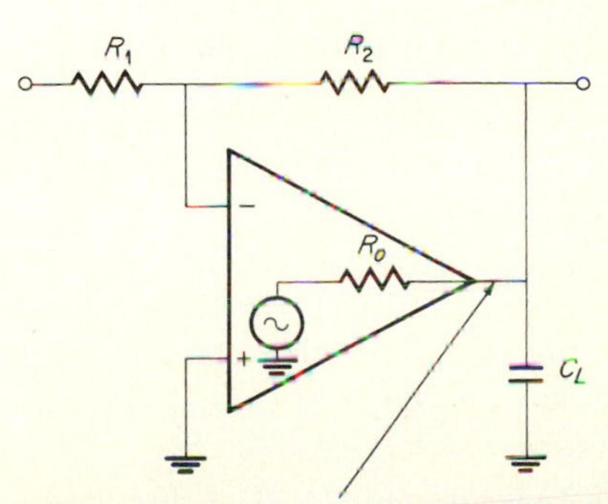

Fig. 4-23 Effect of capacity loading on an operational amplifier with finite output impedance.

4-4-5 EFFECT OF OFFSET AND DRIFT

As mentioned previously, an ideal operational amplifier would have zero output voltage for zero input. In practice there will always be some output voltage for no input, which is called an *offset*. The output offset can arise from two sources, the input-voltage offset and the input-current offset. These two input offsets are independent of each other and contribute to the output offset in a manner depending on the particular circuit used. Offsets in themselves are not a serious problem, since they can always be canceled by equal and opposite compensating signals. The problem with offsets is that they change, primarily with temperature, time, and supply voltages; the changes are called *drifts*. Their effects can be analyzed by means of the equivalent circuit for an operational amplifier, Fig. 4-24, in which the sources of offsets are included.

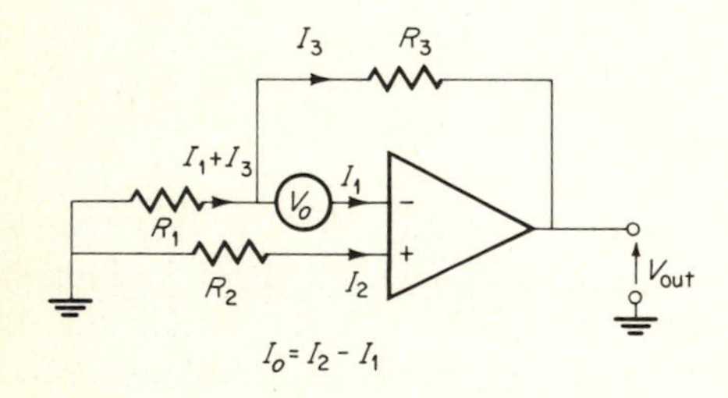

Fig. 4-24 General operational-amplifier circuit with input voltage and current offsets shown.

Offset voltages are always referred to the input, since they are independent of circuit gain and can easily be compared with input signal levels by dividing the measured output signal and offset by the gain. Offset and drift effects are most easily analyzed by grounding both inputs, which is done in Fig. 4-24 through impedances R_1 and R_2, to obtain generalized results, since both the inverting and noninverting modes of operation are thereby covered.

The offset voltage V_o is that voltage which must be applied between the operational-amplifier input terminals to obtain zero output voltage.

The offset current I_o is the difference between the two input currents, I_1 and I_2.

V_o and I_o and the magnitudes of R_1, R_2, and R_3 all affect the resultant output-voltage offset as shown in the following derivation, which refers to Fig. 4-24. Equating the voltages across R_1 and R_3 with V_{out}, we obtain

$$(I_1 + I_3)R_1 + I_3 R_3 = V_{out} \tag{4-70}$$

Summing the voltages around the loop including R_1, R_2, and the offset generator V_o gives

$$(I_1 + I_3)R_1 - V_o - I_2 R_2 = 0 \tag{4-71}$$

In writing Eq. (4-71) it should be noted that there is no voltage drop between the inverting and noninverting terminals of the operational amplifier, because we have removed the offset components from the amplifier and are considering them externally. Notice also that I_1 does not necessarily equal I_2, there being other current paths inside the operational amplifier that permit this to be so. The difference between I_1 and I_2 is the offset current I_o.

Eliminating I_3 from Eqs. (4-70) and (4-71) and solving for V_{out} gives

$$V_{\text{out}} = V_o\left(1 + \frac{R_3}{R_1}\right) + I_2R_2\left(1 + \frac{R_3}{R_1}\right) - I_1R_3 \tag{4-72}$$

If I_2 were equal to or nearly equal to I_1 as in a high-quality operational amplifier, V_{out} could be minimized by making the coefficients of I_2 and I_1 equal, since they have opposite signs. So we choose

$$R_3 = R_2\left(1 + \frac{R_3}{R_1}\right) \tag{4-73}$$

or

$$R_2 = \frac{R_1R_3}{R_1 + R_3} \tag{4-74}$$

Equation (4-74) indicates that the desired condition is that we make R_2 equal to the parallel combination of R_1 and R_3. Doing this, we can rewrite Eq. (4-72) as

$$V_{\text{out}} = V_o\left(1 + \frac{R_3}{R_1}\right) + I_oR_3 \tag{4-75}$$

since $I_o = I_2 - I_1$. Notice the similarity between the quantity $1 + R_3/R_1$ in Eq. (4-75) and the quantity β first introduced in Eq. (4-40).

If the circuit in Fig. 4-24 is used in the inverting mode, its gain is $-R_3/R_1$, so that the input-voltage offset is multiplied by 1 plus the magnitude of the gain. [See Eq. (4-75)]. If the circuit is used in the noninverting mode, its gain is $1 + R_3/R_1$, so that the input-voltage offset is multiplied by the gain.

To determine the effects of offset changes due, for example, to temperature, V_{out}, V_o, and I_o are replaced by ΔV_{out}, ΔV_o, and ΔI_o. For example, an inverting amplifier has, referring to Fig. 4-24, $R_1 = 10\ \text{k}\Omega$, $R_3 = 100\ \text{k}\Omega$, and $R_2 =$ the equivalent resistance of R_1 and R_3 in parallel. The offset-voltage temperature coefficient is 10 μV/°C, and the offset-current temperature coefficient is 1 nA/°C. (In practice the offset-current temperature coefficient is a nonlinear function of temperature.) How will the output change with temperature with the input shorted? From Eq. (4-75),

$$\Delta V_{\text{out}} = 10^{-5}(1 + 10) + (10^{-9})(10^{+5}) = 210\ \mu\text{V/°C}$$

The result obtained is a worst-case answer, since it was assumed that the voltage and current drifts added.

Some general comments regarding drift are relevant at this point. Notice that negative feedback, in spite of its many other beneficial effects, does not improve the drift performance of an amplifier. Drift cannot be separated from a dc input signal, so that although feedback may be employed to reduce gain and hence output drift, it reduces the signal by a similar amount.

Generally speaking, in low-impedance circuits where low values for R_3 in Fig. 4-24 can be used, the major source of drift is the voltage drift. Until recently it was possible to obtain matched pairs of bipolar transistors with appreciably lower offset-voltage temperature coefficients than field effect transistors. Consequently, bipolar transistors would be superior as front ends for operational amplifiers intended for use with low impedances, where the application called for minimum drift. However, JFETs have been considerably improved in this respect of late, and the choice would be made mainly on economic grounds. For high-impedance circuitry the JFET or even the MOSFET is superior from the point of view of temperature because of its low offset current-coefficient.

For both voltage and current offset coefficients, manufacturers often quote an average value since the drift rate varies at different temperatures. This should be borne in mind, since the drift at any particular temperature might be more or less than that specified. A better specification would be to quote the total drift over a given temperature range. For example, an amplifier with a specified drift of 1 mV over a 100°C temperature range would have a drift of more than 10 μV/°C at some temperatures and less than that figure at other temperatures.

Because of the differential nature of operational amplifiers, power-supply voltage variations tend to be common-moded out. However, there are residual effects, and they are usually specified in terms of a voltage-supply rejection ratio which, in high-quality amplifiers, would be measured in microvolts per volt.

Radio-frequency pickup on long input leads is another source of offsets, since it may be rectified by the input stages and appear at the output as a fixed or slowly varying voltage.

There is an interesting type of circuit for which a knowledge of the effects of voltage and current drift is important. It is often used when a circuit design calls for the use of inconveniently large value resistors. Very large value resistors, about 10^{10} Ω and above, tend not to be as stable as smaller values, so circuits using lower, but equivalent value, resistors are desirable provided that their limitations are understood. Consider the two circuits shown in Fig. 4-25. Both are current-to-voltage converters, and both have the same signal gain as measured in volts out per ampere in. Their drift performances are, however, quite different. In Fig. 4-25, I_{in} and E_{out} are input-current and output-voltage signals, respectively, and I_d and E_d are equivalent input drift currents and voltages. Since I_{in} comes from a current generator, which has an infinite output impedance, all I_d will flow in R_1.

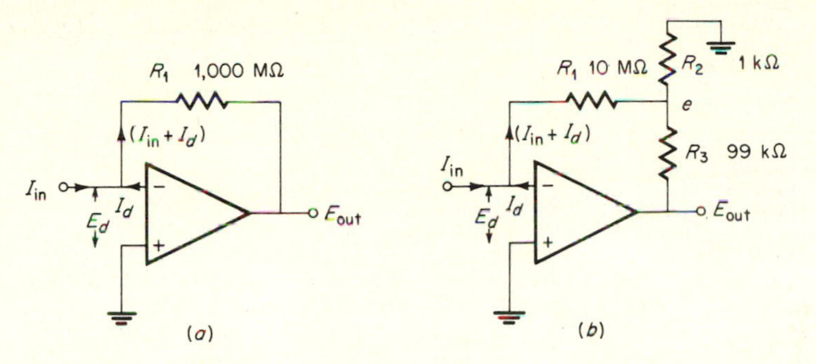

Fig. 4-25 Two current-amplifier circuits with the same signal gain, but with different drift performances.

For circuit *a*:

$$E_{out} = -[E_d + (I_{in} + I_d)R_1] \tag{4-76}$$

$$= -\underbrace{(I_{in} R_1)}_{\text{signal}} - \underbrace{(I_d R_1 + E_d)}_{\text{drift}} \tag{4-77}$$

$$= -10^9 I_{in} - 10^9 I_d - E_d \tag{4-78}$$

For circuit *b*:

$$e = -[E_d + (I_{in} + I_d)R_1] \tag{4-79}$$

$$E_{out} = \left(\frac{R_3 + R_2}{R_2}\right) e = 100e \tag{4-80}$$

$$= -\underbrace{(100 I_{in} R_1)}_{\text{signal}} - \underbrace{(100 I_d R_1 + 100 E_d)}_{\text{drift}} \tag{4-81}$$

$$= -10^9 I_{in} - 10^9 I_d - 100 E_d \tag{4-82}$$

A comparison between Eqs. (4-78) and (4-82) shows that in both cases the input-current signal and drift are multiplied by a factor of 10^9, whereas the voltage drift at the output due to the same voltage generator E_d at the input is 100 times greater for circuit *b* than for circuit *a*.

4-4-6 EFFECTS OF NOISE

Noise may be defined as any spurious signal in the output of an amplifier which is not present in the input. The drifts we have been considering are merely special cases of noise which occur at very low frequencies. As with drift, noise can be considered to arise from two input sources, one a series voltage source, the other a parallel current source. This is illustrated in Fig. 4-26.

e_n is the rms value of the voltage-noise generator per root cycle, so that its contribution to noise over a bandwidth Δf is $(e_n^2 \, \Delta f)^{1/2}$.

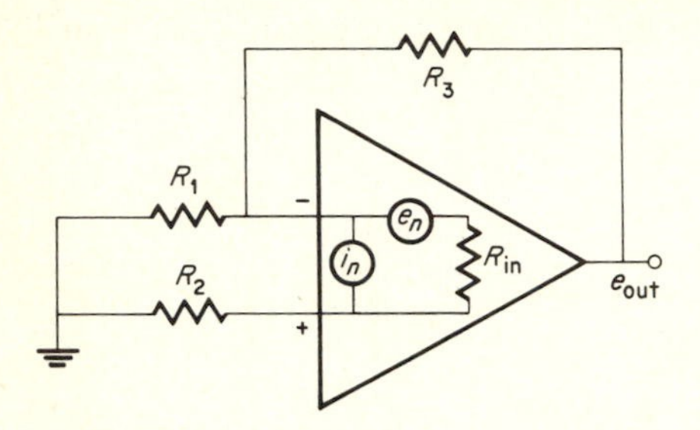

Fig. 4-26 Operational-amplifier circuit showing equivalent input voltage- and current-noise generators.

i_n is the rms value of the current-noise generator per root cycle, so that its contribution to noise over a bandwidth Δf is $(i_n^2 \ \Delta f)^{1/2}$.

e_n is found from the noise measured at the output, with suitable allowance made for gain, with the input shorted. i_n is found from the noise measured at the output, with suitable allowance made for gain, with a source resistance large enough to make the contribution of the voltage-noise generator negligible. Thus, i_n, flowing in the large source resistance, produces a noise IR drop, that is, a noise-voltage generator, large compared with the actual noise-voltage generator. In measuring i_n by this method, the thermal noise generated by the source resistor itself must also be accounted for.

There are several other ways of characterizing the noise performance of an operational amplifier, including noise figure, noise factor, and equivalent noise resistance, but the method of using equivalent voltage- and current-noise generators is the most common. In addition it is used here because of the obvious analogy between the drift circuit of Fig. 4-24 and the noise circuit of Fig. 4-26.

In general the voltage- and current-noise generators are not statistically independent, but because of the manner in which the input transistors in operational amplifiers are operated, the correlation coefficient is low and can often be neglected.

The value of the overall equivalent input voltage-noise generator for the circuit of Fig. 4-26 has three components. The most apparent is e_n. Also i_n flows through the circuit impedances to produce a noise voltage. By inspection, the impedance seen by generator i_n is

$$R = R_2 + \frac{R_1 R_3}{R_1 + R_3}$$

The resulting noise voltage is therefore $i_n R$. This assumes that R_{in} is very large compared with R, which is usually the case. If not, R_{in} can easily be accounted for, since it appears in parallel with R and merely has the effect of reducing the total resistance seen by i_n. The third component is that due to the noise generated in the source and feedback resistors themselves and is the noise that would be obtained from the amplifier if it were ideal from the point of view of noise, that is, noiseless. This third component is the "white" or "Johnson" noise and is equal to $\sqrt{4KTR\,\Delta f}$, where

K Boltzmann's constant = $(1.38)(10^{-23})$ joule/K

T absolute temperature, K

R resistance for which the noise is being defined

Δf bandwidth being considered

The equivalent input voltage-noise generator e_e is obtained by adding the three components, noting that we must first square since we are dealing with rms quantities. No noise power is present due to an $e_n i_n$ product, since we are assuming zero correlation coefficient between the voltage- and current-noise generators. So we have

$$e_e^2 = 4KTR + i_n^2 R^2 + e_n^2 \tag{4-83}$$

Equation (4-83) gives e_e, the equivalent noise generator for unit bandwidth. For example, suppose that an operational amplifier is to be used over a bandwidth of 100 Hz and that the effective circuit resistance R is 100 kΩ. If i_n is 10^{-12} A/Hz$^{1/2}$, and e_n is 10^{-7} V/Hz$^{1/2}$, find the equivalent voltage-noise generator.

$$4KT = (1.66)(10^{-20}) \qquad \text{at room temperature}$$

For a 1-Hz bandwidth

$$\begin{aligned} e_e^2 &= (1.66)(10^{-20})(10^{+5}) + (10^{-24})(10^{10}) + (10^{-14}) \\ &= (2.166)(10^{-14})\ \text{V}^2 \end{aligned}$$

For a 100-Hz bandwidth

$$\begin{aligned} e_e^2 &= (100)(2.166)(10^{-14}) = (2.166)(10^{-12})\ \text{V}^2 \\ e_e &= \sqrt{2.166} = 1.47\ \mu\text{V} \end{aligned}$$

4-5 TYPES OF OPERATIONAL AMPLIFIERS AVAILABLE

The ideal operational amplifier can never be built, and the compromises necessary in the design of practical amplifiers have resulted in the evolution of a number of different types, each with its own advantages and disadvantages in that it deviates to a greater or lesser degree from the ideal.

The number of different operational amplifiers available is measured in the hundreds, and it is convenient to divide the types available into two main categories.

The first is general-purpose, low-cost, moderate-performance operational amplifiers. The 741, which is made by several manufacturers, is typical of this first group. The 741 is supplied in a number of different models, some with better specifications than others. It also comes as a dual or a quad unit, that is, with two or four completely separate amplifiers in a single IC package. Table 4-2 summarizes the important specifications of typical 741-type operational amplifiers and compares these specifications both with those of an ideal operational amplifier and with those of the operational amplifiers which comprise the second category.

The second category covers a wide range of operational amplifiers in which at least one specification is made to approach the ideal more closely than in the 741. This improvement in a particular specification is achieved at the expense of cost and may be accomplished by using a JFET or MOSFET front end, by using chopper stabilization, or by exercising extreme care in the manufacturing process to achieve matching. In addition to the ideal operational amplifier and the general-purpose 741, six other amplifiers each of which has at least one exceptionally good specification, are listed in Table 4-2.

4-6 MISCELLANEOUS OPERATIONAL-AMPLIFIER CIRCUITS

4-6-1 OFFSET CANCELLATION

Although the ideal operational amplifier has zero current and voltage offset, both are nonzero in practical amplifiers. The equivalent input voltage produces an output-voltage offset which depends on the closed-loop gain of the circuit. The equivalent input-current offset produces an output-voltage offset which depends on the impedance levels and the closed-loop gain of the circuit. The total output offset may be canceled by using the circuit of Fig. 4-27. R_1 and R_2 are the gain-determining resistors. R_4 is equal to the parallel impedance of R_1 and R_2 to minimize current-offset effects. [See Eq. (4-73).] R_3 is usually very much larger than R_1 or R_2, so its effect on the total parallel impedance that R_4 must balance is negligible. In the absence of R_3 and potentiometer RP_1, any offset current will flow into R_1 and R_2, thereby producing an offset voltage at the

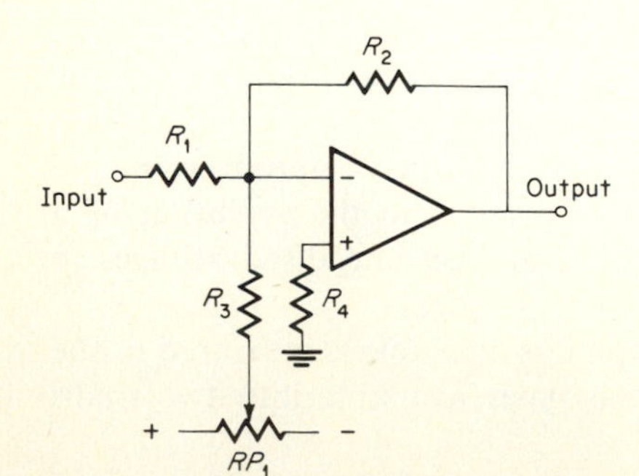

Fig. 4-27 Offset cancellation for an inverting amplifier.

TABLE 4-2 Comparison of types of operational amplifiers

Specification \ Amplifier		Ideal	Typical 741	Intersil ICH8500A	PMI Mono-OP-7A	Nat. Semi LH0032	Datel AM-490-2	Ferranti ZN417
Bias current	nA	0	500	0.00001				
Offset voltage	mV	0	5		0.01			
Drift	μV/°C	0	2		0.2			
Bandwidth	MHz	∞	1			500		
Slew rate	V/μS	∞	0.7			70		
Voltage gain	—	∞	$(5)(10^4)$				$(5)(10^8)$	
Output swing	V	±Supply	±14					225

amplifier output which will add to or subtract from that produced by the voltage input offset. With R_3 and RP_1 in the circuit, the output voltage can be set to zero for zero input to R_1. Notice that this type of compensation will zero the output at only one temperature. This emphasizes that for accurate dc amplifiers employing operational amplifiers, the critical factor is not the offset but the temperature coefficient of the offset, or drift.

A circuit for offset cancellation in a noninverting amplifier is given in Fig. 4-28. The principle of operation is similar to that for the circuit in Fig. 4-27, with one exception. The zeroing circuit for the noninverting circuit also changes the gain, although by an appropriate choice of resistor values the effect can be made negligible for any particular accuracy requirement. This involves making R_3 plus the effective resistance of RP_1 large compared with R_2 in parallel with R_1.

4-6-2 MULTIPLE INPUTS

Operational amplifiers may be used to sum a number of inputs, the equation for the circuit of Fig. 4-29 being, for an ideal amplifier,

$$e_o = -\left(\frac{R_f e_1}{R_1} + \frac{R_f e_2}{R_2} + \cdots + \frac{R_f e_n}{R_n}\right) \tag{4-84}$$

Thus, in the special case in which $R_1 = R_2 = R_3 = R_n$,

$$e_o = -\frac{Rf}{R_n}(e_1 + e_2 + \cdots + e_n) \tag{4-85}$$

and the output is equal to a constant $-R_f/R_n$ times the sum of the inputs.

Two factors that affect the performance of multiple-input circuits must be considered in their design. The first concerns drift, which is measured by grounding all inputs and measuring the output drift. Under these circumstances all the input resistors are in parallel and, if equal, have an effective resistance of R_n/n. The circuit gain is then nR_f/R_n, and the equivalent input-voltage drift is amplified by a factor of $1 + nR_f/R_n$, whereas signals are amplified only by R_f/R_n. If $n = 10$, $R_f = 100\ \text{k}\Omega$, and $R_n = 10\ \text{k}\Omega$, the drift will be amplified by a factor of 101 while signals are only amplified by a factor of 10. The drift performance of an operational amplifier with multiple inputs tends therefore to be worse than when used with a single input.

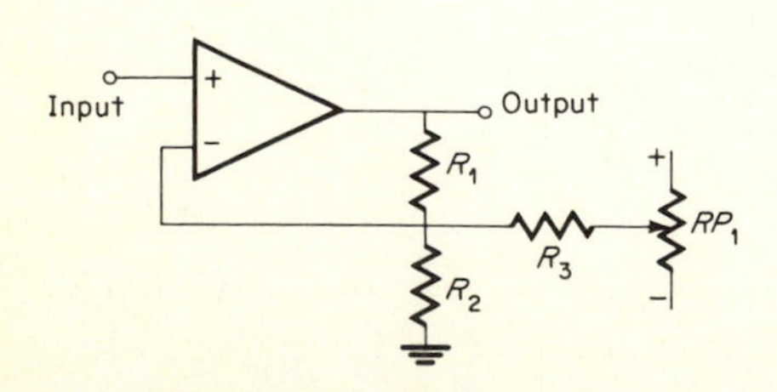

Fig. 4-28 Offset cancellation for a noninverting amplifier.

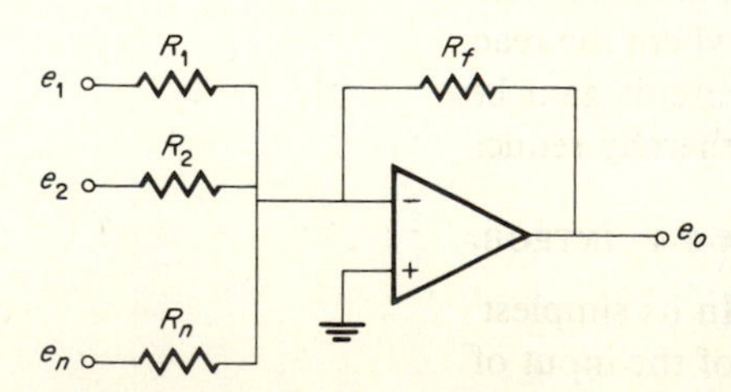

Fig. 4-29 Operational amplifier with multiple inputs.

The second effect that occurs with multiple-input circuits is a reduction in the loop gain, which is important because, as we have previously indicated, the improvement in circuit performance due to feedback is in many instances proportional to loop gain. For the circuit in Fig. 4-14, we defined in Eq. (4-40)

$$\frac{1}{\beta} = 1 + \frac{R_2}{R_1}$$

For the circuit of Fig. 4-29, with all input resistors equal, it can be shown (see Prob. 4-24) that $1/\beta = 1 + nR_f/R_n$, so that the loop gain $A_o\beta$ will be reduced.

4-6-3 DIFFERENTIATOR NOISE

In its simplest form the differentiator is usually an unsatisfactory circuit, because one effect of the linear increase in gain with frequency which is characteristic of this circuit is poor noise performance. Since the circuit gain increases with frequency, high-frequency noise is amplified more than lower-frequency noise. If the amplifier bandwidth is high, very high levels of output noise may be realized. In fact, the output noise can easily be larger than the amplified signal, even though the input signal is larger than the equivalent input noise. One technique for limiting the high-frequency response of an operational-amplifier circuit, and hence the output noise, is illustrated in Fig. 4-30.

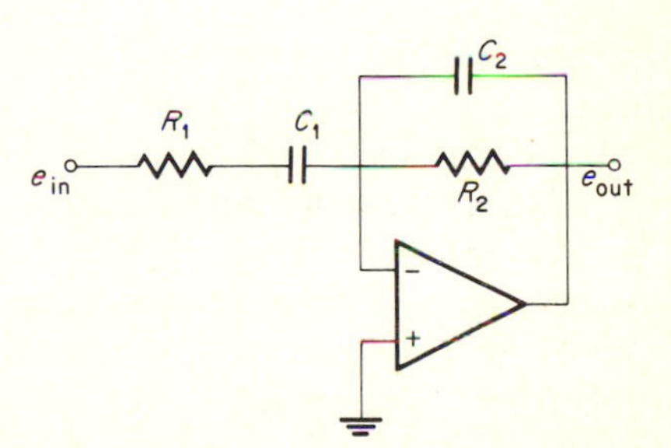

Fig. 4-30 Practical differentiator.

Over the frequency range in which it is desired to perform a differentiation, the reactance of C_1 is chosen to be much larger than R_1, and the reactance of C_2 is chosen to be much larger than R_2. Thus, over the frequency range of interest, the circuit acts as a differentiator with time constant C_1R_2. At higher frequencies, where the reactances of C_1 and C_2 become small compared with R_1 and R_2, the circuit gain becomes the reactance of C_2 divided by R_1 and approaches zero, thereby reducing the amplification of high-frequency noise.

4-6-4 INTEGRATOR DRIFT

In its simplest form the integrator is often an unsatisfactory circuit because part of the input offset current will charge the feedback capacitor, resulting in a constantly changing output even with no input. One technique for minimizing this problem is shown in Fig. 4-31.

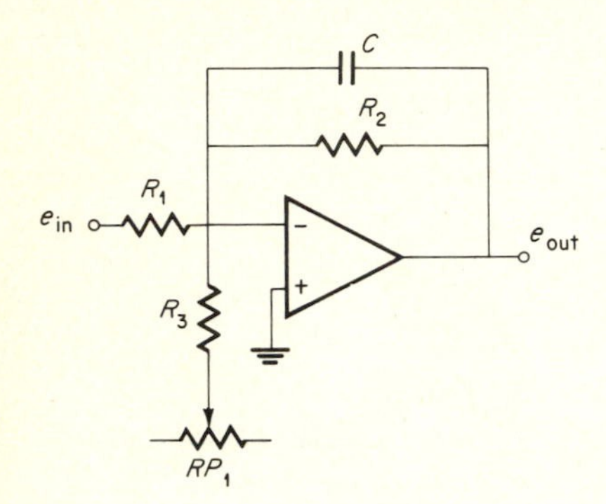

Fig. 4-31 Practical integrator.

This is a current-balancing scheme, as may be deduced by a comparison between Figs. 4-27 and 4-31. R_2 is made large by comparison with the reactance of C over the range of frequencies at which it is desired to integrate. The offset current through R_2 is minimized by adjustment of RP_1, and the gain of the circuit to dc is limited to $-R_2/R_1$, thereby preventing the output from eventually drifting into saturation, which it would do if R_2 were not present.

4-6-5 NONLINEAR FUNCTION GENERATORS

There is a class of circuit which employs nonlinear impedance elements in feedback loops to obtain a nonlinear relationship between the input and the output. The elements are nonlinear in the sense that current flow in response to an input voltage is not proportional to that input. Many types of elements are used, ranging from diodes used as switches to change an impedance level as an input voltage is increased, to devices such as varistors which are inherently nonlinear, to pulse-modulation circuits. It is outside the scope of this book to analyze the inner workings of such devices, and we will confine ourselves to a particular case in which such elements are used in conjunction with an operational amplifier to generate a desired function. Suppose, given time-varying voltages e_1 and e_2, that it is desired to obtain the square root of the sum of the squares of these two voltages. Assume an element is available which has the property that its output voltage is 0.01 times its input squared. The 0.01 is called a *scaling factor* and is commonly found in analog computing circuits (and especially in squarers) to prevent output voltages from becoming too large when squared. Consider the circuit of Fig. 4-32, which consists of one operational amplifier, three squaring circuits marked N^2, and three equal resistors.

The three squaring elements have as outputs $e_1{}^2/100$, $e_2{}^2/100$, and $e_3{}^2/100$, as shown. Summing currents at the inverting input, which we may assume is a virtual ground, we have

$$\frac{e_1{}^2}{100R} + \frac{e_2{}^2}{100R} = -\frac{e_3{}^2}{100R} \tag{4-86}$$

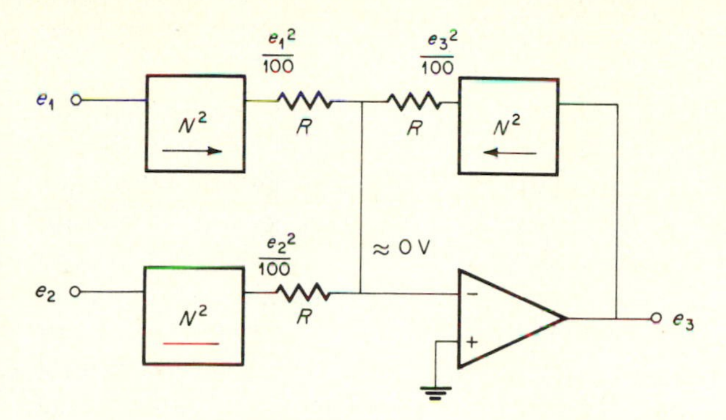

Fig. 4-32 Operational-amplifier circuit for computing the square root of the sum of the squares of two time-varying voltage inputs.

From Eq. (4-86), by multiplying through by $100R$, and taking the square root,

$$e_3 = -\sqrt{e_1^2 + e_2^2} \tag{4-87}$$

as desired.

4-6-6 LOGARITHMIC AMPLIFIERS

A special type of nonlinear amplifier that takes advantage of certain properties of both operational amplifiers and transistors is the logarithmic amplifier shown in its basic form in Fig. 4-33. It can be shown that if a transistor is operated with nearly zero collector-to-base voltage, the relationship between the collector current and the base-to-emitter voltage may be expressed in the form

$$V_{BE} = K \log I_c \tag{4-88}$$

where K is a constant which includes temperature, the charge on the electron, Boltzmann's constant, and the effects of leakage currents. Notice that the transistor base is grounded and that the collector is at a virtual ground, which satisfies the need for nearly zero collector-to-base voltage. $e_1 = I_c R$, and since V_{BE} is the circuit output voltage e_2, we can write, from Eq. (4-88),

$$e_2 = K \log \left(\frac{e_1}{R}\right) \tag{4-89}$$

Thus we have the desired logarithmic relationship between e_2 and e_1.

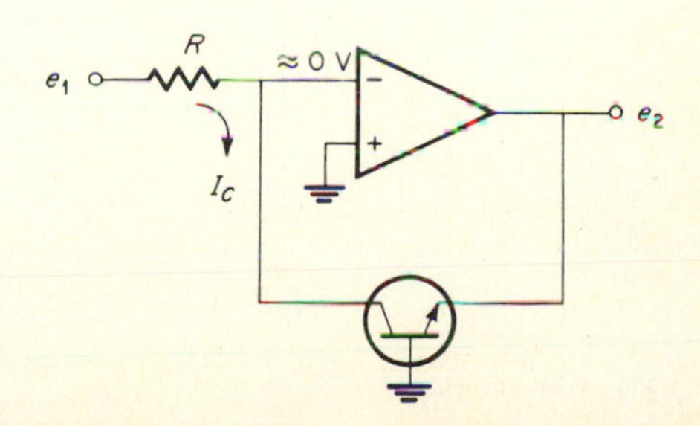

Fig. 4-33 Logarithmic amplifier.

4-6-7 THE GYRATOR

The gyrator is an interesting type of commercially available circuit that may eventually contribute to the elimination of inductors from filters and tuned circuits, a significant factor in the design of integrated circuits, where the fabrication of sizable inductors is impractical. A gyrator is a two-port network, one of the properties of which is that if a capacitor is connected across one port, the impedance seen looking into the other port is inductive. Thus if a gyrator can be made from a circuit not containing inductors, a means exists for making an effective inductor from integratable components. Figure 4-34 shows a circuit for a gyrator with capacitor C connected across one port. That Z_{in} is inductive can be shown in the following manner:

$$Z_{\text{in}} = \frac{e_1}{i_1} = \frac{e_1}{i_2 + i_3} \tag{4-90}$$

But

$$i_2 = \frac{e_1 - e_2}{R} \tag{4-91}$$

and

$$i_3 = \frac{e_1 - 2e_1}{R} = -\frac{e_1}{R} \tag{4-92}$$

The $2e_1$ appears at the output of the upper amplifier because it is set up to be a noninverting amplifier with a gain of 2. So we have

$$i_2 + i_3 = -\frac{e_2}{R} \tag{4-93}$$

To obtain an expression for e_2, we can write

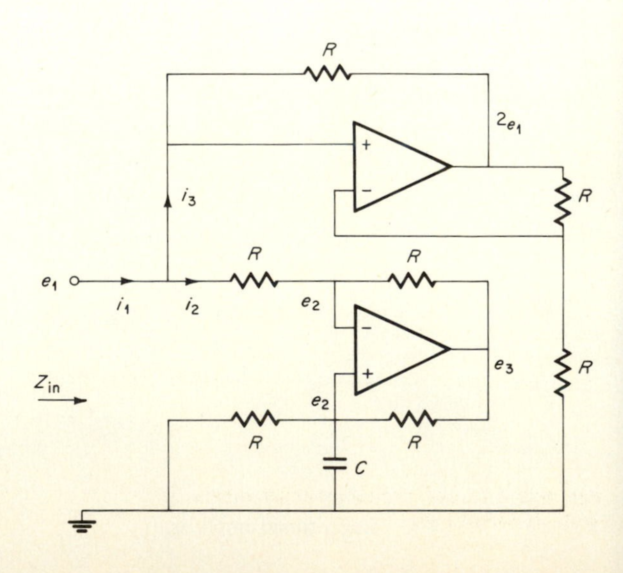

Fig. 4-34 Gyrator.

$$\frac{e_1 - e_2}{R} = \frac{e_2 - e_3}{R} \tag{4-94}$$

by equating the currents at the inverting input of the lower amplifier. Also, by equating currents at the noninverting input of the lower amplifier,

$$-\frac{e_2}{R} = \frac{e_2}{1/pC} + \frac{e_2 - e_3}{R} \tag{4-95}$$

Solving Eqs. (4-94) and (4-95) gives

$$e_2 = -\frac{e_1}{pCR} \tag{4-96}$$

Equation (4-93) can be rewritten, using Eq. (4-96), as

$$i_2 + i_3 = \frac{e_1}{pCR^2} \tag{4-97}$$

Equation (4-90) can be rewritten, using Eq. (4-97),

$$Z_{in} = pCR^2 \tag{4-98}$$

Z_{in} is positive, equal to a constant (CR^2) multiplied by p, which is $j\omega$ for the steady state, and is therefore equal to an inductance of value $L = CR^2$. If $R = 10\ \text{k}\Omega$ and $C = 0.01\ \mu\text{F}$, $L = 1$ H.

4-6-8 OPERATIONAL-AMPLIFIER PULSE CIRCUITS

All the commonly used pulse circuits, such as astable (free running), bistable (flip-flop), and monostable (one-shot) multivibrators, can be realized using operational amplifiers as the active element. This might be useful if one needed such a circuit, had a spare operational amplifier available, and did not want to add a digital circuit. A feature of such circuits is ease of design, since the timing- or level-sensing elements can be external to, and considered separately from, the active element. To illustrate the use of operational amplifiers in pulse circuits, consider Fig. 4-35, which shows a circuit for a monostable multivibrator, a

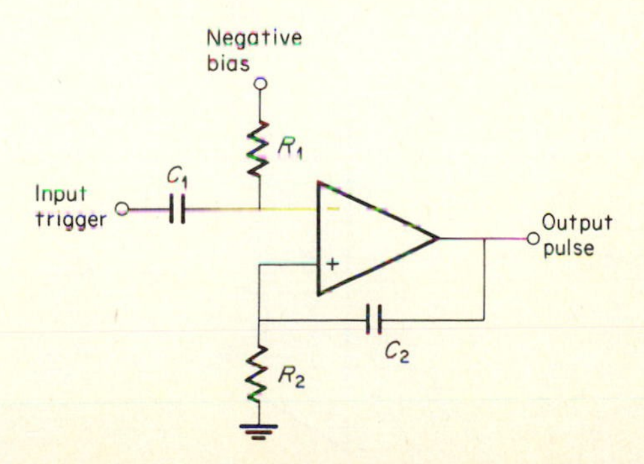

Fig. 4-35 An operational amplifier monostable multivibrator.

circuit which has as its output a pulse of fixed length and amplitude, independent of the input trigger.

Circuit operation is as follows. Since the inverting input is returned to a negative bias and since there is no dc feedback, the output will be saturated to a positive level. The noninverting input is at ground potential. Suppose now that a positive trigger is applied to the inverting input via C_1, of sufficient magnitude so that the inverting input is driven positive with respect to the noninverting input. The output will then saturate to a negative level; this output change is coupled through C_2 to the noninverting input, driving it negative also. Thus, even when the trigger is no longer present, the output will remain negative because the noninverting input is negative with respect to the inverting input. This situation will remain until C_2, which discharges exponentially toward ground through R_2, permits the noninverting input to reach the same level as the inverting input, at which time the output will again saturate positive and stay there until the receipt of another trigger. C_1 is therefore a trigger path, R_1 is a biasing resistor, and $C_2 R_2$ is the time constant which determines the length of the negative output pulse. The operational amplifier is the gain element used to obtain the regenerative switching action necessary for monostable multivibrator operation.

In this chapter we have analyzed a wide variety of operational-amplifier circuits. No operational-amplifier circuit collection will ever be complete, however, because new circuits and new variations are constantly being introduced. Fortunately, all these depend on certain basic principles, and it has been the intention here to give readers understanding and insight into these general principles so that they can understand any circuit they might come across, or design their own.

4-7 SUMMARY OF OPERATIONAL-AMPLIFIER FORMULAS

In this section exact and approximate formulas are given for the two most common operational-amplifier configurations, the inverting and noninverting amplifiers. The formulas given in this section refer to Figs. 4-36 and 4-37, which are

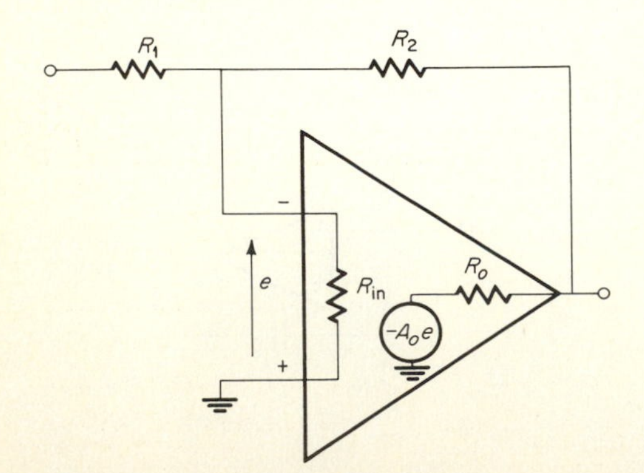

Fig. 4-36 General inverting-amplifier circuit.

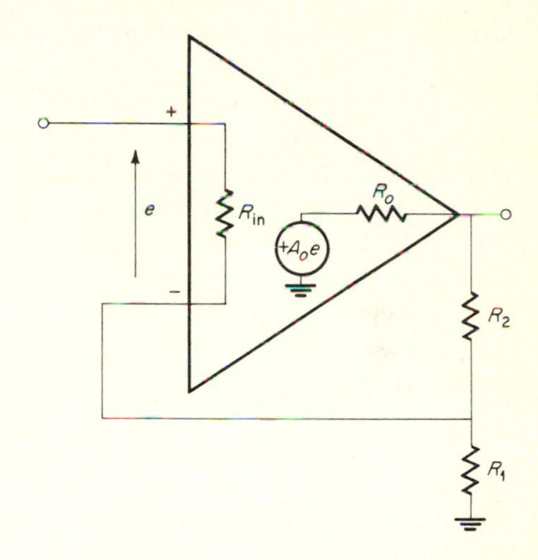

Fig. 4-37 General noninverting-amplifier circuit.

general circuits for the inverting and noninverting configurations, respectively, and which include the effects of finite open-loop gain, finite input resistance, and nonzero output resistance.

The approximate formulas are derived from the exact formulas by assuming that A_o and R_{in} are very large and that R_o is very small. It is possible for a particular term to dominate, especially since very high input impedance operational amplifiers are available, where R_{in} as a number may be larger than A'_o. In doubtful cases one should insert the numerical values into the exact formula, to determine which terms are in fact negligible.

The following definitions apply to the formulas in Tables 4-3 and 4-4:

- A_F Closed-loop gain
- R'_{in} Input impedance with feedback
- R'_o Output impedance with feedback
- A_o Open-loop gain
- A_o/A_F Loop gain

TABLE 4-3 Summary of inverting-amplifier formulas (see Fig. 4-36)

Exact	Approximate
$A_F = \dfrac{\dfrac{R_o}{A_oR_2} - 1}{\dfrac{1}{A_o}\left(\dfrac{R_o}{R_2} + \dfrac{R_oR_1}{R_{\text{in}}R_2} + \dfrac{R_1}{R_{\text{in}}} + \dfrac{R_1}{R_2} + 1\right) + \dfrac{R_1}{R_2}}$	$-\dfrac{R_2}{R_1}$
$R'_{\text{in}} = R_1 + \dfrac{(R_2 + R_o)R_{\text{in}}}{(R_2 + R_o + R_{\text{in}}) + A_oR_{\text{in}}}$	R_1
$R'_o = \dfrac{R_o[R_1(R_2 + R_{\text{in}}) + R_2R_{\text{in}}](A_oR_{\text{in}}R_2 - R_oR_{\text{in}})}{A_oR_{\text{in}}R_2[R_1(R_2 + R_o + R_{\text{in}}) + (R_2 + R_o)R_{\text{in}} + A_oR_{\text{in}}R_1]}$	$R_o\left(\dfrac{1 + R_2/R_1}{A_o}\right) = \dfrac{R_o}{\text{loop gain}}$

TABLE 4-4 Summary of noninverting-amplifier formulas (see Fig. 4-37)

Exact	Approximate
$A_F = \dfrac{\left(1+\dfrac{R_2}{R_1}\right)+\dfrac{R_o}{A_oR_{\text{in}}}}{1+\dfrac{1}{A_oR_{\text{in}}R_1}[R_{\text{in}}(R_2+R_1+R_o)+R_1(R_o+R_2)]}$	$1+\dfrac{R_2}{R_1}$
$R'_{\text{in}} = R_{\text{in}} + R_1\left(\dfrac{R_2+R_o+A_oR_{\text{in}}}{R_1+R_2+R_o}\right)$	$\dfrac{A_oR_{\text{in}}}{1+R_2/R_1} = R_{\text{in}}(\text{loop gain})$
$R'_o = \dfrac{R_o[(R_1+R_2)R_{\text{in}}+R_1R_2][R_1R_o+A_oR_{\text{in}}(R_1+R_2)]}{[A_oR_{\text{in}}(R_1+R_2)][R_{\text{in}}(R_1+R_2+R_o)+R_1(R_2+R_o)+A_oR_{\text{in}}R_1]}$	$R_o\dfrac{1+R_2/R_1}{A_o} = \dfrac{R_o}{\text{loop gain}}$

4-8 ANALYSIS OF OPERATIONAL-AMPLIFIER CIRCUITS BY INSPECTION

Many operational-amplifier circuits are susceptible to analysis by mathematics which can be performed mentally or by inspection, rather than by the solution of simultaneous equations on paper. A written description of the steps involved tends to be rather lengthy, but this should not obscure the fact that the method is usually quicker and gives more insight into the operation of the circuit than a purely mathematical approach.

An example of the use of this method has already been given in Sec. 4-3-2, where the gain and input impedance of the voltage follower of Fig. 4-3 were found.

Now consider the circuit in Fig. Prob. 4-9. The answer obtained to Prob. 4-9 by conventional solution of simultaneous equations may be obtained more simply as follows. Suppose input 2 is grounded and a signal e_1 is fed into input 1. The output will be $+2e_1$ because the upper amplifier is a noninverting amplifier with a gain of 2. This in turn is because the output of the lower amplifier is at 0 V and is essentially at ground potential. Now suppose input 1 is grounded and a signal e_2 is fed into input 2. $2e_2$ will appear at the output of the lower amplifier because it is a noninverting amplifier with a gain of 2. Examination of the upper amplifier will reveal that with its input (input 1) grounded, it acts as an inverting amplifier with a gain of -1 to signals appearing at the output of the lower amplifier. Since $2e_2$ exists at the output of the lower amplifier, the upper-amplifier output will be $-2e_2$. Finally, suppose that e_1 and e_2 are applied simultaneously to inputs 1 and 2, respectively. Since the circuit is linear, we can use the principle of superposition and say that the output will be the sum of the outputs obtained when they are applied separately, which is $2e_1 - 2e_2$ or twice the difference between the two inputs.

For another demonstration of this method, refer to the two circuits shown in Fig. 4-25, and consider why they have equal input-current-to-output-voltage gains. If E_{out} for circuit *a* is 1 V, the voltage across R_1 will be 1 V because the inverting input is a virtual ground. The current fed back through R_1 will be $\frac{1}{1000}$ μA. If E_{out} for circuit *b* is 1 V, voltage *e* will be 0.01 V, owing to the voltage-divider action of R_2 and R_3, and the voltage across R_1 will also be 0.01 V. The current fed back through R_1 will therefore be $0.01/10 = \frac{1}{1000}$ μA, which is the same as for circuit *a*. If I_d is small in both cases compared with $\frac{1}{1000}$, which it would be if the amplifier is usable from the point of view of signal-to-noise ratio, the input current in both cases is $\frac{1}{1000}$ μA. Since the same input current produces the same output voltage of 1 V, the current-voltage gains are the same.

PROBLEMS

4-1 Find the gain of the inverting and noninverting amplifiers of Table 4-1 if $R_1 = R_2 =$ 10 kΩ.

4-2 For the adder-subtractor of Table 4-1, find e_5 if $e_1 = 1$ V, $e_2 = 2$ V, $e_3 = 3$ V, and $e_4 = 4$ V.

4-3 For the integrator of Table 4-1, $e_1 = 0$ at $t \leq 0$. If $R = 1$ MΩ and $C = 1$ μF, what is e_2 after 1 sec if a +1-V signal is applied to the input and held for $t > 1$ sec?

4-4 For the differentiator of Table 4-1, what is the input impedance if $C = 1000$ pF and the frequency is 10 kHz?

4-5 Show that the circuit of Fig. Prob. 4-5 has an output of

$$e_2 = +\frac{2}{RC}\int e_1\,dt$$

and is therefore a positive-gain integrator.

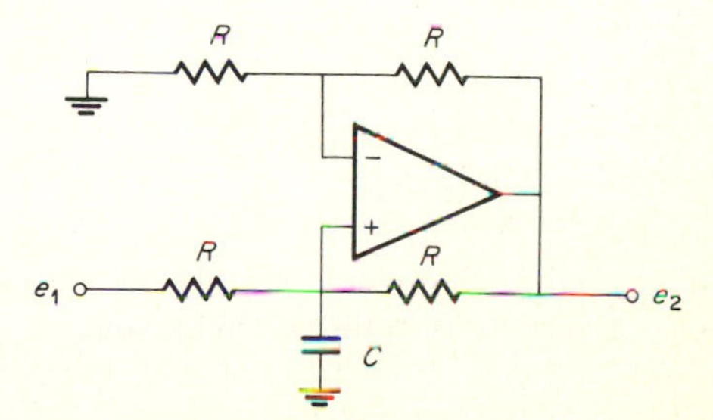

Fig. Prob. 4-5

4-6 Calculate the output drift of the current amplifier shown in Fig. Prob. 4-6 for the following values: $R_1 = 1$ MΩ, $R_2 = 100$ Ω, $R_3 = 10{,}000$ Ω, $e_d = 10$ μV/°C, and $i_d = 10$ nA/°C. Assume that the current and voltage drifts add.

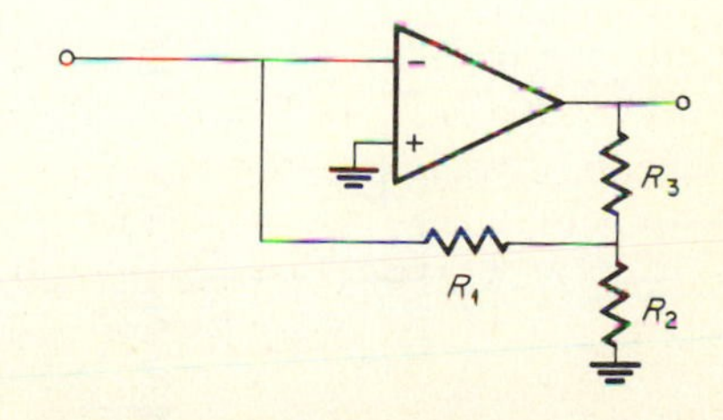

Fig. Prob. 4-6

4-7 For the circuit of Fig. Prob. 4-7, prove that

$$e_5 = (e_1 + e_2) - (e_3 + e_4)$$

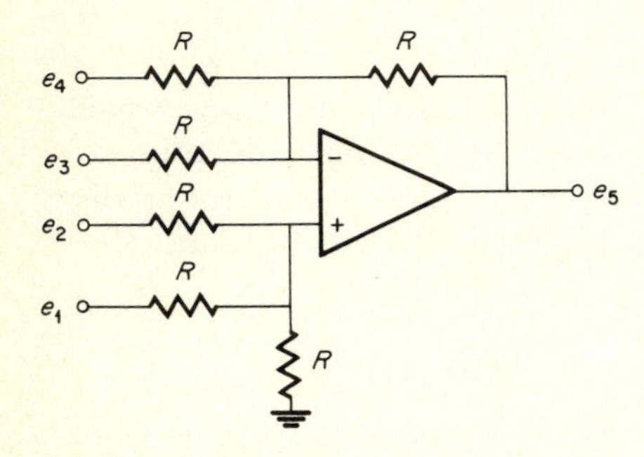

Fig. Prob. 4-7

4-8 Show, for the circuit of Fig. Prob. 4-8, that the instantaneous current passing through the meter is unidirectional and equal to $e_1/3R$, where e_1 is the instantaneous value of the input voltage, which may be positive or negative.

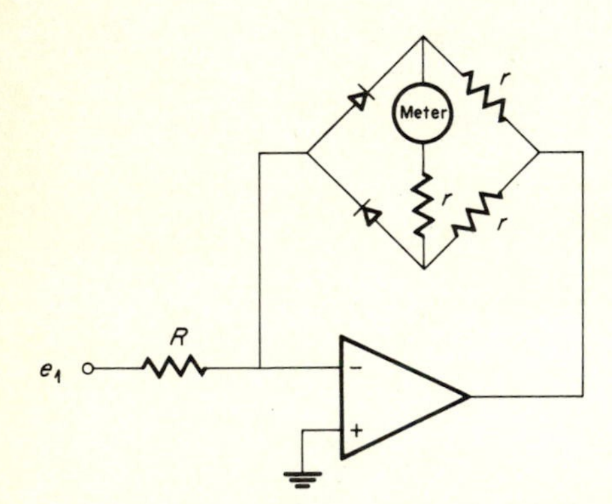

Fig. Prob. 4-8

4-9 Show that the output of the high-input-impedance circuit of Fig. 4-9 is twice the difference between the two input voltages.

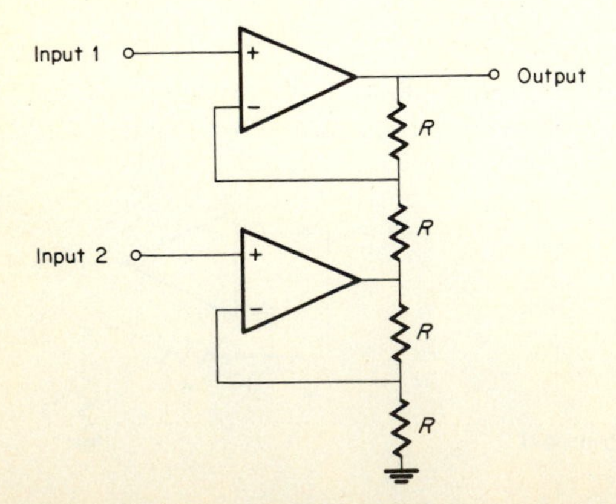

Fig. Prob. 4-9

4-10 Show that the circuit of Fig. Prob. 4-10 generates, through the floating load R_4, a current that is independent of R_4 and equal to e_1R_2/R_1R_3. Assume $R_2 \gg R_3$.

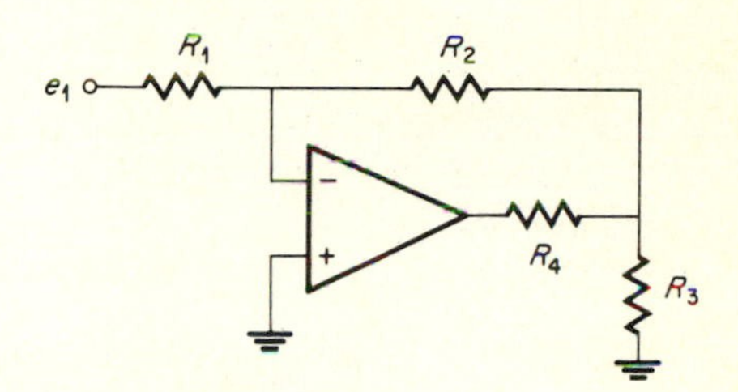

Fig. Prob. 4-10

4-11 For the circuit of Fig. Prob. 4-11, prove that a current flows through R_2 which is independent of R_2 and equal to $-e_1/R_1$.

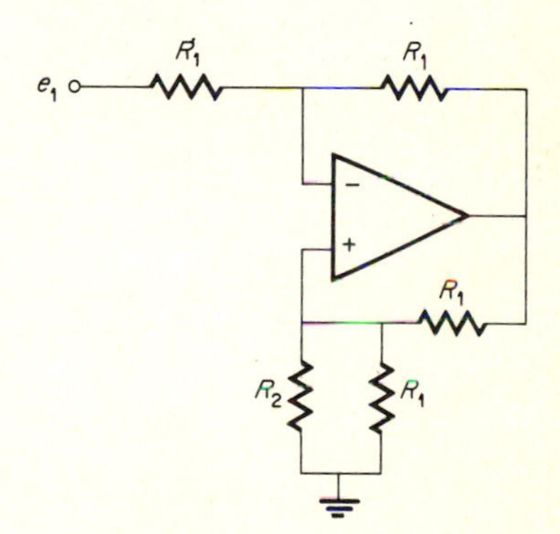

Fig. Prob. 4-11

4-12 The impedance looking into terminals 1 and 2 of the circuit shown in Fig. Prob. 4-12 is a pure inductor isolated from ground. Show that its value is CR^2. Use the voltage and current notation of Fig. 4-12.

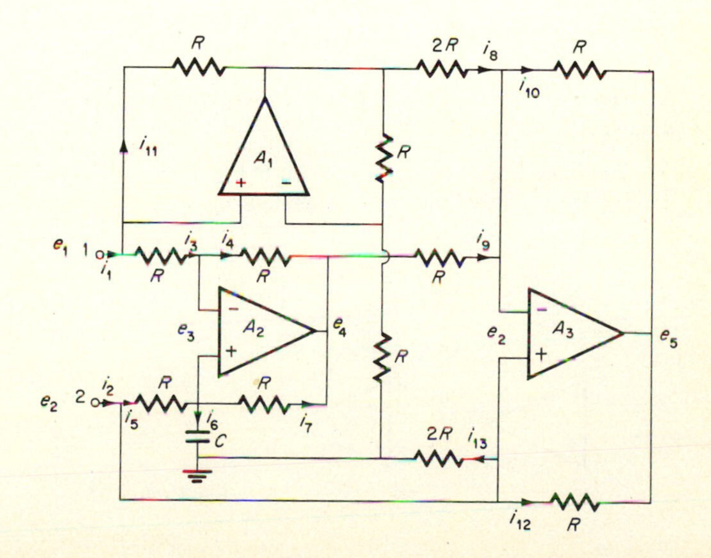

Fig. Prob. 4-12

4-13 The circuit of Fig. Prob. 4-13 is a high-pass filter. Show that its transfer function is

$$\frac{e_2}{e_1} = \frac{Kp^2}{p^2 + [(2 + C - K)/\sqrt{C}]p + 1}$$

where $K = (R_2 + R_1)/R_1$.

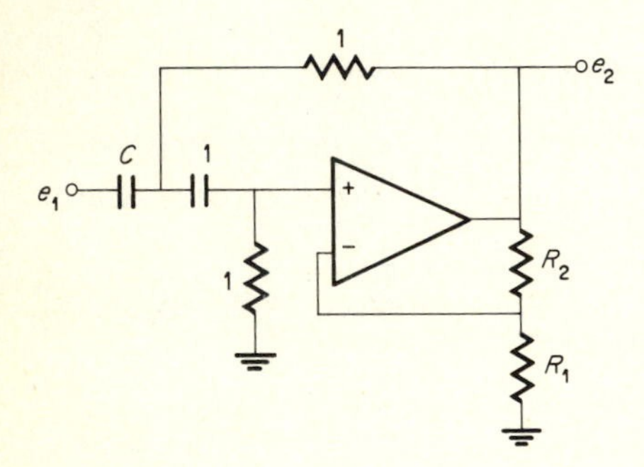

Fig. Prob. 4-13

4-14 Show that if C_f equals the sum of input and stray capacity C_s, the amplifier in Fig. Prob. 4-14 has a flat frequency response to infinity, when fed from a signal source having an output resistance of R. Assume an ideal operation amplifier.

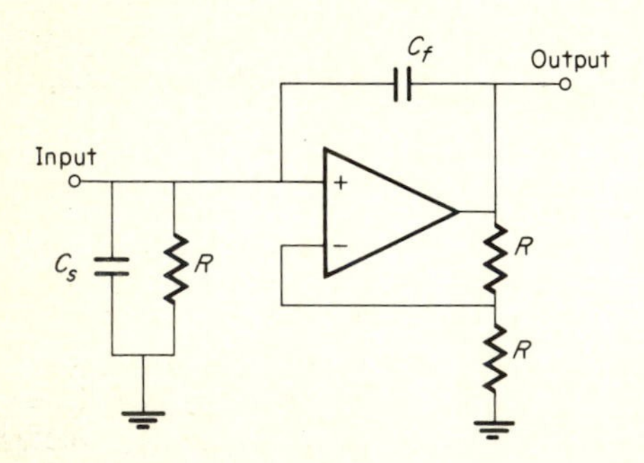

Fig. Prob. 4-14

4-15 The input impedance of the circuit in Fig. Prob. 4-15 can be varied by adjustment of the resistor marked $(n - 1)R$. Show that the input impedance is Z/n.

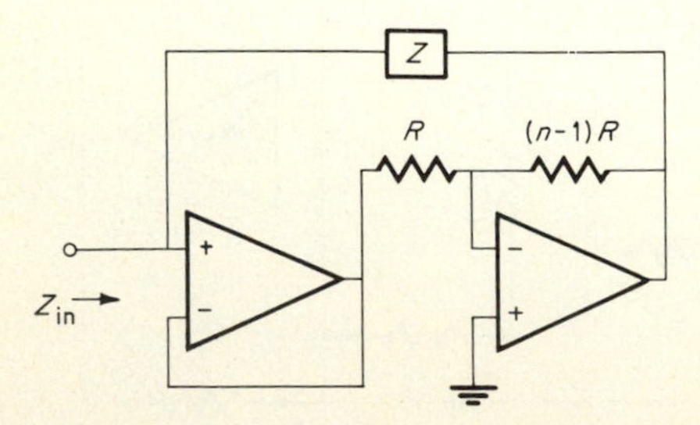

Fig. Prob. 4-15

4-16 Prove that the circuit of Fig. Prob. 4-16 has an infinite input impedance.

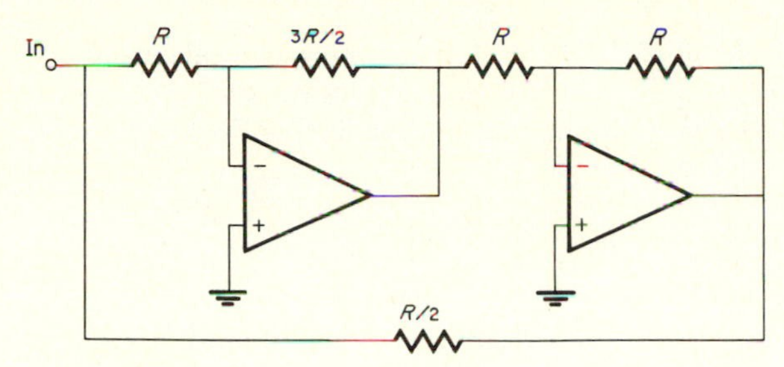

Fig. Prob. 4-16

4-17 The circuit in Fig. Prob. 4-17 acts as a charge amplifier. Show that its output is $-\Delta q/C$, where Δq is a change in input charge.

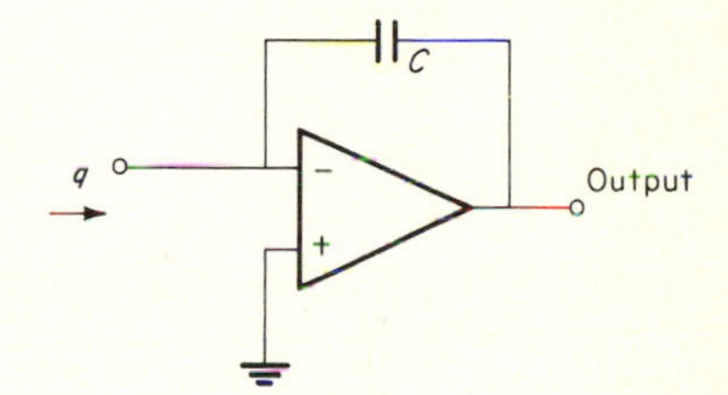

Fig. Prob. 4-17

4-18 When an operational amplifier used in a low-gain, closed-loop configuration is coupled to its load by means of a long length of high-capacity cable, a resistor is often connected in series with the amplifier output. Which end of the cable is it connected to? Why?

4-19 If a unity-gain operational amplifier has a dc open-loop gain of 10^8, the errors introduced by all but the very best feedback resistors will swamp any errors due to changes in the open-loop gain. Under what circumstances would such a high open-loop gain be necessary if the amplifier were to be used with moderate-stability resistive feedback elements? Explain your answer using a Bode plot.

4-20 Referring to Figs. 4-36 and 4-37, assume an operational amplifier has the following characteristics: $A_o = 10^6$, $R_o = 1\ \text{k}\Omega$, $R_{\text{in}} = 100\ \text{k}\Omega$, $R_1 = 1\ \text{k}\Omega$, and $R_2 = 100\ \text{k}\Omega$. For both the inverting and noninverting circuits, find the closed-loop gain, the loop gain, the closed-loop input impedance, and the closed-loop output impedance.

4-21 Show that the input impedance of the noninverting amplifier in Fig. Prob. 4-21 is approximately

$$\frac{R_{\text{in}} A_o}{1 + (R_2/R_1)}$$

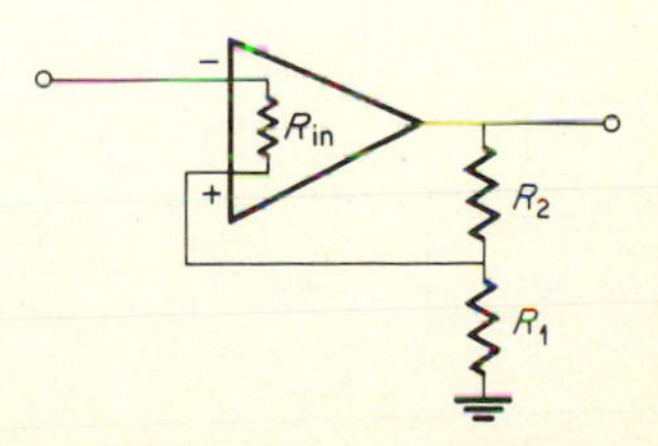

Fig. Prob. 4-21

Assume an ideal operational amplifier except for finite input impedance R_{in} and finite open-loop gain A_o.

4-22 Find the loop gain for the circuit of Fig. Prob. 4-21 assuming an ideal operational amplifier except for finite open-loop gain A_o and finite input resistance R_{in}.

4-23 Show that the output impedance of the noninverting amplifier of Fig. Prob. 4-21 is

$$\frac{R_o}{A_o/(1 + R_2/R_1)}$$

Show that this reduces to approximately R_o/loop gain. Assume an ideal operational amplifier except for nonzero R_o and finite open-loop gain A_o. Calculate the closed-loop output impedance for $R_o = 1000\ \Omega$, $A_o = 10^5$, $R_2/R_1 = 10$.

4-24 Show that for the circuit of Fig. 4-29 the loop gain is

$$\frac{A_o}{1 + (nR_f/R_n)}$$

if $R_1 = R_2 = \cdots = R_n$.

4-25 Explain why the common mode rejection of the circuit in Fig. 4-13 is better than that of the circuit in Fig. 4-7. For Fig. 4-13, $r = 1\ \text{k}\Omega$, $R = 100\ \text{k}\Omega$, and $R' = 10\ \text{k}\Omega$. If $e_1 = +10$ mV and $e_2 = +4$ mV, what is e_5?

chapter 5

Digital Integrated Circuits

5-1 INTRODUCTION

In the two previous chapters we covered the subject of linear integrated circuits. We saw how certain types of linear circuits are especially suited to integration, primarily because they are required in large numbers. The operational amplifier was a particularly good example of a circuit suited to integration, because it is a basic "building block" around which many other types of circuit can be built. This chapter introduces digital circuits, and these are extreme examples of circuits which can take full advantage of integration as a fabrication technique.

As indicated in Chap. 3, integrated circuits (and especially monolithic integrated circuits) can be very inexpensive when made in large quantities, are very reliable owing to the reduction in hand-made interconnections, and are best suited to applications where close tolerances on components are not required. Digital systems, no matter how complex, are composed of large numbers of a few basic circuits, many of which are described in Sec. 5-2. Because the numbers required are large, a fabrication technique capable of producing large numbers of similar circuits at low cost is needed. To achieve reliability in the overall system, high reliability in the individual circuits is a must. Also, although digital systems are capable of great accuracy, they achieve this without the use of close tolerances or selected parts. Monolithic integrated circuits meet all these requirements and, in addition, can greatly reduce the size of a package performing a given function.

The two primary features of digital systems are speed and accuracy. Both are important in such diverse fields as finance, where the 0.01 percent accuracy considered good for analog circuits is inadequate, and in space flight. Without computers in finance to handle the vast amount of paperwork generated by a modern civilization, there would be chaos, disorganization, and collapse, and the need for accuracy in this field does not have to be explained. If, in a space flight, problems of celestial navigation had to be solved without the aid of a digital computer, the flight would often be over before calculations involving a mid-course correction, for example, could themselves be completed. Also, the accuracy required to place a spacecraft in orbit a few hundred miles above a planet that is separated from the earth by perhaps 100 million mi [161 million km] is obvious.

Comparing the speed and accuracy achievable with analog and digital techniques is difficult, because such a comparison depends on a variety of factors. However, some reasonable generalizations can be made. Analog computations can be performed in milliseconds with accuracies of about 1 percent. Greater speeds are possible, and 0.1 percent accuracies can be achieved, but microsecond calculations and 0.01 percent accuracies at the same time are not routinely feasible. Digital computations, on the other hand, can be made in submicrosecond times, and relatively low-cost electronic digital desk calculators are available with accuracies of typically 14 significant digits. Although faster and more accurate, digital computers have the disadvantage that they tend to be more complex. This, in turn, makes them more costly than analog computers.

In addition to their use in computers, digital techniques are used in instrumentation (volt, current, and ohm meters, counters), telemetry (transmission of scientific data and photographs from spacecraft), and data processing (handling of business information).

Digital systems are constructed by means of a series of fabrication levels of increasing complexity. Before integrated circuits became practical, the first level consisted of one or more basic buildings blocks (flip-flops, gates—see Sec. 5-2), made with discrete components on a printed-circuit board. Many such boards would be plugged into a rack to form a subsystem—perhaps an analog-to-digital converter—and many racks in several cabinets would comprise the complete system. The coming of integrated circuits has resulted in greater complexity at the lower levels. Thus, a printed-circuit board may now contain 20 integrated-circuit modules, with each module being a complete subsystem such as a counter or arithmetic unit. The complete system obviously contains far fewer printed-circuit boards than the discrete-component version would, and therefore is much smaller. However, the development of very complex integrated circuits has had a greater impact than just reducing size. It means that circuit design engineers, who were previously designing discrete-component circuits and concerning themselves with such topics as transistor bias stabilization, are now working more as systems engineers. This in turn means that an individual can now produce systems of much greater complexity than before, in the same amount of time. These complex building blocks or subsystems are the result of what is called medium- and large-scale integration. Medium- and large-scale integration are discussed in detail later in this chapter.

Monolithic digital integrated circuits are currently in very widespread use, and in this chapter we shall be discussing basic digital principles, fundamental building blocks, types of logic element, memories, subsystems, MSI, LSI, and applications, as these apply to monolithic integrated circuits. For a good introduction to digital principles and applications, Ref. 1 at the end of this chapter, is recommended.

5-2 BASIC BUILDING BLOCKS

The purpose of a digital system, be it a calculator, a telemetry link, or whatever, involves the manipulation of numbers. The most familiar number system is the decimal system, which was adopted by early man because his most obvious calculation aid, the digits on his hands, number 10. Figure 5-1*a* shows a simple

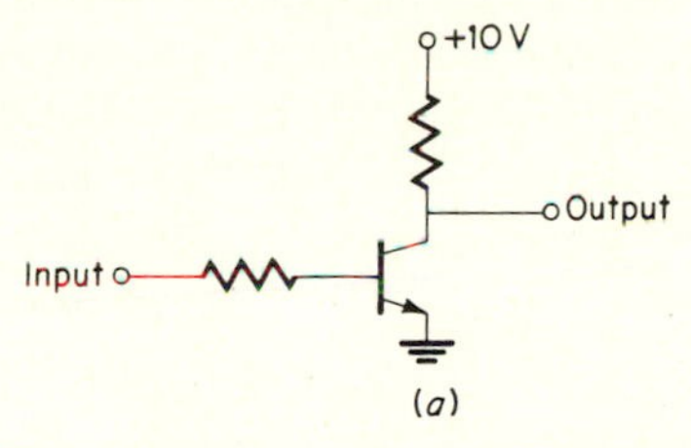

Number of objects	*Decimal number*				*Binary number*					
	10^2	10^1	10^0		2^5	2^4	2^3	2^2	2^1	2^0
	100	10	1		32	16	8	4	2	1
Zero			0							0
I			1							1
II			2							0
III			3							1
IIII			4					1	0	0
IIIII			5					1	0	1
IIIII IIIII IIIII		1	5				1	1	1	1
IIIII IIIII IIIII IIIII IIIII		2	5			1	1	0	0	1

(*b*)

Fig. 5-1 (*a*) Circuit for indicating numbers; (*b*) comparison of decimal and binary numbers.

transistor circuit which could be used to indicate numbers from 0 to 10. Application of the appropriate input level would result in some output voltage, for example, 5 V, which would represent the number 5. Unfortunately, the circuit has the disadvantage that different output voltages might well be obtained for the same input, if the transistor were changed. Also, accurate resistors would have to be used, since the output voltages are separated by only 10 percent of full scale. Even if the circuit had been set up with a selected transistor and accurate resistors, the output voltage would change unduly with temperature, easily more than 10 percent.

Fortunately, other number systems are available which enable the circuit in Fig. 5-1*a* to be used without transistor selection, with resistors having large tolerances, and over a wide range of temperatures. The most important of these number systems is the *binary*, which requires the use of only two symbols, in contrast to the 10 required by the decimal system. The number of symbols required by a number system is called the radix or base. The radix of the decimal system is 10 (symbols 0 through 9); and of the binary, 2 (symbols 0 and 1).

Figure 5-1*b* shows how numbers may be represented in the binary system and compares the binary representation with the corresponding decimal number. Just as decimal numbers are composed of units, tens, hundreds, and so on, so binary numbers are composed of units, twos, fours, eights, and so on. For example, consider the number 25 in Fig. 5-1*b*, for which the decimal and binary notations are abbreviations as follows:

$$\text{Decimal:}\quad 25 = (2 \times 10^1) + (5 \times 10^0)$$

$$\text{Binary:}\quad 25 = (1 \times 2^4) + (1 \times 2^3) + (0 \times 2^2) + (0 + 2^1) + (1 \times 2^0)$$

Note that 10^0 and 2^0 both equal 1.

Thus in the binary system the number 25 is represented by 11001, a result which is extremely important both with respect to the simple circuit in Fig. 5-1*a* and to digital integrated circuits in general. Notice that although we used 1 and 0 in Fig. 5-1*b*, we could have used any two symbols, for example, ON and OFF. Notice also that it is very easy to design the circuit in Fig. 5-1*a* to be ON or OFF without transistor selection, without accurate-value resistors, and over a wide range of temperatures. These are the kinds of requirements liked by designers of digital integrated circuits. For example, with an input of 0 V, the transistor in Fig. 5-1*a* will be OFF and the output will be at about +10 V. With an input of +10 V, the transistor will be ON and the output will be almost 0 V. Notice that the input and output levels are compatible—that is to say, the input levels are around 0 and +10 V and the outputs are, too. The input voltages do not have to be very accurate to produce the desired outputs of 0 and +10 V. Also, since outputs are used as inputs for other, similar building blocks, the 0- and +10-V outputs do not have to be very accurate either. We can easily design a circuit such that 0 is represented by a voltage between 0 and +1 V, and 1 is represented by a voltage between +9 and +10 V.

The significant point here is that most of the disadvantages of the circuit of Fig. 5-1*a* disappear when it is used to represent binary instead of decimal numbers—the circuit itself is unchanged. We can choose 0 V to represent 0 and +10 V to represent 1, in which case we have adopted what is called a *positive logic system*, the 1 being more positive than the 0. Or we could choose 0 V to represent 1 and +10 V to represent 0, in which case we would have adopted a *negative logic system*, the 1 being more negative than the 0.

A 1 and 0 or ON and OFF system of numbers is called *binary* because combinations of two states can represent any number we wish. Other names for the two states are also used, for example, *true* and *false*, or *high* and *low*, or even 0 and +10 V. From now on, for the sake of brevity and because they are most universal, we shall mainly use the two digits 1 and 0.

Binary numbers can be added, subtracted, multiplied, divided, and otherwise manipulated just as can decimal numbers. The arithmetic is just as easy, though not as familiar. A very clear review of the rules governing binary arithmetic is given in Ref. 1.

Having seen how electronic circuits work best with a binary system of numbers, it nevertheless remains true that humans, through familiarity, still prefer to work with decimal numbers. Thus, although a computer may do its "thinking" in binary numbers, it usually accepts instructions and presents answers in decimal form. This interfacing is performed with the help of number systems such as the *binary-coded decimal* (BCD). BCD codes are based on the fact that any number from 1 to 9 can be represented by combinations of four numbers such as 8, 4, 2, and 1 or 4, 2, 2, and 1. In BCD, each decimal digit is represented by its binary equivalent, and four binary digits are sufficient for each decimal. For example, the decimal number 183 may be expressed in 8421 BCD forms as follows:

1	8	3
0001	1000	0011

Thus, 183 in 8421 BCD is 0001 1000 0011. The 8421 code and several other codes and their manipulation in binary arithmetic are discussed at length in Ref. 2.

Figure 5-1*a* introduces a basic binary circuit or building block capable of being combined with other similar circuits to build a digital system. In drawing circuit diagrams of digital systems, the various functional elements (such as the circuit in Fig. 5-1*a*) are represented by symbols, instead of by diagrams showing resistors and transistors. For example, Fig. 5-1*a* is an inverter (or NOT circuit), since a 1 (+10 V) input results in a 0 (0 V) output, and a 0 (0 V) input results in a 1 (+10 V) output, if positive logic is assumed. (If the input is a 1 the output is *not*, and vice versa.) There are many other circuits that will perform this function, but they can all be represented by the same symbol, that for the inverter circuit in Table 5-1. The inverter circuit is not the only digital circuit that inverts; when such inversion does occur, it is frequently indicated on the symbol by a small circle such as the one on the right-hand apex of the inverter triangle in Table 5-1. Inversion is common in digital logic because many logic circuits use the common-emitter or common-source configuration. Notice that the input and output of the inverter are called A and $\bar{A}$, respectively. The bar over the A simply means that if A is a 1, $\bar{A}$ is a 0; and if A is a 0, $\bar{A}$ is a 1. This permits the function of the inverter to be defined independently of whether its input is in a 1 state or a 0. $\bar{A}$ is called the *complement* of A.

Other basic elements commonly used in digital systems are the AND, OR, NAND, NOR, EXCLUSIVE OR, EXCLUSIVE NOR, and three-state logic gates (three-state logic is discussed in Sec. 5-3-11). The symbols and truth tables for these gates are shown in Table 5-1. (A truth table is a listing of all the possible input-output possibilities for a logic circuit.)

Logic systems may be synthesized and analyzed by means of Boolean algebra, named after its inventor George Boole. Boole died in 1864, never to know that one day his invention would be a powerful and widely used tool in the design of digital systems. Boolean algebra is especially useful for digital-

TABLE 5-1 Basic logic building blocks

Function	*Symbol*	*Truth table* Inputs		Output	*Comments*
INVERTER	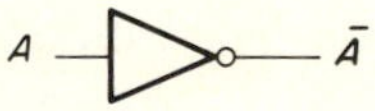 A → $\bar{A}$	A 1 0		$\bar{A}$ 0 1	Inverts
AND	A, B → AB	A 0 0 1 1	B 0 1 0 1	AB 0 0 0 1	No inversion; mechanical equivalent is: 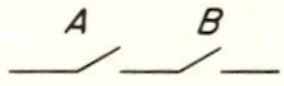 A B
OR	A, B → $A+B$	A 0 0 1 1	B 0 1 0 1	$A+B$ 0 1 1 1	No inversion; mechanical equivalent is: 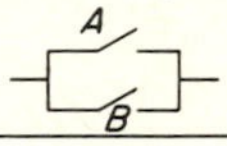 A B
NAND	A, B → $\overline{AB}$	A 0 0 1 1	B 0 1 0 1	$\overline{AB}$ 1 1 1 0	Same as AND but inverts
NOR	A, B → $\overline{A+B}$	A 0 0 1 1	B 0 1 0 1	$\overline{A+B}$ 1 0 0 0	Same as OR but inverts
Exclusive OR	A, B → $A\bar{B}+\bar{A}B$	A 0 0 1 1	B 0 1 0 1	$A\bar{B}+\bar{A}B$ 0 1 1 0	Output is a 1 when *A or B* but not *both* are 1; implemented by: 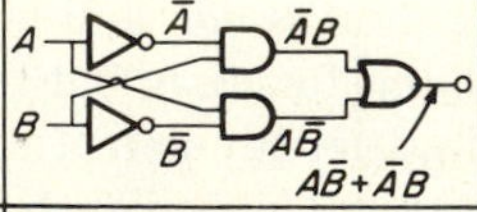 A, $\bar{A}$, $\bar{A}B$, B, $\bar{B}$, $A\bar{B}$, $A\bar{B}+\bar{A}B$
Negated OR	A, B → $\overline{AB}$	A 0 0 1 1	B 0 1 0 1	$\overline{AB}$ 1 1 1 0	Implemented by: 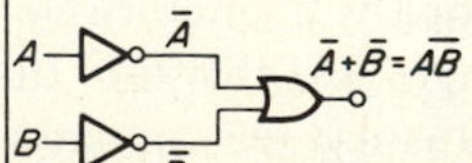 A, $\bar{A}$, B, $\bar{B}$, $\bar{A}+\bar{B}=\overline{AB}$
Three state logic	B, A → C	A 1 0 Any	B 0 0 1	C 0 1 Hi Z	With B low, circuit is an inverter. With B high, output is high impedance, permitting wire-ORing.

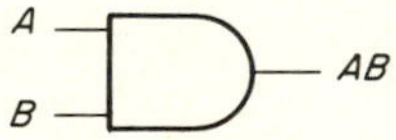

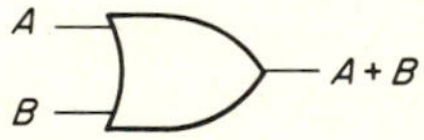

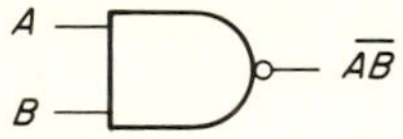

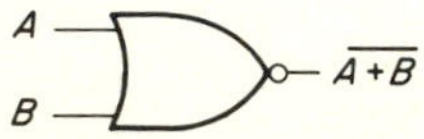

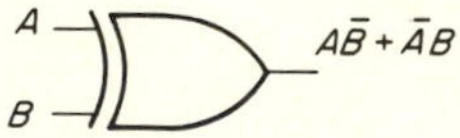

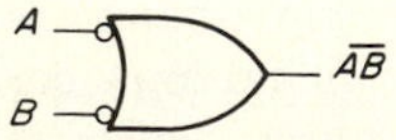

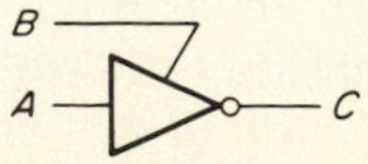

computer design because it permits, by the manipulation of symbols, crude forms of reasoning and logic to be performed mathematically. A complete review of Boolean algebra is outside the scope of this chapter, although the basic identities are given in Table 5-2.

Boolean algebra, like ordinary algebra, involves the manipulation of symbols for which numbers can be substituted. In Boolean algebra, though, only the numbers 1 and 0 can be substituted for symbols. Thus, in Table 5-1 the symbols A and B may each be replaced in the truth tables by 1 and 0.

TABLE 5-2 Boolean algebra identities

Operations

OR (+)	AND (× or ·)
$0 + 0 = 0$	$0 \times 0 = 0$
$0 + 1 = 1$	$0 \times 1 = 0$
$1 + 1 = 1$	$1 \times 1 = 1$
$A + 0 = A$	$A \times 0 = 0$
$A + 1 = 1$	$A \times 1 = A$
$A + A = A$	$A \times A = AA = A$
$A + \bar{A} = 1$	$A \times \bar{A} = A\bar{A} = 0$

Laws

$$\left.\begin{aligned} A + B &= B + A \\ AB &= BA \end{aligned}\right\} \text{Commutative laws}$$

$$\left.\begin{aligned} (A + B) + C &= A + (B + C) \\ (AB)C &= A(BC) \end{aligned}\right\} \text{Associative laws}$$

$$\left.\begin{aligned} A(B + C) &= AB + AC \\ A + BC &= (A + B)(A + C) \end{aligned}\right\} \text{Distributive laws}$$

$$A + AB = A$$

$$A + \bar{A}B = A + B$$

$$\left.\begin{aligned} \overline{ABCD} &= \bar{A} + \bar{B} + \bar{C} + \bar{D} \\ \overline{A + B + C + D} &= \bar{A}\bar{B}\bar{C}\bar{D} \end{aligned}\right\} \text{De Morgan's laws}$$

It is possible to write expressions in Boolean algebra to symbolize the action of the various gates in Table 5-1. For example, if an AND gate has two inputs A and B and an output X, the Boolean expression relating A, B, and X is

$$X = AB \tag{5-1}$$

Equation (5-1) indicates that an AND-gate output is analogous to multiplying the inputs, since in ordinary algebra AB means A multiplied by B. Although not strictly a multiplication, the results are the same. Notice that Eq. (5-1) is to be read "X equals *A and B*," not A times B.

Another way of arriving at the same truth table is to note simply that an AND gate gives an output only when there is a 1 on all inputs, in this case *A and B*. Thus, only the input combination 1 and 1 gives a 1 out; all other input combinations give a zero out.

The expression symbolizing the action of the OR gate in Table 5-1 is

$$X = A + B \tag{5-2}$$

where X is the output and A and B are inputs. Equation (5-2) means X equals *A or B*, not A plus B. Thus, an input 1 on A or B or both results in an output 1.

The NAND and NOR gates shown in Table 5-1 are similar to the AND and OR gates, except that the outputs are inverted. Inversion in Boolean algebra is symbolized by a bar above the inverted quantity. Thus if A is 1, $\bar{A}$ is 0.

Another very useful logic gate is the exclusive-OR shown in Table 5-1. This differs from the simple OR gate in that it has a 1 output when *A or B* but *not both* are 1. Thus, with A and B both 1, an OR gate has a 1 out, whereas an exclusive-OR has a 0 out. Comparison of the truth tables in Table 5-1 will further clarify the difference.

If both inputs of an OR gate are supplied through inverters, the resulting circuit is called a negated-OR. As shown in Table 5-1 the symbol for a negated-OR circuit is usually simplified to that of an OR gate with small circles on each input to represent the inversion. A comparison of the truth tables for a negated-OR and a NAND gate show that the functions of the two circuits are identical.

Let us now see how Boolean expressions and their manipulation can lead to circuit simplification and hence cost savings in the design of digital systems.

Referring to Fig. 5-2, assume that we are required to perform a logic operation described by the equation

$$F = AB\bar{C} + ABC + \bar{A}BC + \bar{A}B\bar{C} \tag{5-3}$$

Figure 5-2*a* shows how we might implement this expression by using two NOT gates, four AND gates, and one OR gate. However, Boolean expressions can be simplified, just as can ordinary algebraic equations, often with the result that a much simpler expression—and hence circuit—is obtained. Two of the most common methods are simple algebraic manipulation, as illustrated in Fig. 5-2*b*, and mapping, as in Fig. 5-2*c*.

In Fig. 5-2*b* we first factor out B. Terms are then rearranged to give give $(A + \bar{A})(C + \bar{C})$ as shown. But $(A + \bar{A}) = (C + \bar{C}) = 1$ from Table 5-2, so that $F = B$. Thus, the circuit can be simplified to that in Fig. 5-2*d*, simply taking the output off the B line.

Figure 5-2*c* shows what is called a *Karnaugh map*. It permits simplification of Boolean expressions without actually performing the algebra. Karnaugh mapping involves three steps, which are:

1. Multiply out all terms of the function so that each term contains each variable not more than once.
2. Draw the map.
3. Simplify the expression.

The function in Fig. 5-2 already satisfies step 1, so we can immediately draw the map. A Karnaugh map has 2^n squares, where n is the number of variables in the function. In this case $n = 3$ so eight squares are required, and they are numbered 1 to 8 in Fig. 5-2*c* for purposes of explanation. The rows

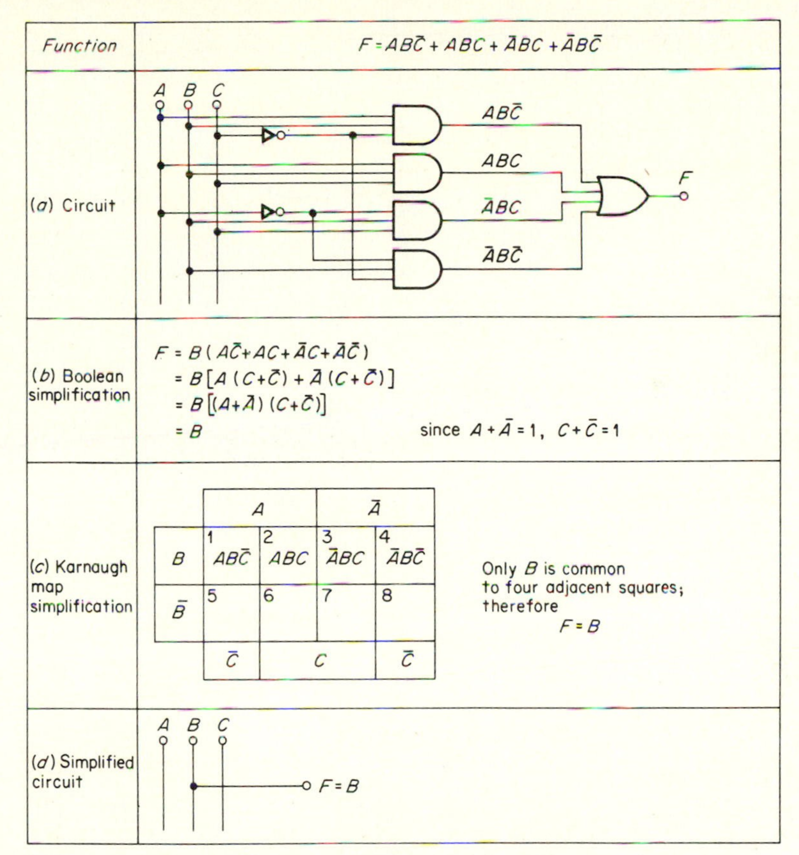

Fig. 5-2 Circuit simplification by algebra and mapping.

and columns of squares are marked with A, B, C, $\bar{A}$, $\bar{B}$, $\bar{C}$, so that any combination of these quantities can be represented by a square. For example, in Fig. 5-2*c*, square 1 represents $AB\bar{C}$, and since this occurs in the function F, we write $AB\bar{C}$ in square 1. Also, square 5 represents $A\bar{B}\bar{C}$, but we do not write this in square 5 because it does not occur in F. All four terms that appear in F are marked on the map as shown. Notice that *any* function with three variables will have a Karnaugh map of the form of that in Fig. 5-2*c*, although the squares filled in will depend on the function.

The final step is to simplify F by using the map. To do this, look for groupings of one, two, four, or eight squares. For example, four squares can form a 2-by-2 square, or, as in the case of Fig. 5-2*c*, a 4-by-1 rectangle. Having found a grouping, look for a factor that is common to all squares in the group. In Fig. 5-2*c*, such a factor is B. Only B is common to squares 1, 2, 3, and 4. No other squares are marked, so $F = B$ is the simplification.

The function in Fig. 5-3,

$$F = AC + ABC \tag{5-4}$$

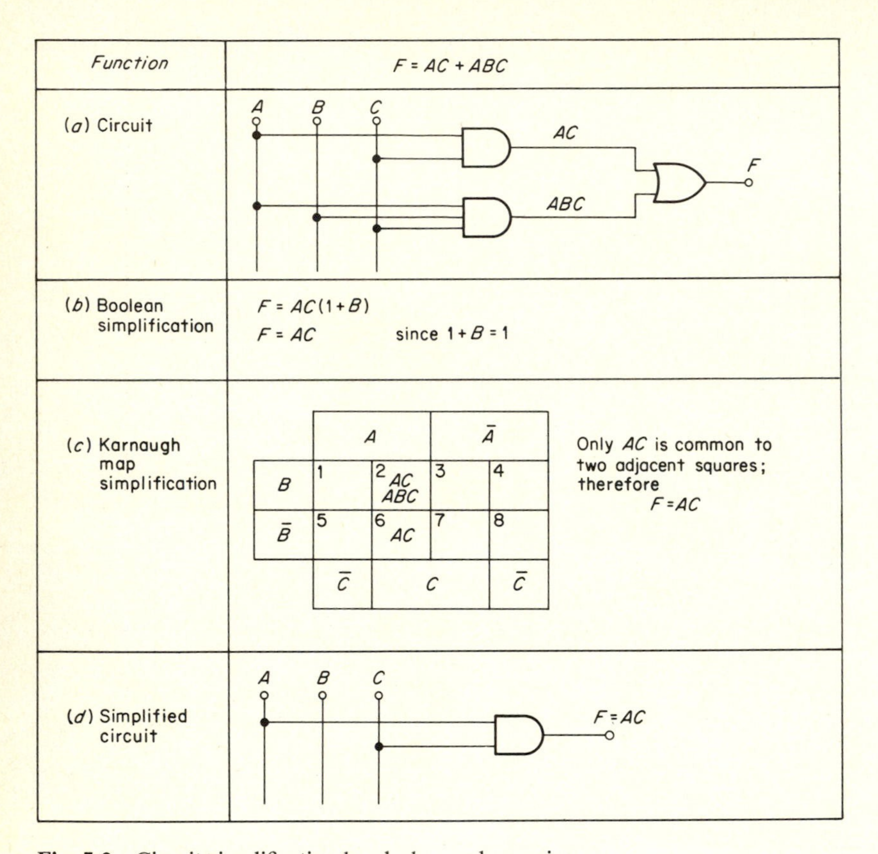

Fig. 5-3 Circuit simplification by algebra and mapping.

illustrates some other points concerning algebraic manipulations and mapping. In Fig. 5-3*b*, AC is first factored out. Noting from Table 5-2 that $(1 + B) = 1$, we are left with $F = AC$. Thus the circuit of Fig. 5-3*a* simplifies to that in Fig. 5-3*d*.

To simplify by mapping we notice that three variables are involved, so that eight squares are required. The form of Fig. 5-3*c* is the same as that of Fig. 5-2*c*. Notice that AC occupies two squares, those numbered 2 and 6. This is because AC does not contain B or $\bar{B}$. AC is AC whether B is a 1 or a 0. AC is independent of the condition of B. Therefore AC must occupy both the B and $\bar{B}$ horizontal rows. In contrast, ABC occupies only square 2. Again we look for groupings of one, two, four, etc., and find a group of two. AC is a common factor in the three terms in the two squares, so $F = AC$.

The function in Fig. 5-4 is

$$F = ABCD + ABC\bar{D} + \bar{A}BC\bar{D} + \bar{A}\bar{B}C\bar{D} + \bar{A}B\bar{C}\bar{D} + \bar{A}\bar{B}\bar{C}\bar{D} \tag{5-5}$$

and Fig. 5-4*a* shows how the circuit could be implemented before simplification.

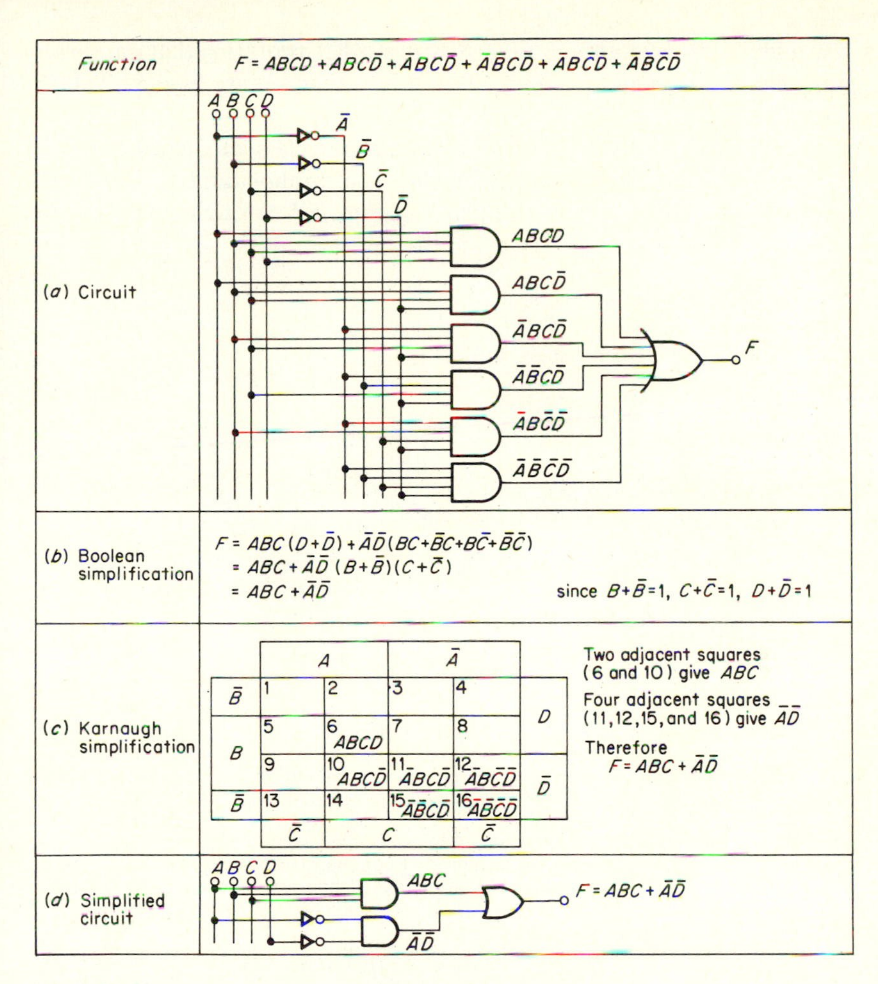

Fig. 5-4 Circuit simplification by algebra and mapping.

To simplify algebraically, ABC is factored from the first two terms, and $\bar{A}\bar{D}$ from the last four. Further factoring and noting that

$$(B + \bar{B}) = (C + \bar{C}) = 1$$

results in $F = ABC + \bar{A}\bar{D}$ as in Fig. 5-4*b*.

To simplify by mapping, note that since we have four variables, $2^4 = 16$ squares are required for the map. The map is drawn, and the six terms in the function are noted in their appropriate squares. Thus in Fig. 5-4*c*, squares 6, 10, 11, 12, 15, and 16 are filled. Now look for groupings of one, two, four, etc. Squares 6 and 10 form a 2-by-1 rectangle, and 11, 12, 15, and 16 form a 2-by-2 square. The common factor in squares 6 and 10 is ABC. That in squares 11, 12, 15, and 16 is $\bar{A}\bar{D}$. Therefore $F = ABC + \bar{A}\bar{D}$. Either algebra or mapping results in the circuit of Fig. 5-4*a* being simplified to that in Fig. 5-4*d*.

A few additional facts are required to solve other problems that can arise in algebraic manipulation and mapping. The most important are listed in Tables 5-2 and 5-3.

In addition to gates, there is another important class of digital circuits, namely, the *flip-flop* or bistable multivibrator. The symbols and functions of these and some other basic circuits are given in Figs. 5-5 through 5-11 and are summarized in Table 5-4.

TABLE 5-3 Karnaugh mapping

$F = D(AB\bar{C} + \bar{A}B\bar{C} + \bar{A}\bar{B}C)$

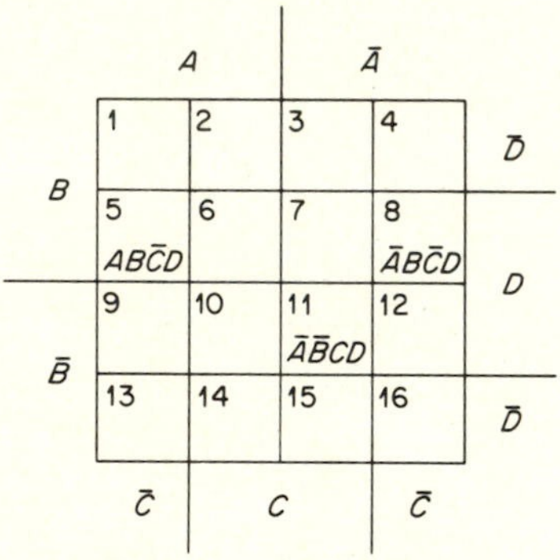

	A	A	$\bar{A}$	$\bar{A}$	
B	1	2	3	4	$\bar{D}$
B	5 $AB\bar{C}D$	6	7	8 $\bar{A}B\bar{C}D$	D
$\bar{B}$	9	10	11 $\bar{A}\bar{B}CD$	12	D
$\bar{B}$	13	14	15	16	$\bar{D}$
	$\bar{C}$	C	C	$\bar{C}$	

1. Draw chart with 2^n squares, where n is the number of variables. ($n = 4$ in above example, since there are 4 variables A, B, C, and D.)
2. Multiply out the function to be mapped, so that each term contains each variable only once. For example, $F = D(AB\bar{C} + \bar{A}B\bar{C} + \bar{A}\bar{B}C)$ should be expressed as $F = (AB\bar{C}D + \bar{A}B\bar{C}D + \bar{A}\bar{B}CD)$.
3. Locate each term on the map. For example, $AB\bar{C}D$ occupies square number 5 above.
4. Look for groupings of 1, 2, 4, 8 squares and extract common factors. Notice that 1 and 4, 1 and 13, 1 and 2, and 1 and 5 are considered groupings. 1 and 6 are not since they form a diagonal. 1, 2, 3, 4 is a 4 × 1 rectangular group and 1, 2, 5, 6 is a 2 × 2 square group. 1, 2, 13, 14 also forms a 2 × 2 square group.
5. For the 3 terms in the example above, squares 5 and 8 form a group of 2. $B\bar{C}D$ is a common factor in this group. Also, square 11 is a group of 1, so $\bar{A}\bar{B}CD$ is a common factor. Thus,

$$F = B\bar{C}D + \bar{A}\bar{B}CD$$

Unlike the gates in Table 5-1, a flip-flop is a logic circuit which contains feedback, so that the output state is a function not only of the input, but also of the previous history of the input. Thus a flip-flop may be used as a *memory* element. For example, a flip-flop has two outputs, often called Q and $\bar{Q}$ to indicate that, if the Q output is a 1, the $\bar{Q}$ output must be a 0, and vice versa. A momentary input signal will cause the outputs to flip, so that the Q output becomes a 0 and the $\bar{Q}$ a 1. The two outputs remain in these states even though the input signal is removed. A flip-flop is therefore analogous to a toggle switch whose "output" (its switch position) is a function of its last "input" (the mechanical switching action).

The simplest flip-flop is the triggering or T flip-flop shown in Fig. 5-5. Its outputs change state each time a triggering pulse (T) is applied to the input. Notice in Fig. 5-5 that the outputs are referred to as Q and $\bar{Q}$, not 1 and 0,

TABLE 5-4 Flip-flop and other pulse-output circuits

<table>
<tr><th>Circuit</th><th>Symbol</th><th>Remarks</th></tr>
<tr><td>(a) T flip-flop</td><td>T; Q, $\bar{Q}$</td><td>Each input pulse toggles the outputs, that is, causes them to change state
<table><tr><th>T</th><th>Q</th><th>$\bar{Q}$</th></tr><tr><td>n</td><td>0</td><td>1</td></tr><tr><td>n + 1</td><td>1</td><td>0</td></tr><tr><td>n + 2</td><td>0</td><td>1</td></tr></table></td></tr>
<tr><td>(b) RS flip-flop</td><td>R, S; Q, $\bar{Q}$</td><td>R input causes reset, S input causes set, both regardless of prior state
<table><tr><th>R</th><th>S</th><th>Q</th><th>$\bar{Q}$</th></tr><tr><td>0</td><td>0</td><td colspan="2">No Change</td></tr><tr><td>1</td><td>1</td><td colspan="2">Not defined</td></tr><tr><td>0</td><td>1</td><td>1</td><td>0</td></tr><tr><td>1</td><td>0</td><td>0</td><td>1</td></tr></table></td></tr>
<tr><td>(c) RST flip-flop</td><td>R, T, S; Q, $\bar{Q}$</td><td>Combines RS and T functions</td></tr>
<tr><td>(d) D flip-flop (latch)</td><td>T (CP), D; Q, $\bar{Q}$</td><td>The data, 1 or 0, on the D input is transferred to the Q output upon receipt of a clock pulse (CP) on the T input</td></tr>
<tr><td>(e) JK flip-flop</td><td>S; J, T (CP), K; Q, $\bar{Q}$; R</td><td><table><tr><th colspan="2">If J and K are:</th><th colspan="3">The next trigger (CP) pulse will:</th></tr><tr><th>J</th><th>K</th><th></th><th>Q</th><th>$\bar{Q}$</th></tr><tr><td>0</td><td>0</td><td>Have no effect</td><td colspan="2">No change</td></tr><tr><td>1</td><td>0</td><td>Set</td><td>1</td><td>0</td></tr><tr><td>0</td><td>1</td><td>Reset</td><td>0</td><td>1</td></tr><tr><td>1</td><td>1</td><td>Toggle</td><td colspan="2">Both change state</td></tr></table></td></tr>
<tr><td>(f) Monostable multivibrator</td><td>Input — MMV — Output</td><td>Output pulse width is fixed</td></tr>
<tr><td>(g) Astable multivibrator</td><td>AMV — Output</td><td>Free-running, square-wave oscillator</td></tr>
<tr><td>(h) Schmitt trigger</td><td>Input — ST — Output</td><td>Bistable, but with no memory; gives output only as long as input is above a fixed dc level</td></tr>
<tr><td>(i) Comparator</td><td>Input, Ref — Output</td><td>Similar to Schmitt, but more accurate and more easily adjusted switching level</td></tr>
</table>

since the Q output can be a 1 or a 0, and $\bar{Q}$ will always be the complement of Q. A series of T flip-flops can be used as a simple counter as in Fig. 5-6*a*. Suppose each flip-flop requires an *input transition from a* 0 *to a* 1 *to trigger it*, and that all $\bar{Q}$ outputs are initially 0, so that all Q's are 1s. Now apply a train of 1 pulses to the input. The first pulse flips the first flip-flop so its Q and $\bar{Q}$ outputs become 0 and 1. *B* does not change, because its input comes from Q_c which has just gone from a 1 to a 0. *A* does not change, because its input has not seen a transition at all. Thus, for one pulse in, the $\bar{Q}$ outputs of *A*, *B*, and *C* will be 001 as in

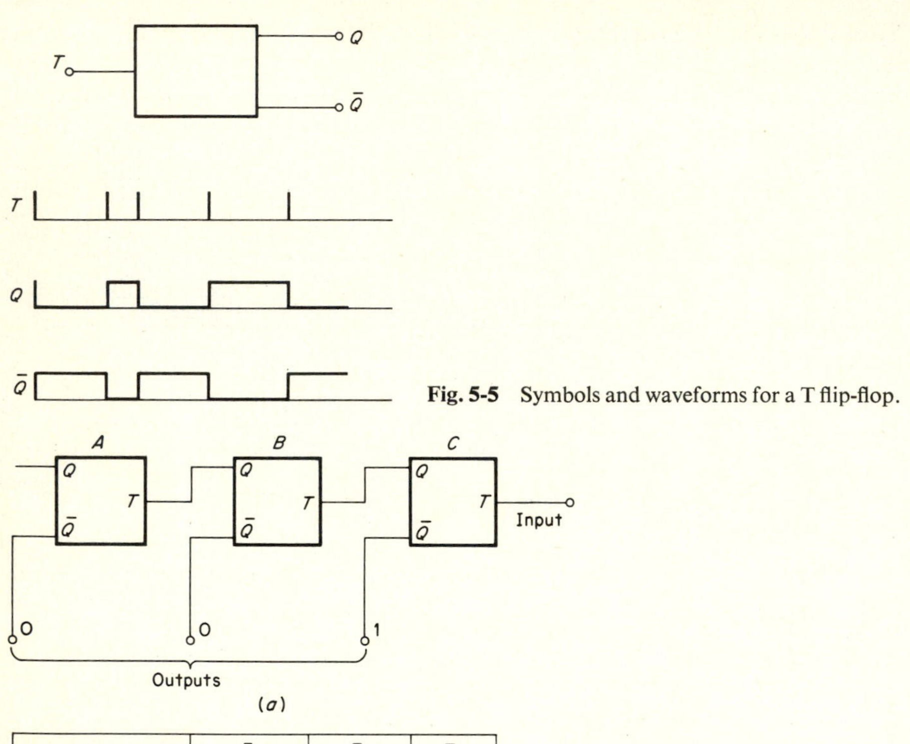

Fig. 5-5 Symbols and waveforms for a T flip-flop.

No. of input pulses	$\bar{Q}_A$	$\bar{Q}_B$	$\bar{Q}_C$
0	0	0	0
1	0	0	1
2	0	1	0
3	0	1	1
4	1	0	0
5	1	0	1
6	1	1	0
7	1	1	1

(b)

Fig. 5-6 (*a*) Simple T flip-flop counter; (*b*) typical output conditions for a train of input pulses.

Fig. 5-6*a*; this corresponds to the number 1 in binary code. Further inputs will result in the $\bar{Q}$ outputs changing to represent the total number of input pulses as illustrated in Fig. 5-6*b*, and the circuit acts as a counter.

A second important type of flip-flop is the RS flip-flop shown in Fig. 5-7. RS means *reset-set*, terms which describe the functions of the two inputs. A logic input on the reset input will reset the outputs so that Q and $\bar{Q}$ are in certain specified states, say 0 and 1. This will happen regardless of what states

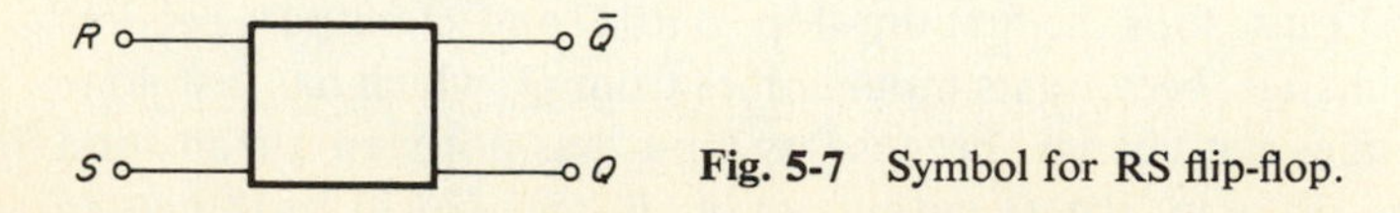

Fig. 5-7 Symbol for RS flip-flop.

Q and $\bar{Q}$ were in before application of the reset pulse. Similarly, a logic input on the set input will set the outputs so that the Q and $\bar{Q}$ states are opposite to those created by the reset pulse. Again this is independent of the states of Q and $\bar{Q}$ before application of the set input. The RS flip-flop has the disadvantage that its outputs are undefined for both inputs at 1, although with both inputs at 0, Q and $\bar{Q}$ remain unchanged. (See Table 5-4*b*.)

The RST flip-flop in Fig. 5-8*a* is a combination of the RS and the T flip-flops; it can reset, set, and trigger. In the circuit of Fig. 5-6*a*, which used T flip-flops, no provision was made for initially setting the three outputs all to 0. Figure 5-8*b* shows how this could be done using an RST flip-flop. The toggle

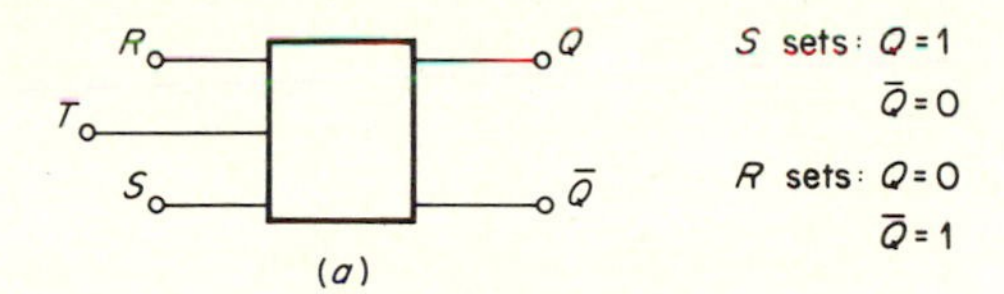

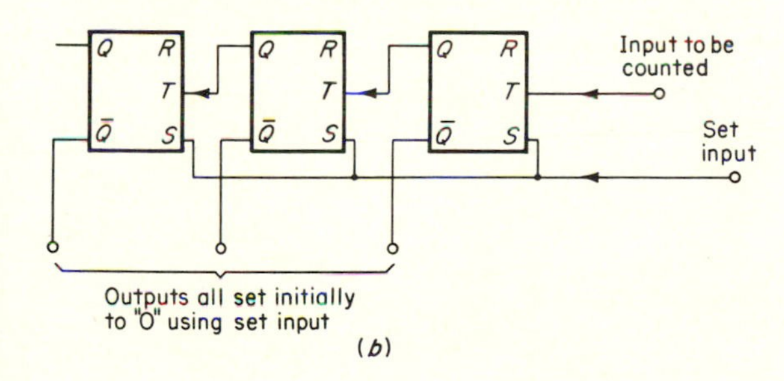

Fig. 5-8 (*a*) Symbol for RST flip-flop; (*b*) resettable counter using RST flip-flops.

action is again achieved through the use of the T input. However, the S, or set, input is used to initially set all outputs to 0, which it will do regardless of the previous count stored.

A fourth kind of flip-flop is the D type or *latch*, shown in Fig. 5-9*a*. Data in the form of a 0 or 1 on the D input is transferred to the Q output when a

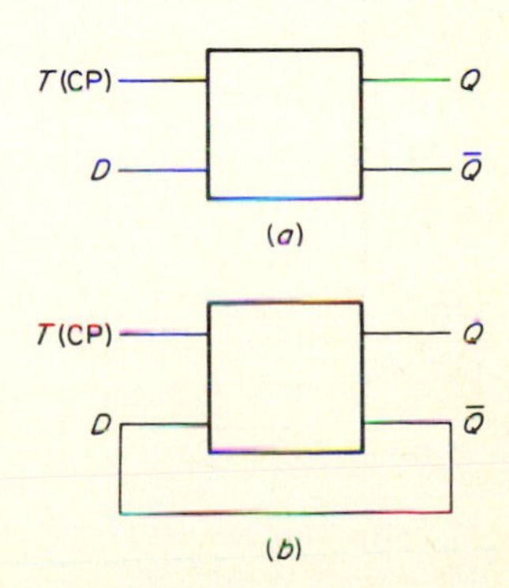

Fig. 5-9 (*a*) Symbol for D flip-flop; (*b*) D flip-flop connected as a T flip-flop.

trigger T (also called a clock pulse, CP) occurs at the T (CP) input. An array of such flip-flops may be used to store a digital word by parallel transfer. In such a case there would be one D flip-flop for each bit of the word. Each bit is applied to its own D input and, upon receipt of a clock pulse common to all flip-flops, the digital word is transferred to the appropriate Q outputs.

Many D flip-flops have set and reset inputs also, which perform the same functions as in an RS flip-flop.

A D-type flip-flop may be converted to a T type or toggle by connecting the $\bar{Q}$ output back to the D input and applying clock pulses to the T (CP) input as in Fig. 5-9*b*. If Q is a 1, $\bar{Q}$ and D are both 0. A clock pulse transfers the 0 on D to Q, making Q a 0 and $\bar{Q}$ a 1. Each succeeding clock pulse causes the Q and $\bar{Q}$ outputs to change as in a toggle flip-flop.

A very important type of flip-flop is the one shown in Fig. 5-10*a*, the JK flip-flop. It is a more universal flip-flop because it combines the functions of the T, RS, RST, and D flip-flops. The action of a JK flip-flop is summarized in Fig. 5.10*b*. The outputs Q and $\bar{Q}$ may or may not change state (flip or toggle) when a trigger (clock pulse) is applied to the T (or CP) input, depending on the states of the JK inputs. If both J and K are 0 ($J = K = 0$), Q and $\bar{Q}$ will be unchanged by a trigger. If $J = 1$ and $K = 0$, the flip-flop will be set by a trigger; whereas if $J = 0$ and $K = 1$, it will be reset. In the set and reset conditions, Q and $\bar{Q}$ are defined as, say, 1 and 0 for set and 0 and 1 for reset. With $J = K = 1$, the next trigger pulse will cause the flip-flop to toggle, so that the state of Q and $\bar{Q}$ interchange.

Many JK flip-flops also have the R and S inputs shown in Fig. 5-10*a*.

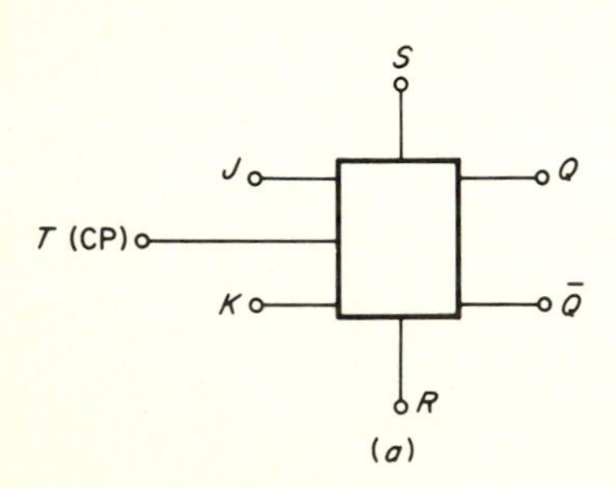

(*a*)

If J and K are:		*The next trigger pulse will:*	*Outputs*	
J	*K*		*Q*	*Q̄*
0	0	Have no effect	No change	
1	0	Set	1	0
0	1	Reset (clear)	0	1
1	1	Toggle	Both change state	

(*b*)

Fig. 5-10 (*a*) Symbol for JK flip-flop; (*b*) truth table for JK flip-flop.

Notice that we have just described R and S modes of operation for the JK flip-flop, with $J = O$, $K = 1$ and $J = 1$, $K = 0$. However, these modes require a trigger input. The R and S inputs permit reset and set operations independent of a trigger input.

Another important type of flip-flop is the *master-slave* (MS) flip-flop. This consists of two flip-flops, one called a master and the other the slave, and several logic gates. When all these functions are combined on a single chip, the result is an example of a simple form of medium-scale integration. An MS flip-flop is often used to overcome the problem of "racing" in ordinary flip-flops. Racing occurs when inputs to a flip-flop are simultaneous, so that the output is not defined (Ref. 1).

The flip-flops we have discussed so far are all variations of one member of the family of circuits called multivibrators. They are, in fact, all *bistable multivibrators*, the bistable meaning that they have two stable states. Thus a flip-flop can remain indefinitely with its two outputs at 1 and 0 (one state) or at 0 and 1 (a second state). Both states are stable, and the circuit is bistable. Two other types of multivibrators are used in digital systems, and these are shown in Fig. 5-11.

As the name implies, the *monostable multivibrator* (also called a *one-shot*) shown in Fig. 5-11*a* has only one stable state. When triggered, it goes into the

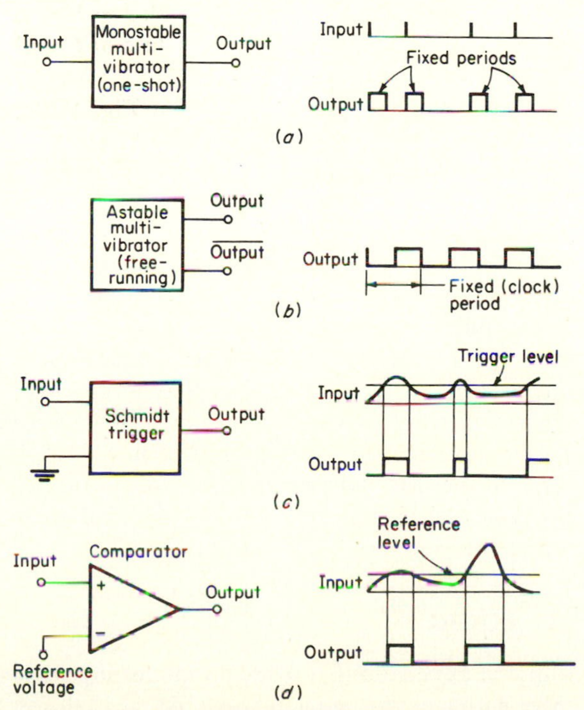

Fig. 5-11 (*a*) Monostable multivibrator (one-shot); (*b*) astable multivibrator (free-running); (*c*) Schmitt trigger; (*d*) comparator.

unstable state for a fixed amount of time and then returns to the stable state. In other words, the length of time that a one-shot spends in its unstable state is stable. The one-shot can therefore be used to generate fixed time delays from variable-length pulses; it is also used for gating, pulse reshaping, and changing the width of other fixed-width pulses.

The *astable* (free-running) multivibrator of Fig. 5-11*b* has no stable states; it switches back and forth continuously between its two unstable states. Thus it has no input (except perhaps for a synchronizing input) and two outputs either or both of which may be used in digital systems. When used in a digital system, the free-running multivibrator output is often called a clock pulse (CP). (The JK flip-flop in Fig. 5-10 has a CP input.) The word "clock" is appropriate because the free-running multivibrator can be used to provide synchronizing pulses to coordinate the operation of a complete system containing many flip-flops and gates. Thus it acts as a *master clock*.

The *Schmitt trigger* of Fig. 5-11*c* is both a bistable circuit and a voltage comparator. It differs from the bistable, astable, and free-running flip-flops in that it has no memory. These three circuits remain flipped after the triggering pulse is removed. The Schmitt trigger remains flipped only as long as a trigger is present. To trigger a Schmitt circuit, a voltage level greater than some predetermined level, called an *upper trip point*, is required. To change the circuit back to its original state, the input level must fall to another predetermined level called the *lower trip point*. The upper- and lower-trip-point voltages are not the same, and the difference between them is called the *circuit hysteresis*. In Fig. 5-11*c* the hysteresis is shown as zero. Because of its voltage-sensing action, a Schmitt trigger can be used to "tell" when some voltage has risen above a given level; that is, it acts as a comparator. This is useful in analog-to-digital converter circuits.

A Schmitt trigger is really a high-gain dc amplifier with positive feedback. A related circuit is the operational-amplifier type of comparator shown in Fig. 5-11*d*. It may be an operational amplifier (see Sec. 4-3-1), or it may be a similar type of circuit in which features necessary for linear operation have been sacrified to achieve speed and low hysteresis. This type of comparator is more universal in that the voltage level required to trigger it is more easily adjusted and the hysteresis can be made lower (1 mV) and more stable than with a Schmitt trigger. The various types of multivibrators are summarized in Table 5-4.

In this section we have described the most important basic circuits used to build digital systems. There are many variations and combinations of these circuits, descriptions of which are gladly supplied by their manufacturers.

5-3 INTEGRATED-CIRCUIT LOGIC

Integrated circuits intended for digital applications provide an interesting study of how circuits evolve to keep pace with a changing technology. As we shall see, logic circuits are classified by abbreviations which indicate the type of circuit used. Thus RTL means resistor-transistor logic, DTL means diode-transistor logic, and so on. The first integrated logic circuits were simply integrated-circuit versions of the old-discrete-circuit types of logic. Recall from our discussions on integrated-circuit fabrication in Chap. 3 that to simply copy the circuit for a discrete-component design, but use monolithic integrated-circuit fabrication, usually does not result in optimizing the performance-price ratio. A discrete-component-fabrication philosophy involves an attempt to minimize the number of components used and to use a low ratio of active (transistors) to passive (resistors) parts. In integrated-circuit fabrication it costs little to add a few transistors to a circuit, and replacing resistors with transistors is actually desirable from a space-saving point of view.

There are five main types of logic circuits in use: transistor-transistor logic (TTL), Shottky TTL, emitter-coupled logic (ECL), complementary metal oxide semiconductor (CMOS), and integrated-injection logic (I^2L). These will all be discussed and compared later in this chapter, but we shall first review some of the earlier types of logic to indicate how the later forms evolved.

5-3-1 DCTL

Figure 5-12 shows a three-input, direct-coupled transistor logic (DCTL) NOR gate, which is one of many types of DCTL gate. The name *direct-coupled* comes from the fact that the inputs are coupled directly to the transistor bases.

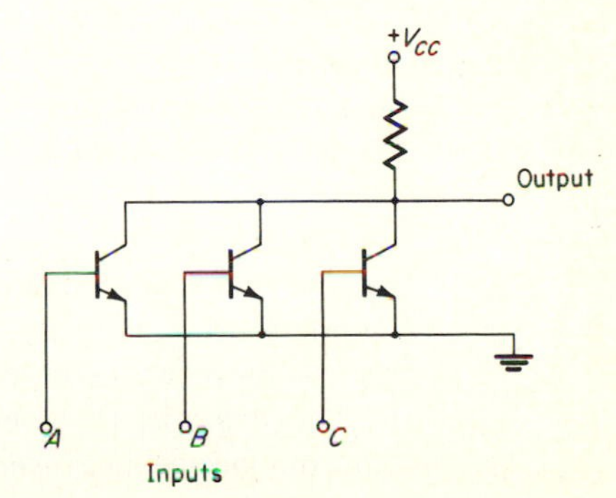

Fig. 5-12 Direct-coupled transistor logic (DCTL) NOR gate.

A 1 input on *A or B or C* will turn ON the transistor whose base has the 1 on it (assuming positive logic), and this will cause the output to be low or a 0. When all inputs are 0, all three transistors will be OFF and the output will be a 1. From Table 5-1, this is the condition for a NOR gate.

DCTL gates are susceptible to what is called "current hogging." Suppose the circuit in Fig. 5-12 is loaded with several other DCTL NOR gates. This means that several base-emitter junctions are being driven from the same point and, due to inevitable differences in their base-to-emitter characteristics, one junction will turn on first. It is quite possible for this first junction to "hog" sufficient current to prevent some of the other junctions from turning ON.

5-3-2 RTL

The problem of current hogging can be most easily solved by placing a resistor in series with each input base. The result is the resistor-transistor logic (RTL) NOR circuit of Fig. 5-13. Current hogging is avoided because the resistors isolate the bases from the common driving point, thereby permitting the base-emitter voltages to individually adjust to the levels necessary for turn-ON. The addition of the input series resistors also increases the circuit input impedance, so that the fan-out of the driving circuit is increased.

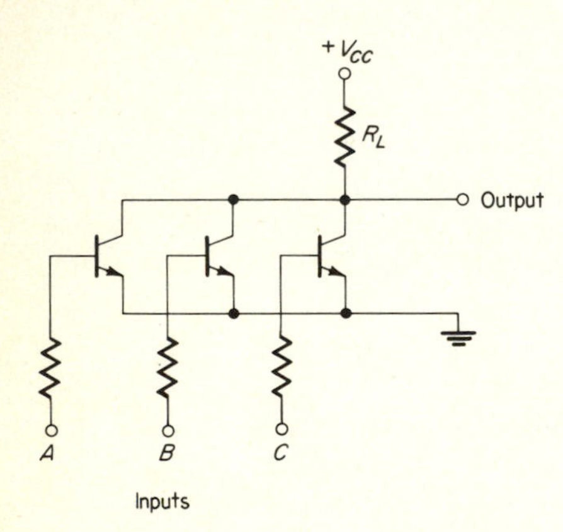

Fig. 5-13 Resistor-transistor logic (RTL) NOR gate.

RTL circuits were the first type of logic to be integrated in the early 1960s, and there are several reasons for this. The basic circuit had been proved in discrete-circuit form and was readily adaptable to integrated-circuit fabrication. Thus, designers were familiar with RTL, and lots of reliability and evaluation data were available.

RTL circuits have three main disadvantages: relatively low speed, low fan-out, and temperature sensitivity. The low speed arises because the external series base resistor combines with the input capacity of the transistor to form a low-pass filter. This degrades the rise and fall times of any input pulse.

The low fan-out occurs because the input current to a given transistor is limited by having to flow through R_L (in Fig. 5-13) and the series base resistor of the next gate.

5-3-3 RCTL

Speed in RTL circuits is a function of the size of the resistors used, since inherent transistor and stray capacities must be charged and discharged through collector and base resistors. To increase speed, resistors can be reduced in value, but this increases the circuit power requirements. To achieve high speed with less power, RTL circuits can be modified to the RCTL (resistor-capacitor-transistor-logic) form shown in Fig. 5-14. Here, the base resistors have been

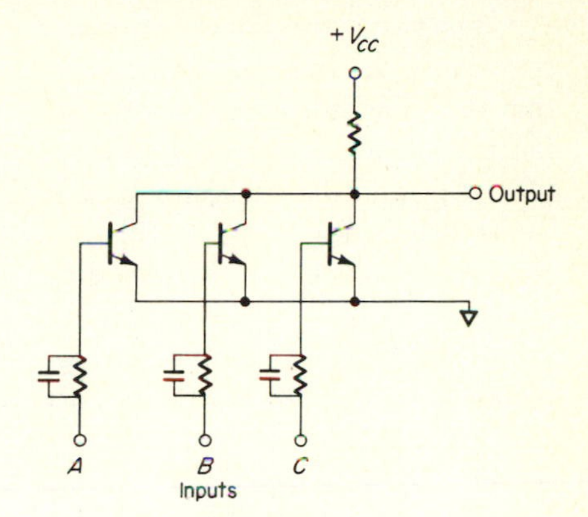

Fig. 5-14 Resistor-capacitor-transistor logic (RCTL) NOR gate.

paralleled by speed-up capacitors, so that fast rise and fall can be achieved, even with relatively large base resistors. Apart from a higher ratio of speed to power, RCTL is not significantly different from RTL and has not become widely accepted.

5-3-4 DTL

Another type of logic which was very popular as a discrete circuit and which was quickly translated into integrated form is DTL (diode-transistor logic). As shown in Fig. 5-15, this usually consists of an input-diode AND gate (D_1, D_2, D_3, and D_4) followed by a transistor inverter which results in a NOT-AND or NAND gate. If any of inputs *A*, *B*, *C*, and *D* are low (a logic 0), point *X* will be at approximately +0.7 V, and the transistor will be turned OFF because its V_{BE} will be less than 0.7 V owing to D_5 and D_6. The output is therefore high (a logic 1). However, if inputs *A*, *B*, *C*, *and D* are *all* high (logic 1), the four input diodes all turn OFF, and the values of the two base bias resistors are such that the transistor turns ON. Its output is therefore low (a logic 0) only when all inputs are high (1). From Table 5-1, this is the condition for a NAND gate.

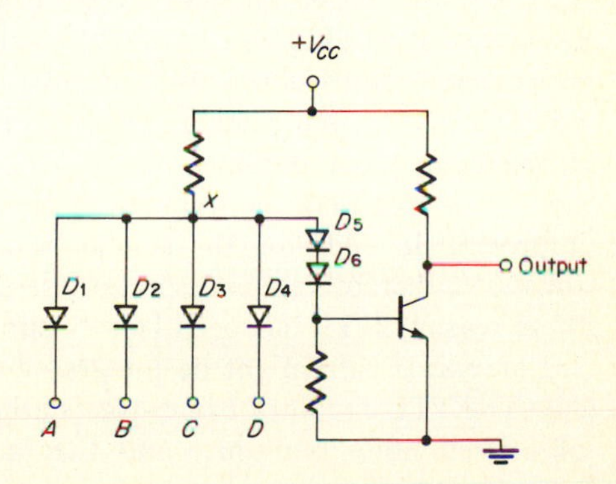

Fig. 5-15 Diode-transistor logic (DTL) NAND gate.

The two extra diodes, D_5 and D_6, serve two purposes. With any input low they cause the base of the transistor to be well below +0.7 V, which ensures that the transistor is solidly off. Also, all inputs must rise above +1.4 V before the transistor can turn ON, because point X in Fig. 5-15 is three diode drops (D_5, D_6, and the transistor base emitter) or $3 \times 0.7 = 2.1$ V above ground as the transistor turns ON. At this time the four input diodes are one diode drop closer to ground, that is, two diode drops or $2 \times 0.7 = 1.4$ V above ground. Thus the transition between the transistor being ON and OFF occurs at inputs of +1.4 V for the DTL circuit in Fig. 5-15, compared with about +0.7 V for the circuits in Figs. 5-12 to 5-14. This means that +1-V noise spikes would not give a false output from a DTL gate, whereas they might for DCTL, RTL, or RCTL.

DTL is well suited to monolithic integrated-circuit fabrication. No capacitors are required, and component values are not critical. Also, some monolithic versions of the DTL NAND gate in Fig. 5-15 replace D_5 with a transistor as in Table 5-5. This improves performance and yield because the extra transistor supplies the drive necessary to turn ON the output transistor, thereby reducing the current demand on the previous stage. Replacing a diode with a transistor in a discrete circuit would involve wiring one extra lead and the use of a more costly component (a transistor instead of a diode). Per gate, the cost would be small, but, for a large system, costs would multiply. Replacing a diode with a transistor in a monolithic circuit would be a very minor addition, would not increase costs, but instead would improve performance.

5-3-5 TTL

A form of logic that is related to and has evolved from DTL is shown in Fig. 5-16*a*. It is called *transistor-transistor logic*, which is abbreviated either to TTL or T^2L. In the DTL circuit of Fig. 5-15, D_4 and D_5 form an *np-pn* combination. In TTL this combination is replaced with an *npn* transistor. To obtain the multiple-input AND gate similar to the four-diode AND gate in the DTL circuit, TTL circuits use multiple emitters on the *npn* transistor. The logic action of the TTL circuit can be understood by comparison with the explanation for the DTL circuit. TTL is an excellent example of the application of monolithic integrated-circuit technology to improve what is basically a discrete-component type of circuit. Even when a DTL circuit is integrated, it cannot take advantage of the technique in the same way as TTL can. Since the multiemitter transistor is smaller in area than the number of diodes it replaces, the yield from a wafer is improved. In addition, the smaller area results in a lower capacitance to the substrate, thereby reducing circuit rise and fall times and increasing speed. For these reasons DTL has been largely superseded by TTL. Fig. 5-16b shows how a TTL NAND gate might be integrated.

The TTL circuit of Fig. 5-16*a* is almost never used in its basic form because of limited noise immunity and the fact that higher speeds and fan-out are

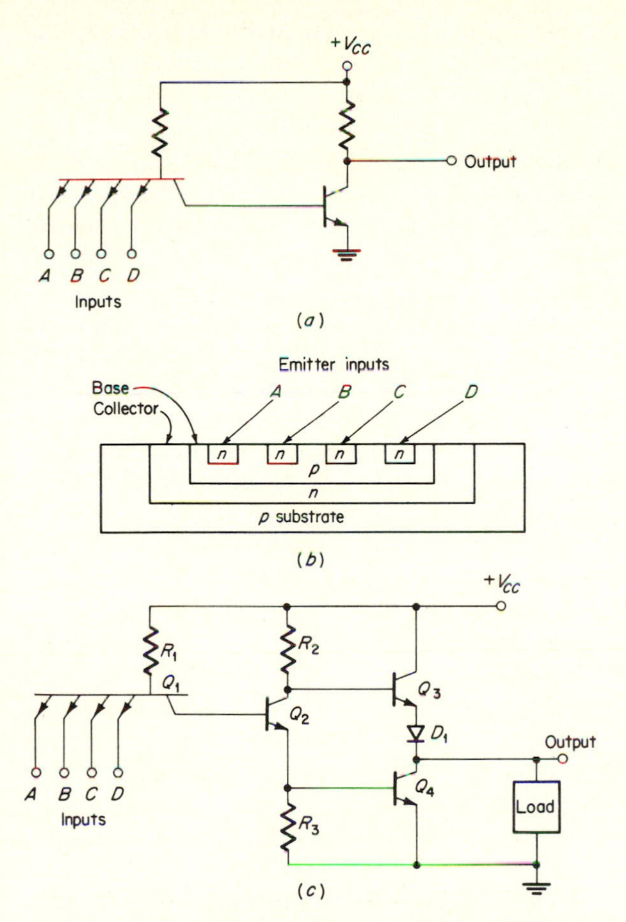

Fig. 5-16 (*a*) Transistor-transistor logic (TTL or T^2L) NAND gate; (*b*) integrated-circuit form of TTL gate; (*c*) TTL NAND gate with totem-pole output.

possible with a modified version such as that shown in Fig. 5-16*c*. The multi-emitter-transistor AND gate is still used, the principal difference being in the output circuit. TTL input circuits require higher drive currents than DTL. This is why TTL circuits usually have high-power output stages. Notice in Fig. 5-16*c* that most of the extra components are active elements, that is, diodes and transistors. This is in keeping with monolithic fabrication philosophy in which transistors or diodes are preferred to resistors because they are smaller, resulting in higher yields and lower costs.

The output circuit in Fig. 5-16*c* is called a *totem-pole* output, because the three output components Q_3, D_1, and Q_4 are stacked one on top of another in the manner of a totem pole. Circuit action is as follows. If any input *A*, *B*, *C*, or *D* is low, Q_1 turns ON and Q_1 collector potential is low. This turns OFF Q_2, grounding its emitter and the base of Q_4, so that Q_4 is also turned OFF. Under these circumstances, the net result is a circuit consisting of R_2, Q_3, D_1, and the load. Thus, the output is high; Q_3 acts as an emitter follower with R_2 as its base resistor and with the load and D_1 as its emitter impedance.

If *A, B, C, and D* in Fig. 5-16*c* are *all* high, the base and collector of Q_1 form a forward-biased diode, which permits current to flow from $+V_{CC}$ through R_1 to the base of Q_2, turning Q_2 ON. When Q_2 turns ON, the drop across R_3 is sufficient to forward bias the base-emitter diode of Q_4, turning it ON. When Q_4 is ON, its collector potential (the output) is nearly that of its emitter, so the output is low.

The function of D_1 in Fig. 5-16*c* is to prevent both Q_3 and Q_4 from being ON simultaneously. If both were ON at the same time, they would represent a very low impedance across the supply, which would draw excessive current and put large noise spikes on the supply line .

The reason for using the totem-pole arrangement in TTL circuits is that it provides both "pullup" and "pulldown." The circuits in Figs. 5-12 to 5-15 and 5-16*a* have pulldown only, and this arises as follows. All five circuits have a transistor with a collector resistor as the output stage. In addition, there will always be some output capacity from the output to ground. When the output transistor turns ON, it acts like a shorting switch and rapidly discharges any output capacity. Thus, the output is rapidly pulled down. However, when the output transistor turns OFF, it acts like an open switch, and the output capacity must charge to the logic high through the collector resistor. Thus the output is not pulled up.

By contrast, in Fig. 5-16*c*, Q_4 pulls the output down to a logic low, and Q_3, acting as an emitter follower with a low output impedance, pulls the output up to the logic high. Another way of stating this is to say that the totem-pole output presents a low impedance in both the high and low states. A consequence of this is that noise pickup is reduced, since the noise must couple through some impedance (such as interwiring capacity) and is divided down potentiometrically by the low totem-pole output impedance.

5-3-6 SCHOTTKY TTL

Standard TTL, like all the forms of logic discussed so far, uses transistors in a saturating mode. When a transistor in a TTL circuit is turned on, there is no clamping to prevent the collector voltage from becoming sufficiently close to the emitter voltage to allow the transistor to saturate. When the transistor is turned OFF, storage-time effects prevent the collector from rising immediately, thereby causing a delay between the turn-off signal on the base and the corresponding collector voltage swing. Gate speeds cannot therefore be optimized; in digital systems, where gates (and hence gate delays) are cascaded, this can be a significant problem.

Two types of bipolar-transistor logic have evolved to overcome the problem of delays caused by storage-time effects in turned-ON transistors. Both employ circuits which do not allow the transistors to become saturated. One is Schottky TTL, and the other is emitter-coupled logic or ECL (Sec. 5-3-7).

Schottky-TTL logic circuits are similar to the standard TTL circuit shown in Fig. 5-16*c*, except for the addition of one component, the Schottky diode, to every transistor that would otherwise saturate. The diode is connected from collector to base as shown in Fig. 5-17. When the collector of the transistor is high, the diode is reverse-biased and has no significant effect. However, when the collector is low, so that the transistor is ON, the Schottky diode has a very significant effect; this is due to two important properties of such diodes. First,

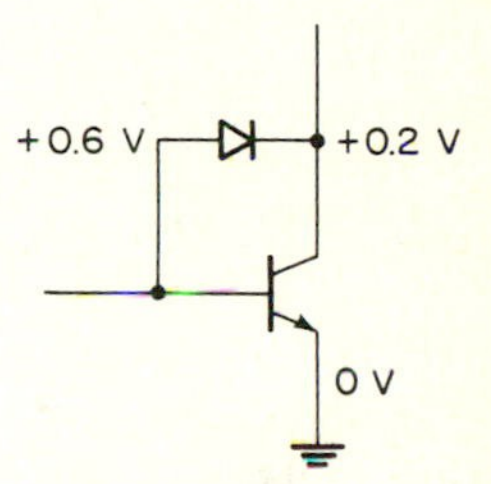

Fig. 5-17 Schottky-diode clamping to prevent transistor saturation.

the Schottky diode has a metal-to-silicon junction, not a silicon-to-silicon junction as in ordinary *pn*-junction diodes. There are therefore no minority carriers and no stored charges. (Stored charges are minority carriers in ordinary *pn*-junction diodes.) The absence of stored charges means that the diode can switch faster than diodes which do have stored charges. The second desirable property of the Schottky diode is that its forward voltage drop is less than that of a forward-biased silicon diode. With the transistor in Fig. 5-17 ON, its base will be at typically +0.6 V. Since the forward drop across the Schottky diode is only about 0.4 V, the collector can go no lower than, and is clamped at, +0.2 V. The transistor therefore can never saturate and can be turned OFF more rapidly than if it did saturate.

It is difficult to compare standard TTL with Schottky TTL numerically, because of the different types of each available, although the fact that Schottky is more expensive than standard is a significant factor. Schottky TTL comes, as does standard TTL, in two power levels—low and standard power. Low-power Schottky TTL requires one-fifth the power of standard TTL, yet its speed is comparable to that of standard TTL. Standard-power Schottky TTL takes the same power as standard TTL, but is two to three times faster. For moderately fast circuits up to about 50 MHz, standard TTL would probably have the edge for many requirements because of its lower cost. If low power were important, however, low-power Schottky TTL could be used. For faster circuits either standard-power Schottky TTL or emitter-coupled logic (ECL) would have to be used. ECL is discussed in the next section.

5-3-7 ECL

ECL, which is sometimes referred to as current-mode logic (CML), is explained with reference to the basic circuit of Fig. 5-18. This is a differential amplifier of the type discussed in Chap. 3. Before proceeding with the description of its operation as it applies to digital circuits, we should recall that this type of circuit is particularly suited to monolithic fabrication techniques. Logic levels are a function of resistor ratios, and the practical circuit (see Fig. 5-18*b*) has a high ratio of active to passive elements.

Suppose, for the circuit in Fig. 5-18*a*, that the input is at 0 V. Both emitters will be at -0.7 V, and the emitter current will be $4.3 \text{ V}/4.3 \text{ k}\Omega = 1$ mA. By symmetry this current will divide equally into Q_1 and Q_2, so that 0.5 mA flows

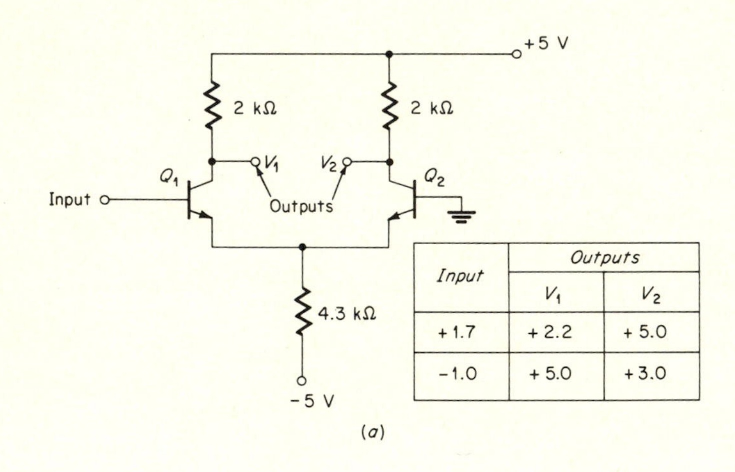

Input	Outputs	
	V_1	V_2
+1.7	+2.2	+5.0
−1.0	+5.0	+3.0

(*a*)

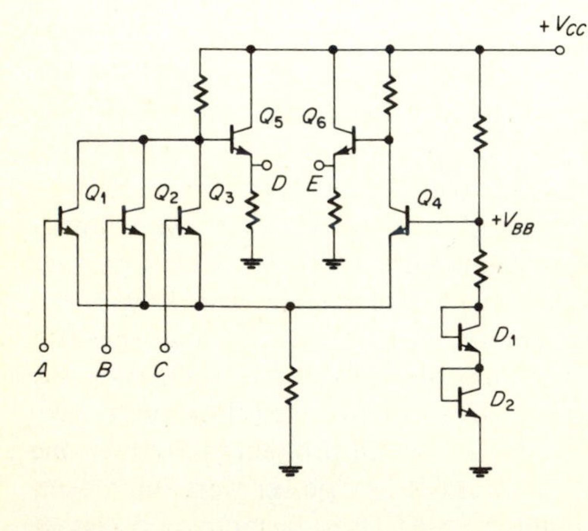

Inputs			Outputs	
			NOR	OR
A	B	C	D	$E = \overline{D}$
0	0	0	1	0
0	0	1	0	1
0	1	0	0	1
0	1	1	0	1
1	0	0	0	1
1	0	1	0	1
1	1	0	0	1
1	1	1	0	1

(*b*)

Fig. 5-18 (*a*) Differential amplifier used in ECL (CML) and voltage levels; (*b*) practical ECL (CML) NOR-OR gate and truth table.

in each 2-kΩ load, resulting in a voltage drop of 1 V. Each collector is therefore at 5 − 1 = +4 V. The situation described so far is for a differential pair used in a linear amplifier. For a digital application, the circuit is operated with the input either high or low. Suppose the high logic level is +1.7 V. The base of Q_1 will be at +1.7 V, and by follower action its emitter will be 0.7 V more negative than this, or +1 V. The emitter of Q_2 will also be at +1, V and Q_2 will be turned OFF, because its base is at 0 V (ground). Q_2 collector voltage will be +5 V, since there is no current to produce a drop across its collector load. Also, all the current from the common-emitter resistor will flow in Q_1 collector load. This current is 6 V/4.3 kΩ = 1.4 mA, producing a voltage drop of (2 kΩ) (1.4 mA) = 2.8 V across a 2-kΩ resistor. Q_1 collector voltage is therefore 5 − 2.8 = +2.2 V.

Now suppose that the input goes to the low logic level of −1.0 V. Since the common emitters always "follow" the most positive base (for *npn* transistors), the emitters assume a potential of −0.7 V. This is 0.7 V below Q_2 base which is at 0 V. Q_1 is therefore turned OFF and its collector voltage is +5.0 V. Q_2 now takes all the current supplied by the emitter resistor. This current is 4.3 V/4.3 kΩ = 1 mA, causing a 2-V drop across Q_2 collector resistor, so that Q_2 collector is at +3.0 V.

These numbers are summarized in Fig. 5-18*a* so that we can analyze their significance. First of all, Fig. 5-18*a* explains the terms ECL and CML. The circuit obviously consists of emitter-coupled transistors; hence the term *emitter-coupled logic* (ECL). Also, circuit operation is such that the *current* in the emitter resistor is switched from one transistor to the other, according to whether the input is high or low. This *current*-switching action accounts for the other name, *current-mode logic* (CML). The table of collector voltages in Fig. 5-18*a* reveals the primary advantage of ECL, which is that neither of the two transistors *saturates*. It is for this reason that ECL is the fastest type of logic available, with *propagation delay times* (PDT) of the order of a couple of nanoseconds. Standard TTL has PDTs of about four times that of ECL. Notice that the input swing is 2.7 V and that the collector swings are 2.8 and 2.0 V, respectively. This means that the circuit can be designed such that either output is adequate to drive other, similar circuits after appropriate level shifting. ECL also has the unique feature that *two* outputs are available, one of which is always the complement of the other. This provides additional flexibility to the logic-circuit designer.

ECL can be designed to have a logic function by the addition of transistors in parallel to the input transistor, as shown in Fig. 5-18*b*. Q_1, Q_2, and Q_3 are the logic input transistors, and Q_4 is the common reference transistor. Notice that the base of Q_4 is taken to a supply $+V_{BB}$ and that the common emitter resistor is grounded. This is in contrast to the circuit in Fig. 5.18*a* and is done to avoid the use of two power supplies. V_{BB} is derived from V_{CC} and is temperature-compensated by the use of diode-connected transistors D_1 and D_2. Emitter followers are used at the outputs to lower the output impedance and to provide a degree of level shifting. Since there are two outputs, D and E, each the complement of the other, the circuit functions as either a NOR gate or an OR gate using positive logic, depending on the output point selected. For example, a high on *A*, *B*, or *C* turns on Q_1, Q_2, or Q_3, making *D* low. This is the condition for a NOR gate. Similarly, a high on *A*, *B*, or *C* turns OFF Q_4, making *E* high. This is the condition for an OR gate. The truth table in Fig. 5.18*b* shows the two outputs for all possible input combinations.

5-3-8 I^2L

One of the more interesting types of bipolar logic is *integrated-injection logic* (I^2L). This is a form of logic from which resistors used for biasing and loading have been eliminated, except for one resistor used to bias a whole chip. Resistor elements require appreciable power and chip space, so their elimination results in higher-density circuits operating at reduced power. Where speed is essential, as in a large computer, chips containing 1000 or more I^2L gates can operate with less power than 100-gate TTL chips and at almost the same speed. Where high packing density is more important, as, for example, in a digital wristwatch, I^2L chips are capable of microwatt power dissipation, yet can provide high currents when necessary to drive light-emitting-diode displays.

An important feature of I^2L is that it is a *new circuit technique* rather than a new technology, which is in contrast to MOS logic when it was introduced. Production of I^2L is therefore relatively easy, since existing facilities for the manufacture of bipolar logic can be used.

I^2L results from rearranging the DCTL logic discussed in Sec. 5-3-1 by using a complementary transistor-equivalent circuit. We can follow the transition from DCTL to I^2L by referring to Fig. 5-19.

Figure 5-19*a* shows part of a logic system implemented with NAND gates, and Fig. 5-19*b* shows the circuit details when DCTL logic is used. Point *A* on Fig. 5-19*a* corresponds to point *A* on Fig. 5-19*b*, and so on. In Fig. 5-19*c*, one part has been taken from each of the three NAND gates of Fig. 5-19*b*, and the three parts, R_1, Q_1, and Q_2, have been combined into a single circuit. Notice that the transistors have common emitters and common bases so that Q_1 and Q_2 can be replaced with a single transistor (Q_3 in Fig. 5-19*d*) which has one emitter and one base. Q_3 is a dual-collector transistor so that two collectors are available as with Q_1 and Q_2. In Fig. 5-19*c*, resistor R_1 has two functions, one as a source of current for the bases of Q_1 and Q_2, and the other as a load resistor for previous stages. In Fig. 5-19*d*, *pnp* transistor Q_4 performs the same functions. One way in which the current-source function of R_1 might be implemented would be to connect a resistor from a standard +5-V power supply to the emitter of Q_4. This resistor would be common to many gates, and the current it *injects* into Q_4 would set the system operating current—high current for high speed, and low current for low power. The injection of current in this manner is the reason for the name integrated-injection logic.

Figure 5-19*d* reveals some of the main features of I^2L. There are no resistors in the circuit. (The single current-injection resistor is common to many circuits.)

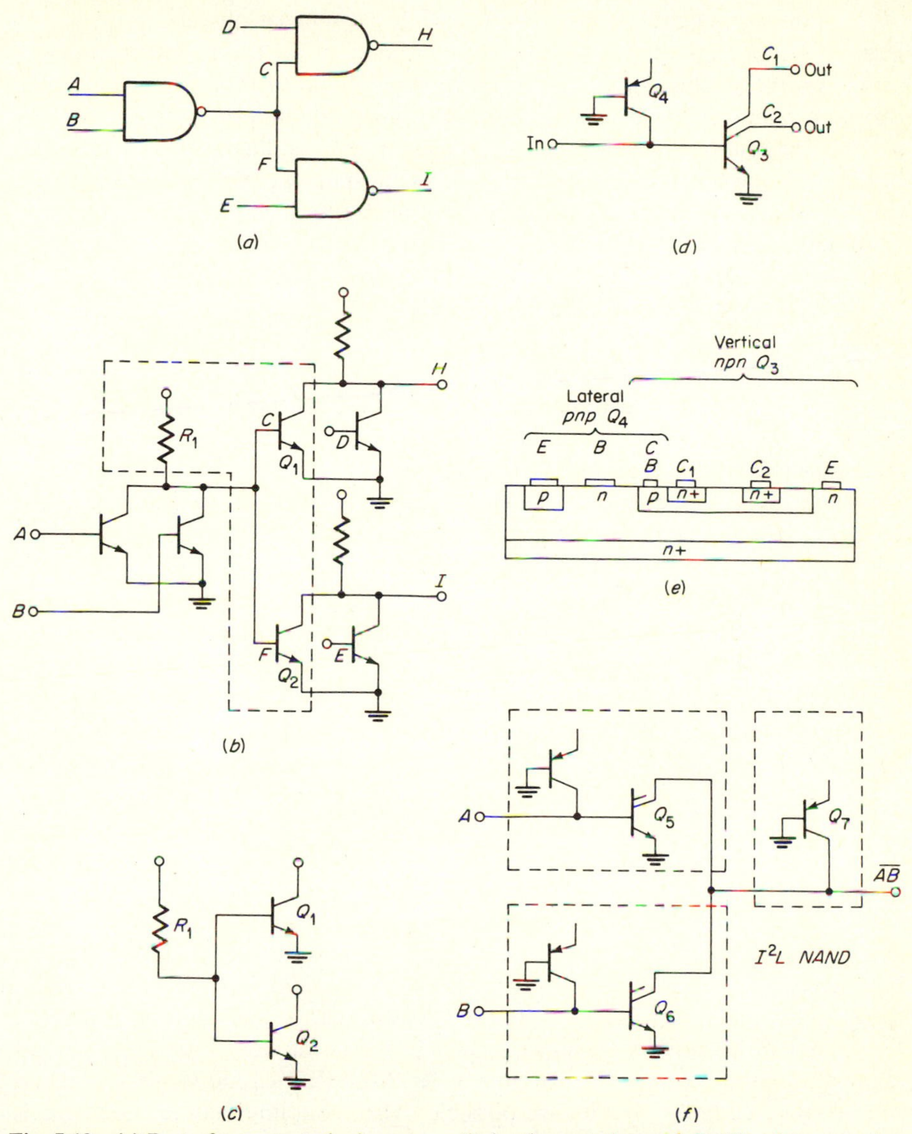

Fig. 5-19 (*a*) Part of NAND-gate logic system; (*b*) implementation with DCTL; (*c*) separation of DCTL elements; (*d*) I^2L equivalent of part *c*; (*e*) I^2L fabrication; (*f*) NAND gate using I^2L.

The circuit requires *npn* and *pnp* transistors on the same chip, but this is an established technique so that no new processing steps are required. Current hogging is not a problem as it was with DCTL, because the different bases and emitters are combined into a single base-emitter junction. Thus in Fig. 5-19*c*, Q_1 and Q_2 might draw different base currents from R_1, and one transistor might draw so much more than its share (owing to differences in Q_1 and Q_2) that the other transistor might not have enough base current to operate correctly. In Fig. 5-19*d*, Q_3 has only one base-emitter junction, so the problem does not arise. Another feature of I^2L is that it is very simple to fabricate, because Q_4 can be a lateral *pnp* transistor (*p* to *n* to *p* goes horizontal) and Q_4 can be a vertical *npn* (*n* to *p* to *n* goes vertical) transistor. This is illustrated in Fig. 5-19*e*. Notice that the base of the *pnp* transistor is the same *n* region as the emitter of the *npn* transistor, which is as required in Fig. 5-19*c* since the base and emitter are common. Also, the collector of the *pnp* transistor is the *p* region of the base of the *npn* transistor, which again is as desired. The existence of common regions and the absence of certain interconnection are some of the reasons why I^2L is easy to manufacture and accounts for the less commonly used name, *merged-transistor logic*.

Figure 5-19*d* also illustrates that I^2L is a new circuit technique since, unlike all other logic gates discussed so far, it has one logic input (excluding the current-bias injection required) and two logic outputs, although other I^2L gates might have more than two outputs through having more than two collectors on Q_3. In any case it has been usual to work with gates that have several inputs and one output. (ECL has two outputs, but one is merely the complement of the other.) However, the various NAND, NOR and flip-flop functions can be implemented by appropriate interconnection of the basic I^2L gates. In MSI/LSI applications, this interconnection is of course done on the chip. As an example, an I^2L NAND gate is shown in Fig. 5-19*f*. Q_5 and Q_6 are turned ON or OFF by A and B. Q_7 is the common collector load giving rise to the $\overline{AB}$ NAND output, as shown.

5-3-9 P AND NMOS LOGIC

All the logic circuits discussed so far have employed bipolar transistors as the amplifying element. It was explained in Sec. 2-6 that MOSFETs are very suitable for integrated-circuit logic. They can act as both amplifier and load resistor, thereby eliminating the need for ordinary load resistors, as is the case with I^2L. Compared with ordinary resistors, MOSFETs take up very little area, so that many more circuits can be formed on a wafer. MOSFET circuit yields are higher, so that more complex circuits are possible. Also, coupling between elements is simplified because it is direct.

There are two kinds of digital MOS transistor circuits. The first uses transistors of one polarity only. A gate or other logic element would therefore contain either all *p*-type or all *n*-type MOS transistors, but not both on the same chip. We can refer to these two subgroups as *p*-type or *n*-type MOS logic. The second kind of logic employs both *p*- and *n*-type MOS transistors on the same chip and is called *complementary MOS* (CMOS) logic. Although CMOS is slow compared with other logic families, it has its own considerable advantages and has emerged as one of the predominant families. We shall first discuss noncomplementary MOS logic, which contains either *n*- or *p*-type MOS transistors on a chip but not both.

Figure 5-20 shows a possible configuration for a *p*-type MOS NAND gate. Notice that there are no resistors in the circuit. Instead the gate consists of three MOS transistors as logic elements, with a fourth as a load resistor. Current hogging is not a problem as it is with DCTL, because the MOS inputs draw virtually no quiescent gate current, so that input resistors are not required, Enhancement-mode devices are preferred over depletion because the enhancement types can be turned ON or OFF with the supply voltage or common as inputs, so that dual-polarity supplies are not necessary. This is why JFETs, which are depletion devices, are not used in digital circuits. If $-V_{DD}$ in Fig. 5-20 is -12 V, then -12 V will turn the transistors ON and 0 V will turn them OFF. With positive logic 0 V would be a 1 and -12 V would be a 0, since 1 is assigned to the most positive voltage. If *A*, *B*, or *C* in Fig. 5-20 is at -12 V (a logic 0), one or more of the three input transistors will be ON, thereby having a low resistance from drain to source, causing the output to be nearly 0 V (a

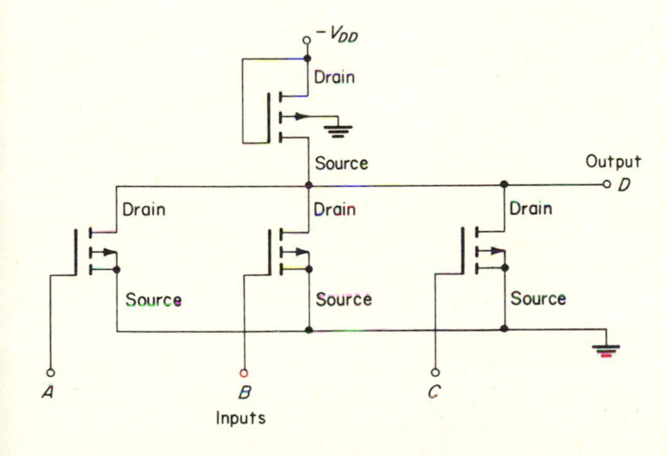

Inputs			*Output*
A	*B*	*C*	*D*
0	0	0	1
0	0	1	1
0	1	0	1
0	1	1	1
1	0	0	1
1	0	1	1
1	1	0	1
1	1	1	0

For positive logic: 1 = 0 V (Turns *p*-channel enhancement MOS off)
0 = -12 V (Turns *p*-channel enhancement MOS on)

Fig. 5-20 PMOS NAND gate and truth table.

logic 1). Only when *A*, *B*, and *C* are all at 0 V can the output be high. The truth table in Fig. 5-20 shows that the circuit is a positive-logic NAND gate (or a negative-logic NOR gate, since the two are identical).

5-3-10 CMOS LOGIC

Although MOS devices used in place of resistors increase circuit density, the load transistor still dissipates the same amount of power. Space is saved, but not power. When *both p and n-type* MOS transistors are used on the same chip, the two types of devices are arranged in a manner which results in a vast reduction in the amount of power consumed. This low power consumption is a primary advantage of CMOS digital circuits. How the power reduction is achieved is explained with reference to Fig. 5-21.

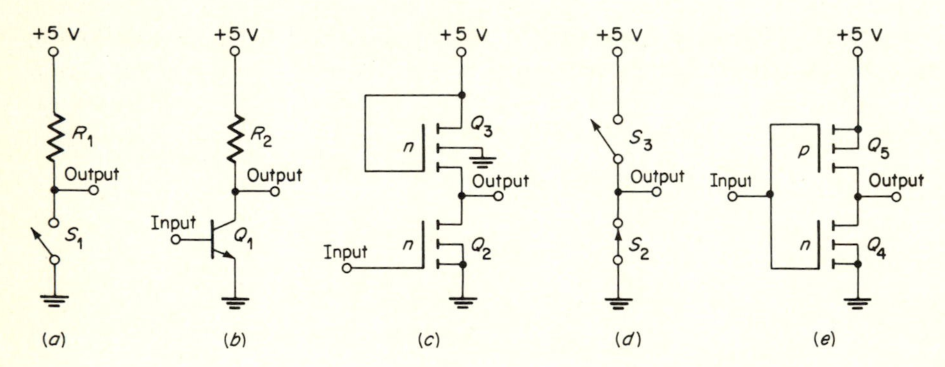

Fig. 5-21 Digital switches.

One purpose of any gate, flip-flop, or other element in a digital system is to produce an output-voltage change whose magnitude is known within certain limits, in response to some standardized input-voltage level change. In Fig. 5-21 the outputs for all five circuits will change from 0 to +5 V or from +5 V to 0, depending on the input. In Fig. 5-21*a* the switching is accomplished mechanically by opening or closing S_1. With S_1 open, the output is +5 V; with S_1 closed, 0 V. In Fig. 5-21*b* switch S_1 has been replaced with bipolar transistor Q_1, but the effect is the same. In Fig. 5-21*c* the bipolar transistor has been replaced with an *n*-channel enhancement-mode transistor Q_2, and the load resistor R_2 with a similar transistor Q_3, which is used as a resistor. Again, appropriate inputs produce 0- or +5-V outputs. Notice that circuits *a*, *b*, and *c* have a common feature. With S_1 closed, power is dissipated in R_1. With Q_1 ON, power is dissipated in R_2. With Q_2 ON, power is dissipated in Q_3. With S_1 open and Q_1 and Q_2 OFF, all three circuits consume negligible power.

The circuit of Fig. 5-21*d* may also be used to switch voltage levels. With S_2 open and S_3 closed, the output is at +5 V; and with S_2 closed and S_3 open, the output is at 0 V. The significant feature of the circuit is that *no power is* *consumed* with the output either high or low. The same is true of the complementary MOS (CMOS) circuit of Fig. 5-21*e*. With the input at 0 V, Q_4 is OFF and Q_5 is ON, so that the output is at +5 V. Since Q_4 is OFF, no current can flow from the plus supply to common, so no power is dissipated. With the input at +5 V, Q_4 is ON and Q_5 is OFF, so that the output is at 0 V. Again, since Q_5 is OFF, the circuit consumes no power except that due to leakage currents. However, when the gate switches from one level to another some power is consumed, because both transistors are partly on at the same time. For this reason the power used by CMOS is a function of frequency: the higher the frequency, the more power required. Because of duty-cycle effects it is impossible to give hard-and-fast rules, but at around 1 MHz, CMOS begins to lose its advantage in power economy over TTL.

Notice that the CMOS circuit of Fig. 5-21*e* is an inverter circuit, as are those of Fig. 5-21*b* and *c*.

Figure 5-22*a* shows how a CMOS NOR gate may be implemented. Q_1 and Q_2 are *n*-type MOS transistors, while Q_3 and Q_4 are *p* type. Inputs *A* and *B* switch between $+V_{DD}$ (logic 1) and ground (logic 0). For Q_1 and Q_2 to be ON,

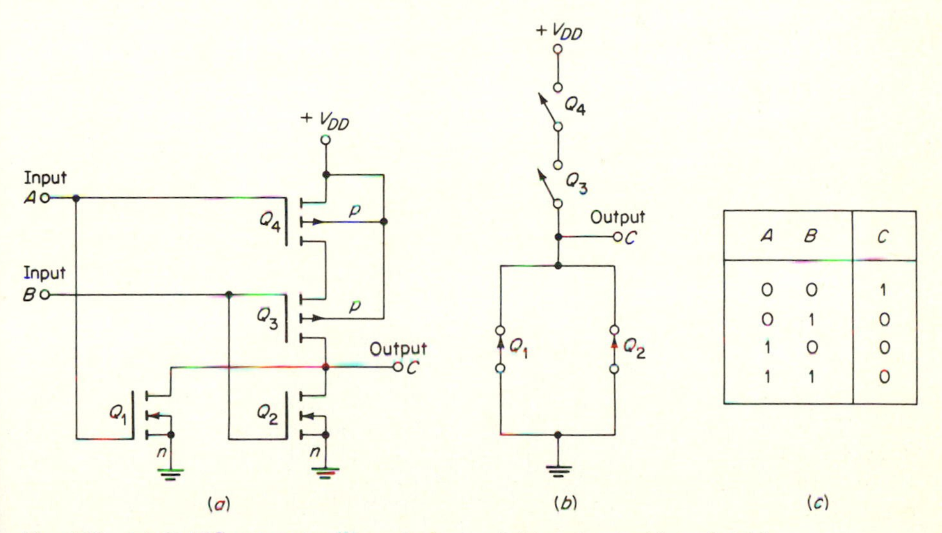

A	B	C
0	0	1
0	1	0
1	0	0
1	1	0

Fig. 5-22 (*a*) CMOS NOR gate; (*b*) equivalent switching circuit; (*c*) truth table.

that is, to act as closed switches as in Fig. 5-22*b*, *A* and *B* must have logic 1 inputs. Logic 1 inputs also result in Q_3 and Q_4 being OFF so that they act as open switches. Thus, if $A = B = 1$, then $C = 0$. Applying combinations of inputs to *A* and *B* produces switch settings that result in the truth table of Fig. 5-22*c*, which is that for a NOR gate. Notice that in all possible situations described by the truth table there is at least one open switch in the path from $+V$ to ground, so that the gate draws only leakage currents from the supply for any static state.

In addition to the advantage of requiring low power, CMOS is capable of operation over a wide range of supply voltages, typically 3 to 18 V, although some devices can operate at 1.1 V. The 3- to 18-V devices can be used at 5 V with other logic families, if appropriate interfacing circuits are used. They may also be used at 15 V in applications in which digital circuits are mixed with such analog devices as operational amplifiers powered from ± 15-V supplies. Since CMOS logic switches faster when powered from 15-V supplies than when the supply is $+5$ V, higher supply voltages are often used when speed is important.

So far we have mentioned the two main advantages of CMOS, namely, very low power requirements and high packing density for a given chip area. The high density has made possible complex functions in a single-chip device. We have also mentioned the main disadvantage of CMOS, that of being relatively slow. However, variations in basic fabrication technology have resulted in greatly improved speed performance. In silicon-on-sapphire MOS (SOS/MOS) devices, the process begins with an epitaxial layer of silicon being grown on a wafer of sapphire. The insulating properties of the sapphire have the effect of reducing the parasitic capacitances always found in integrated circuits. The result is an increase in speed of from 2 to 4 over standard CMOS.

5-3-11 THREE-STATE LOGIC

In many digital systems the wiring and package count can be considerably reduced by a technique called *wire-ORing*. Figure 5-23 shows two RTL NOR gates whose collectors have been shorted together, as indicated by the dotted line, to provide a single common output. A logic 1 on any or all of the four inputs results in a logic 0 output. The 2 two-input NOR gates have therefore been combined to produce a single four-input NOR gate. The important thing to observe in Fig. 5-23 is that neither Q_1 nor Q_2 will be damaged by any combination of input logic levels. This is not true of logic families which have active pullup as well as active pulldown. The CMOS NOR gate of Fig. 5-22*a* is an example of a circuit with active pullup and pulldown. If the outputs of two such gates are connected together, it is apparent from Fig. 5-22*b* that certain input

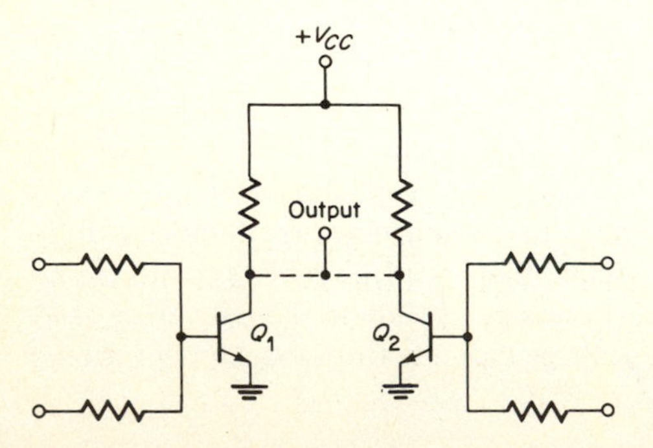

Fig. 5-23 RTL wire-ORed.

combinations will result in a short circuit, or at least a very-low-resistance path, across the power supply. In practice the shorting current will be limited by the ON resistances of the turned-on transistors or perhaps by current limiting in the power supply. Damage to the transistors may or may not occur, but power consumption would rise and the output logic level, instead of being at $+V_{DD}$ or ground, would rest at some undetermined point between the two, usually about halfway. Such a logic level is unusable.

These remarks apply to any logic family with active pullup and pulldown. Apart from CMOS, TTL is the most common of such families. These two families, like all the logic discussed so far, have two output states in which, for positive logic, a 0 means that the output is coupled via a low impedance to ground and a 1 means that the output is coupled via a low impedance to the positive supply. However, it is possible to redesign active pullup circuits so that they have a third state in addition to the usual high and low states. The purpose of having this third state is to allow wired-ORing of active pullup circuits. In this third state the output, instead of being coupled to $+V_{DD}$ or ground via a low impedance (or closed semiconductor switch), is coupled to both via high impedances (or open semiconductor switches). Gates with this third state may have their outputs wire-ORed, that is, shorted together, without damage since there is no short across the supply. In the other two states the active pullup and pulldown can operate as usual with the corresponding increase in speed that such circuits have over, for example, RTL with its passive pullup.

The truth table for a three-state inverter is given in Fig. 5-24*a*, and one CMOS implementation in Fig. 5-24*b*. The symbol is shown in Fig. 5-24*c*.

Q_2 and Q_3 in Fig. 5-24*b* form an inverter identical to that shown in Fig. 5-21*e*. As described earlier, connecting the outputs of two such gates together could cause excessive power dissipation, damage to the circuit, and unusable logic levels. The addition of Q_1 and Q_4 overcomes these problems. A suitable signal on the DISABLE input will open switches Q_1 and Q_4 so that the output is isolated from both power-supply lines by a high impedance. The inverter shown symbolically in Fig. 5-24*b* is necessary to provide the correct polarity of switching voltages. (It is shown as a symbol rather than in transistor form to simplify the explanation. It would be a conventional CMOS inverter, with one *p* and one *n* MOS transistor.)

From the truth table of Fig. 5-24*a*, with the ENABLE/DISABLE input low, the gate functions as a conventional inverter since Q_1 and Q_4 are always ON, regardless of the other input levels. With the ENABLE/DISABLE input high, the gate output impedance is always high, regardless of other inputs, and the output logic level will be determined by whatever logic level is on some other gate wire-ORed to the inverter of Fig. 5-24*b*. For example, 10 three-state inverters might have their outputs wire-ORed, that is, connected together. Nine would be disabled so that their outputs were high impedances, and one would be enabled so that it behaved as an ordinary inverter. The output voltage levels of all nine disabled gates would be the same as that of the single enabled one.

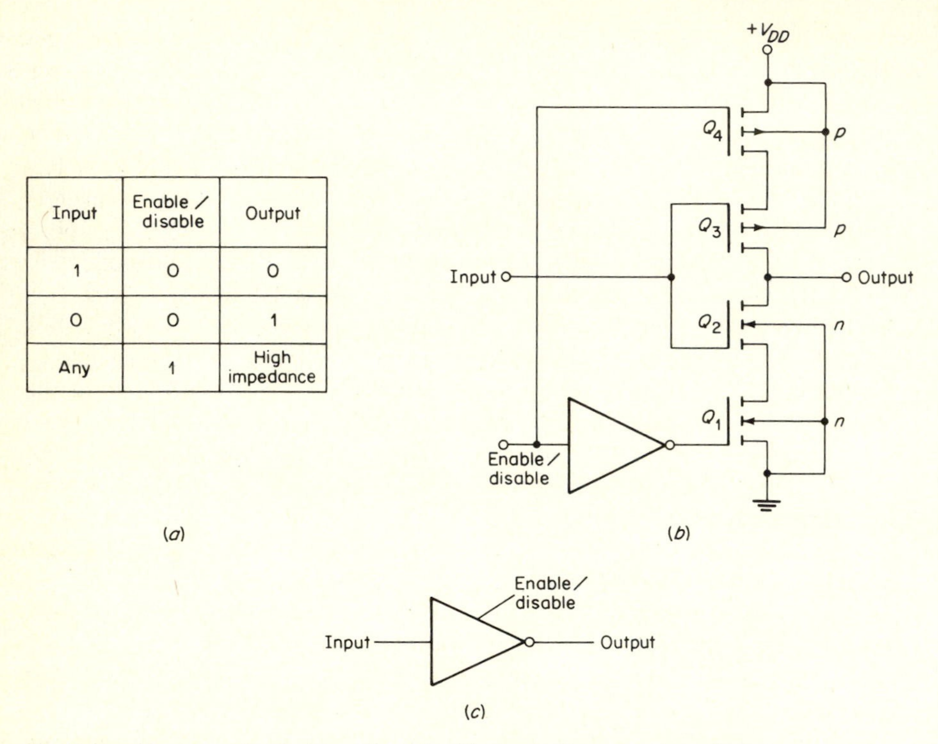

Input	Enable / disable	Output
1	0	0
0	0	1
Any	1	High impedance

Fig. 5-24 Three-state logic: (*a*) truth table; (*b*) circuit diagram; (*c*) symbol.

Notice that the term wire-ORing is used rather loosely. OR gates are not the only circuits that can be wired in this manner. NOR, NAND, and many other types of circuits can be wired similarly, but the result is still often referred to as wire-ORing. Three-state logic is used for multiplexing, data bussing, data routing, and memory expansion. It is also applicable to analog multiplexing.

Table 5-5 summarizes typical circuits for the most common types of logic and indicates some important features of each circuit.

Table 5-6 provides some *typical* performance data for the most commonly used logic lines. The word "typical" must be stressed here because all the numbers in Table 5-6 can vary, depending on the manufacturer and the specific circuit.

Speed can often be increased by reducing load-resistor values, since this in turn reduces circuit *RC* time constants. However, for a given supply voltage, each gate will draw more current so that speed is attained at the expense of power. Table 5-6 illustrates this tendency.

5-3-12 PRACTICAL CONSIDERATIONS

As with all electronic circuits, there are certain practical considerations that must be taken into account if digital logic families are to be applied successfully. For example, when working with TTL it is often stated that +5 V represents a

TABLE 5-5 Typical logic circuits

Type of logic	*Abbr.*	*Typical circuit*	*Remarks*
Direct-coupled transistor logic	DCTL	NOR	Not used much owing to current hogging
Resistor-transistor logic	RTL	NOR	Base resistors avoid current hogging, but reduce gate speed owing to *RC* input filter effect
Resistor-capacitor transistor logic	RCTL	NOR	Same circuit as RTL except for addition of speed-up capacitors; not good for monolithic *IC* since too many passive (*R* and *C*) components
Diode-transistor logic	DTL	NAND	Has reasonably high active-to-passive component ratio which is good for monolithic *IC*
Transistor-transistor logic	TTL or T^2L	NAND	Multiemitter transistor very suited to monolithic *IC*; very fast; totem-pole output has pullup and pulldown. Schottky TTL has fast clamping diodes to prevent transistor saturation. Result is faster circuits.
Current-mode or emitter-coupled logic	CML ECL	D=NOR E=OR	Differential circuit whose operation depends on parameter ratios, and which has high ratio of active-to-passive elements; therefore highly suited to monolithic *IC*. Very fast since transistors do not saturate. Two outputs available
PMOS transistor logic	MOSL	NAND	No resistors used so can be made very small; hence is adaptable to MSI and LSI. High noise margin owing to high threshold voltages

TABLE 5-5 (cont.)

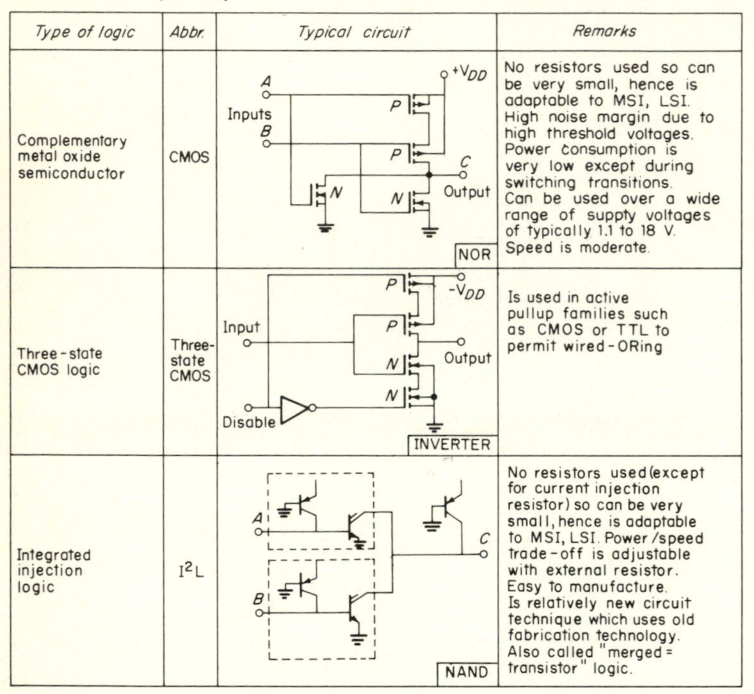

Type of logic	*Abbr.*	*Typical circuit*	*Remarks*
Complementary metal oxide semiconductor	CMOS	$+V_{DD}$, A, Inputs, B, P, P, N, N, C, Output — NOR	No resistors used so can be very small, hence is adaptable to MSI, LSI. High noise margin due to high threshold voltages. Power consumption is very low except during switching transitions. Can be used over a wide range of suppty voltages of typically 1.1 to 18 V. Speed is moderate.
Three-state CMOS logic	Three-state CMOS	P, $-V_{DD}$, Input, P, N, Output, N, Disable — INVERTER	Is used in active pullup families such as CMOS or TTL to permit wired-ORing
Integrated injection logic	I²L	A, B, C — NAND	No resistors used (except for current injection resistor) so can be very small, hence is adaptable to MSI, LSI. Power/speed trade-off is adjustable with external resistor. Easy to manufacture. Is relatively new circuit technique which uses old fabrication technology. Also called "merged = transistor" logic.

logic 1 and 0 V a logic 0, but in practice 1 and 0 are represented by *ranges* of voltage levels. These ranges are different for inputs and outputs to ensure that any output is always more than adequate to drive any input under worst-case conditions. For example, for TTL the *output* range is 0 to 0.4 V for a logic 0 and 2.4 to 5 V for a logic 1, while the *input* range is 0 to 0.8 V for a logic 0 and 2.0 to 5 V for a logic 1. Hence a worst-case output of +2.4 V is greater than the minimum of 2.0 V required to ensure a 1 input.

TABLE 5-6 Comparison of performance of various forms of IC logic. Numbers are typical only.

Type of Logic	Standby Power mW	Flip-flop Toggle Freq. MHz	Propagation Delay per Gate nS
RTL	5	8	12
DTL	8	10	30
Standard TTL	10	35	10
Low power Schottky TTL	2	70	5
CMOS at 5V	.00001*	8	30
CMOS at 15V	.00001*	10	20
SOS/MOS at 15V	.00001*	20	10
PMOS or NMOS	5	2	250
ECL	70	1000	1
I²L	Depend on injected current, see text		

* CMOS and SOS/MOS power dissipation approaches that of standard TTL at a few megahertz.

The high and low ranges define *a noise margin*, which is the difference between the worst-case input and output voltages for that level. For example, for the above TTL levels:

$$\text{The high (1) noise margin} = 2.4 - 2.0 = 0.4 \text{ V}$$

$$\text{The low (0) noise margin} = 0.8 - 0.4 = 0.4 \text{ V}$$

Similar considerations apply to all logic families, and the specific numbers must be extracted from the appropriate data sheet.

Many logic families, such as TTL, require large amounts of current, in the range of amperes, if many ICs are used. These currents, with large transients superimposed upon them, are often routed through long cables to reach the cards and racks in the system, resulting in large voltage spikes on the power lines. To avoid false triggering of gates and flip-flops by these spikes, it is important not only to use an adequately regulated supply, but also to employ local decoupling on each printed-circuit card. Often several separate decoupling networks are necessary on a single card.

Depending on the logic family used and the speeds required, special *interface circuits* called *line driver/receivers* must be employed to carry signals over distances of more than a few feet. Without these driver/receivers, time delays, reflections, and capacitive and inductive loading can degrade a system's noise performance. In the case of TTL these interface circuits convert TTL signals to or from other signals suitable for transmission lines. They are available in many forms. The simplest is a single-ended transmitter/receiver with one signal line and a common ground return. For lines greater than a few hundred feet in length, in noisy environments, balanced differential transmitter/receivers with a twisted pair of signal wires plus a common ground return are recommended. The twisted pair cancels electromagnetically induced noise currents, and much noise is common-moded out by the differential nature of the circuits.

Another problem occurs when interfacing different logic families, since one family may have inadequate voltage and current outputs to drive another. Or, as in the case of 15-V CMOS driving 5-V TTL, for example, the voltage drive may be too much while the current drive may be too little. Each case must be considered on its merits, and appropriate interfacing circuitry either bought or designed.

5-4 INTEGRATED-CIRCUIT MEMORIES

Many digital systems and all digital computers require memories. The purpose of a memory is to store both the data relevant to a problem and the instructions necessary for its solution. In addition, the memory must be so designed that access to the stored information is possible. Thus, a memory system is characterized by its *storage capacity* and its *access time*. A large storage capacity means that more data can be handled and more complex operations performed. Rapid access to the information in the memory means that complex problems can be solved in a reasonable time.

The need for records or data libraries requiring billions of bits of capacity is usually met by devices called *peripherals*, employing such magnetic recording devices as tapes, drums, and discs. The need for internal memories to hold and disperse information rapidly while calculations, compiling, and other data processing is going on is typically met by either core memories or integrated-circuit memories. IC memories therefore are used where access time must be fast while bit capacity is *relatively* small. For this application they must compete with core memories, not only on technical grounds but also for reasons of cost. Many types of memories are available to the digital systems designer. The most common are summarized in Tables 5-7 and 5-8, along with their most important characteristics.

Digital memories use elements that are capable of storing bits of information. In the binary system a bit may be a 1 or a 0, so a device with two stable states is required. Among the most commonly used are magnetic devices, which

TABLE 5-7 Comparison of various types of memory. Numbers are typical only

Type of Memory	Volatile	Erasable	Access Time, μsec	Bit Capacity	Remarks
Punched card	No	No	100/CARD	10^2	External storage Low speed
Paper tape	No	No	10^6	10^4	External storage Low speed
Magnetic core	No	Yes	1	10^5	Internal storage High speed
Magnetic drum	No	Yes	10	10^7	Internal storage Medium speed
Magnetic tape	No	Yes	10^6	10^8	External storage High capacity Low speed
Magnetic scroll tape	No	Yes	10^6	10^{11}	
Magnetic plated wire	No	Yes	1	10^5	Internal storage High speed
Magnetic film	No	Yes	0.5	10^5	Internal storage High speed
Magnetic disc (incl floppy disc)	No	Yes	10	10^7	Internal storage High speed
Tunnel diode	Yes	Yes	10^{-1}	10^4	High speed Temporary storage
Delay line	Yes	Yes	1	10^3	High speed, low capacity, temporary storage
Integrated circuit bipolar	Yes	Yes	10^{-2}	10^4	High speed, low capacity, temporary storage
Integrated circuit MOS	Yes	Yes	10^{-1} 3×10^{-2}	10^4	High speed, low capacity, temporary storage
Integrated circuit CMOS	Yes	Yes	10^{-2}	10^4	Low power, so standby batteries can be used to solve volatility problem.
Electrically alterable nonvolatile MNOS	No	Yes	0.5	10^3	Retains data with power off. No clocks required

have two states because they can be magnetized in either of two directions. As indicated in Table 5-7, several types of magnetic memory system are used, but the most popular small or medium-sized system is the magnetic core. This consists of a toroid made of ferromagnetic material, which can be magnetized by passing a current through a wire looped several times around the toroid. The core can be magnetized in either direction by reversing the direction of current flow through the wire. Each magnetic direction can be arbitrarily defined as a 1 or a 0, so that each core can store one bit of information. To read the information out of a core, that is, to find out if a 1 or a 0 is stored, involves some complications. A readout wire is passed through the core, but a readout current can be induced in the wire only by a changing magnetic field. However, since this changes the magnetic state, the stored information is no longer stored. For example, suppose the core is magnetized in the 1 direction. To read out, we could magnetize the core to the 0 direction, and a current would be induced in the readout wire, thereby indicating a 1. If the core had been orginally magnetized in the 0 direction, trying to magnetize in the zero direction would produce no change in flux and no readout current, thereby indicating a 0. But regardless of whether a 1 or a 0 has been originally stored, a 0 will be finally stored. Thus, the readout is *destructive* unless extra circuitry is added to remagnetize the core to its original state by sensing and storing the original data.

The cores used are extremely small—18 mils outside diameter is typical—so that many bits of memory can be packaged in a reasonable volume. In addition to improving production techniques to reduce costs, core manufacturers are taking advantage of integrated-circuit developments to minimize the number of parts used in the electronics associated with core memories. Thus, the often-predicted demise of the magnetic-core memory may still be a long way off. Since cost is always an important parameter to consider in the selection of any device, it is worth noting that the stringing of the cores is now the most costly item.

The fact that magnetized materials have two stable states is used in several other types of memory system. The magnetic material can be deposited on drums, tapes, disks, plated wires, thin films, etc. Drums have high storage capacity and serial or cyclic access. Tape reels have the highest storage capacity, but access is

TABLE 5-8 Summary of various types of integrated circuit memory

(*a*) Types of Memory Access	
Random access	Any address can be accessed at random, that is, without going through other addresses first. Data retrieval time is relatively fixed.
Serial access	Typified by shift register or charge-coupled-device (CCD) memories. Data retrieval is serial in a fixed order. All data ahead of required data must be read first.
(*b*) Static versus Dynamic Storage	
Static storage	Data is stored in flip-flops or other memory cells in which the data does not deteriorate with time.
Dynamic storage	Data is stored in leaky capacitors so that refresh circuitry is required to prevent data loss.

TABLE 5-8 (cont.)

(*c*) **Types of Integrated Circuit Memories**

Memory	Features	Access	Volatile*
Read/write†	Data can be read or written with equal ease. Used whenever data is needed fast and often, and is frequently changed.	Random or serial	Yes
Read-only† (ROM)	Data is stored permanently by a masking step during manufacture. Used whenever data is not to be changed as in fixed tables, constants or computer instruction sets.	Random	No
Programmable read-only (PROM)	Can be programmed after IC is manufactured. Data is written by opening fusible metal links or pn junctions with high currents. Can only be written into once. Same use as a ROM, but is cheaper in small quantities.	Random	No
Reprogrammable read-only	Are ROMs that can be written into many times. Are really "read-mostly" rather than "read-only" memories. Two main types: (a) Those in which writing is by voltage application and erasing (all data at once) by exposing chip to ultra-violet radiation through a window on the IC and (b) Those in which reading and writing is electrical. Data remains stored even with no power applied through the use of MNOS transistors. Used where frequent or at least more than one change in a program is required as in debugging a program.	Random	No
Content-addressable (CAM)	This extracts all data in an address when a part of the contents of that address match a specified number. Used in associative memories to obtain stored data related in some way to the input data.	Random	Yes
Charge-coupled-device (CCD)	Digital input is converted to charge and stepped through a shift register. Requires refresh circuitry to prevent data loss.	Serial	Yes
Programmable logic array (PLA)	Is a memory structure that is mask or field (FPLA) programmed as is a ROM. However, it implements logic functions. Outputs are functions of the input variables and the logic operation stored in the address. Is more flexible than random (hard-wired separate IC) circuits because PLAs and FPLAs are programmable.	Random	No

* Volatile means data in memory is lost when power is removed.
† Note that RAMs and ROMs are not two separate classifications. A ROM can also be a RAM. RAM does not refer to read/write memories only. However, in popular usage "RAM" is often used to designate a read/write memory and "ROM" a read only memory.

serial and relatively slow. Disks have high capacity, and access is rapid.

Many other types of memory exist, apart from those employing magnetic principles. Examples are punched cards, punched paper tape, tunnel diodes, delay lines, and integrated circuits. These are compared and summarized in Table 5-7. We shall now turn our attention to integrated-circuit memories.

As can be seen from Table 5-8, integrated-circuit memories can be classified in a number of ways. One relates to the method of *access*, which can be *serial* or *random* as indicated in Table 5-8*a*. In a serial-access memory, such as a reel of magnetic tape or an IC shift-register memory composed of flip-flops, data retrieval is serial and in a fixed order. All data ahead of the desired data must be read first. In contrast, in a random-access memory (RAM) any address (or memory location) can be accessed at random, without reading other data first.

Integrated-circuit memories may also be *static* or *dynamic* as in Table 5-8*b*. With static storage, information may be fed in, left for any length of time without further processing, and read out at some later time. The data are changed only when new information is fed in, or lost when the power is turned off. Dynamic storage uses capacitors, usually the gate capacitors of MOS transistors, to store bits of data, one bit per capacitor. Such memories are called dynamic since the capacitors must be periodically recharged to prevent data loss due to leakage. The recharging is called *refreshing*.

A type of integrated-circuit memory which can be static or dynamic is the shift register. A static shift-register memory consists of a number of flip-flops connected in series. Data are entered one bit at a time (serial entry) at the input of the chain of flip-flops, or all bits at one time (parallel entry). After entry of the data, the shift register can be left for any period of time without refreshing and with no loss of data. Readout may also be serial or parallel. A dynamic shift-register memory uses a number of MOS circuits in series, with gate capacitors again as the storage elements, with one capacitor needed per bit.

Figure 5-25*a* shows the typical organization of a RAM. Each square represents a flip-flop, which can contain one bit of information, a 1 or a 0. The 1 or 0 can be placed into the flip-flop (WRITE operation) or extracted from it (READ operation). To read or write, the particular flip-flop (into which the bit is to be written or from which the bit is to be read) must be located. This is done by means of an address. In Fig. 5-25*a* there are $N \times N$ or N^2 flip-flops; each must have its own unique address, so there must be N^2 addresses also. In a 1024-bit memory for example, $N = 32$ and there are 1024 flip-flops and 1024 addresses. It is impractical to run 1024 address lines into an IC RAM chip (even 28 pins is slightly unwieldy for a single IC). Fortunately, this is unnecessary, since a binary word 10 bits long can represent 1024 binary numbers and hence 1024 addresses. Figure 5-25*b* shows how 10 address lines may be split into five row address lines and five column address lines, where "row" and "cloumn" have the meanings indicated in Fig. 5-25*a*. Typical addresses might be as shown in Fig. 5-25*c*, where the row and column addresses have five bits each. In the arrangement shown, the flip-flop addressed is the one whose number equals the binary number represented by all 10 bits appropriately weighted plus 1. From Fig. 5-25*c*, we have:

Binary number	Decimal number	Flip-flop addressed
0000000000	0	$0 + 1 = 1$
0000000001	1	$1 + 1 = 2$
0111111111	511	$511 + 1 = 512$
1000000000	512	$512 + 1 = 513$
1111111111	1023	$1023 + 1 = 1024$

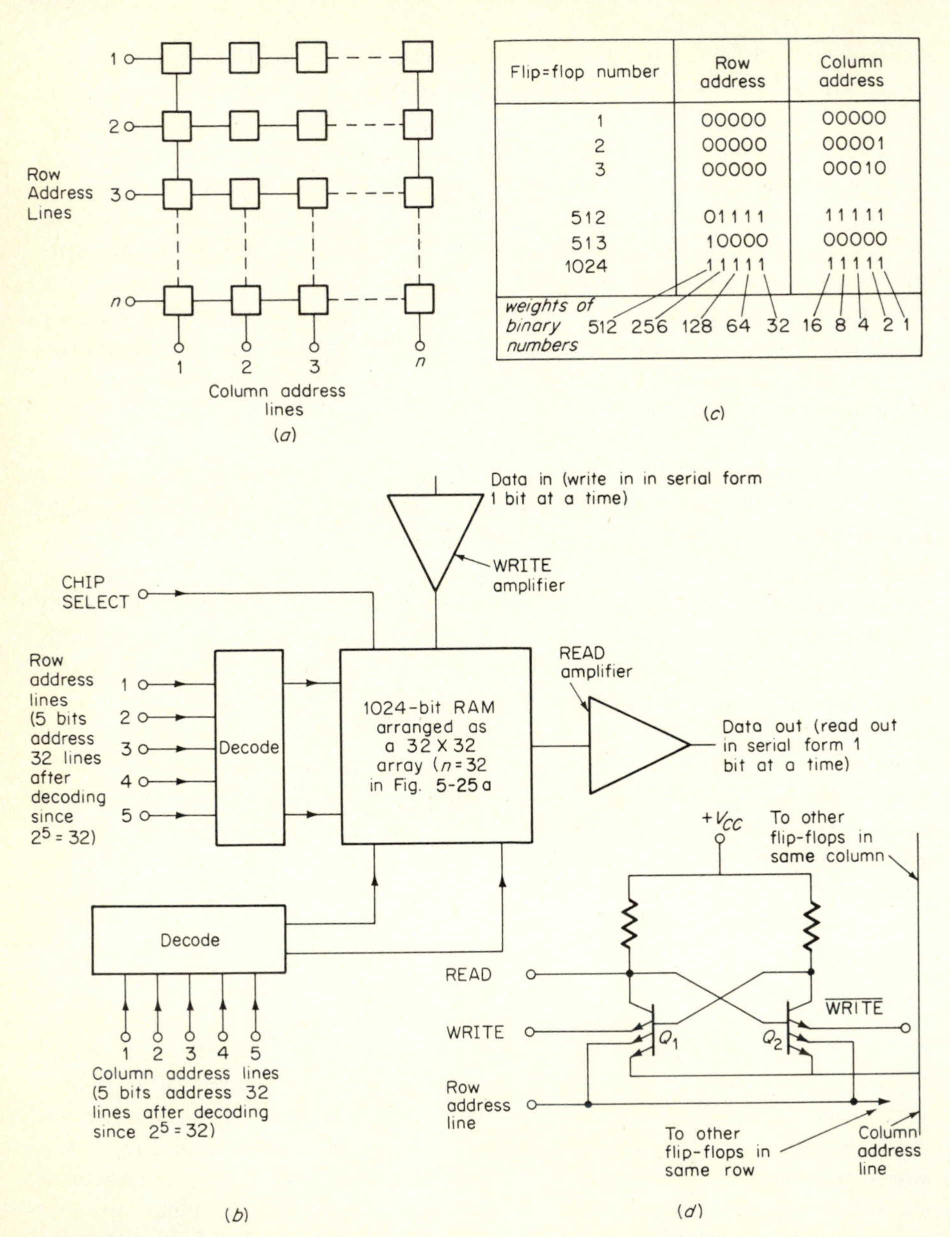

Fig. 5-25 (*a*) Arrangement of IC memory cells in an array; (*b*) 1024-bit RAM with coincidence selection; (*c*) examples of coincidence addressing; (*d*) flip-flop memory cell using multiemitter bipolar transistors.

When the particular flip-flop has been addressed, it remains first to write a 1 or 0 into it and later to read the bit from it. The bit is fed into the addressed flip-flop via a write amplifier which often has two outputs, a WRITE output and a $\overline{\text{WRITE}}$ output, to force the flip-flop to the desired state as in Fig. 5-25*d*. Data go in on the write line one bit at a time, and are taken out later via a read amplifier one bit at a time. Usually, it is not possible to write and read at the same time. Note the use of a multiemitter transistor in the memory cell in Fig. 5-25*d*, which is consistent with IC technology.

If a memory larger than 1024 bits is required, say 10,240, then ten 1024-bit memories can be used. Any one flip-flop out of the 10,240 can be addressed by first selecting the chip on which it lies using the CHIP SELECT input from Fig. 5-25*b*, and then selecting the flip-flop using the two 5-bit row and column addresses as before.

In most practical memories the addressing is arranged so that a single address will retrieve from the memory not just a single bit, but a whole word of perhaps 4 or 16 bits, but usually 8 bits, which is often called a *byte*. For example, to specify the memory locations of 256 eight-bit words, or 256 bytes, would only require 256 addresses, not $256 \times 8 = 2048$ addresses.

In addition to bipolar-transistor memory cells, such as were shown in Fig. 5-25*d*, MOSFETs can also be used as static cells for RAMs. Figure 5-26 shows an RS flip-flop using *p*-channel enhancement-mode transistors. Notice that no resistors are used, MOSFETs being employed as load resistors instead. The use of enhancement-mode devices permits direct coupling from drains to gates, thereby eliminating coupling resistors and capacitors. This flip-flop is *static* in the sense that, once set, it will store a bit indefinitely, so long as power is not removed or a set or reset pulse is not applied.

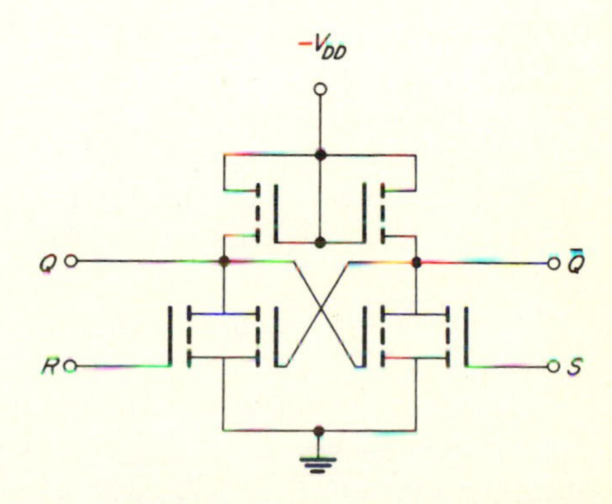

Fig. 5-26 MOSFET flip-flop for static storage.

Figure 5-27 shows how a MOSFET shift register can be used as a dynamic storage device. The storage is again achieved by using a MOSFET as a sample-and-hold device as in Fig. 5-27*a*. Information is set into the cell (sampled) by closing S_1 while an input voltage (bit) is present. When S_1 opens, C_1 will not discharge rapidly (held), owing to the high gate-to-channel resistance of the

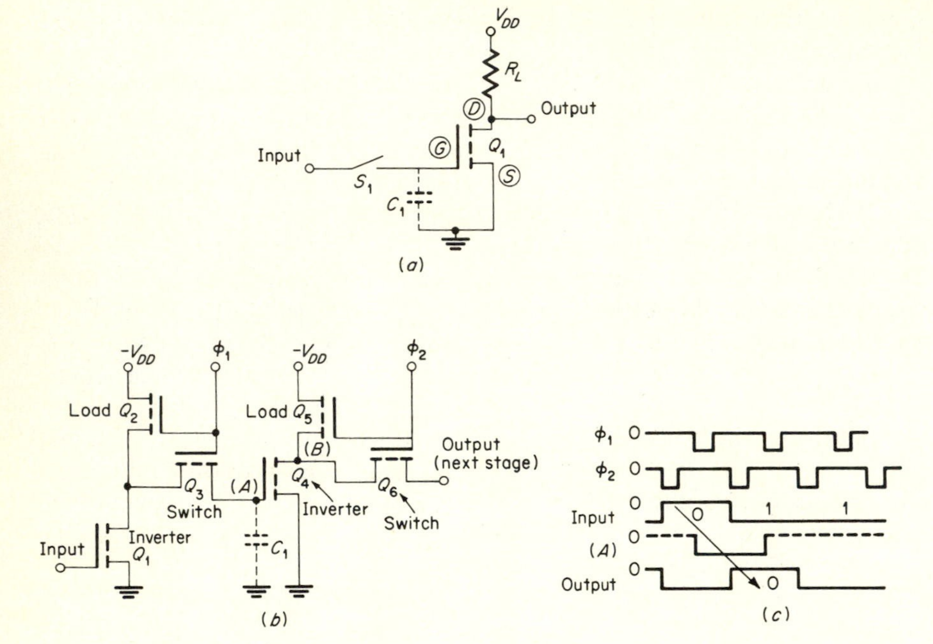

Fig. 5-27 (*a*) Principle of MOSFET dynamic storage; (*b*) one stage of a PMOS dynamic shift register; (*c*) waveforms for part *b*.

MOSFET, which approaches 10^{15} Ω. This dynamic storage capacitance can thus be used to construct very simple MOS memory cells as shown in Fig. 5-27*b* and *c*.

The operation of one stage of a dynamic PMOS shift register is as follows. Notice that there are two out-of-phase negative-going clock inputs. ϕ_1 clocks the signal through Q_3 to C_1, and ϕ_2 then shifts the information to the next stage. The negative ϕ_1 pulse simultaneously turns Q_2 and Q_3 ON, and ϕ_2 activates Q_5 and Q_6. Q_2 and Q_5 act as MOS load resistors, and Q_1 and Q_4 are inverters, their input capacity providing the temporary storage. Q_3 and Q_6 are bidirectional switches which allow data to shift through the cell when the clock pulse turns them ON. To illustrate the progression of data through the cell, consider the waveforms in Fig. 5-27*c*. Assume negative logic. When ϕ_1 goes negative, Q_2 and Q_3 are turned ON. Q_1 remains OFF because its gate-to-source voltage is not adequate for turn-ON with its gate at 0 V. The drain of Q_1 thus approaches $-V_{DD}$ (logic 1). This voltage is transferred through Q_3 to the input capacity C_1 of Q_4, where it is temporarily held even after ϕ_1 goes to zero. Then ϕ_2 goes negative, and Q_5 and Q_6 are turned ON. Since the gate of Q_4 is adequately negative with respect to its source, it is turned ON, providing a low-impedance path to ground from the drain. This pulls point *B* to ground, and hence a 0 is transferred through Q_6 to the next stage. Notice that the 0 has been shifted by one ϕ_2 time period, thus demonstrating the shift-register or delay capability of the circuit. The two-phase clock cannot be constructed from a simple multivibrator because of the necessity that ϕ_1 and ϕ_2 do not *change* at the same time.

A disadvantage of such dynamic storage is that the maximum storage time is generally less than 0.2 ms; therefore the input data must be replenished at a rate of 5 kHz. The storage time is limited owing the size of C_1 in Fig. 5.27*a* and the leakage resistance of an imperfect switch (S_1 is an MOS) and Q_1.

RAMs can also be dynamic in that they use capacitor storage and require refresh circuitry. Figure 5-28 shows a typical MOS dynamic RAM cell. Q_2 is the MOS transistor whose gate capacitance C is used to store charge. With Q_1 ON, the 0 or 1 on the write data line is stored. With Q_1 OFF and Q_3 ON, the 0 or 1 bit is read out into the read data line.

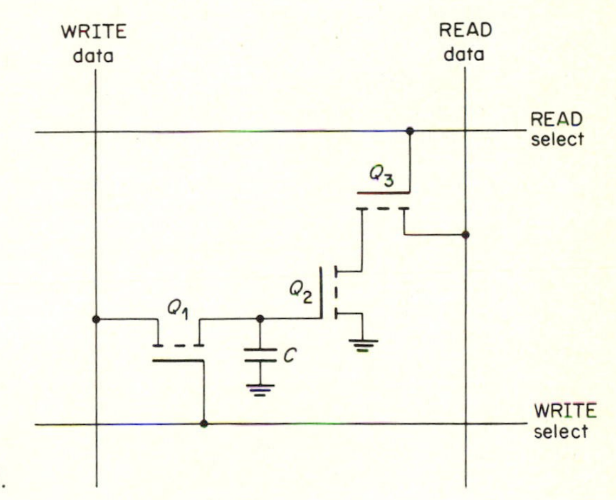

Fig. 5-28 Dynamic RAM cell.

Figure 5-29 is an interesting circuit because it takes advantage of both the temporary MOS input capacity storage *and* the flip-flop characteristics. During data storage ϕ_2 and ϕ_3 are negative, so that Q_2 and Q_3 are ON and the cross-connection paths are complete. During this time ϕ_1 is high so that Q_5 is OFF, thus isolating the memory cell from the rest of the circuit. When data are to be transferred, ϕ_2 and ϕ_3 go high and turn Q_2 and Q_3 OFF so that the cell halves are isolated. The output data are temporarily stored in the input capacity of Q_4. While Q_2 and Q_3 are still OFF, ϕ_1 goes negative long enough to allow entry of

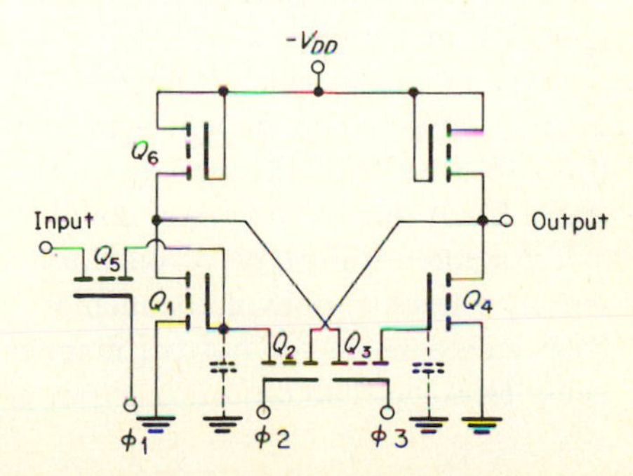

Fig. 5-29 Typical MOS static shift-register stage.

new data through Q_5, into the input capacity of Q_1. After ϕ_1 returns to zero, ϕ_2 turns on somewhat faster than ϕ_3 to ensure that data are shifted to the right.

Table 5-8*c* (page 215) lists the types of IC memory available. The first two listed are read/write and read-only. It is worth mentioning some confusion that has arisen in distinguishing between these two types of memory. Notice that *both* can be *random* access, so we would be justified in referring to either as a RAM. Moreover, it has become common to refer to read/write memories which have random access as RAMs, and read-only memories which have random access as ROMs. For example, microprocessors, to be discussed later in this chapter, are invariably designed to be used with what are referred to on data sheets as RAMs and ROMs, even though the ROMs are also RAMs. The solution is to interpet "RAM" as being a read/write random-access memory, and "ROM" as being a read-only random-access memory.

A ROM contains information that is permanently stored or at least requires special procedures to erase it, such as exposure to ultraviolet light. ROMs are used to store constants such as π and e, to store tables of data to be looked up some time during some program run, and to store instructions to be followed by a microprocessor or minicomputer.

The reprogrammable ROMs (PROMs) listed in Table 5-8*c* operate through the use of memory cells using MOS devices with floating gates. The gate is isolated in an insulating layer of silicon dioxide, where it can still control the device channel to be ON or OFF. It is charged by a process called *avalanche injection*, and since there is nowhere for the charge to go, the MOS transistor remains ON or OFF indefinitely, years in practice. However, illumination with ultraviolet light produces a photocurrent which allows the gate charge to be bled off and the data erased. The PROM is then ready for reprogramming.

ROMs are organized in a manner similar to read/write RAMs in that they consist of bit storage cells arranged in arrays such as that shown in Fig. 5-25*a*, so that random access is achieved by addressing of the appropriate bit lines. The bit storage cells are not flip-flops because it is not necessary to change the state of a particular cell in a ROM. Instead they can be circuits such as that in Fig. 5-30, which are prewired to output either a 1 or a 0.

The ROM can be prewired to the desired bit pattern either during the manufacturing process or by the customer. In the manufacturing process the ROM is programmed during fabrication by means of a mask having a customer-specified bit pattern. A customer who wishes to program a ROM must obtain the type of ROM which is intended for this application. Initially all bits are 1s; 1s are converted to 0s where appropriate by passing a high enough current through a fusible link, such as that shown in Fig. 5-30, so that that particular transistor is open and cannot conduct.

An interesting type of memory with some unusual properties is the electrically alterable, nonvolatile memory. Data can be written into and read out of such memories in the normal manner by correctly addressing the appropriate cells. However, unlike other types of electrically alterable cells, such as the

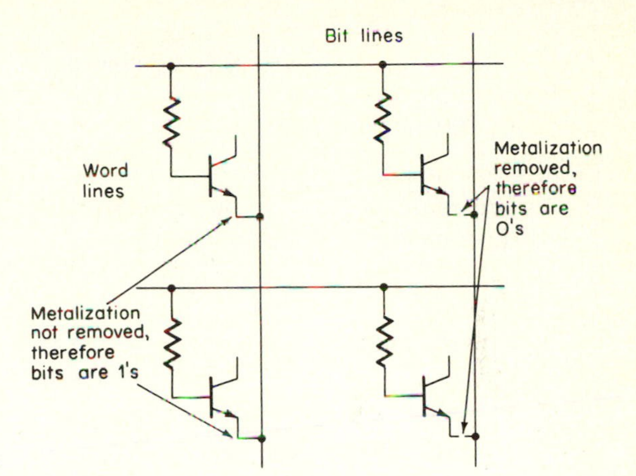

Fig. 5-30 Typical ROM cell.

flip-flop, data are not lost when power is removed. This is because the data are stored in metal nitride oxide silicon (MNOS) transistors, which suffer a shift in their turn-on voltage thresholds when a write pulse is applied. The shift is induced by the deposition of charges in energy traps in the nitride–silicon-dioxide gate insulation of the MNOS transistors. The data can be stored indefinitely with no power applied, so that long-term power consumption can be zero. Readout is nondestructive, and the data can be erased electrically if desired and new data entered. No refresh circuitry is required, as with dynamic memories.

As usual, the choice between integrated-circuit and other memories depends on cost, as well as such technical factors as performance and reliability. For reasons of cost it turns out that integrated circuits are most applicable for relatively small memories. For example, in large core memories the necessary associated driving and readout circuitry can be shared among many cores by appropriate design. For memories of 10^4 or fewer bits, the ratio of driving and readout circuitry to the number of cores is increased. This can be seen by considering the extreme and impractical example of a one-bit memory. Here, one driver and one readout circuit would be necessary. But if transistors are used as drivers and readout devices, they might as well be used as storage flip-flops, and the core would be unnecessary. Thus it is that flip-flops are more economical storage devices than cores for smaller memories.

Integrated-circuit read/write RAMs have the disadvantage that they are volatile since the stored information is lost if power is removed. (This is not the case with IC ROMs.) However, CMOS and I^2L circuits can be designed to run on such low power that a standby battery is feasible; this makes the memory effectively nonvolatile, so that such memories can compete with core memories on this score.

IC memories do have the advantage that they are compatible with integrated-circuit logic. It is easy to drive a memory cell made from a transistor flip-flop directly, while the same is not true of magnetic cores, where the logic and cores must be interfaced with special circuitry.

5-5 OTHER INTEGRATED-CIRCUIT SUBSYSTEMS

5-5-1 INTRODUCTION

In the previous subsections we covered basic digital theory and building blocks. In this subsection we shall discuss several of the more common (nonmemory) digital subsystems which are made up of combinations of the previously mentioned circuits. The discussion will progress from very simple subsystems such as the MS flip-flop (merely two flip-flops and two gates appropriately connected) to semicomplex functions such as counters, decoders, shift registers, multiplexers/demultiplexers, encoders/code converters, digital-to-analog (and vice versa) converters, plus the arithmetic operations of comparison, addition, subtraction, multiplication, and division.

Medium-scale integrated circuits (MSI) are introduced here because many of these subsystems are actually available on a single chip of silicon. Hence if we define MSI as a level of sophistication in which more than 12 and fewer than 100 equivalent gates are contained in one integrated-circuit package, we find that most of these subsystems have been integrated on a chip with an area of a few thousand square mils. For example, the advent of MSI has reduced the size of a complete decade counter by a factor of 5 or more, compared to the discrete-component or conventional individual flip-flop integrated-circuit counterpart. The cost is also reduced substantially and the reliability is increased due to the reduction of interconnections. [Section 5-6 covers an even greater level of digital sophistication, wherein an entire watch or calculator can be fabricated on one or two *large-scale integration* (LSI) chips.]

The progression from fabricating discrete diodes, transistors, resistors, etc., to complete digital subsystems on one chip is possible because the active devices in a digital integrated circuit are operated in an ON-OFF fashion, in contrast to a linear integrated circuit which must operate stably over a range of levels. Hence, it is not surprising that the yield of digital integrated circuits is better than that of linear integrated circuits. Also, there are only a few different types of circuits, even in the most complex digital computer. [In the linear world there is a much greater variety of different circuits (Chaps. 3 and 4).] Hence, once a digital-integrated-circuit manufacturer is capable of fabricating the basic digital building blocks with good yield, it is only natural to extend this capability to more complex arrays.

It is beyond the scope of this text to cover the multitude of digital subsystems, and the many different techniques available to accomplish the various digital operations. As an example, we shall discuss only the most popular digital counter techniques and concentrate on those available in MSI form.

5-5-2 COUNTERS

One of the functions most often required in almost any digital system is the ability to count. As discussed in Sec. 5-2, such counting is not done in the familiar decimal system but rather in the binary system of 1s and 0s, since this is more convenient to the flip-flops of Secs. 5-2 to 5-4. The simplest type of counter is the binary *ripple* counter shown in Fig. 5-6. That counter contains three cascaded flip-flops, with the output of each flip-flop triggering the next, with a total count capacity of 0 to 7. There are many variations of this basic binary counter, such as elimination of the cumulative ripple delay of each flip-flop by applying the clock (synchronously) to all the trigger inputs simultaneously. Also, the maximum count can be extended to 15, for instance, by adding one more flip-flop. Certain decimal numbers can also be skipped by adding appropriate gating to eliminate particular states from occurring. The total number of allowed states is defined as the *modulus* of the counter. For example, one that is capable of counts 0, 1, 2, 3, 4, 5, 6, and 7 is a *mod-8* counter as in Fig. 5-6. A similar counter with one disallowed state (say it skips 4) is a mod-7 counter.

The *decade counter* has a base or modulus of 10. The basic binary counter of Fig. 5-6 can be converted to a decade counter by adding one more flip-flop and eliminating 6 of the 16 possible states. Any six states can be eliminated, and the elimination can be accomplished by several different methods. As an example let us use the BCD code where each decimal digit is represented by four binary digits (Sec. 5-2). One of the most popular methods of eliminating states is to leave off the last six counts so that the remaining 10 counts have the same weighting as the binary counter of Fig. 5-6. Referring to the waveforms and truth table of Fig. 5-31, we see that to limit the binary counter to a total of 10 counts we must stop at the count of 9 instead of advancing to 10. The status of the flip-flops at the counts of 9 and 10 would normally be

Count	(8) D	(4) C	(2) B	(1) A
9	1	0	0	1
10	1	0	1	0

If we are to return the counter to 0 after the count of 9, we must

1. Prevent flip-flop B from setting to 1.
2. Reset D to 0.

The basic decade counter is similar to the binary counter of Fig. 5-6 except for: the extra stage, the fact that *master-slave* (MS) RST flip-flops are used to prevent *race problems* (Fig. 5-32), and the fact that interconnection changes are made to stop the count at 9. With reference to Fig. 5-31*a*, the operation of the decade counter is as follows. When the counter reaches 9, that is,

D	C	B	A
1	0	0	1

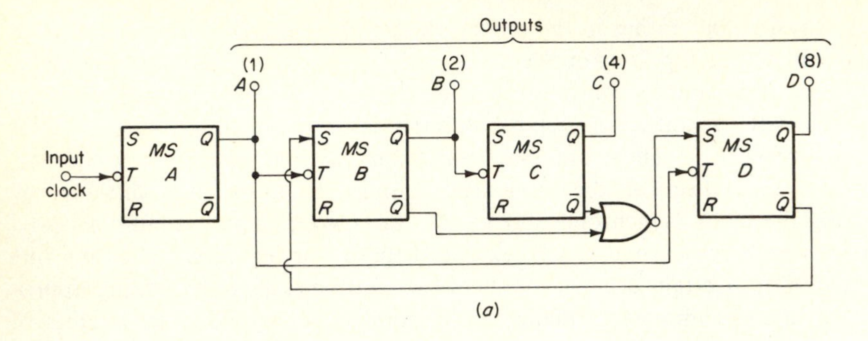

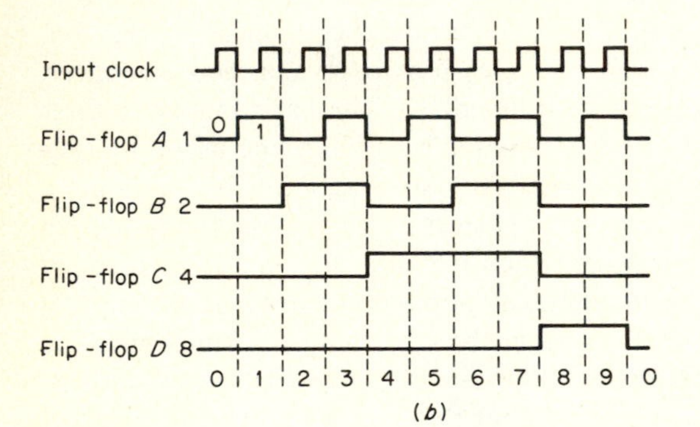

Count	Flip-flops			
	D (8)	C (4)	B (2)	A (1)
0	0	0	0	0
1	0	0	0	1
2	0	0	1	0
3	0	0	1	1
4	0	1	0	0
5	0	1	0	1
6	0	1	1	0
7	0	1	1	1
8	1	0	0	0
9	1	0	0	1
0	0	0	0	0

(c)

Fig. 5-31 One method of constructing a decade counter: (*a*) block diagram; (*b*) waveforms; (*c*) truth table.

$\bar{Q}_D = 0$, and this is fed back to the set input of flip-flop *B*. Since flip-flop *B* is 0, our first requirement is satisfied. At the count of 9, $\bar{Q}_B$ and $\bar{Q}_C = 1$ and the NOR-gate output is thus 0. The NOR gate is connected to the set input of flip-flop *D*, and, since R_D is normally high, flip-flop *D* is reset to 0 after the count of 9.

The waveforms of Fig. 5-31*b* demonstrate the inherent frequency-division capability of digital counters. That is, the input or clock frequency is divided in half at each successive flip-flop. (Figures 5-51 and 5-56 show counters in frequency-division and voltage-to-time conversion applications). Reference 2 demonstrates an extremely versatile MSI *variable-module* counter which can be operated in a variety of counting or frequency-division modes, depending on the "programming" of two mode-select inputs.

Figure 5-31 serves as an excellent example of a digital circuit which would be susceptible to the race problem if MS flip-flops were not employed. "Racing" occurs when two or more inputs to a gate are applied almost instantaneously, as to the NAND gate in Fig. 5-32*a*. Notice that we say almost, because any event must take a finite time to happen. In this case the times are the rise and fall times shown in Fig. 5-32*b*. At time t_1, *A* is 0 and *B* is 1, so *C* must be a 1 for

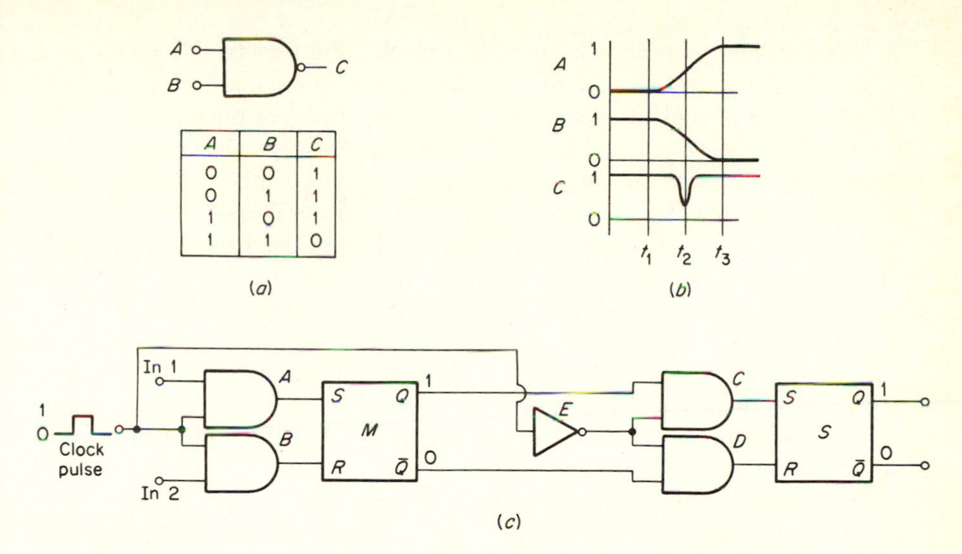

A	B	C
0	0	1
0	1	1
1	0	1
1	1	0

Fig. 5-32 Master-slave flip-flop. (*a*) NAND gate and truth table; (*b*) waveforms showing race problem; (*c*) master-slave flip-flop circuit.

NAND-gate action. At time t_3, A is 1 and B is 0, so again C is a 1. However, at time t_2, A and B are neither high nor low and C is indeterminate, because the output of a NAND gate is not defined for this condition. The waveform of C will depend on which of A or B wins the "race" to change its level. If the output C were to drive a flip-flop, the "glitch" in the waveform for C in Fig. 5-32*b* could cause false triggering. The false triggering can be avoided by combining two flip-flops in such a way that racing cannot occur. One flip-flop is called the *master* and the other the *slave*, and the combination is referred to as a *master-slave flip-flop*. This circuit is shown in Fig. 5-32*c*. First we shall consider how the circuit works and then see how it helps solve the race problem. Circuit operation is best understood by considering the conditions that exist before the clock pulse (clock input a 0), during the clock pulse (clock input a 1), and after the clock pulse (clock input a 0 again). The circuit consists of a master flip-flop M fed from two AND gates A and B, a slave flip-flop S fed from two AND gates C and D, and an inverter E.

When the clock pulse is 0, both A and B outputs will be 0, regardless of the logic levels of inputs 1 and 2, because *both* inputs of an AND gate must be 1 for the output to be a 1. A and B are therefore *disabled*. Thus, inputs 1 and 2 can be racing, and the state of the master flip-flop will be unchanged. Suppose the master flip-flop output is as shown in Fig. 5-32*c*, that is, $Q = 1$ and $\bar{Q} = 0$. AND gate C is *enabled by the* 1 output of inverter E and 1s on both inputs. Its output, and therefore the set input of the slave, is thus 1. D output and the slave reset are both 0. Let us assume that the slave outputs are as shown in Fig. 5-32*c*, that is, $Q = 1$ and $\bar{Q} = 0$.

Now consider what happens when the clock pulse goes to a 1. First of all, the output of inverter E is a 0 and AND gates C and D are disabled, so that the SET and RESET inputs of the slave cannot change. Therefore the slave Q and $\bar{Q}$ cannot change. On the other hand, AND gates A and B are *enabled.* If input 1 is a 1 and input 2 is a 0, the master SET and RESET inputs will be 1 and 0. If input 1 is a 0 and input 2 is 1, the master SET and RESET will be 0 and 1.

Summarizing so far, we can say that with the clock pulse at a 0, the slave will be set or reset, depending on the state of the master, but the master cannot be changed. With the clock pulse a 1, the slave cannot be changed but the master can.

Now let the clock go to a 0 again. Once again, the master is *disabled* and its contents are transferred to the slave, which is *enabled.*

The important thing to notice from all this is that the output of the MS flip-flop cannot change state until after the clock pulse has gone low.

The decoded output of a single *decade counter unit* (DCU) represents only one decimal digit (varying from 0 through 9); therefore additional DCU stages must be cascaded to increase the capacity to numbers greater than 9, i.e., tens and hundreds. A typical three-digit DCU capable of counting to 999 is shown in Fig. 5-33. Notice that a carry line connects the highest-order BCD output of each DCU to each succeeding higher-power DCU. The carry is necessary to advance the count from one power to the next; i.e., after the units DCU has progressed through the count of 9, the tens DCU must start at 0001.

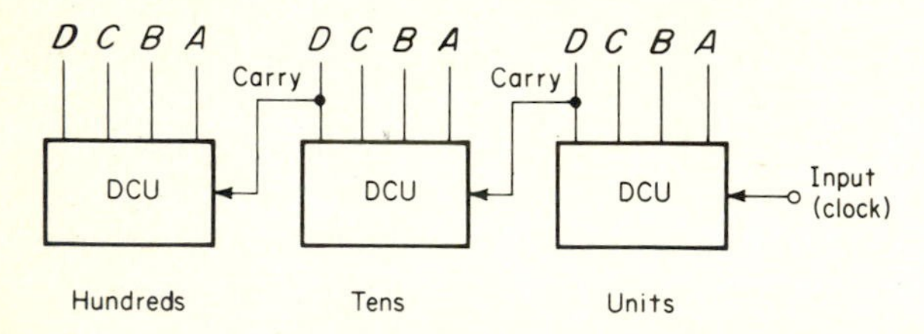

Fig. 5-33 Three cascaded decade counter units (DCU) capable of a maximum count of 999.

So far, we have discussed only serial or *asynchronous* counters in which the signal "ripples" through each flip-flop, resulting in an inherently low speed. This disadvantage can be overcome by a parallel or *synchronous* counter in which all the flip-flops are triggered simultaneously. (Another important advantage of synchronous versus asynchronous counting is the elimination of troublesome decoding spikes since all the flip-flops change state simultaneously.)

We have only discussed counters which count up, i.e., from zero toward some higher number. However, it is often desirable to count in either direction —up or down. The implementation of an *up-down counter* is rather simple because a normal up counter can be converted to a down counter by triggering each flip-flop from the $\bar{Q}$ side of the previous flip-flop instead of the Q side. Typical MSI synchronous up-down BCD counters have an up-down control input which determines whether the flip-flops are interconnected via their Q or $\bar{Q}$ outputs. (Figure 5-49 shows an up-down counter application in an analog-to-digital converter.)

5-5-3 SHIFT REGISTERS

The next digital subsystem topic is the shift register. Its circuit is similar to that of the multiple flip-flop counters of Figs. 5-6 and 5-31, but its function is somewhat different. Registers are a very important part of most digital systems. They are used to temporarily store binary information, especially before or after conversion or encoding/decoding operations. They allow a simple means of converting from serial to parallel (or vice versa) formats. Registers are also fundamental to the basic arithmetic operations such as mutliplication, division, complementation, and analog-to-digital (and vice versa) conversion, as shown in Secs. 5-5-7 and 5-5-8. Since a flip-flop can store only one binary number (1 or 0), a shift register must contain one flip-flop for each bit of a binary number. There must also be some provision for entering (shifting) the binary number into, out of, and from one stage to another of the shift register. This brings up the two general methods of shifting data into shift registers, serial and parallel. The first involves serially shifting one bit at a time into the register, whereas all the data are entered at the same time in a parallel shift register. (The more universal MSI shift registers are designed so that data can be entered serially or in parallel so that the register can be used for a variety of sequential operations.)

Figure 5-34 demonstrates the basic serial shift-register technique. This is a four-stage (four-bit) serial shift register. Notice that each Q output is directly coupled to the J input of the adjacent flip-flop, and the triggering inputs are all in parallel to provide simultaneous shifting of data from A to B, etc. After all flip-flops are initially reset to 0, the J and K inputs of the first flip-flop (A) determine whether a 1 or a 0 is inserted into the register. At each clock pulse the state ($Q = 0$ or 1) of flip-flop A is shifted to the next-higher-order stage (B), and so on until the four-bit word is entered in the register. Four more clock pulses will then shift the word completely out of the register.

Figure 5-35 shows a versatile MSI four-bit shift register and some of the functions it can perform. (During any one of these operations all other modes must be inhibited.)

The operation of the MSI shift register of Fig. 5-35*a* is as follows. The flip-flops are TTL clocked RS multivibrators, and positive logic is used. The mode-control input determines whether the outputs of the flip-flops are serially coupled to the inputs of the succeeding flip-flops or whether these connections are inhibited and the parallel inputs are enabled. When a 0 is applied to the mode control, all the X AND gates are enabled, and their outputs will go high once the associated flip-flop Q outputs go high. The Y AND gates are disabled. In this mode the Q output of each flip-flop is coupled to the input of the succeeding flip-flop, as shown by the logic states and signal flow arrow in Fig. 5-35*a*.

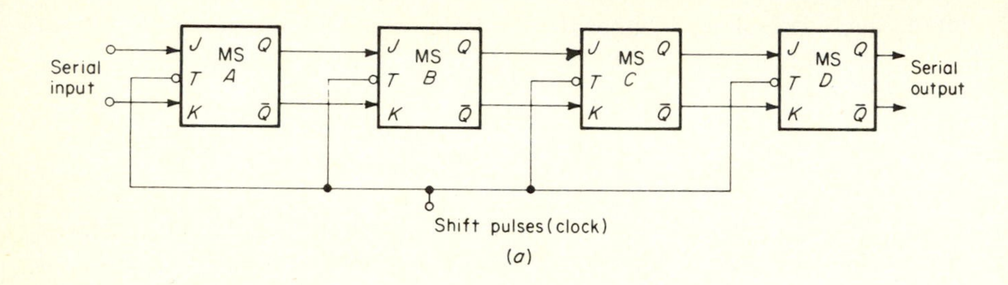

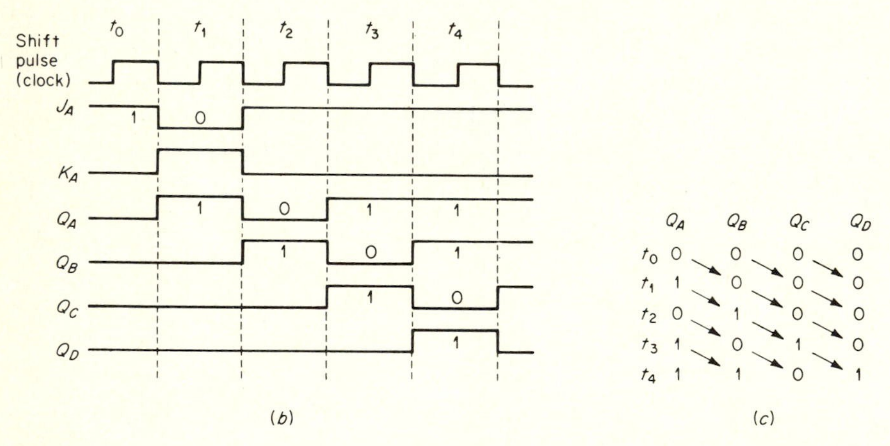

Fig. 5-34 Basic serial shift register: (*a*) block diagram; (*b*) waveforms; (*c*) truth table showing data progression.

The serial data are then entered at the serial input and shifted right when the clock pulses are applied to clock 1. Notice that the disabled *Y* AND gates inhibit the parallel inputs (and the clock AND gates inhibit clock 2). When a 1 is applied to the mode control, all the *Y* AND gates are enabled, while the *X* gates are disabled. In this condition the succeeding flip-flop output-input connection is inhibited, and data can flow through the parallel inputs and clock 2. One of two functions can be performed in this mode. Parallel data can be loaded *or* left shift can be accomplished by cross-connecting the parallel outputs and inputs as shown in Fig. 5-35*d*. In this case the output of *D* is coupled back to the input of *C*, etc., and serial data are entered in input *D* and shifted left if clock 2 is applied, or shifted right if the clock is applied to clock 1. Although Fig. 5-35 shows only a four-bit shift register, such registers can be connected in series to form 8, 12, 16, . . . -bit shift registers.

An interesting application of shifting left or right is that multiplication or division by multiples of 2 can be performed. In the following illustration, shifting 001 (=1) one digit to the left results in 010 (=2) or an effective multiplication by 2. Shifting left by another stage is equivalent to multiplying by 4,

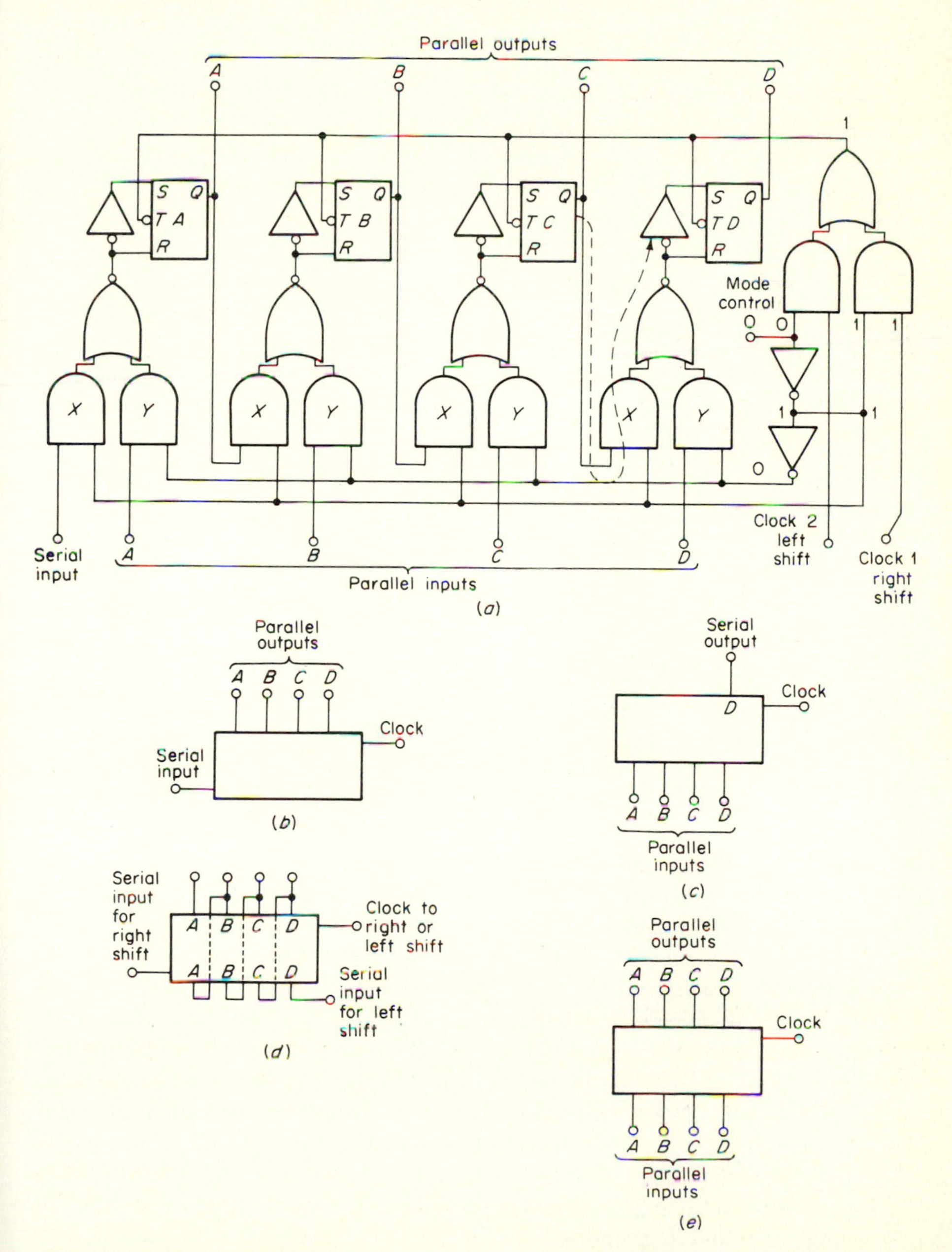

Fig. 5-35 (*a*) MSI four-bit bidirectional (right or left) shift register, used as: (*b*) serial-to-parallel converter; (*c*) parallel-to-serial converter; (*d*) right or left shift register; (*e*) parallel-in, parallel-out register. *Texas Instruments.*

and this process can be carried on to provide any range of binary (2, 4, 8, 16, 32, . . .) multiplication or division.

Binary number	*Decimal equivalent*		
Original			
001	1		
Shifted left 1 bit		× 2	
010	2		× 4
Shifted left 2 bits			
100	4		
Original			
011	3		
Shifted left 1 bit		× 2	
110	6		
Original			
100	4		
Shifted right 1 bit		÷ 2	
010	2		÷ 4
Shifted right 2 bits			
001	1		

Figure 5-35 is an excellent example of how MSI integrated circuits have eliminated the need for interconnecting individual integrated-circuit flip-flops and gates to perform many functions. To duplicate the shift register of Fig. 5-35*a* would require at least six individual integrated-circuit packages of the same size as the MSI package plus pc-board interconnections, or approximately 8 in^3 [131 cc]. The cost would be approximately double that of the MSI package. The speed and power consumption of the MSI shift register of Fig. 5-35 are 20 MHz and 250 mW, respectively, and these figures are probably typical of a conventional multi-integrated circuit register of the same logic family. [16-k bit charge-coupled-device (CCD) shift registers are also available.]

A logical extension of the basic serial shift register of Fig. 5-34 is the *ring counter*. A ring counter is merely a serial shift register with direct feedback from the outputs of the last stage to the inputs of the first stage. The data circulate in the ring counter; thus the designation *circulating register*, which the ring counter is sometimes called. The interesting aspect of the ring counter is that the waveforms can be made to be exactly equivalent to those required from a decoder used to convert the binary output of a decade counter to the more useful decimal system. Hence, if such waveforms are desired, the ring counter can provide the conversion directly without the additional complexity of a decoder.

By crossing the feedback connections of the ring counter instead of feeding them back directly, an even more versatile shift-register variation can be attained. Such a subsystem is called a *shift counter*. The main advantages of the shift counter over the ring counter are:

1. The modulus of the shift counter is $2N$, versus $1N$ for the ring counter (where N is the number of stages).
2. The shift counter is self-starting, and proper operation is easier to maintain than in the ring counter.

5-5-4 MULTIPLEXING

Figure 5-36*a* shows the basic *multiplexer/demultiplexer* (mux/demux) principle. As shown, a multiplexer selects one of several input data sources (lines) to transmit over a single output line. Prior to the advent of semiconductor switching, such data sampling was accomplished by mechanical multiposition or stepping switches which mechanically sampled each of the inputs in a given sequence. Figure 5-36*a* also shows that *demultiplexing* is the inverse of multiplexing. Typical mux/demux applications include computers, remote process control, and communication systems where it is desirable to reduce the number of wires between subsystems. In such applications multiple data inputs are multiplexed to time-share a single data transmission line to the distribution point, where they are demultiplexed and routed to their individual destinations (such as displays or receivers).

Figure 5-36*b* shows one-half of a typical MSI dual four-input multiplexer. The two-bit input address determines which input is routed (or selected) to the output as shown by the truth table of Fig. 5-36*c*. Such addresses are generally derived from binary counters, and in this case a mod-4 counter would provide the proper 00, 01, 10, and 11 address sequence. The example of Fig. 5-36*b* shows the B input routed to the output; the reader is encouraged to study other input and address conditions to verify the truth table of Fig. 5-36*c*.

(In multiplexer applications such as *time-division multiplexing*, multiple measurements are sent over a common channel by dividing the available time intervals among the measurements to form a composite pulse train.)

5-5-5 DECODING/DEMULTIPLEXING

A decoder identifies or recognizes a code, and there are two general categories of decoders—logic decoders and display decoders. Examples of digital-display decoder/drivers are discussed in Sec. 6-3. Typically, display decoders decode binary BCD information so as to drive the appropriate elements of a decimal display. Logic decoders are frequently used to demultiplex (Fig. 5-36*a*), route data or clock signals, and address memory subsystems (Figs. 5-25 and 5-39).

Figure 5-36*a* shows that the role of the *demultiplexer* is to route the serial multiplexed data from one common line (or bus) to one of several output lines.

Figure 5-37 shows half of a typical IC dual one-of-four decoder. Figure 5-37*b* shows that the XY address (or output selection) is the same as the multiplexer of Fig. 5-36, and that for each state of the address one and only one of the outputs is activated (low). For demultiplexing applications the serial input data are applied to the enable input; the right-hand side of Fig. 5-37*b* shows to which

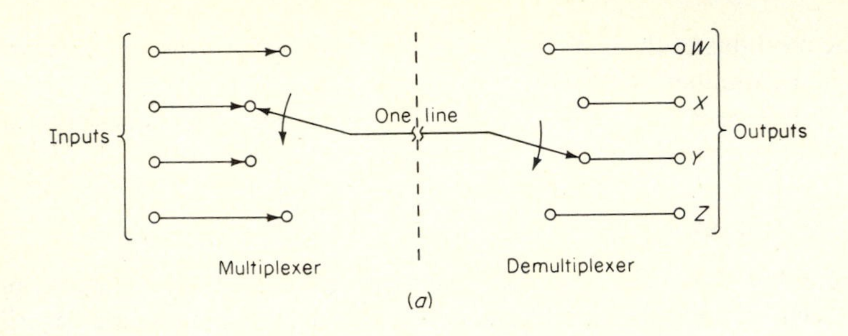

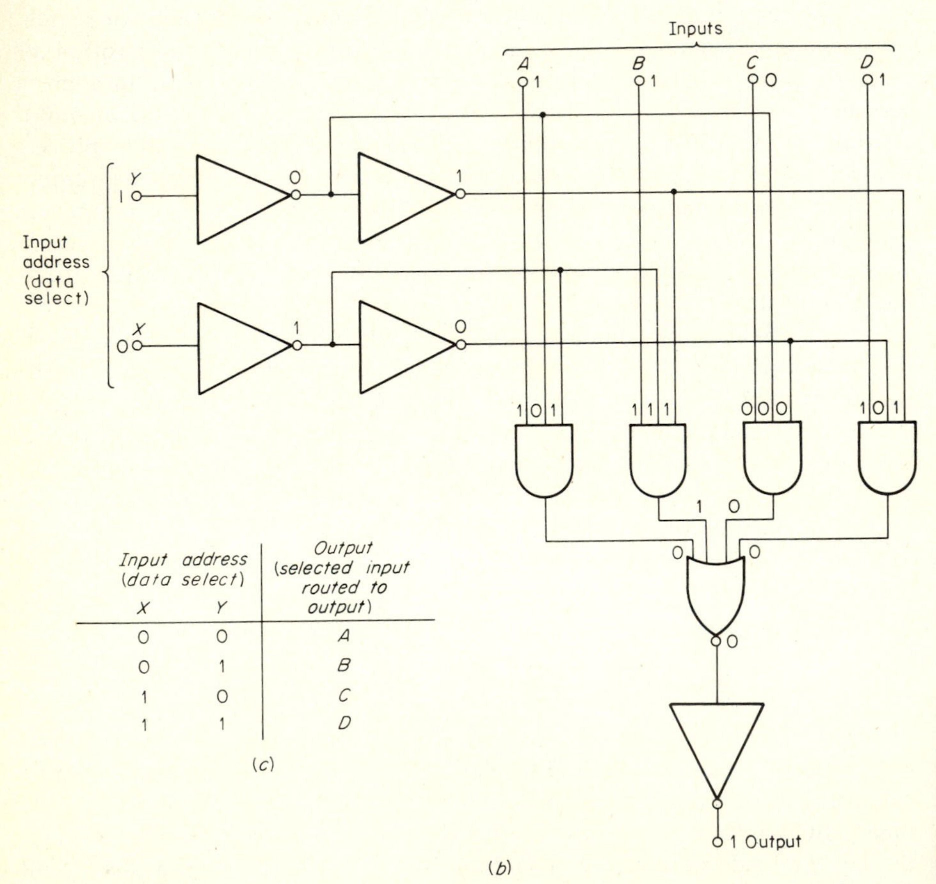

Input address (data select) X	Y	Output (selected input routed to output)
0	0	A
0	1	B
1	0	C
1	1	D

(c)

Fig. 5-36 Multiplexer: (*a*) stepper-switch analogy; (*b*) schematic for one-half of a typical MSI dual four-input multiplexer; (*c*) truth table.

output the input data are routed for each of the possible addresses. (MSI logic decoders are available with up to 16 outputs, and most can be cascaded to accommodate more bits.)

Figure 5-38 shows the block diagram and schematic for a typical MSI *binary* (*BCD*)-*to-decimal* 1-of-10 (or 4-line-to-10-line) decoder. Such a decoder

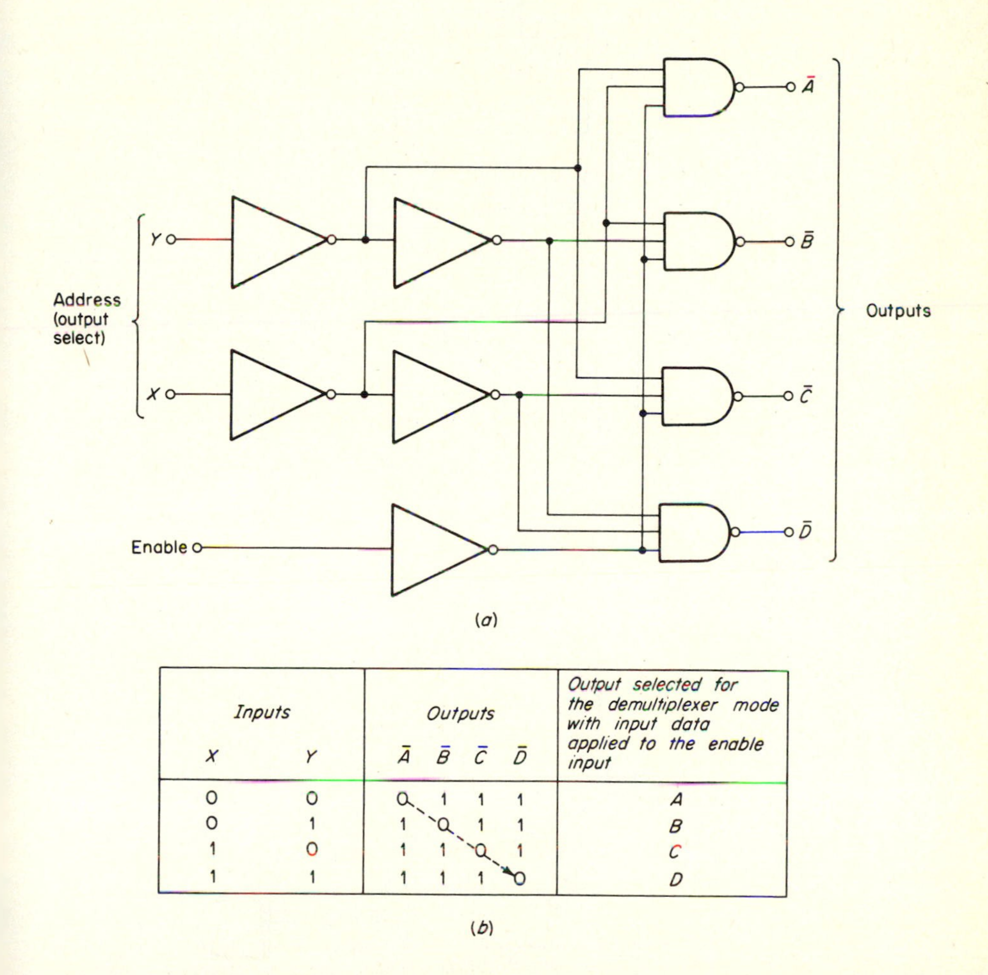

Inputs		*Outputs*				*Output selected for the demultiplexer mode with input data applied to the enable input*
X	*Y*	$\bar{A}$	$\bar{B}$	$\bar{C}$	$\bar{D}$	
0	0	0	1	1	1	*A*
0	1	1	0	1	1	*B*
1	0	1	1	0	1	*C*
1	1	1	1	1	0	*D*

(b)

Fig. 5-37 One-half of a dual one-of-four decoder: (*a*) schematic; (*b*) truth table, including the simplified demultiplexer mode. *Fairchild Semiconductor*.

is frequently required to convert the (machine language) BCD output of a decade counter such as shown in Fig. 5-31 to the familiar decimal system.

Since each decimal equivalent has a unique set of outputs from the decade counter, a four-input AND gate can be set up to respond to each BCD output. For instance, the decimal number 5 is represented by a decoder input of

(8)	(4)	(2)	(1)
D	*C*	*B*	*A*
0	1	0	1

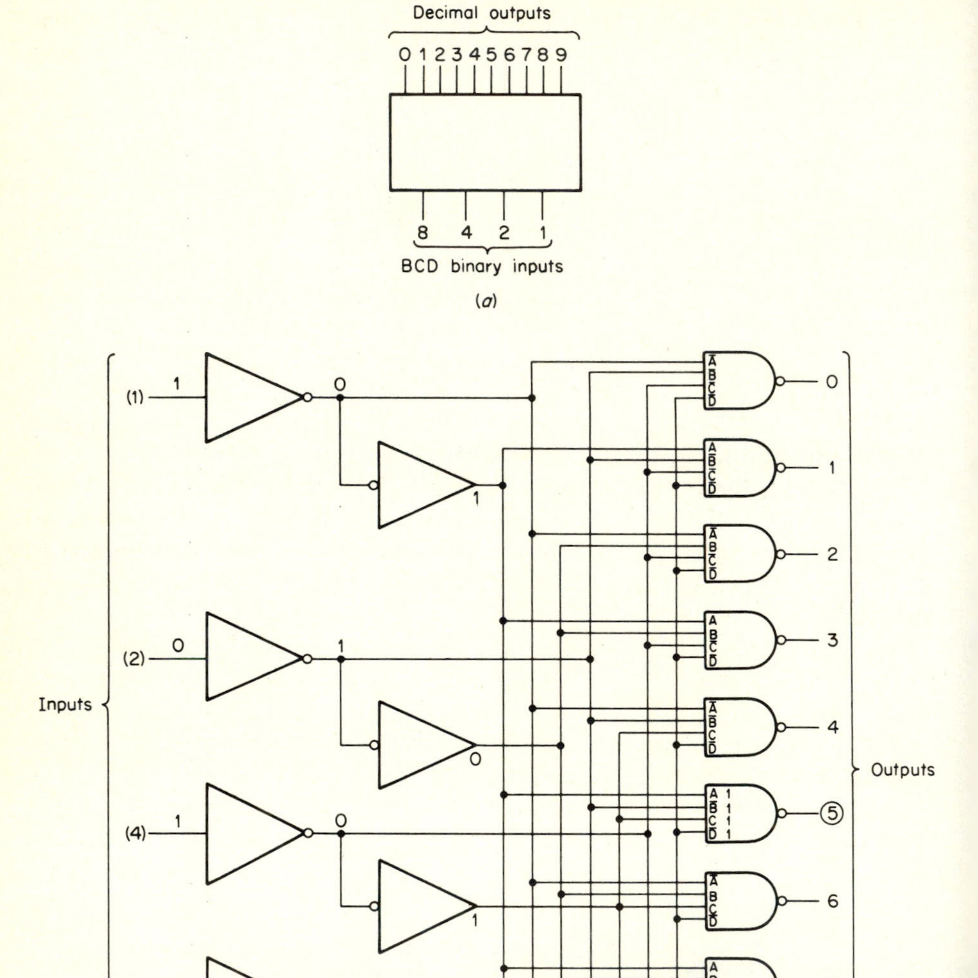

(a)

(b)

Fig. 5-38 Binary (8421 BCD) to decimal decoder: (*a*) block diagram; (*b*) schematic showing an example of decimal 5 decoding. *Texas Instruments*.

Notice that for the output 5, the NAND-gate inputs are $\bar{D}$, C, $\bar{B}$, and $\bar{A}$, and that these inputs are all high (1s) only for the case of a binary 5 (0101) input to the decoder. This gate output then uniquely represents the number 5.

5-5-6 ENCODERS AND CODE CONVERSION

Encoding is the inverse of decoding (Sec. 5-5-5) and may involve multiple input and output lines. Generally only one input is excited, and the encoder generates an output code depending on which input is high. (Notice that this differs from multiplexing, where only a single output line is involved.)

Before discussing encoders or code converters, let us briefly review some of the more popular binary *codes*. Since humans are generally more familiar with the decimal (base 10) number system, the inputs and outputs of digital systems must be decimal. However, inside a digital system a binary code (Fig. 5-1) is generally used because it is simpler and more natural to the digital arithmetic operations (Sec. 5-5-7). In the binary-coded decimal (BCD) system of Sec. 5-1, each decimal digit is represented by a group of binary bits, thus providing a relatively simple interface/conversion between the decimal and binary number systems.

The three most-used three-bit BCD codes are the 8421, excess-3, and Gray codes shown in Table 5-9. Table 5-9 also summarizes the key advantages and disadvantages of these codes. Since two of the three codes are not suitable for arithmetic operations yet have other unique input/output advantages, the need for digital code conversion is obvious.

A typical *priority encoder* for transforming a decimal number into the binary 8421 BCD code is shown in Ref. 3. (See Fig. 5-38 for an 8421 BCD-to-decimal *decoder*.) The word *priority* means that the appropriate output is obtained as long as the correct decimal input line is activated, regardless of the state of the rest of the inputs. Priority encoders provide a hardware priority interrupt and assignment capability.

TABLE 5-9 (a) 8421, Excess-3, and Grey BCD codes

Decimal	Binary	8421	Excess-3	Gray	Decimal
0	0 0 0 0	0 0 0 0	0 0 1 1	0 0 0 0	0
1	0 0 0 1	0 0 0 1	0 1 0 0	0 0 0 1	1
2	0 0 1 0	0 0 1 0	0 1 0 1	0 0 1 1	2
3	0 0 1 1	0 0 1 1	0 1 1 0	0 0 1 0	3
4	0 1 0 0	0 1 0 0	0 1 1 1	0 1 1 0	4
5	0 1 0 1	0 1 0 1	1 0 0 0	0 1 1 1	5
6	0 1 1 0	0 1 1 0	1 0 0 1	0 1 0 1	6
7	0 1 1 1	0 1 1 1	1 0 1 0	0 1 0 0	7
8	1 0 0 0	1 0 0 0	1 0 1 1	1 1 0 0	8
9	1 0 0 1	1 0 0 1	1 1 0 0	1 1 0 1	9
10	1 0 1 0	0 0 0 1 0 0 0 0	0 1 0 0 0 0 1 1	1 1 1 1	10
11	1 0 1 1	0 0 0 1 0 0 0 1	0 1 0 0 0 1 0 0	1 1 1 0	11
12	1 1 0 0	0 0 0 1 0 0 1 0	0 1 0 0 0 1 0 1	1 0 1 0	12

TABLE 5-9 (b) 8421, Excess-3, and Grey BCD codes

Code	Advantages	Disadvantages
8421	• Weighted (each bit position has a fixed value) • Conversion to decimal numbers easy because only the binary numbers 0 through 9 are needed for each decimal digit	• Rules of binary addition do not apply • Not efficient since only 10 of 16 possible combinations are utilized
Excess-3	• Self complementing (1's complement = 9's complement of the digital number) • Rules of binary addition apply	• Unweighted
Gray Code	• Each progressive number differs by only one bit; therefore, useful for input/output devices such as shift encoders	• Unweighted • Rules of binary addition do not apply

As shown in Ref. 4, several code converters can be assembled with standard integrated circuits and/or MSI ICs; however, LSI ROMs are generally utilized in code-conversion applications which involve more than ≈ 50 outputs or more than 2000 bits of memory. Figure 5-39 shows a typical example of a (MOS LSI) ROM which converts the ASCII computer code (American Standard Code for Information Interchange) to the IBM Selectric typewriter code, and vice versa. Such converters are frequently used to interface Selectric typewriters to computers. (The ROM code converter is similar to the RAM of Fig. 5-25, except that the ROM has no writing capability and the code conversion is permanently stored in the memory cells.)

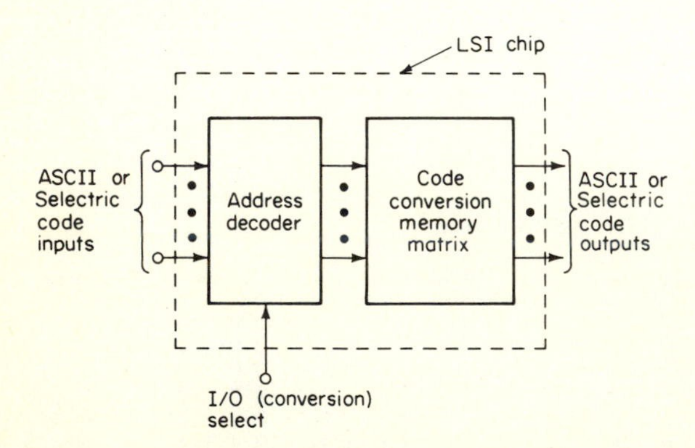

Fig. 5-39 Typical MOS LSI ROM ASCII-to-IBM Selectric typewriter code-converter block diagram.

5-5-7 ARITHMETIC OPERATIONS

The simplest digital arithmetic circuit is the *digital comparator* or *magnitude comparator*. This circuit indicates whether one binary number A is equal to, greater than, or less than another number B, that is, whether $A = B$, $A > B$, or $A < B$. The exclusive-NOR gate of Table 5-1 functions as an *equality detector* because its output is 1 only if $A = B$ (if both equal 1 or both equal 0) as shown by the equation

$$D = \overline{\bar{A}B + A\bar{B}} = \begin{cases} 1 & \text{if } A = B \\ 0 & \text{if } A \neq B \end{cases}$$

A simple AND gate with one input inverted can check for the condition "A less than or greater than B" since

$$C = \bar{A}B = 1 \text{ iff } A < B \qquad (A = 0 \text{ and } B = 1)$$

and

$$E = A\bar{B} = 1 \text{ iff } A > B \qquad (A = 1 \text{ and } B = 0)$$

(The reader is encouraged to verify these equations by referring to the basic logic blocks and Boolean-algebra identities of Table 5-1 and 5-2.)

Figure 5-40 shows a one-bit digital magnitude comparator with the $A = B$ output provided by an exclusive-NOR gate and the less-than/greater-than outputs derived from an inverter/AND-gate combination. Several manufacturers offer four-bit IC magnitude comparators utilizing 4 one-bit comparators as shown in Fig. 5-40, plus additional gating to provide the following decisions for four-bit words:

$$A_3 A_2 A_1 A_0 = B_3 B_2 B_1 B_0$$

$$A_3 A_2 A_1 A_0 < B_3 B_2 B_1 B_0$$

$$A_3 A_2 A_1 A_0 > B_3 B_2 B_1 B_0$$

Words of greater length can be compared by cascading two or more four-bit comparators with the outputs of the LSB-stage input to the MSB of the next stage. [In computer applications the arithmetic logic unit (ALU) usually provides the comparison functions.]

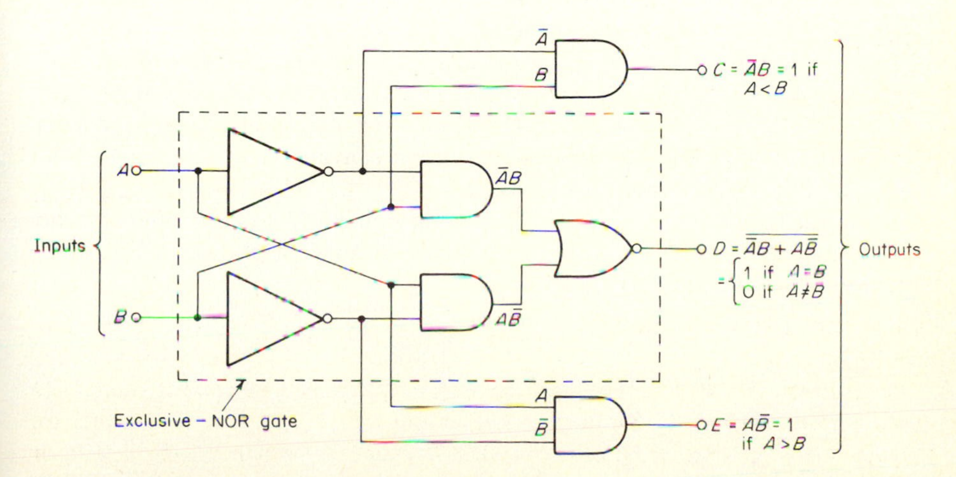

Fig. 5-40 One-bit digital magnitude comparator.

The processes of binary addition, subtraction, multiplication, and division are carried out in much the same manner as in the decimal system. As an example, consider the following addition rules and examples (+ = plus):

$$0 + 0 = 0$$

$$0 + 1 = 1$$

$$1 + 0 = 1$$

$$1 + 1 = 0 + \text{carry of } 1$$

Binary number		Decimal equivalent
(MSB) (LSB)		
1 0 1	LSB column $1 + 0 = 1$	5
+1 1 0	Second column $0 + 1 = 1$	+6
1 0 1 1	MSB column $1 + 1 = 0 +$ carry of 1	11
$(2^3)(2^2)(2^1)(2^0)$		
1 1 1	LSB column $1 + 1 = 0 +$ carry of 1	7
+1 0 1	Second column $1 + 0 + 1 = 0 +$ carry of 1	+5
1 1 0 0	MSB column $1 + 1 + 1 = 1 +$ carry of 1	12

The conventional integrated-circuit addition element is the full adder, and MSI-level adders generally contain dual full adders. A half-adder will add two binary digits, but half-adders are rarely used by themselves because we frequently need additional capacity whenever two binary numbers result in a carry or more than two binary numbers are added. (The above three-digit examples show this.) In such instances the full adder is employed, and it generally consists of two half-adders plus additional gating as shown in Fig. 5-41*a*.

The circuit of Fig. 5-41*a* will add three digits at a time and produce a sum-and-carry output. The example shown in Fig. 5-41*a* and *c* shows the correct output for 1 + 1 + 0, which is a sum of 0 and a carry of 1. A three-digit (full-adder) truth table is shown in Fig. 5-41*b*. One of the many methods of producing a full adder is shown in Fig. 5-41*c*. (This circuit is available in pairs as an MSI dual full adder or as a four-bit full adder.)

Several three-input full adders are often connected in series to add two binary numbers of four or more digits. Such an adder is shown in Fig. 5-42. The two binary numbers to be added are

$$\begin{array}{rrrrr} & & & & \text{(LSB)} \\ & A_4 & A_3 & A_2 & A_1 \\ + & B_4 & B_3 & B_2 & B_1 \\ \hline C & S_4 & S_4 & S_2 & S_1 \end{array}$$

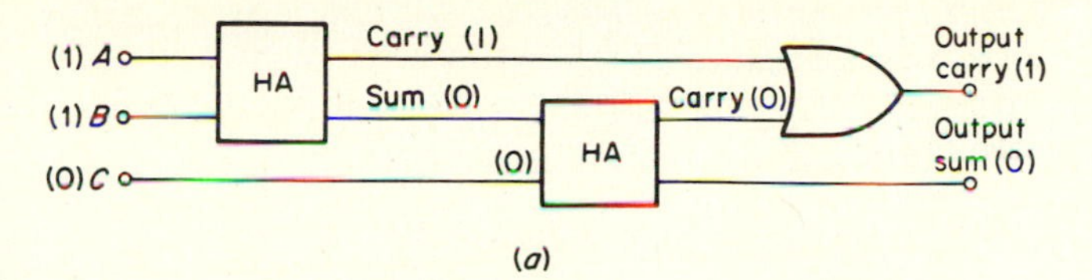

(a)

Inputs			Outputs	
A	B	C	Sum	Carry
0	0	0	0	0
0	0	1	1	0
0	1	0	1	0
0	1	1	0	1
1	0	0	1	0
1	0	1	0	1
1	1	0	0	1
1	1	1	1	1

(b)

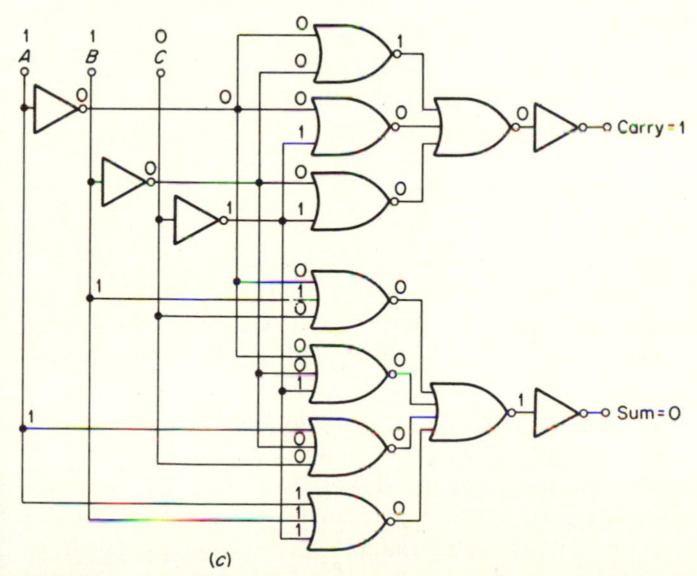

(c)

Fig. 5-41 Full adder: (a) block diagram; (b) truth table; (c) one-half of an MSI dual full adder. *Motorola.*

and the answer appears at the full-adder sum outputs. Notice that this is a parallel adder; that is, the A and B numbers are shifted from the A and B registers, respectively, and the complete output appears as soon as the carry propagates down the line. Parallel addition is thus very fast but requires more hardware. The capacity of Fig. 5-42 would be two 1111 numbers or 15 plus 15 = 30 (plus a carry in of 1); however, more full adders can be cascaded to increase the capacity. (Notice that if there were no carry input, the LSB would require just a half-adder, because only two inputs are added at that stage.) Digital systems operating in the 8421 BCD binary code need an adder that

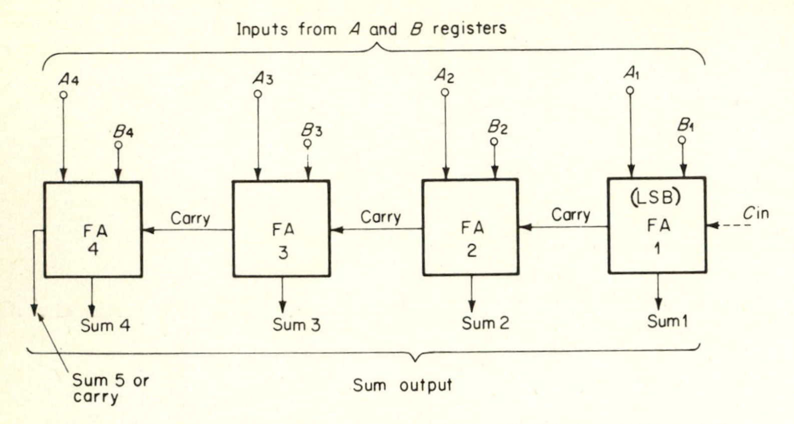

Fig. 5-42 Parallel binary adder with a capacity of 1111 plus 1111, or 30.

operates in the same code. Recall from the BCD counter of Fig. 5-31 that the 8421 BCD code uses only 10 of the 16 possible four-bit combinations. Thus such a BCD adder must be reset to 0 after the count of 9. Reference 5 shows a method of accomplishing this.

The operation of a binary subtractor is very similar to that of decimal subtraction; the rules are summarized in the truth table of Fig. 5-43*a*. Notice that the only time a "borrow" is necessary is when 1 is subtracted from 0, giving a difference of 1 and a borrow of 1 from the next column. Like the half-adder, the half-subtractor is adequate for only two bits; hence a full subtractor (as shown in Fig. 5-43*b*) is required for three-input subtractions. A parallel binary subtractor can be constructed by cascading several full subtractors, but more often such a subsystem is constructed from the more readily available full-adder integrated circuits, as shown in Fig. 5-43*c*. This technique takes advantage of the fact that binary subtraction can be done by taking the 2s complement of the subtrahend and adding that to the minuend. As shown by the "7 minus 3" example of Fig. 5-43*c*, the 2s complement is generated by inverting each bit of the number and adding 1 to the LSB. Notice that the only difference between the full subtractor and the full adder is that the B (subtrahend) inputs are inverted and the final carry is fed back. These two changes provide the 2s complement for subtraction.

The *arithmetic logic unit* (ALU) is an MSI digital IC which combines the comparison, addition, and subtraction functions previously discussed with several other arithmetic and logic functions. Basically, the more sophisticated ALU ICs provide all the 16 possible logic operations for 2 four-bit words, i.e. inversion, AND, OR, NAND, NOR, exclusive-OR, exclusive-NOR, and logical 1 for the two possible combinations of two variables. Arithmetically, ALUs add, subtract, compare, double, pass, increment/decrement, and add or subtract 1 from the various combinations of two variables. ALUs do not mutiply or divide. The specific operation which an ALU performs at any one time is determined by the

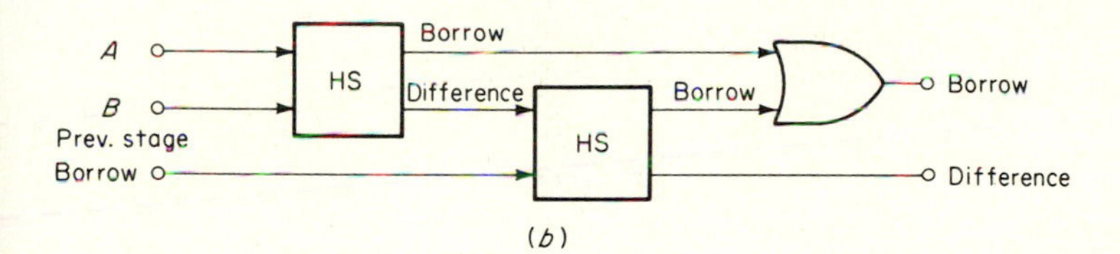

A	$-B$	Difference ($A-B$)	Borrow
0	0	0	0
0	1	1	1
1	0	1	0
1	1	0	0

(*a*)

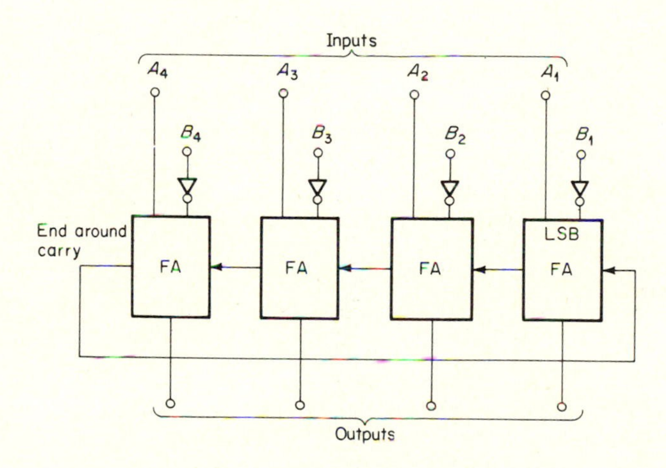

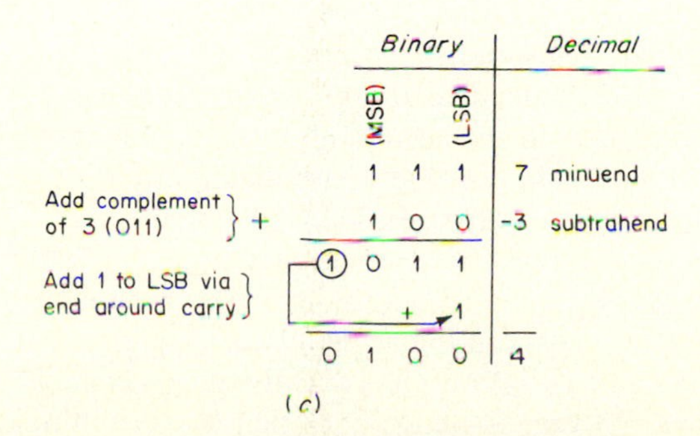

Fig. 5-43 Binary subtraction: (*a*) truth table; (*b*) full-subtractor block diagram; (*c*) example of 7 minus 3, using full adders and the 2s-complement technique.

state of the function-select inputs. The ALU is particularly useful in applications where several functions must be performed in sequence. Reference 6 describes a typical MSI parallel four-bit ALU.

A digital *accumulator* is an LSI IC that contains an ALU and a bidirectional shift register (Fig. 5-35).

The processes of binary multiplication and division can be accomplished by successive addition or subtraction, or shift-and-add (subtract) operations, respectively. Hence the full adder or full subtractor forms the basis of all arithmetic operations. In binary multiplication, anytime either the multiplicand or the multiplier is 0 the product is also 0. Hence, only 1 × 1 gives a 1 answer. As an example, consider the product of 7 × 5.

Binary							Decimal
					(LSB)		
			1	1	1	multiplicand	7
			×1	0	1	multiplier	×5
			1	1	1	partial products	35
		0	0	0			
	1	1	1				
1	0	0	0	1	1	product = sum of partial products	
32 +	0 +	0 +	0 +	2 +	1 = 35		

The process involves multiplication of the multiplicand by the multiplier LSB and so on to form the three partial products. Each partial product is shifted one digit to the left of the preceding partial product, and all three are added to give the final product. One can imagine that shift registers will be employed to shift the partial products, and full adders to add them.

Figure 5-44*a* is a block diagram of the basic binary multiplication process. The operation of Fig. 5-44 proceeds as follows. Ignoring the sign bit for the moment, the accumulator is reset, and the multiplicand and multiplier are shifted into their appropriate registers. If the multiplier LSB is 1, the contents of the multiplicand register are added to those of the accumulator by the full adders, and the product appears in the accumulator. If the multiplier LSB is 0, no addition is performed, just as in the case of decimal multiplication. Next the accumulator and multiplier registers are shifted one place to the right. The second multiplier LSB is then at the far right of the multiplier register, and the previous processes are repeated. After one more addition (if the multiplier LSB is 1) and right shift, the product appears in the accumulator. The signs are checked, and, if they coincide, the sign is + ; if not, it is negative as in decimal multiplication. An example is shown in Fig. 5-44*b*. (Notice that binary multiplication or division requires significantly more hardware than the other arithmetic operations.)

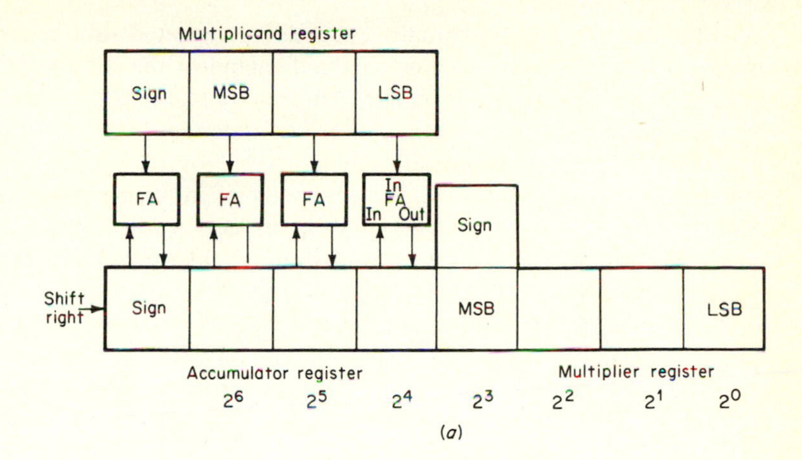

(a)

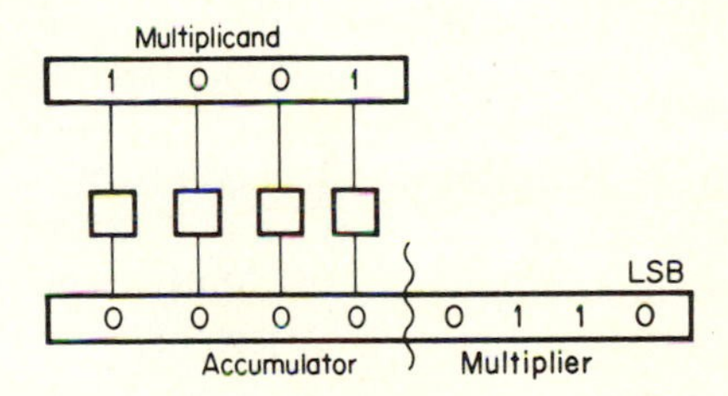

Multiplier LSB = 0. Don't add; shift accumulator and multiplier right:

1 0 0 1
0 0 0 0 } 0 0 1 1

Multiplier LSB = 1. Add multiplicand and accumulator:

1 0 0 1 } 0 0 1 1

Shift accumulator and multiplier right:

1 0 0 1
0 1 0 0 } 1 0 0 1

Multiplier LSB = 1. Add multiplicand and accumulator:

1 1 0 1 } 1 0 0 1

Shift accumulator and multiplier right:

0 1 1 0 } 1 1 0 0

Multiplier LSB = 0 Don't add, shift right:

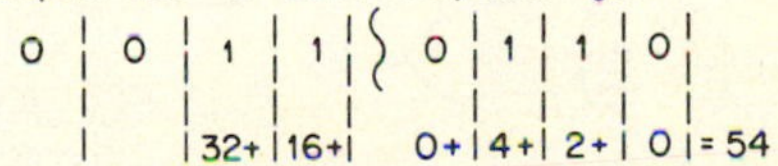

(b)

Fig. 5-44 (a) Binary multiplier. *Adapted from A. Malvino and D. Leach, "Digital Principles and Applications," p. 387, McGraw-Hill Book Company, New York, 1969.* (b) Example of 9 times 6.

Binary division is essentially a process of repeated subtractions, with the divisor being subtracted from either the dividend or the remainder. Hence, the process can be implemented by a system exactly like that of Fig. 5-44*a*, except that full subtractors are used instead of full adders.

Digital data are continually being moved or transferred within digital systems; hence there is always the possibility of introducing an error into the system from noise, intermittent failures, or race problems (Fig. 5-32). (Such an error is typically manifest as an undesired bit change within a multibit word.) Hence, some means of checking for such errors must be incorporated into the system so that the erroneous data can be discarded and a retransmission initiated. The simplest and most popular means of checking for single-bit errors in digital words is the addition of an extra bit, called a *parity bit*. *Even parity* is defined as the condition in which the total number of 1s in a word, including the parity bit, is even. For example, the four-bit word plus parity bit 10111 has even parity since there are an even number of 1s (four of them). *Odd parity* exists when the number of 1s is odd; i.e., if the parity bit in the previous example happened to be 0, there would be three 1s and thus odd parity. (Odd parity is generally preferred because there will always be at least one 1 in any transmission. That is, even if all the word bits are 0, odd parity requires that the parity bit be 1.)

Notice that the parity error-detection scheme requires the transmission of redundant information, and thus lowers the efficiency of transmission. The parity-bit technique will detect only single-bit errors; however, the odds of multiple errors are extremely low. Also, a parity bit cannot correct errors. To include such correction, more redundant information is needed.

Figure 5-45 shows an exclusive-NOR *parity tree* which is frequently used to detect or generate even or odd parity. Recall from Table 5-1 that the output of an exclusive-NOR gate is 0 only when *either* of its inputs is 1, but not when both are 1 (or both are 0). The output of the four-bit parity checker of Fig. 5-45 is 1 when the four input bits have even parity, and 0 for odd parity. Hence, the circuit can determine whether even or odd parity exists. This circuit can also be used as a parity generator because the noninverted output is correct to add the fifth odd parity bit to the normal four-bit word; i.e., the example of Fig. 5-45 shows that the noninverted output is 0 for an odd number of 1s in the input, and a

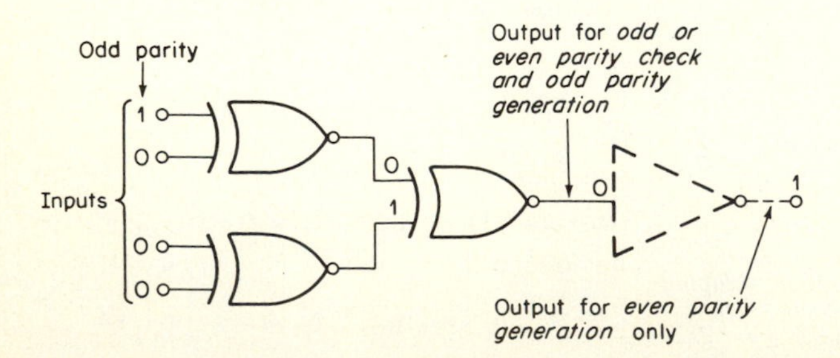

Fig. 5-45 One-half of a typical MSI dual four-bit exclusive-NOR parity tree.

parity bit of 0 is the correct parity bit for an odd-parity word. Likewise, the inverted output provides the proper additional parity bit for even-parity generation. (Eight-bit MSI parity generators/checkers are available with mode-control lines which allow them to perform as either even or odd parity generators or checkers.)

Figure 5-46 shows how a four-bit digital data transmission is verified by generating an odd parity bit at the transmitter and checking for odd parity at the receiver.

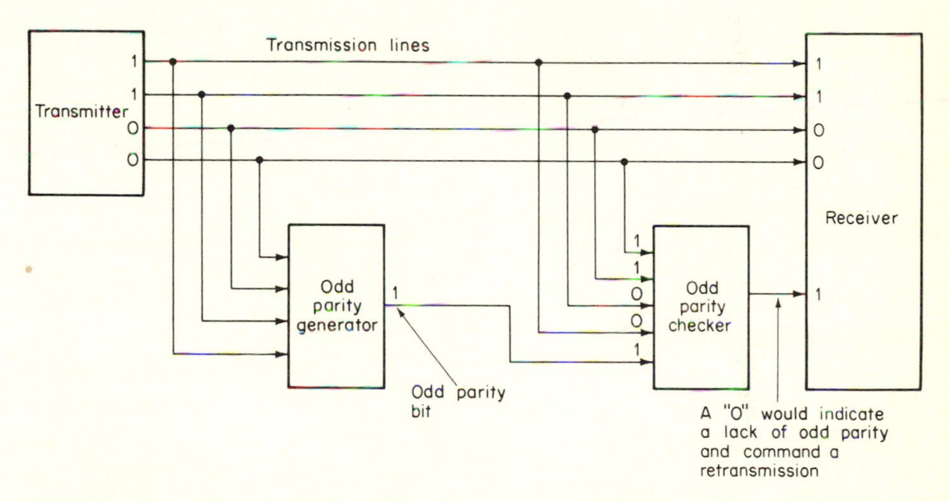

Fig. 5-46 Four-bit data-transmission system with an odd-parity generator at the transmitter and an odd-parity checker at the receiver.

5-5-8 DIGITAL-TO-ANALOG AND ANALOG-TO-DIGITAL CONVERSION

We shall now discuss the important operations of digital-to-analog (D-A) and analog-to-digital (A-D) conversion. The best examples of D-A include conversion from binary machine language to analog signals for plotters, recorders, or meters. A-D conversion is often needed to convert analog transducer outputs to digital form for input into a digital system. (Interestingly enough, the present level of miniaturization as discussed in Sec. 5-6 will allow the designer to add a complete A-D converter right on the back of a transducer-amplifier. In this respect, one IC package could contain the transducer and amplifier with the A-D converter in a similar sized package.)

D-A conversion is much simpler than A-D conversion, and in fact a D-A converter is usually included as part of an A-D converter. For this reason we shall discuss D-A conversion first. The basic problem in D-A conversion is to change a string of digital 0s and 1s to an equivalent analog voltage. For example, consider the binary numbers 0 through 9 and the equivalent weighted analog voltages shown in Fig. 5-47*a*. In this case we have made an arbitrary decision that the maximum analog voltage is 9 V. Since there are 10 different decimal equivalents to express, the lowest (0000) is made equal to 0 V, and the highest (1001) to 9 V, with 1-V increments. Therefore, each binary number is expressed as a discrete analog voltage. Notice that, as we progress from one bit to the next, the equivalent analog voltage doubles; i.e., $2^1 = 2$ V, $2^2 = 4$ V, $2^3 = 8$ V.

Decimal equivalent	Binary number	Bit	Equivalent analog voltage, V
0	0000		0
1	0001	2^0	1
2	0010	2^1	2
3	0011		3
4	0100	2^2	4
5	0101		5
6	0110		6
7	0111		7
8	1000	2^3	8
9	1001		9

(a)

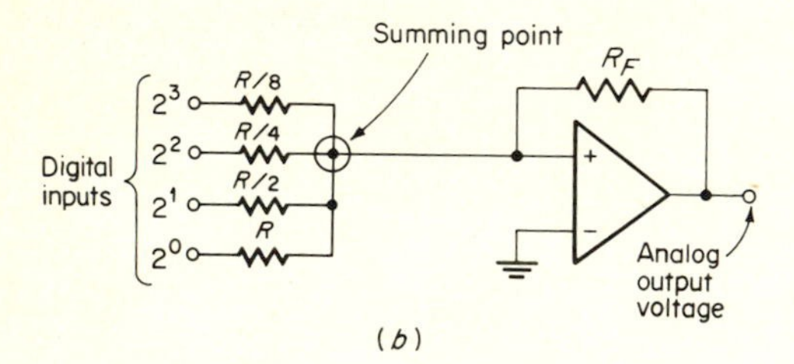

Fig. 5-47 (*a*) One D-A conversion scheme; (*b*) basic binary resistance divider for D-A conversion.

A resistive divider and operational amplifier (Fig. 5-47*b*) is all that is needed to perform the D-A conversion indicated in Fig. 5-47*a*. Notice the common-output summing point and the fact that the weighted resistor values decrease by a factor of 2 for each bit increase. This means that with a high 2^3 input the current through $R/8$ is twice that through $R/4$ with a high 2^2 input. Hence the network fulfills the requirement of doubling the analog current or voltage for every succeeding bit.

Another type of resistive ladder network, known as the R-$2R$ type, is discussed in Ref. 7. This divider serves the same purpose as the circuit of Fig. 5-47*b* with the additional advantages of requiring only two resistor values and providing equal input impedance at all inputs. The length of this text does not permit analyzing these dividers or mentioning all the various types of resistance divider networks available for D-A conversion. For now, let us assume that such dividers will produce the correct weighted analog voltage or current, and that they will be represented by a block entitled *resistive divider*. Such dividers form the heart of all D-A converters so their accuracy, stability, and cost are very important. Other than fabricating dividers from discrete precision resistors, the most popular technique is to deposit an array of thin-film resistors and precision-trim them with a laser or arc source. Such arrays are compatible with integrated-circuit dimensions and prices, possess excellent temperature coefficients (<5 ppm resistance variation per degree Celsius), and can be trimmed to 0.01 percent.

There is more hardware involved in an actual D-A converter than just the precision resistive divider. Figure 5-48 shows a block diagram of a typical D-A converter. Proceeding from the parallel digital input, we see that some form of storage is required for the digital information, plus some means of controlling the data readin and readout. Next, level amplifiers and a precision voltage reference are needed to ensure that the normally imprecise digital signals are equal and constant, regardless of environmental or ladder loading effects. These amplifiers operate in the analog comparator mode (Sec. 4-3-1). That is, they have two inputs: the reference voltage and one of the shift-register outputs. Hence the output of a particular level amplifier is either low or high (≈ 0 or equal to the reference voltage), depending on the state of its input from the shift register. Thus the resistive divider is provided with the proper digital inputs, and it performs the actual digital-to-analog conversion as previously discussed.

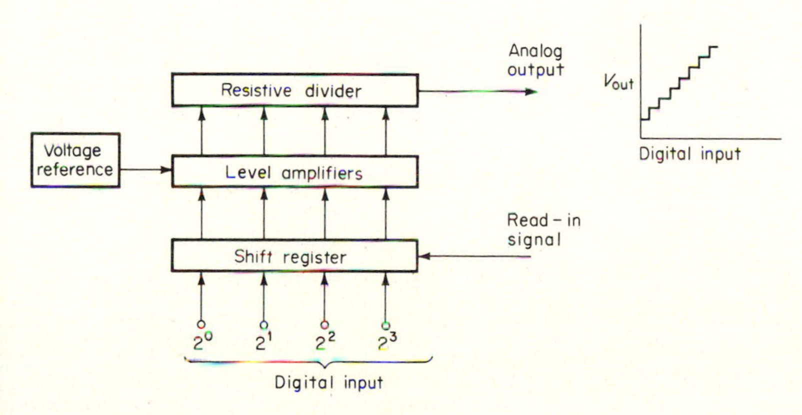

Fig. 5-48 Basic D-A converter block diagram.

The complexity of D-A and especially A-D converters makes them suitable for large-scale integration (LSI), and integrated-circuit D-A converters are available with all the components of Fig. 5-48 included on one chip. (Numerous sources of larger modular D-A converters also exist.)

We shall now discuss the process of A-D conversion, which is more complicated and available in a greater variety of techniques than D-A conversion.

One of the basic building blocks used in A-D converters is the analog voltage comparator. As discussed in Sec. 5-2, this circuit is often an open-loop operational amplifier with one input connected to a reference voltage and the other to an unknown analog voltage. The circuit compares the two inputs and indicates which of the two is greater. As such, the circuit acts like a switch with its inputs arranged so that its output usually remains low when the reference is less than the unknown voltage, but goes high whenever the unknown exceeds the reference. The other basic elements of an A-D converter are a D-A converter and a counter.

Although there are many A-D converter techniques, we shall limit our discussion to the two most popular methods—the counter type (especially the continuous type with up-down counting capability) and the high-speed *successive-approximation* method. The D-A converter is included in an A-D converter to provide a continously variable reference voltage. (Recall that the output of a D-A converter is a staircase analog voltage.)

Figure 5-49*a* shows the essentials of an A-D converter. The circuit is a closed-loop feedback system, and its operation is as follows. The counter is reset so that all its outputs are 0 and hence the D-A converter output is 0 V. Assuming that the analog input voltage is greater than the reference voltage (D-A converter output), the comparator output will be high. The clock is then started, and its pulses pass through the AND gate because both gate inputs are high at every positive-going clock pulse. The clock pulses are thus fed to the counter, and it advances through its normal count sequence. The binary counter outputs cause the D-A converter to provide a continuously increasing staircase-voltage output, as discussed in the previous figure. As shown in Fig. 5-49*b*, this process continues until the staircase reference voltage exceeds the analog input voltage, at which time the comparator output goes low. This stops the clock pulses from passing through the AND gate and thus stops the counter. At this point the output of the counter is the digital equivalent of the analog input voltage, and the conversion is complete. Such a converter must operate rapidly in order to arrive at the correct digital output in a reasonable period of time. In fact, the method of arriving at the final output is the main difference in A-D techniques.

If an up-down counter is used, the conversion time is reduced because such a counter does not have to be reset to zero after each conversion, and the sequence can begin at the last level and hence follow a slowly changing analog waveform better, as shown in Fig. 5-49*c*.

The most universal A-D converter is the *successive-approximation* type, which is similar and only slightly more complicated than the counter type shown in Fig. 5-49. In this type of converter the voltage range is successively divided in half. As shown in Fig. 5-50*a*, the sequence starts with the MSB set in the control register; then this MSB is either taken out or left in, depending on whether the comparator senses that the particular analog input voltage is above or below this level. Next, the second MSB is set in, and the comparator again compares this to the analog input to see if the second MSB should be reset or left in. A typical sequence of trying one bit at time might typically yield 1000 → 0100 → 0110 → 0111, as shown in Fig. 5-50*a*. The conversion time is one cycle per bit, or in this case four times the clock rate (say 1 MHz) or 4 μs. Figure 5-50*b* shows the block diagram for a CMOS LSI 10-bit successive-approximation A-D converter which requires only an external power supply, comparator, and passive clocking components. The thin-film-resistor divider network is deposited on the chip, and tristate logic is utilized in the output stage.

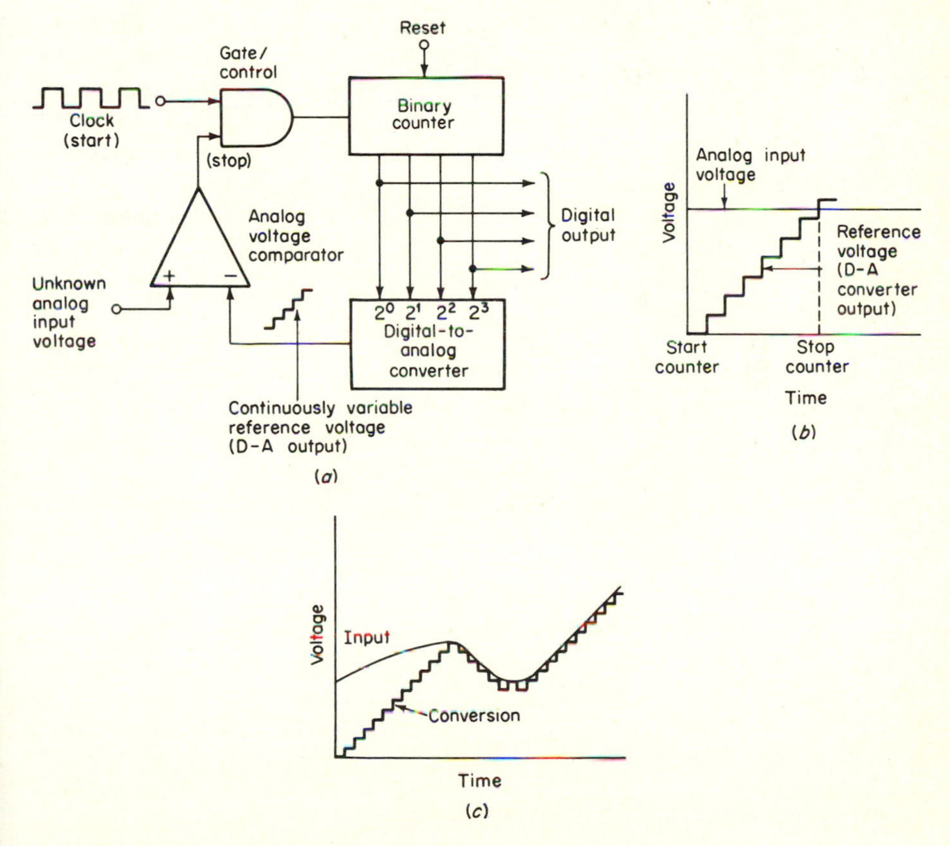

Fig. 5-49 Basic counter-type A-D converter: (*a*) block diagram; (*b*) waveforms; (*c*) rapid-tracking capability with an up-down counter.

We should at least mention the resolution of A-D converters at this point. Basically, the resolution is ±1 part in 2^n, where n = the number of bits in the converter. For example, a seven-bit A-D converter will have a resolution of ±1 part in 128 or ±0.8 percent. If the analog input voltage range is 0 to 5 V, the resolution is 0.008 × 5.0 V, or ±40 mV. A 10-bit A-D converter would have ±1/1024 or ±0.097 percent and ±4.8 mV resolution with the same input voltage range.

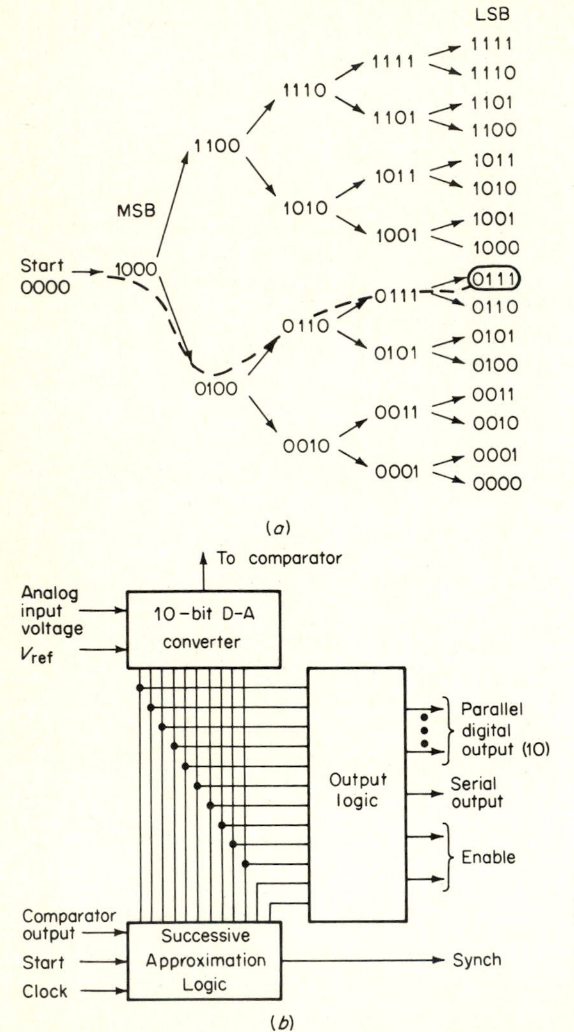

Fig. 5-50 Successive-approximation-type A-D converter: (*a*) typical sequence; (*b*) CMOS LSI 10-bit successive-approximation A-D converter block diagram. *Analog Devices.*

5-6 INTEGRATED-CIRCUIT SYSTEMS

This section is devoted to a level of integrated-circuit sophistication wherein *small systems* (or major portions thereof) are contained on a single chip. This "system-on-a-chip" level of integration is generally referred to as *large-scale integration* (LSI) in constrast to medium-scale integration (MSI) which encompasses subsystems on a chip. MSI is generally defined as a level of integration with fewer than 100 equivalent gates on a chip, whereas LSI involves more than 100 gates. [Actually, three LSI circuits were discussed in Sec. 5-5. They include the ROM ASCII-to-Selectric code converter (Fig. 5-39), the digital accumulator (Sec. 5-5-7), and the analog-to-digital converter (Fig. 5-50*b*).]

The main LSI fabrication techniques are MOS (*p* channel, *n* channel, and complementary MOS), I^2L, and TTL. The reader is referred to Secs. 2-6, 5-3-5, 5-3-6, and 5-3-8 for the inherent advantages of these fabrication techniques for high-density LSI circuits. The theory of operation and basic gate/memory cells for these processes are covered in the above-mentioned sections and in Secs. 2-2, 5-3-9, 5-3-10, and 5-4.

The impact of LSI on the electronics market has been dramatic. Semiconductor manufactuers have gone to LSI for the same reason they went to integrated circuits—the cost is significantly lower. For example, MOS gates cost less than 1 cent each, and such a small system as a calculator or digital voltmeter contains only a few LSI chips, thus reducing the price substantially compared to other techniques. Hence, such complex instruments are beginning to penetrate the consumer market, which has heretofore been unable to afford them.

Electronic-instrument manufacturers have switched to LSI mainly because they can decrease parts count and size while increasing performance and reliability. However, this trend also reduces instrument manufactuers' contributions to the product (and hence their percentage of the profit). This has forced a drastic change in the relationship between semiconductor houses and instrument manufacturers. As an example, consider what has happened in the hand-held-calculator and digital-watch markets. In these cases many of the semiconductor firms have been forced to *manufacture and market the entire product* in order that it be competitive. This is in contrast to their previous practice of producing and selling only the key chips to the companies which traditionally assemble such products. Also, in both cases, a well-established electromechanical design was replaced by high-technology integrated circuits. Another major change brought about by the integrated circuit and LSI is that a larger portion of system design has been transferred from the instrument manufacturer to the integrated-circuit supplier.

There are several examples of LSI circuits being utilized in "smarter" instruments such as oscilloscopes; however, the LSI digital voltmeters, calculator, watch, and microprocessor (Sec. 5-7) are discussed here because they are well known, relatively simple, easily understood, or totally new products made available only with the advent of LSI.

5-6-1 DIGITAL VOLTMETER

The *digital voltmeter* (DVM) has become very competitive with the conventional analog voltmeter, which utilizes a meter-movement readout. The digital voltmeter has not become as inexpensive as a meter-type voltmeter, but the price gap has been reduced to $\approx 2:1$. However, this does not take into account the DVM's increased accuracy and resolution (± 0.02 percent of reading versus ± 2.0 percent), input impedance, speed and ease of reading, and the decrease in measurement error and operator fatigue. DVMs and DPMs are available with a significant portion of their analog and digital circuitry on one or two LSI chips.

Although we shall discuss only the dc DVM [and the related digital panel meter (DPM)], many DVMs are true digital *multimeters* that use various techniques to convert ohms, amperes, and ac to a dc voltage, so that these quantities can also be displayed digitally. Actually, the DVM is a form of A-D converter; however, many conversion techniques employed in DVMs are different and less expensive than those discussed in Sec. 5-5-8. (The particular conversion technique determines the speed, resolution, and noise rejection of the DVM.)

Of the many techniques available for constructing a digital voltmeter, we shall discuss only two, a rather simple low-accuracy and inexpensive method and a more elaborate technique.

Figure 5-51*a* shows a block diagram of a simple ramp-type digital voltmeter. The A-D conversion technique is similar to that of Fig. 5-49 in that a clock, analog comparator, counter, and gate are used; however, the variable-reference-voltage source is a simple ramp generator which would not be used in a more elaborate A-D converter. As shown in Fig. 5-51*a*, a voltage-to-time conversion is made by comparing a linear ramp voltage to the input voltage. The analog-comparator output is saturated low until the ramp voltage very

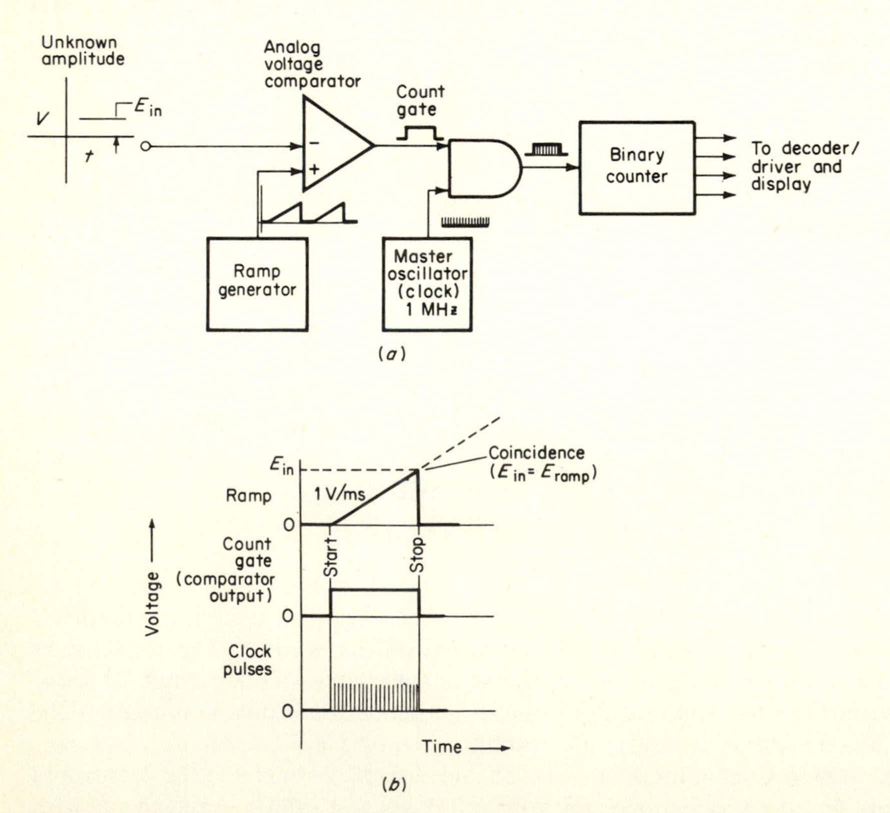

Fig. 5-51 (*a*) Block diagram of a ramp-type digital voltmeter; (*b*) voltage-to-time conversion for the circuit of part *a*.

slightly (typically 1 mV) exceeds the unknown input voltage. At that time the comparator output saturates positively because the noninverting input is more positive than the inverting input (Secs. 4-3 and 5-2). Hence a gate input pulse is generated with a duration proportional to the amplitude of the input voltage, thus completing the voltage-to-time conversion. This count gate enables the AND gate, allowing the counter to count the clock pulses for the duration of the count gate. The number of counts in the counter is then proportional to the input voltage. For example, assuming a ramp slope of 1 V/ms, a total of two hundred 1-μs counted clock pulses would mean that the ramp (or count gate) had been enabled for (200) (1 μs) or 200 μs, thus indicating that coincidence had occurred at 0.2 V. Thus, the input voltage is the solution X of

$$\frac{1\ \text{V}}{1\ \text{ms}} = \frac{X\ \text{V}}{0.2\ \text{ms}}$$

Notice that the A-D converter of Fig. 5-49 is a closed-loop system, and Fig. 5-51*a* is not. In fact, the ramp-type DVM accuracy is limited because of this very fact. In Fig. 5-51 the linearity of the ramp, stability of the master oscillator, and temperature stability of the comparator and ramp generator are all very critical. The system is also susceptible to noise because there is no inherent noise rejection.

A more sophisticated and accurate DVM technique is the dual-slope integrating method of Fig. 5-52. This type of DVM measures the true average of the input voltage over a fixed time period. This is in contrast to the ramp-type DVM which measures the voltage only at one point. Since noise will tend to average to zero (approximately equal + and − excursions), the dual-slope integrating method provides superior noise rejection.

The operation of the DVM of Fig. 5-52 is as follows. The integrator is set to zero by discharging C_1, and S_1 is switched to the input position as shown. The counter is reset to zero. If the unknown input voltage is negative, it is fed directly through the analog switching network, and if E_{in} is positive the input is inverted. As discussed in Sec. 4-3-7, an integrator produces a ramp-type voltage if its input is constant. The integrator output will be

$$-\frac{1}{C_1 R_1}\int -e_{in}\,dt$$

Hence the slope of the integrator output voltage is positive and proportional to E_{in}, as shown in Fig. 5-53. Since the analog-comparator noninverting input is more positive than the inverting input, the comparator output will go high. This enables the AND gate and allows the clock pulses to pass through to the counter. The counter will count up until the MSB goes high, at which time S_1 is switched from the input position to the reference voltage. The analog switching network provides an output reference of the opposite polarity to the input voltage;

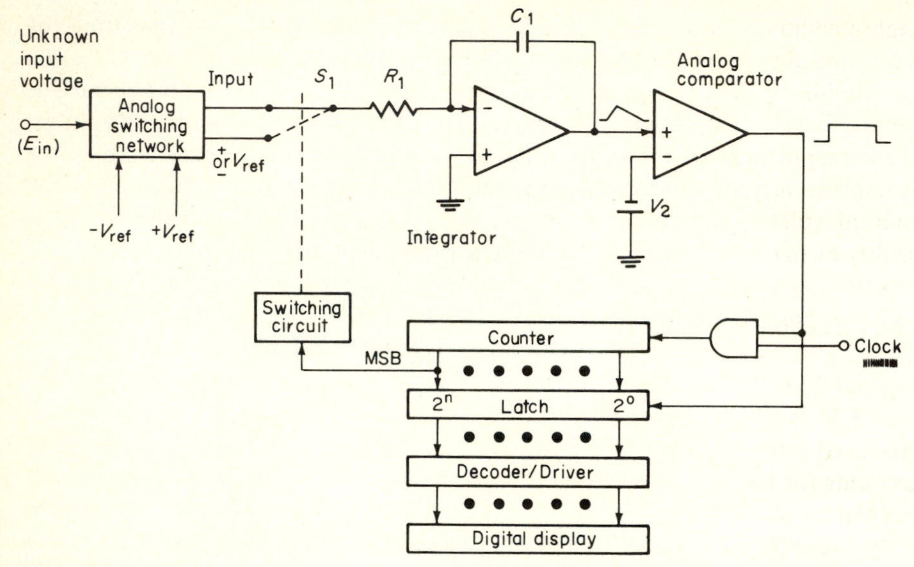

Fig. 5-52 Dual-slope integrating DVM block diagram.

hence, the integrator output polarity will reverse as shown in Fig. 5-53. (Since V_{ref} is constant, the slope of the negative integrator output is also constant.) When the negative-going integrator output reaches zero, the comparator output will go low and inhibit the AND gate, thus stopping the counter. The number of accumulated clock pulses counted during the reference integration time (Fig. 5-53) is then proportional to E_{in}, and the total count is strobed to the display by the negative-going comparator transition. (This particular readout is maintained throughout the next measurement cycle.)

As a numerical example, assume that $E_{in} = 1.000$ V and the clock runs at 1 MHz; hence 1000 counts would be equivalent to 1 ms. The readout might be

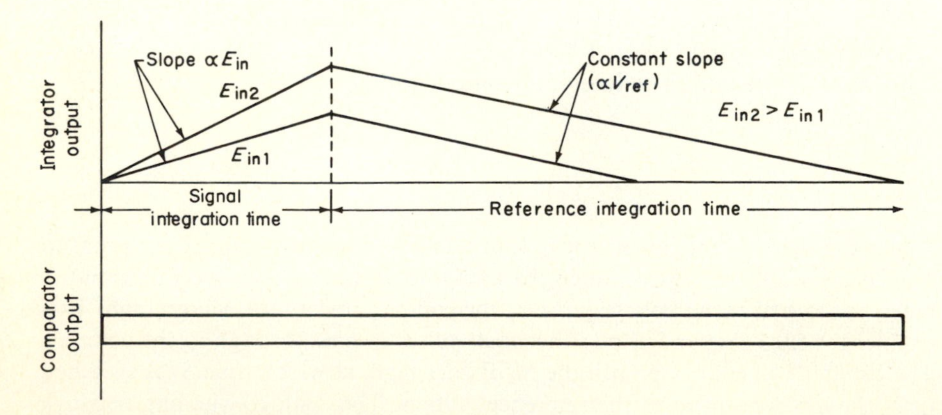

Fig. 5-53 Dual-slope integrating DVM waveforms.

calibrated so that 1000 counts = 1.0 V, so the decimal point would be set after the 1, and the output would read 1.000 V.

Besides its inherent noise-rejection capability, the dual-slope integration method is relatively immune to long-term drift and the temperature sensitivity of the amplifier and C_1, because they will cause errors only if these drifts occur during the very short (few milliseconds) measuring period. The stability of the internal voltage references is critical; however, stable dc voltage sources are readily available.

5-6-2 CALCULATOR

The hand-held calculator is an excellent example of an LSI *system*. As shown in Fig. 5-54, sophisticated keyboard-programmable scientific calculators can be produced with only a handful of "parts." (Extensive use of multiplexing/decoding accounts for the small number of interconnections in Fig. 5-54.) Such miniature calculators started to appear in the early 1970s, and dozens of brands are presently available. They range from the simplest four-function (+, −, ×, ÷) units to scientific calculators with complex mathematical capability, and finally to programmable scientific versions which begin to challenge the minicomputer

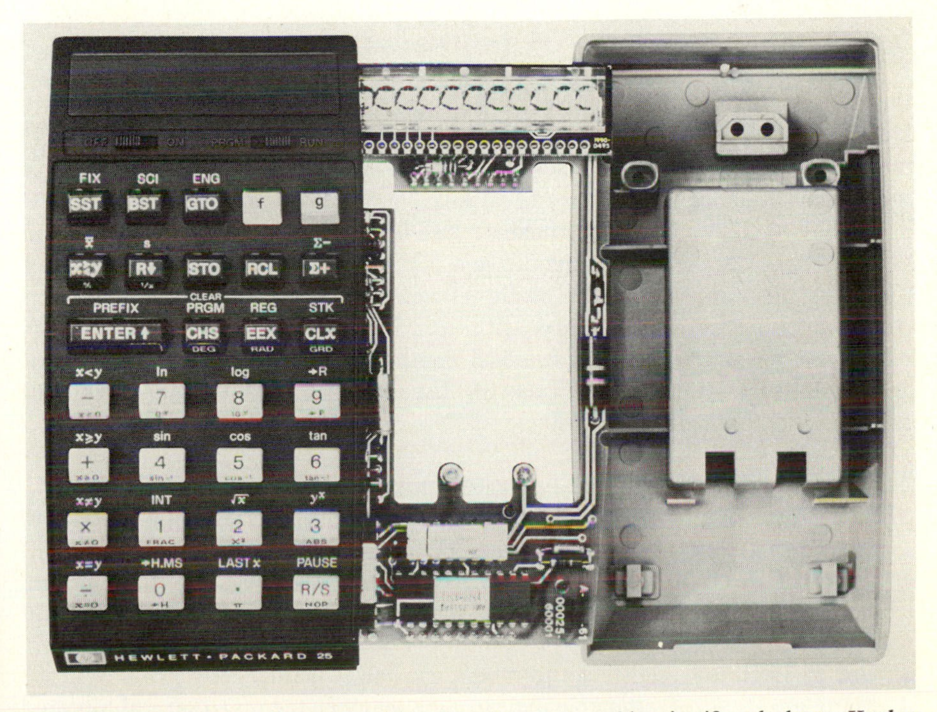

Fig. 5-54 Internal view of a hand-held keyboard-programmable scientific calculator. *Hewlett Packard.*

in computational power. A variety of arithmetic and memory calculator chips are also available from several manufacturers.

Figure 5-55 shows a simplified block diagram for a typical scientific calculator whose basic operation is as follows. The desired function and data are entered via the keyboard; for example, $\pi/4.4 + \sin 34.2°$. (The manner in which the equation is keyed into the calculator depends on the type of mathematics

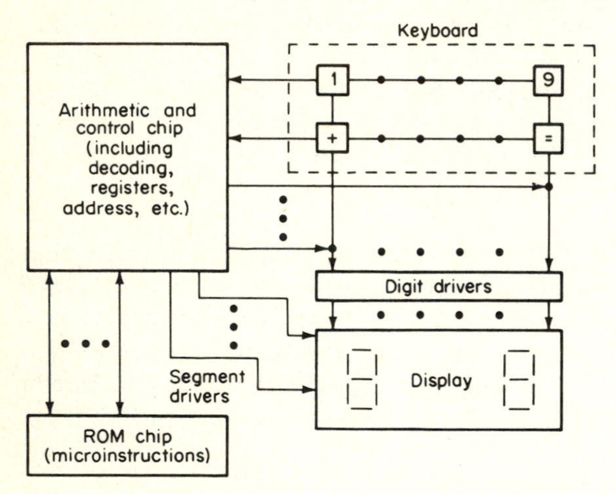

Fig. 5-55 Simplified block diagram of a typical scientific calculator.

the unit is designed for; however, the π constant and sin function frequently require only a single key stroke to call up these microprograms from the ROM table lookup). The keyboard functions and data are fed to the arithmetic/control unit, which addresses a ROM for the desired instructions and then performs the arithmetic operations. The intermediate and final answers are transmitted from the arithmetic/control unit to the display. Notice that a considerable number of functions are lumped into the arithmetic/control unit—i.e., control, timing/clock, encoding, address memory, arithmetic registers, and adders.

As an example of the phenomenal computational power of typical hand-held scientific calculators, the following list identifies some of the capabilities found in many scientific calculators.

1. Up to 10-digit display (plus 2-digit exponent).
2. Single-key commands for (*a*) +, −, ×, ÷; (*b*) all the trigonometric functions in radians or degrees with polar or rectangular coordinates; (*c*) standard log-arithmetic functions; (*d*) $\sqrt{x}$, Y^x, $1/x$. %, π; (*e*) conversions such as those listed in (*b*) plus sign changes; and (*f*) mean and standard-deviation statistical capability.
3. One to 10 storage registers and related exchange features.
4. Many are key-programmable and provide memory for up to several hundred program steps, debugging aids, decision branching, and other functions.

5-6-3 DIGITAL WATCHES AND CLOCKS

Another interesting LSI consumer product is the digital clock/watch. In the late 1960s quartz crystals were introduced as an extremely accurate watch-frequency reference, and this rapidly became an industry standard. In the mid-1970s the digital displays (Sec. 6-3) developed for the hand-held calculator market started to replace the conventional electromechanical watch hands. During this same time, semiconductor manufacturers developed the technology to fabricate the necessary large-module frequency-division counters on a single chip by such LSI techniques as CMOS and I^2L (Sec. 5-3). These and other technological developments resulted in a whole new digital watch/clock concept with one or two LSI chips and a light-emitting diode (LED) or liquid-crystal display (LCD) replacing the maze of precision electromechanical parts normally found in such products.

Figure 5-56 is an example of a typical digital watch. Depending on the type of display, all the subsystems of Fig. 5-56 except the crystal oscillator, battery, and display are contained on a single LSI chip. The watch chip operates as follows. The standard 32.768 kHz quartz crystal provides the stable frequency reference, and this frequency is divided down to 1 Hz by a chain of binary

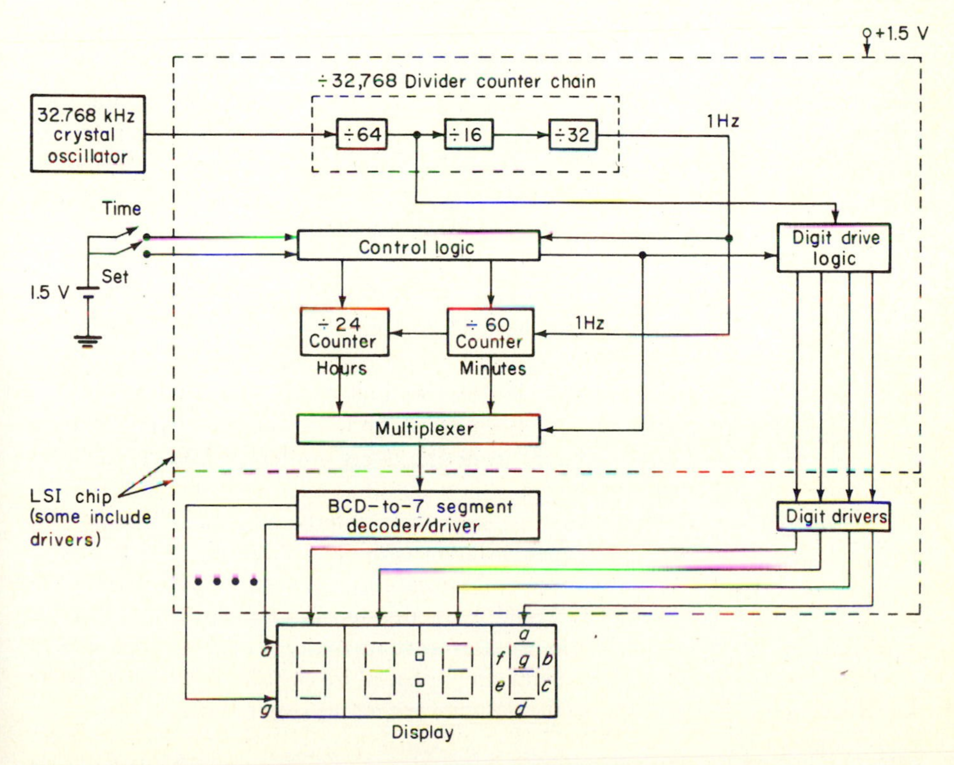

Fig. 5-56 Block diagram for a typical digital watch consisting of one or two LSI chips, LED or LC display, battery, and crystal oscillator.

counters (Sec. 5-5-2). The 1-Hz signal is then fed to a series of $\div 60$ and $\div 24$ counters which count the 1-Hz pulses and provide an output every second, minute, hour, and even day (in some versions). As discussed in Sec. 6-3, the available volume and power in a typical wristwatch requires that a single display decoder/driver be time-shared or multiplxed as shown in Fig. 5-56. Hence, the segment decoder/driver output sequences through the seven possible segments, and the digit driver drives each of the four display digits at a rate above the human eye response. Provision is also made for advancing or setting the watch.

Such watches can operate for over a year with one or two small 1.5-V batteries and maintain accuracies of a few seconds per month (significantly better than their electromechanical counterparts). The reader is referred to Sec. 6-3 for a more detailed description of the LED and LCD digital displays.

5-7 MICROPROCESSORS

5-7-1 FUNCTION OF A MICROPROCESSOR

A microprocessor performs the control and processing functions of a small computer. Data are moved around, calculations are performed, decisions are made, and logic is implemented. These functions are not new. Previously they would have been performed by random logic using gates, flip-flops, registers, and so on, usually in TTL form. Let us make a comparison with such a TTL system, since this will be useful in understanding what *is* new about a microprocessor.

5-7-2 MICROPROCESSOR DESIGN COMPARED WITH RANDOM-LOGIC DESIGN

Figure 5-57 outlines the steps necessary to design a small control system or minicomputer. Every design starts with some kind of requirement, followed by a flow chart if the system has moderate or great complexity. A flow chart shows the flow of signals from input to output as well as signal flow between internal units. It is after the flow chart has been drawn that a design involving random logic becomes significantly different from one using a microprocessor.

With random logic the next step is a detailed circuit design showing the individual gates, flip-flops, registers, and their interconnections. During this design phase there would be frequent references to parts catalogs, with the design being adjusted to meet the realities of availability of parts. With the detailed design complete, a wiring list would be generated, and parts could then be assembled into hardware, perhaps after intermediate breadboarding and/or prototyping stages. Finally a debugging procedure would produce working hardware and perhaps reveal to the designer where his original design procedure had gone wrong.

With a microprocessor the step following the production of a flow chart involves not hardware but software. *Hardware* involves parts, interconnected to

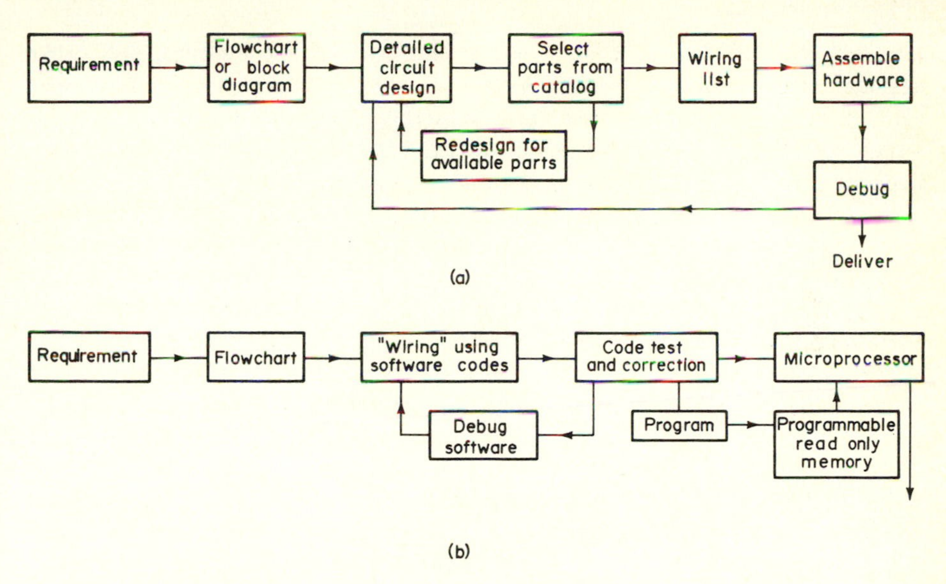

Fig. 5-57 Comparison of design approaches using (*a*) TTL logic; (*b*) a microprocessor.

perform some function. *Software* is a sequence of instructions telling a set of generalized hardware how to combine to perform some particular function. In Fig. 5-57*b* the software converts the system requirements into a form that can be used to program a ROM. When the ROM is correctly programmed for the particular task in mind, that is, has a set of instructions in the correct sequence, gates, registers, and other building blocks that exist inside the microprocessor (just as they do in the random-logic system) are in effect "wired" appropriately to perform the desired function. As the instruction sequence from the ROM is stepped through, the microprocessor is "rewired" at each step so that it carries out each consecutive instruction correctly. To complete the system, it may be necessary to add a read/write memory (usually called a RAM) and some input/output interfacing circuitry so that the microprocessor system can communicate with the outside world of sensors, keyboards, displays, and so on. A significant feature of a microprocessor is that a single type can be used in an almost limitless variety of applications by "rewiring" it using a differently programmed ROM. This is in contrast to the random-logic approach in which a system, once wired, can be used only for the single purpose for which it was designed. In this sense the microprocessor was the first general-purpose LSI chip. It is general-purpose because its interconnections are effectively controlled by external elements (ROMS and read/write memories) which themselves may be interconnected in a limitless number of ways. This solved, for the integrated-circuit manufacturer, the problem of making IC chips of great complexity which could perform only a single function and which therefore had a more limited market than a more general-purpose device. A single type of microprocessor can be used in machine control, game-playing machines, automatic test and control systems, control and monitoring of automobiles, and other applications. Microprocessors are used in such test equipment as oscilloscopes (HP1722A) to read out voltage, time interval, and frequency, and in bridge circuits (Boonton 76A capacitance bridge) to read out dissipation factor, series capacitance, parallel resistance, and Q.

5-7-3 MICROPROCESSOR SOFTWARE

The production of software to produce a practical microprocessor system involves techniques unfamiliar to many circuit designers. A computer programmer would be more at home designing such a system than would a hardware circuit designer. As microprocessor technology develops, there is likely to be an emphasis on simplifying software so that microprocessors can be applied with a minimum of programming experience.

All microprocessors have an *instruction set*, which is a list of instructions telling the microprocessor what to do. A particular microprocessor might have an instruction set with, say, 50 different instructions, and it could be set up to perform some given task by controlling it with a sequence of instructions called a *program*. The program might contain 100 or 1000 instructions, depending on the complexity of the system being designed, so that a given instruction would be used more than once. The microprocessor can only understand and respond to these instructions. A typical instruction sequence would tell the microprocessor to fetch data from one storage register, multiply it by data from another register, and place the result in a third register. The design of a microprocessor system involves writing and implementing a program that will supply the device with the correct sequence of instructions. However, some of the instructions are *conditional*, which means that the direction in which the sequence proceeds depends on whether or not certain conditions are met. This decision-making capability means that the instruction sequence, or program, proceeds in a manner that depends on external conditions. A feature of a microprocessor is, therefore, that it can decide on a course of action (depending on the circumstances) and select a response accordingly.

We shall briefly mention what is involved in programming. This requires the use of a certain amount of computer jargon, which we shall explain as we go along.

First, we must understand that the microprocessor, being a digital device, can respond only to 1s and 0s in the form of high or low voltage levels. An instruction must therefore consist of 1s and 0s, and microprocessors exist in which the instructions are grouped as 4, 8, 12, or 16 bits. A four-bit microprocessor might have eight-bit instructions, but they would occur consecutively in two groups of bits on four parallel lines.

If the instructions written by the programmer were in the form of 1s and 0s, the resulting sequence of instructions, or program, would of course consist of a long column of 1s and 0s. Such an approach has two main disadvantages.

First, it would be almost impossible to remember the code for each instruction so the programmer would have to be constantly referring to a table to discover for example, that the code for the instruction SUBTRACT is 10011011. Second, it would be extremely easy to make a mistake in writing the program and copying it into the instruction memory or ROM.

To overcome these disadvantages, the straight binary notation of 1s and 0s has been replaced with *hexadecimal* notation, a table for which is given in Table 5-10. Hexadecimal notation is similar to BCD except that it extends

TABLE 5-10 Hexadecimal notation

Hex	Binary
0	0 0 0 0
1	0 0 0 1
2	0 0 1 0
3	0 0 1 1
4	0 1 0 0
5	0 1 0 1
6	0 1 1 0
7	0 1 1 1
8	1 0 0 0
9	1 0 0 1
A	1 0 1 0
B	1 0 1 1
C	1 1 0 0
D	1 1 0 1
E	1 1 1 0
F	1 1 1 1

For example:
A = 1010
2A = 00101010
F2A = 111100101010

beyond the decimal number 9. Four binary digits can represent 16 different quantities. In BCD only 10 quantities are represented, whereas hexademical represents the maximum possible of 16 and therefore permits the more efficient use of four-bit data busses, registers, gates, and other devices. In Table 5-10, F2A is a much more manageable way of writing 111100101010 than the method using 1s and 0s. Notice that a hexadecimal word can represent either an instruction, or data, or an address. Therefore 21 (00100001) could be an instruction such as ADD, or it could represent the number 21 (twenty-one) or the number 33 (32 + 1), or a location in the ROM. The particular meaning is determined by where the hexadecimal word is placed in the microprocessor timing sequence.

To further simplify programming, *mnemonic* codes are employed. EXCH(R_X) might be the mnemonic code for "exchange the contents of register R_X with the accumulator." EXCH would have its own hexadecimal number or letter, say A, and this would be followed by a hexadecimal number or letter designating which register is to have its contents exchanged. The instruction A2(10100010) would instruct the microprocessor to "exchange the contents of register number

2 with the accumulator." Clearly, EXCH(R_2) is much easier to remember than 10100010, and a programmer would soon memorize many hexadecimal instructions to further simplify his work.

A program written in menemonic code is said to be written in *source language*. This must be translated into binary language so that the microprocessor can understand it; the binary language is often referred to as an *object code*.

There are two ways to accomplish this conversion, and the one chosen depends on the complexity of the system being designed, the sophistication of the particular microprocessor instruction set, and the length and subtlety of the program being written.

For the "simpler" systems, which can nevertheless involve programs with over 1000 instructions and which can replace hundreds or thousands of ICs, a hexadecimal keyboard can be used in conjunction with logic to program an erasable ROM directly. For example, in the sequence OFA2, OF would be the address of instruction A2 in the program sequence. OF is an eight-bit address that would place instruction A2 [exchange contents of register 2 with accumulator, the mnemonic for which is EXCH(R_2)] in the sixteenth position in the instruction sequence. It would be the sixteenth position because OF is hexadecimal for 00001111, which is the sixteenth number according to Table 5-9. 10 would be the seventeenth, 11 the eighteenth, 12 the nineteenth, and so on. An erasable ROM or a temporary RAM would be used to permit corrections, the final type of ROM selected depending on the quantities involved.

For more complex systems a computer program called an *assembler* could be used; among other things, an assembler converts the mnemonics into machine language, that is 1s and 0s. Computer programs can also be used to debug or correct the program being written, and to keep track of groups of instructions or subroutines, but it should be realized that a computer cannot actually write the program. The designer must decide what instructions to give the microprocessor and in what order.

When the designer has the ROM correctly programmed, it remains to interconnect the microprocessor, the ROM or ROMs containing the instructions, the RAM or RAMs for temporary storage of data, and peripheral interfaces to connect to keyboards, displays, or other external equipment.

5-7-4 A MICROPROCESSOR APPLICATION

Figure 5-58 shows in outline form how a microprocessor might be used in a specific control application. Suppose the problem involves controlling a set of traffic lights at the intersection of two roads, one major with heavy traffic flow and the other minor with light flow.

Signals from car sensors embedded at appropriate locations in the roads would be fed to the microprocessor via a peripheral interface adapter, an IC which converts signals from devices such as car sensors, teletype keyboards, and

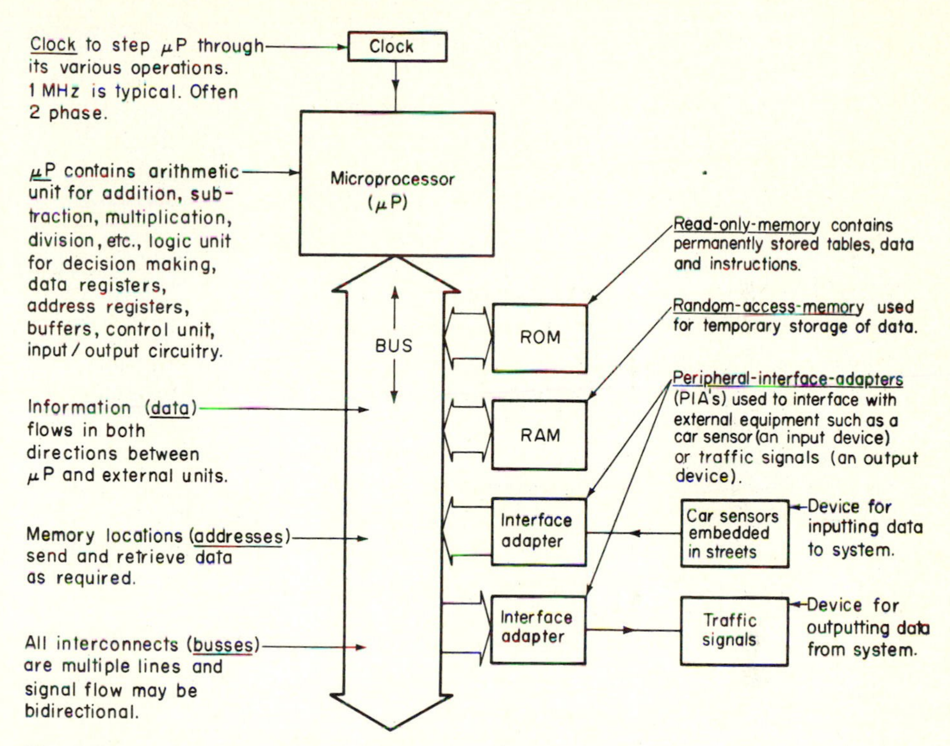

Fig. 5-58 Typical interconnections between microprocessor chip and auxiliary chips and equipment.

analog-to-digital converters into a form compatible with the microprocessor logic being used, perhaps TTL or I^2L.

The microprocessor looks at the interfaced signals from the sensors embedded in the road and continuously decides what to do about the status of the traffic lights. The decision-making involves a number of steps. For example, arithmetic is performed in keeping a running tally of how many cars are where. This tallied information would be stored in the RAM, along with other temporary information such as how long a particular car has been waiting to enter the major road from the minor one, or how many cars have passed through the green light since it turned green. Also, logic is performed based on data received from the ROM, which contains permanently wired instructions. A specific instruction might be as follows: If 100 cars have passed through the intersection along the major road since the light turned green, but none have passed for 15 s, turn the major-road signal to yellow and then red and turn the minor-road signal green. This instruction, along with many others, would be stored in binary-number form in the ROM. Based on information clocked into it from the car sensor via the peripheral interface adapter and on information accumulated in the RAM, the microprocessor would feed signals to the peripheral interface adapter which controls the traffic lights. This would cause the traffic lights to go through the preprogrammed response to the particular situation at the intersection.

We should note that one type of microprocessor could be used at many different intersections with a variety of traffic patterns. Each microprocessor would be programmed by its ROM to respond to the different conditions.

PROBLEMS

5-1 Explain why digital circuits are especially suited to monolithic construction.
5-2 Why are *both* binary and BCD number systems used in digital systems? Express the decimal number 111 in binary and BCD.
5-3 Table 5-1 shows the mechanical equivalents for an AND and an OR gate. Draw a similar mechanical circuit for an exclusive-OR.
5-4 The design of a digital system requires that the following expression be implemented:

$$F = ABCDE + A\bar{B}CDE + \bar{A}BCDE + \bar{A}\bar{B}CDE + CD\bar{E}$$

Draw a circuit using AND, OR, and NOT gates which satisfies the above formula.
5-5 Show, by simplifying by Boolean algebra, that the circuit of Prob. 5-4 can be replaced with one satisfying the expression $F = CD$.
5-6 Repeat Prob. 5-5 using a Karnaugh map.
5-7 A *JK* flip-flop initially has outputs of $Q = 1$, $\bar{Q} = 0$. What will happen to these outputs when trigger pulses are applied to the *CP* input after inputs are applied to the *JK* terminals in the following order:

(*a*) $J = K = 0$
(*b*) $J = 1$, $K = 0$
(*c*) $J = 0$, $K = 1$
(*d*) $J = K = 1$

5-8 ECL is also known by what other designation, and why? Why is ECL so fast? What feature in the output circuit of ECL gives the logic designer extra flexibility not found in other types of logic?
5-9 What is meant by "pullup" and "pulldown" in TTL circuits? What advantage do they provide?
5-10 Figure 5-10 shows an MOS gate which is a NAND gate for positive logic. What is it for negative logic?
5-11 Why are high speed and low power conflicting requirements in logic circuits?
5-12 Give two advantages and two disadvantages of each of the following types of memory:

(*a*) Punched cards
(*b*) Magnetic core
(*c*) CMOS read/write
(*d*) PROM

Give examples of where you would use each type.
5-13 Figure 5-22*a* shows a CMOS NOR gate. Design a CMOS NAND gate.

5-14 Figure 5-25 shows typical flip-flop addresses for a 1024-bit read/write RAM. What are the addresses for flip-flops 100, 256, and 257?
5-15 What logic family can operate with either high speed or low power? How is this accomplished?
5-16 List three ROM applications.
5-17 Name two ways in which ROMs and read/write memories are similar, and two in which they are different.
5-18 List three advantages of MSI-level digital integrated circuits, compared to discrete-component or conventional integrated-circuit counterparts.
5-19 Define MSI and LSI; include examples.
5-20 Sketch a counter and its waveforms for a frequency division of 8. Of 16.
5-21 Briefly describe an MS flip-flop, and explain how it prevents racing.
5-22 Show the block diagram and waveforms for a down counter with a capacity of 16 counts.
5-23 Convert the basic counter of Fig. 5-6 to a mod-5 counter.
5-24 List three applications of digital counters.
5-25 (*a*) Describe the multiplexer/demultiplexer principle.
(*b*) Give a multiplexer/demultiplexer application.
(*c*) What is the main reason for multiplexing?
5-26 Referring to Fig. 5-36*b*, sketch the gate inputs/outputs for the same 1101 ($A\ B\ C\ D$) input for the other three possible address conditions.
5-27 (*a*) What is the difference between a digital decoder and a demultiplexer?
(*b*) What is a 1-of-8 decoder?
(*c*) What is another name for a 4-line-to-10-line decoder?
5-28 Show the inverter inputs and AND-gate inputs/outputs of Fig. 5-38*b* for the decoding of decimal 3.
5-29 (*a*) What is the relationship between the excess-3 and the conventional binary code?
(*b*) List one advantage for each of the 8421, excess-3, and Gray BCD codes.
5-30 Show all the logic levels for the four possible combinations of inputs A and B for Fig. 5-40. (See Table 5-1.)
5-31 What is the difference between an analog and a digital comparator. (See Sec. 4-3-1.)
5-32 (*a*) Show the binary addition of $12 + 9$.
(*b*) Referring to Fig. 5-41*c*, show the logic level of each stage for the addition of $1 + 0 + 1$. ($A = 1$, $B = 0$, $C = 1$.)
5-33 (*a*) Subtract 5 from 7 in binary.
(*b*) Show the binary multiplication 3×6.
5-34 Show each step of the subtraction of 4 from 10 using the 2s-complement technique, assuming full binary adders (but not subtractors) are available.
5-35 Referring to Fig. 5-44, show the logic state of each bit of the accumulator, multiplier, and multiplicand for 5×4.
5-36 Briefly describe an arithmetic logic unit (ALU).
5-37 Briefly describe parity checking, including its purpose and both even and odd parity. Give binary examples.
5-38 Construct a truth table for each stage of Fig. 5-45 for several even- and odd-parity inputs.
5-39 Briefly describe the operation of (*a*) a basic digital-to-analog (D-A) converter and (*b*) a conventional counter-type analog-to-digital (A-D) converter.

5-40 What is the main advantage of a successive-approximation A-D converter?
5-41 How many A-D converter bits are required to achieve a resolution of ± 0.1 per cent of full scale?
5-42 Briefly describe the operation of a dual-slope integrating digital voltmeter.
5-43 What addition is required to Fig. 5-56 to accommodate a day display?
5-44 Explain the following terms as they apply to microprocessors: (*a*) instruction set, (*b*) mnemonic code, (*c*) source language, (*d*) object language, and (*e*) assembler.
5-45 Why are bypass capacitors required in many logic families?
5-46 List three logic families which are found in LSI products.

REFERENCES

1. Malvino, A., and D. Leach: *Digital Principles and Applications*, 2d ed. McGraw-Hill Book Company, New York, 1975.
2. Fairchild Semiconductor: "TTL Applications Handbook," pp. 9-6–9-16, 1973.
3. Texas Instruments Inc.: "TTL Data Book for Design Engineers," pp. 290–291, 1973.
4. Fairchild Semiconductor: *op. cit.*, pp. 5-55–5-61, 1973.
5. Malvino, A., and D. Leach: *op. cit.*, p. 121.
6. Fairchild Semiconductor: *op. cit.*, pp. 5-22–5-31, 1973.
7. Analog Devices Inc.: "Product Guide," pp. 240–242, 1975.

part 3 Optoelectronics

In the last few years a merger of the fields of optics and electronics has produced a new technology generally referred to as *opto-electronics* or *electro-optics*. This has become one of the fastest-growing segments of modern industry, as evidenced by the following list of well-known optoelectronic advances.

1. Photocopy machines
2. Electrical isolation with opto-isolators
3. Digital displays
4. Infrared heat-seeking missiles and night-vision devices
5. Remote sensing aircraft and spacecraft for reconnaissance, weather mapping, and crop-disease and forest-fire detection
6. Laser applications such as detached-retina coagulation, cancer destruction, communications, target identification, aircraft tracking, and holography
7. Industrial photoelectric controls such as counters, conveyer start-stop systems, proximity and level controls, pinhole detection, thickness measurement, food processing, sorting, and position detection and alignment
8. Computer tape and punched-card readers
9. Automatic street lighting and photographic exposure systems
10. Fire, smoke, and intrusion detection
11. Spectrometry
12. Character recognition
13. Solar-energy conversion

This part includes a chapter on light sources and displays, and another on photodetectors and source-detector combinations. Solid-state devices are emphasized throughout, except for comparison purposes or where solid-state equivalents are not presently available. The analysis and design of simple opto-electronic circuits are stressed, and the optics is limited to that necessary for a basic understanding of the photodevices and their applications.

chapter 6

Light Sources and Displays

6-1 BASIC PHYSICS OF LIGHT

Before discussing electronic devices that convert electrical energy to radiant energy and vice versa, it is necessary to cover some of the basic physical principles of light. (It is assumed that the reader is familiar with basic semiconductor physics.) The optics in this chapter may be somewhat unfamiliar to most electronic workers; however, these principles are repeated wherever necessary, to help develop a good working knowledge. It is imperative that one fully comprehend the very basic principles of light in order to successfully apply optoelectronic devices.

Technically, the term *light* means *electromagnetic radiation* of a frequency or wavelength which can be perceived by the human eye. However, a glance at Fig. 6-1 shows that the percentage of the electromagnetic spectrum which is

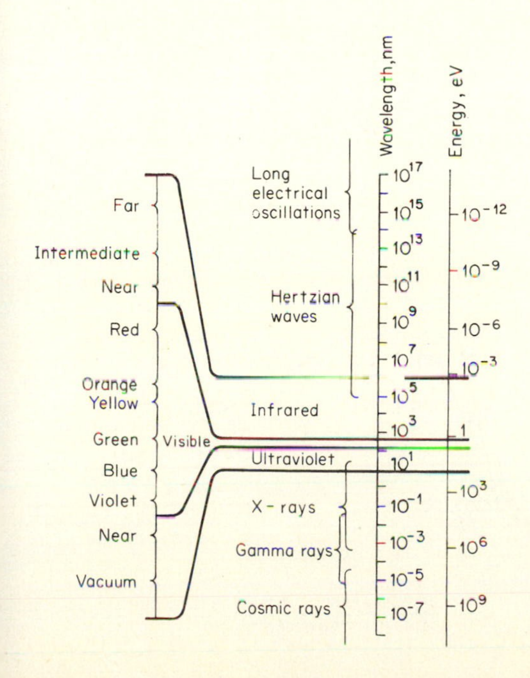

Fig. 6-1 Electromagnetic spectrum. *After J. Mauro, "Optical Engineering Handbook," p. 11-3, General Electric, Syracuse, N.Y., 1966.*

visible is very small. Hence, the term *light* is often used rather loosely to indicate radiant energy of any wavelength.

Two different systems of radiant measurement thus exist, one termed *photometric* and based on the response of the eye, and the other termed (more broadly) *radiometric* and based on the energy of all wavelengths of radiation regardless of physiological phenomena such as visibility. Since most solid-state photodetectors also respond to wavelengths outside the visible spectrum, the radiometric system will be used throughout this section. In some instances non-standardization by manufacturers has resulted in the carryover of photometric terminology into the specifications of many sources and detectors. In such cases it may be necessary to convert to the radiometric system, and Table 6-1 may be used for this purpose. However, the reader is cautioned against indiscriminate use of such conversions in that they are not applicable in all situations or at all wavelengths. The graph of Table 6-1 illustrates this problem. Superimposed on this graph are the typical response of a Si photodetector, the output spectrum of an incandescent lamp, and the visible spectrum. If we were interested in the response of the Si photodetector with this light source and the source were specified in photometric terms (visible spectrum only), we would obviously be in trouble. First, the lamp spectrum is considerably broader than the visible spectrum, and a photometric specification in the visible region only would hardly define the portion of the lamp spectrum to which the photodetector would

TABLE 6-1 Optical conversion table and superimposed spectra of a typical source, detector, and visible spectrum illustrating the problem with the photometric system.

To Convert	Into	Multiply by
candle/cm^2	lamberts	3.142
footcandle	lumen/m^2	10.764
footcandle at standard 2870°K*	milliwatts/cm^2	0.05
lumen	spherical candle power	0.07958
lumen at peak of eye response (555 nm)	watt	0.00148

* *Not true at all wavelengths.*

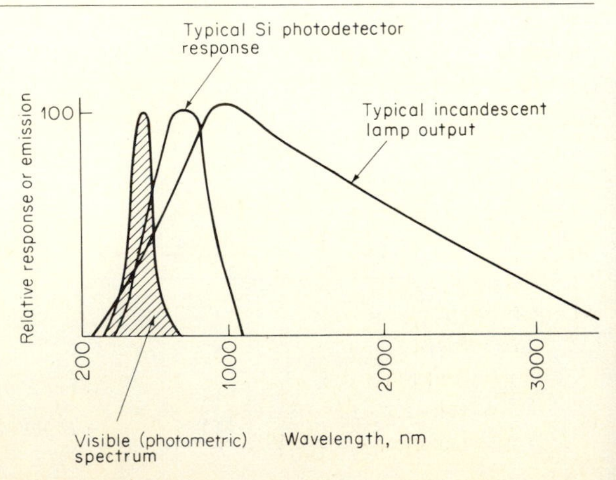

respond. (Of course, the detector does not respond to all of the lamp's radiated power either.) One could perform a simple graphical integration (calculate the areas under the curves) to estimate such conversions; however, Table 6-1 can be used to obtain "ballpark" estimates. A more practical method is to obtain actual sources and detectors with as close a spectral match as possible and then test various combinations to determine the actual responses. This is financially possible because most of the photodetectors and sources mentioned in this section cost only a few dollars or less.

Table 6-2 gives the formulas, in both photometric and radiometric units, for both point and area light-source cases. The point-source formulas of Table 6-2*a* apply when the source-to-detector separation is at least 10 times the diameter of the photodetector or 10 times the diameter of the source, whichever is greater. If this condition is not met, the area-source equations of Table 6-2*b* apply. Notice that the units in the radiometric system are generally watts (W), whereas candles or lumens are used throughout the photometric system.

Table 6-2*a* shows that flux is inversely proportional to the distance from the source squared ($H = I_R/d^2$). In other words, a point source of 1 W radiant intensity at 1 cm distance would be reduced to $1/2^2$ or 0.25 W at 2 cm. In an area-source case, such as using a camera to photograph a scene, increasing the iris opening increases the available steradiancy, or portion of the total light available to the film; hence the increase in exposure with increased iris opening. An optical characteristic shown by Table 6-2*b* is that incident flux density varies directly as a cosine function of the angle θ. As an example, at 45° off axis, an irradiance of 1 W/cm² would be reduced to (cos 45°) (1 W/cm²) or 0.707 W/cm². The use of these equations will become clearer as we proceed with the examples in this section.

There are two related aspects of light that are used together or separately to explain many phenomena associated with light. Optical phenomena such as interference and diffraction patterns are explained by wave theory such as that involved with radio waves. However, most of the phenomena discussed in this chapter are better explained by the quantum theory, in which the particlelike aspects of light are considered. According to this theory, light consists of discrete quanta or packets called *photons*. These are uncharged particles, each having an energy which depends only on their frequency or wavelength as given by

$$E = h\gamma = h\frac{c}{\lambda} \tag{6-1}$$

where E = energy, eV
h = Planck's constant, 4.137×10^{-13} eV·s
γ = frequency, Hz
c = velocity of light, 3×10^8 m/s
λ = wavelength of light, m

TABLE 6-2 (a) Point light source calculations*

(To be used only when the source-to-detector separation (d) is ≧ 10 times the diameter of the source.)

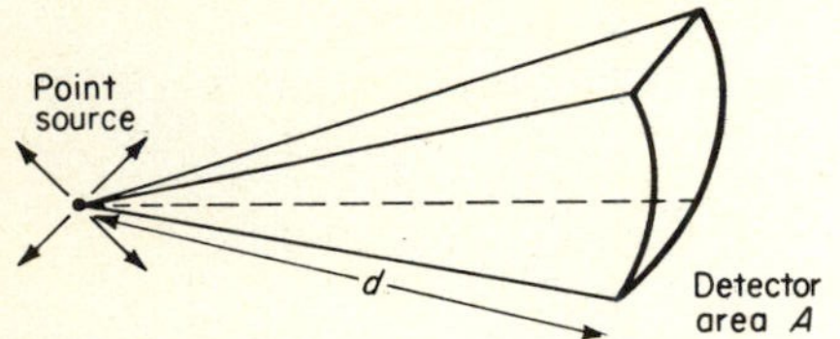

Surface area of a sphere = $4\pi d^2 \approx A$ (where d = the sphere radius). The solid angle or steradian = A/d^2. Therefore a sphere contains 4π sr.

Definition	Radiometric				Photometric			
	Quantity	Symbol	Units	Formula	Quantity	Symbol	Units	Formula
Source intensity	Radiant intensity	I_R	W/sr	$I_R = \frac{P}{4\pi}$	Luminous intensity	I_L	candle (lumen/sr)	$I_L = \frac{F}{4\pi}$
Flux at detector	Radiant flux	P_S	W	$P_S = \frac{I_R A}{d^2}$	Luminous flux	F_S	lumens	$F_S = \frac{I_L A}{d^2}$
Incident flux density at detector	Irradiance	H	W/d^2	$H = \frac{P_S}{A} = \frac{I_R}{d^2} = \frac{P}{4\pi d^2}$	Illuminance	E	(footcandle lumen/ft²)	$E = \frac{I_L}{d^2} = \frac{F}{4\pi d^2}$
Total flux from source	Radiant flux	P	W	$P = 4\pi I_R$	Luminous flux	F	lumens	$F = 4\pi I_L$

* After Howell, E.: The Light Activated SCR, General Electric, *Appl. Note*, 200.34, pp. 7 and 8, 1965.

TABLE 6-2 (b) Area light source calculations*

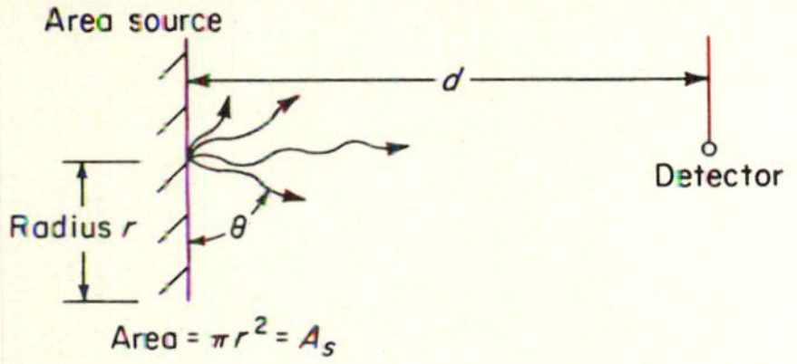

Area source

d

Detector

Radius r

θ

Area $= \pi r^2 = A_s$

Definition	Radiometric				Photometric			
	Quantity	Symbol	Units	Formula	Quantity	Symbol	Units	Formula
Source intensity	Radiance	B_r	W/(cm^2)(sr)		Luminance	B_L	candle/cm^2	
Emitted flux density	Radiant emittance	W	W/cm^2	$W = \pi B_r$	Luminous emittance	L	lumen/cm^2 (lambert)	$L = \pi B_L$
Incident flux at detector	Irradiance	H	W/cm^2	$H = \frac{B_r A_s}{r^2 + d^2}$	Illuminance	E	lumen/ft^2 (footcandles)	$E = \frac{B_L A_s}{r^2 + d^2}$
Total flux from source	Radiant flux	P	W	$P = WA_s$	Luminous flux	F	lumens	$F = LA_s$

* After Howell, E.: The Light Activated SCR, General Electric, *Appl. Note*, 200.34, pp. 7 and 8, 1965.

Notice, from Eq. (6-1), that as the wavelength of light drecreases, its energy increases. This phenomenon is apparent in the electromagnetic spectrum of Fig. 6-1. The exact energy associated with a particular wavelength of light can be calculated by using

$$E(\text{eV}) = \frac{1{,}240}{\lambda(\text{nm})} \tag{6-2}$$

[Units of length will generally be expressed in nanometers (nm). 1 angstrom (Å) = 10^{-8} cm = 0.1 nm = 10^{-4} *microns* or μm, and 1 μ = 1 μm = 2.54×10^{-4} in = 10,000 Å.]

Figure 6-2 will be used to illustrate the important phenomena of emission and absorption of photons from various materials, which generally obey Eqs. (6-1) and (6-2). The outer electron shell of Fig. 6-2 is called the *valence shell*, and the electrons therein are termed *valence electrons*. This outer ring or shell of an atom determines a great many of an atom's characteristics, because the valence electrons are the farthest from the nucleus and hence are the least tightly held to the rest of the atom.

Figure 6-2 demonstrates the two possible valence-electron transitions. The crooked arrow in both cases represents radiant energy. If adequate radiant energy can somehow be added to the system, a valence electron may be "knocked" out of its orbit. This electron can then escape from the attractive force of the nucleus protons and become a free current carrier, thus completing the photon-to-electron (radiant-to-electrical energy) conversion. The radiant energy required to excite the valence electron into conduction (free current carrier) is referred to as the optical *energy gap*, and Table 6-3 shows typical optical energy gaps for several materials used in the fabrication of photodetec-

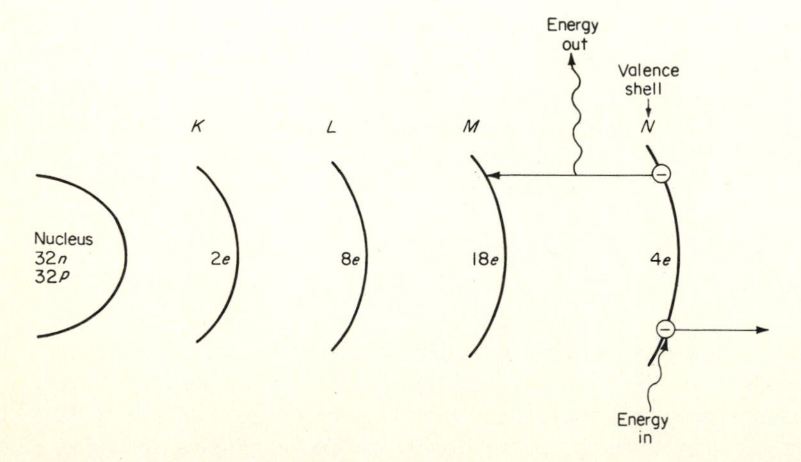

Fig. 6-2 Silicon (Si) atom electron shells, showing two possible valence-electron transitions.

TABLE 6-3 Optical energy gaps of several photodetector materials

Name	Chemical Symbol	Optical Energy Gap at 300°K, eV
Cadmium sulfide	CdS	2.4
Gallium phosphide	GaP	2.2
Cadmium selenide	CdSe	1.7
Gallium arsenide	GaAs	1.4
Silicon	Si	1.1
Germanium	Ge	0.7
Indium arsenide	InAs	0.43
Lead sulfide	PbS	0.37
Lead telluride	PbTe	0.29
Lead selenide	PbSe	0.26
Indium antimonide	InSb	0.23

tors. For example, if radiant energy ≥ 1.1 eV [approximately 1127 nm from Eq. (6-2)] strikes an Si photodetector, a valence electron will be freed. A free hole will also be created, since a hole is defined as the absence of an electron. If these free current carriers encounter an electrostatic field which separates them rather than allowing them to recombine, the free current carriers become a measure of the amount of radiant energy striking the photodetector. This process is the basis of all photodetectors, regardless of the particular material.

The other possible valence-electron transition shown in Fig. 6-2 is an inward transition wherein the amount of radiant energy emitted is proportional to the difference in energy between the two electron shells [Eq. (6-2)]. The radiant output of the light-emitting diodes of Sec. 6-2-4 is a good example of this phenomenon.

6-2 LIGHT SOURCES

This section will be devoted to the various available light sources, with emphasis on those of reasonably small size, and especially the solid-state versions. Although many optoelectronic applications involve the measurement of some natural source of light, a large percentage require some controllable artificial light source. Examples include photographic exposure and processing, position detection, information transfer isolation, transmission and absorption studies, and photodetector calibration.

6-2-1 NATURAL LIGHT SOURCES

Any discussion of light sources must include such natural sources as the sun. A detailed knowledge of the sun's spectral distribution and integrated output power at various places within our galaxy and on the earth's surface is important to those involved in solar-energy conversion or simulation. Figure 6-3 shows a typical solar spectrum and irradiance at various distances from the sun.

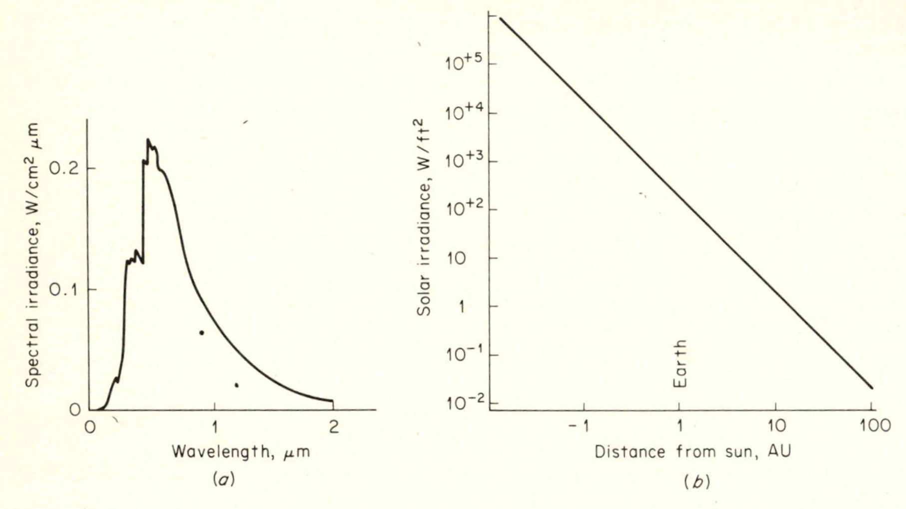

Fig. 6-3 (*a*) Solar spectrum; (*b*) solar irradiance at various distances from the sun. One astronomical unit = the earth-sun separation.

The *solar constant* (or energy per unit area received outside the earth's atmosphere at a distance equal to the earth-sun separation) is approximately 140 mW/cm^2. This brings up the question of how much of this energy is available at the earth's surface after normal atmospheric absorption. Many portions of the solar spectrum are completely absorbed in our atmosphere, and complete cutoff occurs below $\approx$300 nm and above 25 μm. However, on a clear day the solar energy at the earth's surface is $\approx$100 mW/cm^2.

There are several other natural sources such as the sky, moon, stars, aurora, etc., whose brightness is considerably less than that of the sun.

6-2-2 INCANDESCENT LAMPS

Even in a text devoted to solid-state devices, incandescent and gaseous-discharge lamps must be considered because of their widespread use, and for comparison purposes. Although considerable progress has been made in the areas of semiconductor light-emitting diodes (LEDs) and lasers (Secs. 6-2-4 and 6-2-6), there are still several reasons for choosing a conventional incandescent lamp in many applications.

1. Lamps generally offer a much broader spectral output. As an example, incandescent-lamp outputs span the near ultraviolet through the intermediate infrared, whereas the broadest LED output spectrum approaches 100 nm.
2. Lamps have gone through many years of development, andas a result there are many thousands of inexpensive varieties available. Also, the standard calibration light source is a lamp.

The *incandescent lamp* is the most common source of home lighting, and its basic structure of a tungsten filament enclosed in an evacuated glass envelope is well known. One way to explain incandescent-lamp operation is to consider the definition of the word *incandescence*. It is light emission due to the temperature of the source—in other words, thermal radiation. In Sec. 7-1-8 some of the basic concepts of an ideal blackbody will be discussed, and the same general principles apply here. The total radiant power per unit area of a blackbody varies as the fourth power of the temperature, and, as shown in Fig. 7-17, the shape of the blackbody curves remains approximately the same, but the peaks shift upward and toward the ultraviolet with increasing temperature.

No real metal meets the conditions of an ideal blackbody (i.e., absorbing or radiating all wavelengths of radiation), but tungsten filaments behave very similarly to a blackbody except that their radiation at any wavelength is much less. Hence, the light and the color from an incandescent lamp are due to the temperature of the tungsten filament, which varies from 2000 to 3500 K, depending on the particular lamp and input power.

Figure 6-4 shows the important incandescent-lamp characteristics. This figure is important because it serves several purposes. It gives definitions of the terms commonly used in rating lamps, the various equations and graphs expressing candlepower, life, and filament current as a function of operating voltage, plus typical ratings and lifetimes for a few representative subminiature lamps. Notice, from Fig. 6-4*a*, that the luminous intensity from such lamps varies directly as the applied voltage to the 3.5 power, while the life varies inversely to the 12th power of voltage. In other words, the lifetime is very strongly influenced by the operating voltage and can be significantly reduced by overdriving.

The term *mean spherical candlepower* (MSCP) as shown in Fig. 6-4 is the standard means of expressing a lamp's visible output. MSCP refers to the mean photometric candlepower incident on a reflectance sphere which surrounds the entire lamp. In most applications the radiation from the rear and sides of the lamp is lost unless mirrors or lenses are used; therefore only a small portion of the MSCP is actually intercepted by any detector. The filament shape also causes the output radiation to vary with position; however several lamps are available with internal lenses and/or reflectors so that the power in a given spot size is increased tremendously. (Household lamps are specified in terms of average total light output in lumens, which is equivalent to MSCP in candles.)

As an example of the radiant output from a subminiature incandescent lamp, consider the 7152 lamp of Fig. 6-4*a*. The irradiance H at the detector can be estimated by

$$H = \frac{P_{\text{out}}}{4\pi d^2} = \frac{P_{\text{in}}\,\varepsilon}{4\pi d^2} \tag{6-3}$$

[for the case where the source-to-detector separation is at least 10 times the source diameter (Table 6-2*a*)]

Section	Base	Symbols	Lamp number	Design		M.S.C.P.	Life, hr
				Volts	Amps.		
T-1							
	Wire terminal (1.25 in min; 0.125 in max)		7152	5.0	0.115	0.147 ±25%	40,000
			CM8-1175	5.0	0.125	0.22 ±25%	5,000
			CM8-1096	5.0	0.145	0.16 ±25%	10,000
			CM8-1088	10.0	0.027	0.10 ±25%	10,000
			CM8-1089	12.0	0.06	0.15 ±25%	16,000
			CM8-1111	14.0	0.065	0.15 ±25%	12,000
			CM8-1092	18.0	0.026	0.15 ±25%	10,000
			6838	28.0	0.024	0.15 ±25%	16,000

Design voltage

Design voltage represents the voltage at which the lamp, when operated, will yield its rated amperage, candlepower, and life characteristics.

Design amperage

Design amperage represents the value of current flowing through the lamp when operated at design voltage.

M.S.C.P.

Light output of subminiature lamps is generally rated in terms of Mean Spherical Candlepower (M.S.C.P.) which is the average of the total output of the lamp as measured in a photometric reflectance sphere at design voltage.

Life

Lamp life is the average life obtained under static laboratory conditions when the lamp is operated at design voltage. Actual life may differ from laboratory life due to the effects of shock, vibration, voltage variations and environmental conditions encountered in lamp applications.

Lamp rerating

Electrically, subminiature lamps are dynamic devices. Candlepower, current, and life may be varied for all subminiature lamps by operating the lamps at other than design voltage. Rerating may be accomplished by the following formul as:

Rerated M.S.C.P. =

$$\left(\frac{V}{V_1}\right)^{3.5} \times \text{M.S.C.P. at design volts}$$

Rerated life =

$$\left(\frac{V_1}{V}\right)^{12} \times \text{life at design volts}$$

Rerated current =

$$\left(\frac{V}{V_1}\right)^{0.55} \times \text{current at design volts}$$

V = application voltage
V_1 = design voltage

(a)

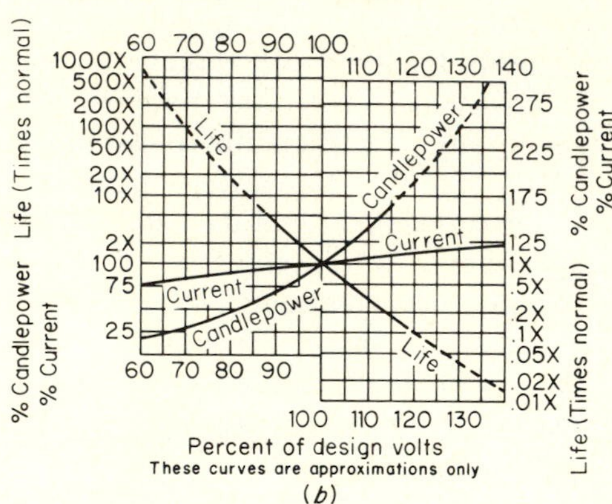

(b)

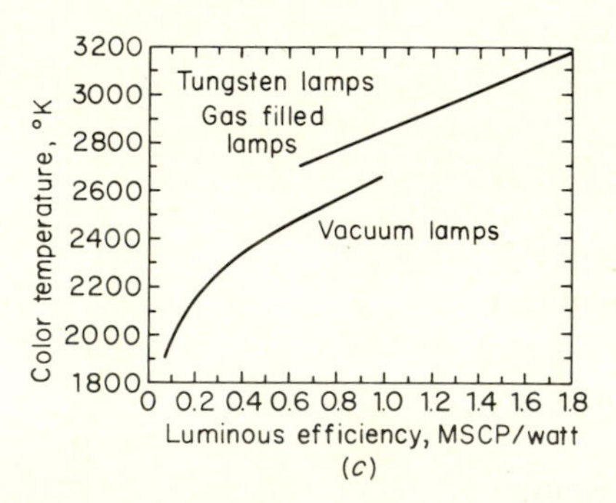

(c)

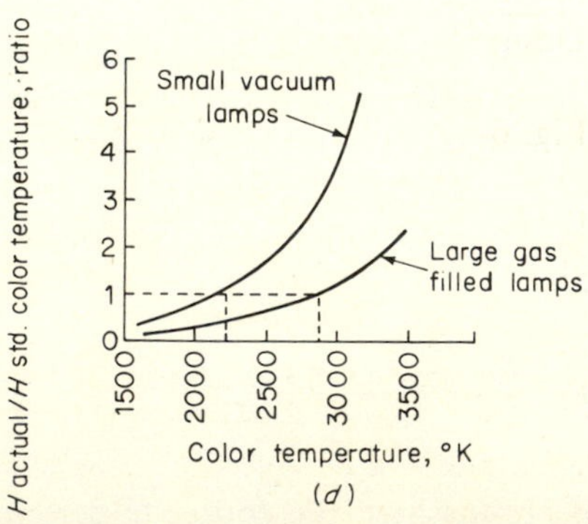

(d)

Fig. 6-4 (*a*) Typical subminiature incandescent-lamp specifications, nomenclature, and formulas. (*Chicago Miniature Lamp Works*. (*b*) Graphical representation of rerated life, current, and candlepower. *General Electric Miniature Lamp Dept*. (*c*) Color temperature versus luminous efficiency. *General Electric Miniature Lamp Dept*. (*d*) $H_{\text{actual}}/H_{\text{standard color temperature}}$ versus color temperature. *Adapted from J. Mauro, "Optical Engineering Handbook," General Electric, Syracuse, N.Y., 1966.*

where P_{out} = radiant output power, mW
P_{in} = electric power into the lamp ($V_{fil} \times I_{fil}$), mW
ε = efficiency of conversion of electric power to radiant power for incandescent lamps (MSCP/watt of Fig. 6-4*c*). Approximately 80 per cent for small vacuum lamps and 90 per cent for larger gas-filled lamps
d = source-to-detector separation, cm
$4\pi d^2$ = surface area of a sphere, in this case the detector in question

For the 7152 lamp operated at design levels and a 1.25-in source-to-detector separation,

$$H = \frac{(5\text{V})(115\text{ mA})(0.8)}{(12.56)[(2.54\text{ cm/in})(1.25\text{ in})]^2} = 3.63\text{ mW/cm}^2$$

To put this number in perspective, recall that the solar irradiance is ≈ 100 mW/cm^2; therefore an irradiance of a few milliwatts per square centimeter from a subminiature incandescent lamp seems reasonable.

If the 7152 lamp were operated at 4 V instead of the 5-V design filament voltage, the changes in life and irradiance at the detector would be as follows. Referring to Fig. 6-4*a*, we see that the rerated lifetime would be

$$\left(\frac{\text{Design voltage}}{\text{Applied voltage}}\right)^{12}\left(\begin{matrix}\text{life at}\\ \text{design voltage}\end{matrix}\right) = \left(\frac{5}{4}\right)^{12}(40 \times 10^3\text{ h}) \approx 582{,}077\text{ h}$$

Alternatively, the graph of Fig. 6-4*b* can be used, and entering the percent-of-design-voltage axis at $\frac{4}{5}$ or 80 percent gives $\approx(16)(40 \times 10^3$ h), or 640,000 h. Figure 6-4*a* shows that at a filament voltage of 4 V the rerated 7152 lamp filament current is

$$\left(\frac{\text{Applied voltage}}{\text{Design voltage}}\right)^{0.55}\left(\begin{matrix}\text{current at}\\ \text{design voltage}\end{matrix}\right) = \left(\frac{4}{5}\right)^{0.55}(0.115\text{ A}) = 0.1\text{ A}$$

From Fig. 6-4*a*, the rerated MSCP is

$$\left(\frac{4}{5}\right)^{3.5}(0.147\text{ MSCP}) = 0.067\text{ MSCP}$$

(Figure 6-4*b* gives approximately the same values for rerated current and MSCP.)

Hence, the rerated efficiency or MSCP/watt is 0.067 MSCP/[0.1 A)(4 V)] = 0.168. Entering Fig. 6.4*c* at this value shows that the rerated color temperature is $\approx$2090 K. (Most subminiature, miniature, and $<$10-W lamps are vacuum lamps.) Figure 6-4*d* can now be used to find the ratio H_{actual}/H_{rated} for this new color temperature. The ratio is ≈ 0.75; hence

$$\begin{aligned} H_{2090\,K} &= (H_{rated})(0.75) \\ &= (3.63\ \text{mW/cm}^2)(0.75) \\ &= 2.72\ \text{mW/cm}^2 \end{aligned}$$

The above example was calculated in radiometric units; however, the photometric equations of Table 6-2*a* ($E = F/4\pi d^2$) and the relationships between MSCP and color temperature shown in Fig. 6-4*c* could be used to calculate the visible output. Chapter 7 contains examples of the outputs of various photodetectors with incandescent lamps and other light sources.

Figure 6-4*c* shows that increasing the lamp's operating temperature increases its luminous efficiency, but this is accompanied by an increased rate of filament evaporation. The evaporation reduces the lamp's output for two reasons: (1) the filament becomes smaller and its lifetime and output are thus reduced, and (2) the evaporated metal will deposit on the inside of the envelope and thus restrict the output illumination. The term *color temperature* as used in Fig. 6-4 refers to the color or spectral distribution of incandescent lamps. An approximate relationship between color and filament temperature is that the color temperature is ≈ 100 K higher than a given filament temperature. The initial incandescent-lamp filament inrush current can be as high as 10 times the normal current. This is caused by the much lower filament resistance at room temperature. Inrush current can be troublesome in many ways. It may damage semiconductor drive devices, generate RF interference, and reduce the life of the lamp. (Notice that the filament of a household incandescent lamp always burns out when the lamp is switched on.)

The lifetimes of lamps can be extended considerably by the addition of an inert gas to suppress filament evaporation. This process involves the collision of the emitted filament atoms with the gas molecules so as to return some of the filament material. This regenerative cycle occurs at a particular temperature at which conditions are favorable for a gas-metal combination. The tungsten-gas molecules then diffuse toward the filament and are dissociated, leaving the tungsten on the filament and the gas ready for further recombination. The most popular form of this type of lamp is the high-temperature quartz-iodine lamp, which is now the most popular standard lamp because of its smaller size and higher ultraviolet output. The increased UV output is due to the higher operating temperature and quartz envelope.

The stability and repeatability of incandescent lamps are fair as long as their operating current or voltage is well regulated. The amount of precision required can be appreciated by noting that the output candlepower changes as the 3.5 power of the applied voltage (Fig. 6-4*a*). The temperature stability of such lamps is excellent because a change in ambient temperature of a few degrees Celsius is obviously negligible compared to the operating temperature.

Tests of miniature-lamp turn-on time and stability have shown that:

1. Operating miniature lamps at 3000 K color temperature ($\approx$two times their rated voltage) increases their efficiency, but the resulting stability degradation makes such operation impractical.
2. Lower-voltage lamps require longer to stabilize, owing to the higher thermal time constant inherent in the larger-diameter filaments used in lower-voltage lamps.
3. Fair stability with time can be achieved at slightly less than design operating conditions.
4. Current regulation produces the best stability at high operating temperatures. Poor stability is achieved with voltage regulation because, as the filament resistance increases with tungsten boil-off, the input electric power decreases as shown by the V^2/R power formula. With constant-current regulation, the input current is held constant when the filament resistance increases; hence the action tends to compensate for the evaporation ($P = I^2R$).
5. Voltage regulation is superior at moderate power levels and produces faster turn-on times because the initial current surge is unlimited.
6. Lamps which have been used turn on faster under current regulation.
7. Turn-on times are less than 1 s for the higher-voltage miniature lamp operated at constant voltage, or at constant current if the lamp is used.

Incandescent lamps are notoriously poor at converting electric power to visible light, as demonstrated by their output of 6 percent visible light and 80 percent infrared.

6-2-3 GAS-DISCHARGE LAMPS

The neon glow lamp and fluorescent lamp are well-known examples of gaseous-discharge lamps. The operation of discharge lamps is explained with the aid of Fig. 6-5. All such lamps consist of a transparent enclosure filled with one of the rare gases, electrodes, and a supply of external electric power. The gas is ionized by collisions with electrons from a heated filament or by field emission from the electrodes. The rate of ionization and thus the conductivity of the gas depends on the discharge current. In fact the conductivity of the gas increases so rapidly that some form of external current limiting is necessary in all discharge lamps to prevent self-destruction. (A series resistor often fulfills this requirement.)

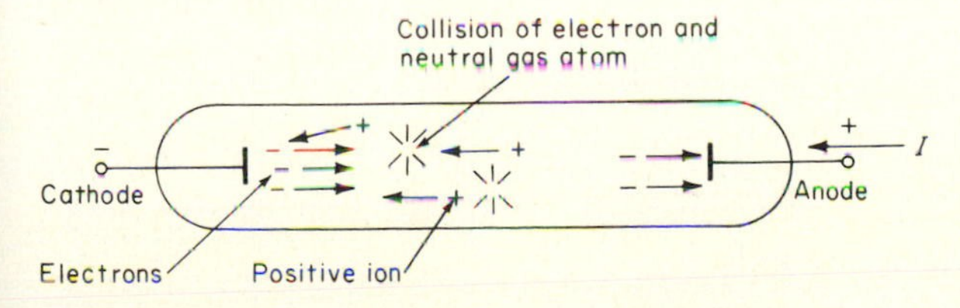

Fig. 6-5 Electric discharge in a gas-discharge lamp.

Light is produced in a gas discharge when the ionized gas atoms return from an excited energy level to lower levels (Sec. 6-1). This process may occur in one or several steps with the emission of wavelengths of light characteristic of the particular energy drops.

In low-pressure discharge lamps the output spectrum consists of various discrete spectral lines such as those shown for the low-pressure mercury lamp in Fig. 6-6*a*. Figure 6-6*b* shows the corresponding energy-level diagram with some

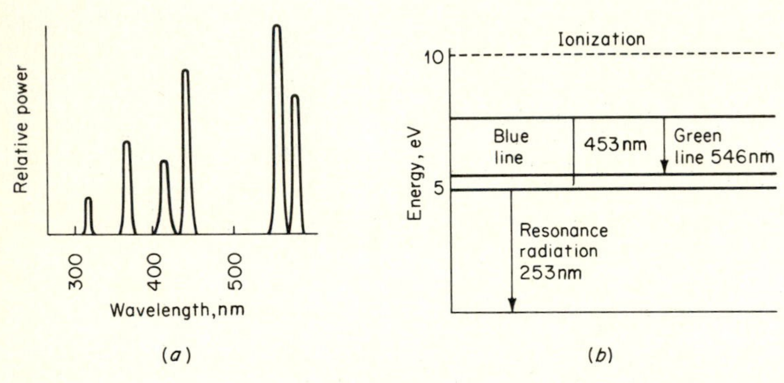

Fig. 6-6 (*a*) Spectral output of a low-pressure mercury lamp; (*b*) energy-level diagram for some of the strongest transitions in low-pressure mercury lamps. *After H. Hewitt and A. Vause, "Lamps and Lighting," pp. 30 and 307, American Elsevier Publishing Company, Inc., New York, 1966.*

of the strongest radiating transitions labeled. The location of such lines is very repeatable; therefore such lamps are often useful in calibrating or checking the wavelength scales of optical instruments. At higher gas pressures the lines are broadened considerably and can in fact become a continuous spectrum much like that of an incandescent lamp (Sec. 6-2-2).

To start the discharge or ionization, it is often necessary to preheat the gas or apply a sudden high voltage to an auxiliary electrode. The gas pressure, electrode material, and electrode spacing can all be varied to produce the desired starting voltage, maintaining voltage, and lifetime.

Figure 6-7 shows excerpts from a neon-glow-lamp data sheet. These lamps have a short electrode separation, and the light is emitted from a glow discharge at the cathode. A typical neon-lamp spectral distribution is shown in Fig. 6-7*a*. As mentioned in Sec. 7-1-2, this source comes closest to matching the response of CdS bulk photoconductors. Figure 6-7*b* gives formulas for determining the lamp current.

Neon glow lamps are used extensively in electronic circuitry for:

1. Overvoltage protection (the lamp is connected in parallel with the sensitive component so that when the voltage exceeds the lamp's breakdown voltage, the lamp provides a path to ground)
2. Low-leakage voltage indicators
3. Sources for photoresistor optical isolators (Sec. 7-2-1)

4. Voltage regulation or level detection
5. Thyristor trigger devices (Sec. 8-2-3)

Neon glow lamps are made with a variety of breakdown voltages ranging from 55 to 150 V, as shown in Fig. 6-7*c*. Ionization times vary with applied voltage but are generally less than 0.5 ms.

As an example of the required neon-lamp external series resistance, consider an A_1A lamp of Fig. 6-7*b* at a desired operating current of 0.5 mA. Solving the formula of Fig. 6-7*b* for the required series resistance (at 115 V) gives

$$R_x = \frac{72\ \text{V}}{0.5\ \text{mA}} - 8\ \text{k}\Omega = 136\ \text{k}\Omega$$

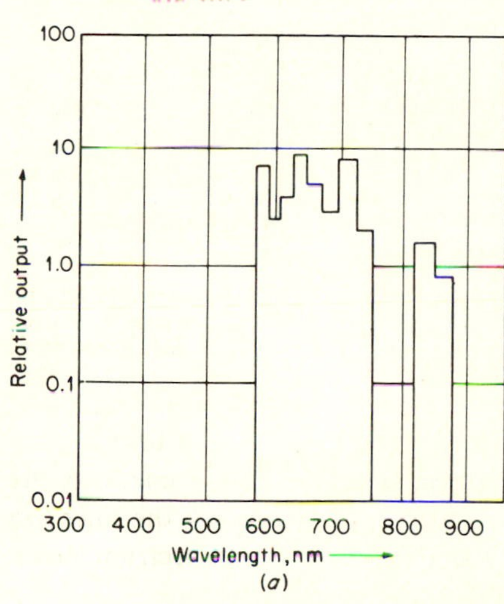

LAMP CURRENT

To determine light output and life, it is necessary to calculate the current which is determined by the supply voltage, series resistance and lamp characteristics. Using this formula lamp current can be calculated:

$$I = \frac{V_S - V_O}{R_I + R_X}$$

V_S = Supply Voltage
V_O = Counter EMF
R_I = Internal Resistance
R_X = External Resistance

AVERAGE AC RMS CHARACTERISTIC VALUES FOR GLOW LAMPS

LAMP TYPES	V_O-VOLTS	R_I-OHMS
A1A, A1B, A1D, A1G, K2A	43	8K
A2B, A9A, C7A	43	4K
B1A, B3A	40	7.5K
A1C, A1H	66	1.8K
C2A, C9A, K1A, K1D, K3A	63	– .5K
B2A	49	.7K
5AH, 8AE	53	.5K

(*b*)

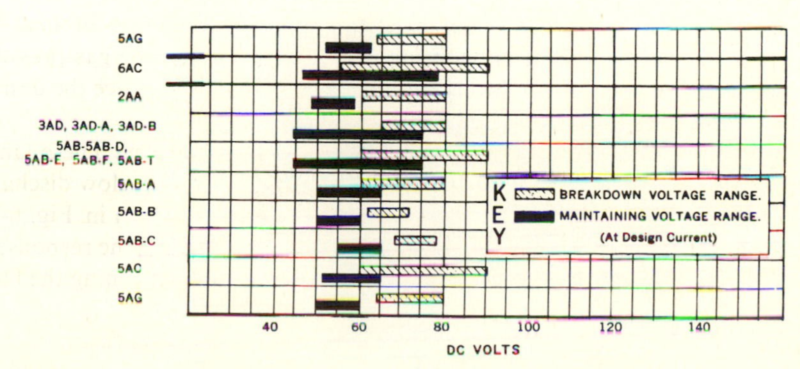

(*c*)

Fig. 6-7 (*a*) Typical high-brightness neon-lamp spectral distribution. *Hewlett-Packard.* (*b*) Neon-lamp current formula plus typical counter-emf and internal-resistance values. *General Electric Lamp Dept.* (*c*) Breakdown and maintaining voltages for neon lamps. *General Electric Lamp Dept.*

This is rather typical of the series resistance that must be added to neon lamps which do not have such a resistance built in.

For optoelectronic applications the following neon-lamp facts should be considered:

1. The radiant output from neon lamps is low (typically <1.0 mW/cm^2); hence the high-brightness versions should be used.
2. Neon-lamp starting voltages vary directly with ambient light level because the cathodes are photosensitive. This problem is significantly reduced in lamps with radioactive additives.
3. The firing and maintaining voltages have a temperature coefficient of ≈ 50 mV/°C.

6-2-4 LIGHT-EMITTING DIODES

As discussed in Sec. 6-1, a junction diode can emit light or exhibit *electroluminescence* (the emission of light from a solid with the excitation provided by an electrostatic field). The emitted light in this case comes from hole-electron recombinations. This becomes clear when one realizes that when a free electron recombines, it may fall all the way from an unbound or higher-energy level to its ground state (Fig. 6-2), releasing a photon of a wavelength corresponding to the energy-level difference associated with this transition [Eq. (6-2)]. In solid-state light-emitting diodes (LEDs), the supply of higher-energy electrons is provided by forward biasing the diode, thus injecting electrons into the n region (and holes into the p region) as shown in Fig. 6-8. The injected holes and electrons then recombine with the majority carriers near the junction. The recombination radiation is emitted in all directions, with most of the external light observed at the top surface, since the amount of material between the junction and that surface is the least. (Electroluminescence is also exhibited in many bulk or undoped materials under high field conditions; however, a doped junction offers a very efficient and well-understood means of generating excess electrons and holes at the proper energy levels.)

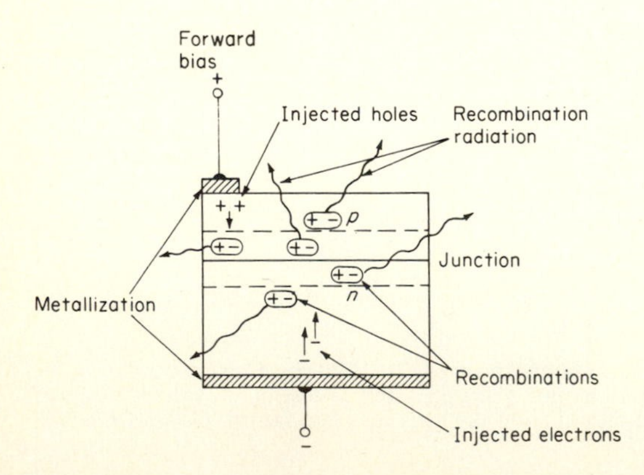

Fig. 6-8 Cross section of a light-emitting diode (LED).

Figure 6-9*a* is a general energy-level diagram which will be used to describe the possible radiative and nonradiative electron transitions involved in LEDs. Figure 6-9*b* shows the spectral output of several of the most common LEDs. Actually, radiative and nonradiative transitions are experienced in any semiconductor under forward bias, but in many cases the nonradiative transitions predominate or the photon losses are too excessive for observation of any external radiation.

Gallium arsenide (GaAs) is one of the first (and best) materials used for *infrared* LEDs (IREDs). One of the main advantages of GaAs is that it exhibits a very high probability for *direct radiative transitions*. That is, a very likely transition is a direct one all the way from the conduction band to the valence

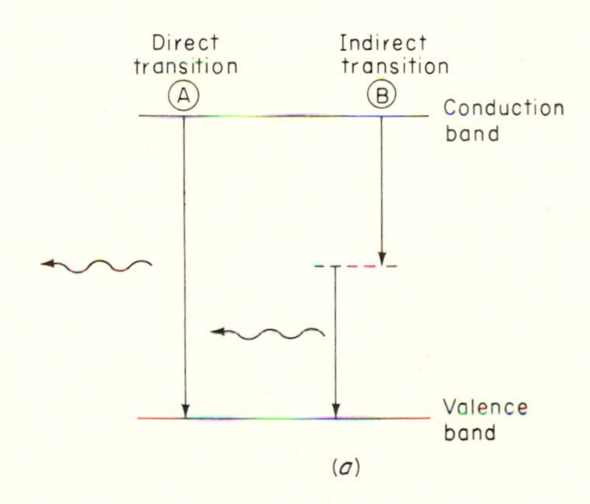

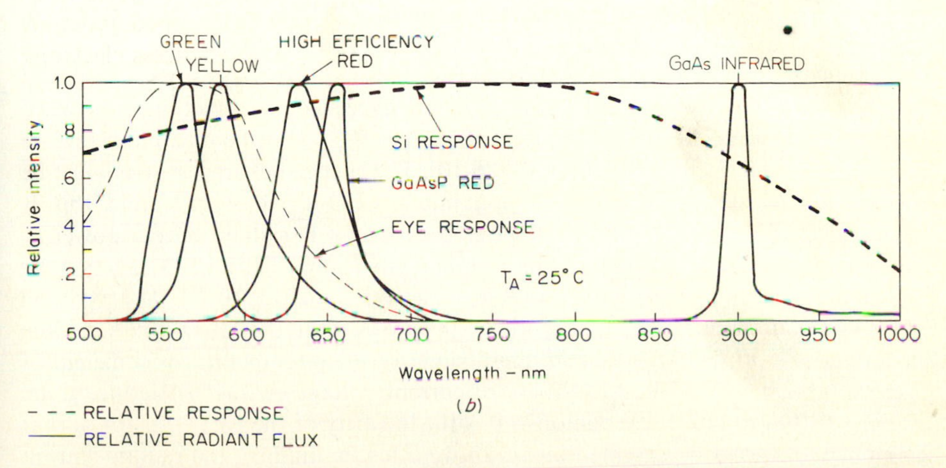

Fig. 6-9 (*a*) LED energy-level diagram; (*b*) relative emission spectra for visible and IR LEDs, including the relative response of the eye and silicon. *Hewlett-Packard.*

band, as shown by *A* of Fig. 6-9*a*. Thus most of the electron's energy is given up in the form of radiant energy approximately equal to the GaAs band gap of 1.4 eV (Table 6-3). This transition yields a wavelength of $\approx$885 nm [Eq. (6-2)], which is near the peak of a silicon (Si) photodetector's spectral response.

Gallium phosphide (GaP) and GaAsP are used to produce *visible* LEDs, and their most probable transition is *B* of Fig. 6-9*a*. GaP is an indirect semiconductor with very poor efficiency. An *indirect transition* occurs when the electron and hole *momenta* (mass times velocity) differ. In this case an intermediate transition must be made, and most of the energy is given up in the form of heat. A mixture of GaAs and GaP with $\leq$45 percent GaP is a direct-band material with an energy gap of $\approx$1.97 eV or an emission wavelength of $\approx$630 nm. As the percentage of phosphide in GaAsP LEDs is increased, the emission wavelength decreases, and GaAsP LEDs can be fabricated to emit anywhere from $\approx$550 nm (the peak of the eye's response) to 650 nm (Fig. 6-9*b*).

The quantum efficiency of conventional infrared LEDs (photons out per electron in) has been increased considerably but is still less than 1 percent for most low-power IREDs and up to 12 percent for a few special units. This can be significantly increased by operating in the laser region, as discussed in Sec. 6-2-6. The luminous efficiency for low-power visible LEDs varies from 50 to 1000 lm/W (lumens per watt). As the energy gap of the LED is increased (emission wavelength decreased), the probability of direct radiative transitions decreases because the crystal flaws provide intermediate energy levels through which two-step radiationless transitions can be made. Three optical phenomena also contribute to the LED's low quantum efficiency. They are: photon absorption due to the opacity of the LED material; reflection from the air/LED material interface due to the differences in index of refraction; and total internal reflection of photons incident at angles greater than the critical angle.

Figure 6-10 shows excerpts from a typical GaAs infrared LED data sheet. We shall use this figure to illustrate the key factors which must be considered in the application of LEDs. Probably the most important LED characteristics are their maximum ratings such as shown in Fig. 6-10*a*. LEDs can easily be destroyed if their rather low maximum drive levels are exceeded [the maximum forward current is 80 mA at a forward voltage of 1.2 V (25°C), and the maximum reverse voltage is -3.0 V for the LED of Fig. 6-10]. The thermal characteristics must be accounted for in determining the maximum LED drive signals, and this is covered in Fig. 6-10*g*. Figure 6-10*b* gives the total LED radiant output power P_o at a given forward current I_F. The nominal efficiency of the LED can thus be calculated from $(P_o/P_{\text{in}})(100) = (P_o/I_F V_F)(100) = [(550 \times 10^{-6})\ \text{W}/(0.05\ \text{A})(1.2\ \text{V})](100) = 0.92$ percent. Figure 6-10*c* shows that the LED emission line width is $\approx$40 nm. (This spectrum has a slight positive temperature coefficient of about $+0.2$ nm/°C.) The LED forward current/voltage characteristics are given in Fig. 6-10*d*. Figure 6-10*e* demonstrates the linearity of the LED radiant output power with forward current; that is, above a few milliamps, the radiant output power approximately doubles as I_F doubles. The LED spectral radiation

ELECTRICAL CHARACTERISTICS (T_A = 25° unless otherwise noted)

Characteristic	Symbol	Min	Typ	Max	Unit
Reverse Leakage Current (V_R = 3.0 V, R_L = 1.0 MΩ)	I_R	—	50	—	nA
Reverse Breakdown Voltage (I_R = 100 μA)	BV_R	3.0	—	—	V
Forward Voltage (I_F = 50 mA)	V_F	—	1.2	1.5	V
Total Capacitance (V_R = 0 V, f = 1.0 MHz)	C_T	—	150	—	pF

OPTICAL CHARACTERISTICS (T_A = 25° unless otherwise noted)

Characteristic	Symbol	Min	Typ	Max	Unit
Total Power Output (I_F = 50 mA)	P_O	200	550	—	μW
Radiant Intensity (I_O = 10 mA)	I_O	—	2.4	—	mW/steradian
Peak Emission Wavelength	λP	—	900	—	nm
Spectral Line Half Width	$\Delta\lambda$	—	40	—	nm

(b)

MAXIMUM RATINGS

Rating	Symbol	Value	Unit
Reverse Voltage	V_R	3.0	V
Forward Current – Continuous	I_F	80	mA
Total Device Dissipation @ T_A = 25° C Derate above 25°C	P_D	120 2.0	mW mW/°C
Operating and Storage Junction Temperature Range	T_J, T_{stg}	-40 to -85	°C

THERMAL CHARACTERISTICS

Characteristic	Symbol	Max	Unit
Thermal Resistance, Junction to Ambient	θ_{JA}	500	°C/W

(a)

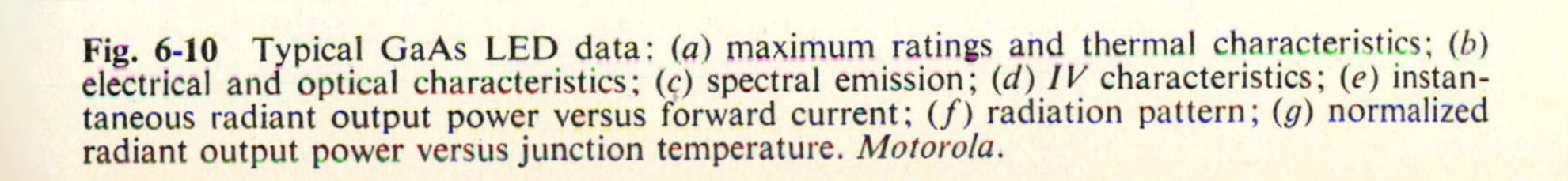

Fig. 6-10 Typical GaAs LED data: (a) maximum ratings and thermal characteristics; (b) electrical and optical characteristics; (c) spectral emission; (d) *IV* characteristics; (e) instantaneous radiant output power versus forward current; (f) radiation pattern; (g) normalized radiant output power versus junction temperature. *Motorola*.

pattern of Fig. 6-10*f* is important because it shows the *divergence angle* θ, which is the total angle for which the relative radiant power output is more than 50 percent or $\pm 15° = 30°$ total for this particular LED.

Figure 6-10*g* (and parts of 6-10*a*) cover the LED thermal characteristics. First, notice from Fig. 6-10*g* that the LED's radiant output decreases as its junction temperature increases. This phenomenon is explained by the fact that, as the LED temperature is increased, more of the electron-hole recombinations become nonradiative.

As an example of the decrease in LED radiant output with increasing temperature, consider the LED of Fig. 6-10 operated at an I_F of 60 mA dc at room temperature. Figure 6-10*d* shows that V_F at this forward current is ≈ 1.2 V. Figure 6-10*e* shows that P_o would be ≈ 0.6 mW at a junction temperature (T_J) of +25°C. To correct P_o for the actual junction temperature, we use Eq. (8-1*a*) of Sec. 8-2-2:

$$T_{J(\text{av})} = T_{\text{ambient}} + \theta_{JA} P_J$$

where P_J is the power dissipated at the junction, or, ($I_F V_F$) (duty cycle), and θ_{JA} is the junction-to-ambient thermal resistance, equal to 500°C/W as shown in Fig. 6-10*a*. For this example

$$T_{J(\text{av})} = 25°\text{C} + (500°\text{C/W})(1.2\ \text{V})(0.06\ \text{A}) = 61°\text{C}$$

Figure 6-10*g* shows that P_o will be reduced by a factor of ≈ 0.7 at a junction temperature of 61°C; hence the actual total power output is (0.6 mW)(0.7) = 0.42 mW. If the ambient air temperature is greater, T_J will be even higher, resulting in a further reduction in P_o. Checking to see if the 120 mW maximum power dissipation at 25°C ambient of Fig. 6-10*a* has been exceeded in this example, we see that the power is $V_F I_F = (1.2\ \text{V})(60\ \text{mA}) = 72$ mW, or well under the allowed dissipation. If a reasonable power-dissipation margin is not available for the desired LED drive and ambient-temperature range, a heat sink may be added (Sec. 8-2-2) to the LED; or the LED can be operated in a pulsed mode; or a higher-power LED may be utilized.

Figure 6-10*e* shows that operation at high pulse forward currents (well above the maximum 80 mA continuous rating of Fig. 6-10*a*) can significantly increase the LED radiant output. Also, as long as the duty cycle is low, such high-pulse-current operation will not exceed the LED's maximum operating temperature. As an example of the advantages of pulsed LED operation, assume a 300-mA, 2-μs-wide I_F pulse, with a period between pulses of 1 ms, which gives a *duty cycle* (pulse width/period) of 2×10^{-3}. Figure 6-10*d* and *e* respectively show that if I_F is 300 mA, then $V_F \approx 1.35$ V, and the instantaneous $P_o \approx 3$ mW at 25°C. Solving Eq. (8-1*a*) gives

$$\begin{aligned} T_{J(\text{av})} &= T_{\text{ambient}} + \theta_{JA}(V_F I_F)(\text{duty cycle}) \\ &= 25°\text{C} + (500°\text{C/W})(1.35\ \text{V})(0.3\ \text{A})(2 \times 10^{-3}) \\ &= 25.4°\text{C} \end{aligned}$$

demonstrating that such low-duty-cycle pulse operation does not cause any significant increase in junction temperature, or decrease in P_o.

Although not shown, data sheets for visible LEDs are very similar to those of the IRED of Fig. 6-10 except that their spectrum is shifted toward the visible (as shown in Fig. 6-9*b*); their efficiency falls as the emission wavelength decreases; and photometric units are used. Section 6-3 contains examples of visible LEDs in display applications.

It is desirable at this point to use the P_o example (Fig. 6-10) to illustrate how to calculate the incident irradiance H at a typical photodetector, and hence show that the total LED output power P_o is not generally available to such detectors. Figure 6-11 shows the typical point-source geometry of a LED irradiating an area A at a distance d and a divergence angle θ. Table 6-2*a* gave the general

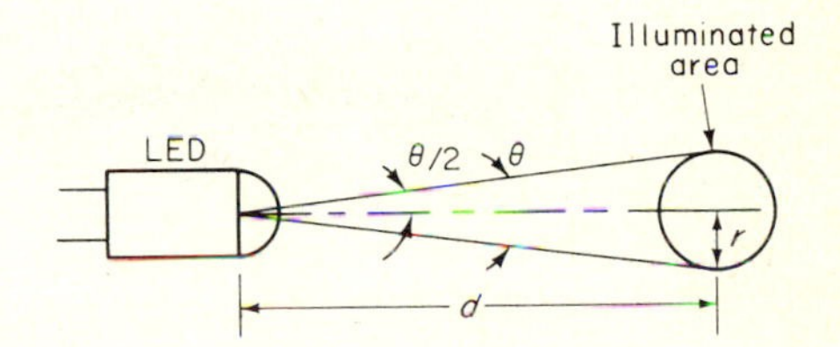

Fig. 6-11 Area illuminated by a LED at a distance d and divergence angle θ.

point-source formula $H = P_o/\text{area}$ in watts per square centimeter (where H is the flux at the detector). For this case the area irradiated by the LED at a distance d can be calculated by noting that

$$\tan\frac{\theta}{2} = \frac{r}{d} \qquad \text{or} \qquad r = d\tan\frac{\theta}{2}$$

Since, for small angles, the tangent is approximately equal to the angle in radians, we have

$$r \approx d\frac{\theta}{2}$$

The area of a circle is πr^2, and substituting the above quantity for r in this equation yields an irradiated area of

$$A = \frac{\pi d^2\theta^2}{4}$$

and hence

$$H = \frac{P_o}{A} = \frac{4P_o}{\pi d^2\theta^2} \tag{6-4}$$

where the various quantities are defined in Fig. 6-11 and θ is in radians, d in centimeters, P_o in milliwatts, and H in milliwatts per square centimeter. (One radian = 57.3°.)

For the LED and drive conditions of the previous dc example we calculated a P_o of 0.42 mW. If the LED-to-detector separation of Fig. 6-11 is 1 in [2.54 cm] and the divergence angle from Fig. 6-10f is 30° (or 0.524 radian), Eq. (6-4) gives

$$H \approx \frac{4(0.42 \text{ mW})}{\pi(2.54 \text{ cm})^2(0.524)^2} \approx 0.32 \text{ mW/cm}^2$$

See Chap. 7 for examples of typical photodetector responses to LED irradiances.

Figure 6-12 demonstrates a LED drive/protection technique utilizing a series current-limiting resistor. Assuming a +5 V input and the LED of

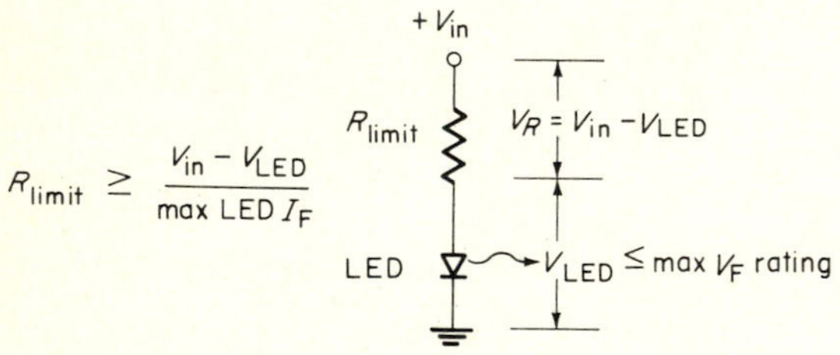

Fig. 6-12 LED current limiting.

Fig. 6-10 with maximum continuous forward current and voltage ratings of ≈80 mA and 1.22 V, respectively, we have

$$R_{limit} \geq \frac{5\text{V} - 1.22 \text{ V}}{80 \text{ mA}} \geq 47\Omega$$

Care must be taken for low supply voltages, when V_{in} approaches V_{LED} because $V_{in} - V_{LED}$ of Fig. 6-12 becomes very small, and small changes in the LED forward voltage due to temperature can cause I_F to vary drastically. Protection from negative-going input voltages which may exceed the reverse-voltage rating of Fig. 6-10a can be provided by adding a diode whose V_{BR} is less than the LED's V_F in parallel with the LED. The LED circuit of Fig. 6-12 can be driven directly from many logic-family gates (such as TTL).

The temperature dependence of LED radiant output (Fig. 6-10g) can generally be neglected in digital applications; however, many linear LED applications require some form of temperature compensation in order to stabilize the LED P_o over the operating-temperature range. Figure 6-10e and g show that both the P_o versus I_F and P_o versus junction temperature characteristics are fairly linear; therefore, one means of temperature compensating a LED is to linearly increase its forward current as the temperature increases. Two arrangements that provide such temperature compensation plus the required drive are shown in Fig. 6-13. In Fig. 6-13*a* the negative temperature coefficient of the LED is canceled by the positive temperature coefficient of the transistor collector current I_C. The value of R_E for such applications is a few ohms, and the optimum value is best found by experimentation. The transistor collector leakage current I_{CO} doubles every 8 or 10°C, and β increases a fraction of a percent per degree centigrade. If the resistance at the base lead is large compared to R_E, all of I_{CO} flows into the base and is multiplied by β. This causes I_C to increase drastically with temperature. Such operation is normally avoided, but in this case it is desirable. I_c also varies as V_{BE} changes with temperature.

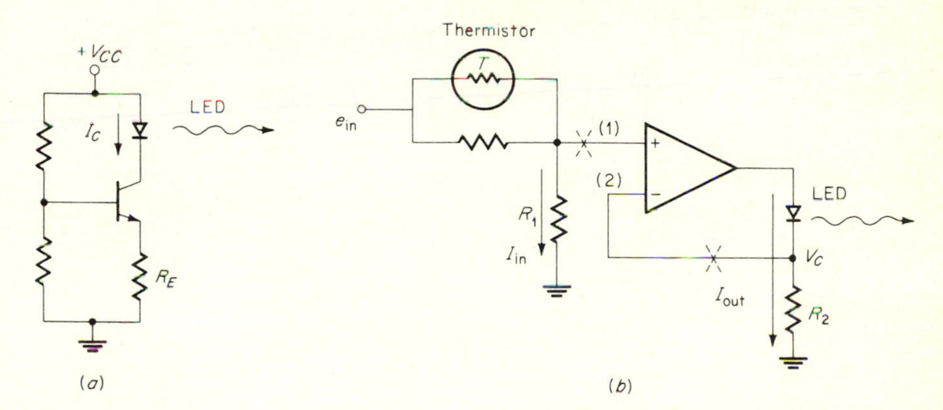

Fig. 6-13 Temperature-compensated drive for LEDs: (*a*) transistor; (*b*) thermistor–operational-amplifier.

Figure 6-13*b* utilizes a linearized thermistor-resistor network and an operational amplifier to achieve LED temperature compensation (Sec. 4-2). Assuming ideal operational-amplifier action (negligible current drawn into the amplifier or voltage difference between input terminals), we see that $V_1 \approx V_2 \approx V_c$; hence the same voltage is impressed across R_1 and R_2, or $I_{in} = V_1/R_1$ and $I_{out} = V_1/R_2$. The current gain of the amplifier is thus

$$\frac{I_{out}}{I_{in}} = \frac{V_1/R_2}{V_1/R_1} \approx \frac{R_1}{R_2} \tag{6-5}$$

Since thermistors have a negative temperature coefficient, or a decrease in resistance with increasing temperature, I_{in} will increase with temperature, and I_{out} will follow. The thermistor-resistor network of Fig. 6-13*b* can be designed with a $\Delta R/\Delta T$ to match the forward-current demands of the LED for constant P_{out} over a limited temperature range (Ref. 1).

To summarize the advantages of LEDs, they possess

1. Extremely high speeds (few nanoseconds)
2. Low cost
3. Long life compared to lamps

4. Linearity of P_o with forward current (above some nominal value of I_F) over a wide range
5. Adaptability to coherent laser operation (Sec. 6-2-6)
6. Low-voltage operation, making them compatible with integrated circuits
7. A variety of spectral output colors

The LED disadvantages include

1. Temperature dependence of radiant output power and wavelength
2. Sensitivity to damage by overvoltage or overcurrent
3. Theoretical overall efficiency is not achieved except in special cooled or pulsed conditions
4. Wide optical bandwidth compared to the lasers of the next section (≈ 10 versus 10^{-2} nm)

One use of LEDs is as a high-speed light source of known wavelength for checking the linearity, speed, etc., of optoelectronic detection circuits. Figure 6-14*a* shows a typical inexpensive setup in which a function of *light* into the megahertz range is available.

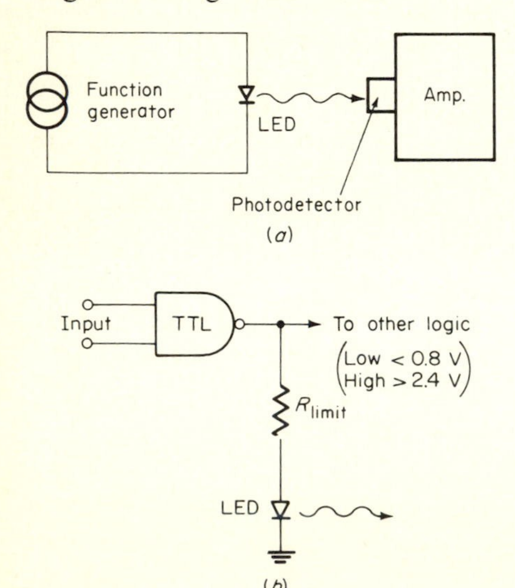

Fig. 6-14 LED applications: (*a*) simple radiometric instrument calibration circuit; (*b*) logic-state indicator.

Figure 6-14*b* shows another typical LED application in which a visible LED indicates the state of a digital logic gate. This circuit maintains the original logic levels; that is, the LED is ON when the logic is in the high state, and the LED is OFF during the low state. LEDs are often used to monitor/indicate voltage levels or polarities. The unijunction transistor oscillator of Fig. 9-11 can be converted to a LED pulser by substituting a LED for R_1.

Applications of LED source/detector combinations, better known as optical isolators, are covered in Sec. 7-2-2.

6-2-5 GAS AND RUBY LASERS

One of the most promising light sources and general scientific tools is the laser. Laser technology is having appreciable impact on the industrial and scientific community. The laser is presently being used to weld, communicate, measure distance, direct tunnel drilling, fuse detached retinas, irradiate cancerous growths, direct military weapons, track aircraft, and improve many optical instruments. The main source of the laser's power stems from the fact that laser light is coherent, with its wave packets all in phase, traveling in one direction, and of essentially one wavelength or color (*monochromatic*). For this reason, a beam of laser light can travel many miles without appreciable divergence or spreading. This is in contrast to conventional incoherent light sources which emit at many wavelengths, with broad variations in phase and direction.

There are many types of lasers, but the basic principle is essentially the same for all. The term *laser* is an acronym for *light amplification by stimulated emission of radiation*. We have already mentioned the *absorption* and *spontaneous-emission* processes in which atoms absorb energy and are excited to a higher-energy level and then emit light during the transition back to the lower-energy state (Sec. 6-1). The probability of occurrence of both of these processes is directly proportional to the number of atoms in the energy level of interest. There is also a second type of radiative transition referred to as *stimulated emission*, which is defined as a radiative transition caused by other photons. In this case the occurrence rate is proportional not only to the number of atoms in the excited state, but also to the radiation or photon density.

Absorption and both types of radiative transitions are demonstrated in Fig. 6-15. Figure 6-15*a* shows an atom absorbing a photon with a consequent

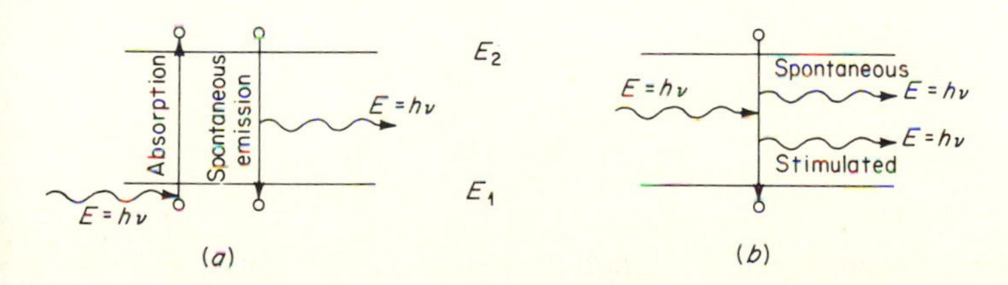

Fig. 6-15 Spontaneous and stimulated emission examples.

rise in electron energy level from E_1 to E_2. The electron eventually spontaneously decays back to the lower-energy level and radiates an energy equal to the difference in energy levels. Stimulated emission is portrayed in Fig. 6-15*b*. In this case the electrons are initially at the excited energy level, and emission is stimulated sooner than in the spontaneous case because the excited atom is struck by a photon of exactly the same energy as the photon it is to emit. Hence, the absorption of the incident photon (under the proper conditions) causes the emission of two photons—one spontaneous and one stimulated. (The two photons leave together and travel in the same direction and phase.) The concept of light amplification thus derives fiom the emission of two photons with an input of only one photon.

The process of getting a large percentage of the atoms into an excited state, as shown by Fig. 6-15*b*, is termed *population inversion* and is one of the keys to laser action. If a large number of atoms can be inverted or excited to the upper energy level, then the probability of stimulated emission and hence light amplification becomes greater.

These concepts are probably best illustrated by a description of a typical laser. Figure 6-16*a* shows a cutaway view of a pulsed ruby laser. The operational

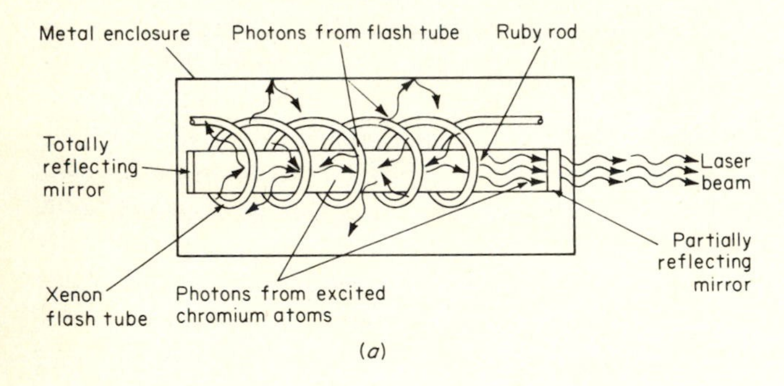

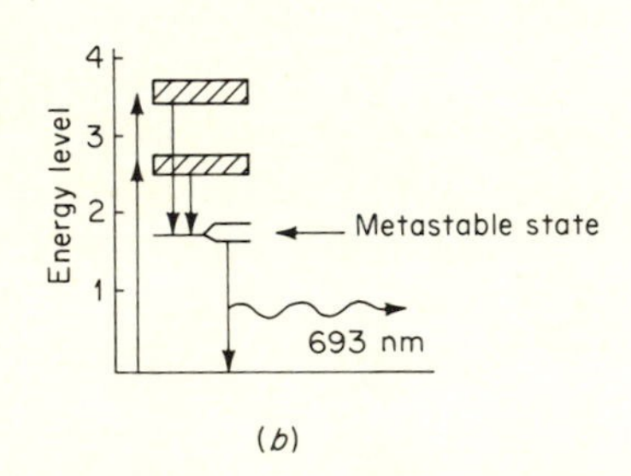

Fig. 6-16 (*a*) Essentials of a ruby laser; (*b*) energy-level diagram for a ruby laser, showing the main chromium-ion transition at 693 nm.

sequence starts with the igniting of the xenon flash tube. Chromium atoms in the ruby rod are then energized or inverted by absorption of the energetic photons from the flashtube. When the excited chromium atoms' electrons fall back to their normal energy states, photons are given off by spontaneous emission. Some of these photons escape from the ruby rod, but many oscillate or bounce back and forth in the length of the rod, being reflected by the mirrors at each end. The ruby rod thus serves as the optical cavity in which a high preference exists for axial photon travel, the other directions being suppressed by losses from the system. A high photon density is then built up in a particular direction so that stimulated emission increases. Each chromium-atom photon absorption occurs in the same phase and direction as the overall wave, because the photon actually comes from the wave. Thus the induced photons also have the same phase and direction as the wave. This cumulative process of flashtube photons exciting chromium atoms, which in turn emit photons in the same preferred direction and phase, continues until the coherent laser beam penetrates through the partially reflecting mirror on one end of the rod. The bulk of the laser light thus originates from stimulated emission, because that is the preferential resonant mode of the optical cavity. The flashtube has thus performed the function of *optically pumping* or inverting, and its light is effectively converted to produce the far more coherent and concentrated laser beam. Figure 6-16*b* shows the actual transitions undergone in a ruby laser. Note that the laser light is not produced by the decay from the highest-energy state, but rather from a stimulated dump from a nearly stable, long-lifetime energy state at 1.78 eV. The optical bandwidth of the final laser beam is often less than 0.001 nm, owing to the selectivity and feedback just described.

Another very popular type of laser is the gas laser. The reasons for its popularity include its low cost, simplicity, and extremely broad range of useful emission wavelengths (0.6 to 100 μm depending on the gas). The helium-neon, (He-Ne) laser will be used as an example of a typical gas laser. This particular laser has been produced in large quantities and can now be found in almost every classroom and laboratory. Figure 6-17 gives the essentials of an inexpensive He-Ne laser. One of the reasons for its low cost is that the excitation is much the same as that of a fluorescent lamp (Sec. 6-2-3). That is, a filament provides a

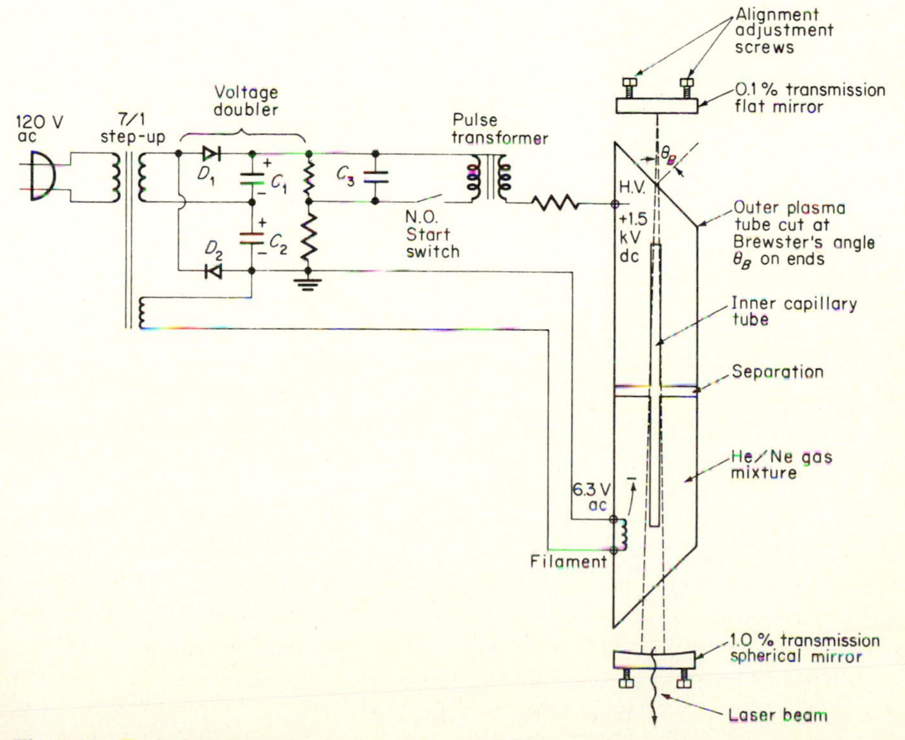

Fig. 6-17 Typical He-Ne laser.

source of electrons which are in turn accelerated by a high voltage, and finally the He-Ne gas mixture is ionized by electron impact. Referring to Fig. 6-17, we see that the potentials for the plasma-tube electrodes are provided by the tapped transformer, voltage doubler, and pulse transformer. Such voltage-multiplying circuits are frequently encountered in optoelectronic equipment, where one can accept a poorly regulated dc voltage at two, four, eight, etc., times the peak ac voltage. In other words, doubler stages can be cascaded several times, but their use is restricted to low-current applications in which the main objective is the creation of a high dc voltage. The output of the voltage doubler of Fig. 6-17 is approximately twice the peak of the transformer secondary voltage because C_1 and C_2 are charged to the peak voltage on alternate half cycles, and the capacitors are connected so that their outputs are additive. The step-up transformer and doubler combine to give a dc potential of ≈ 1.5 kV, which is adequate to sustain a glow discharge in the plasma tube. The small capillary tube is added to provide a region of high plasma density to ensure stimulated emission. The laser is initially started by applying a 5.0-kV pulse to the normal high-voltage line. This is provided when the normally open start switch is closed, discharging C_3 through the pulse transformer.

In many lasers of this type the end reflectors or mirrors are an integral part of the tube itself, but the more universal and versatile arrangement of separate mirrors is shown in Fig. 6-17. These mirrors define the optical cavity, and in this case we have what is known as a *hemispherical resonator* defined by the spherical 1.0-percent transmission mirror on the output end and the flat 0.1-percent transmission mirror on the back. The small transmissions are necessary because the probability of one photon stimulating another in one pass through the tube is only a few percent; therefore the beam must be reflected several times to be self-maintaining. The radius of curvature of the spherical mirror is equal to the mirror separation so that the intracavity radiation pattern focuses at a point on the flat mirror, as shown by the dashed lines. This type of resonator configuration is popular because the mirror separation is easily adjusted to provide the desired stable mode, and angular misalignments are noncritical. The word *mode* refers to the same terminology as used in microwaves, and in this case the lowest-order TEM_{oo} provides the maximum output power and spot uniformity.

There are several emission lines available from the He-Ne system, and one can make any particular one predominate by choosing mirrors which are highly reflective at the wavelength of interest. This causes that specific wavelength to dominate within the cavity; hence the output beam consists mainly of the same preferential wavelength. The end reflectors are adjusted for separation, so that an integral number of wavelengths can exist between the mirrors. The mirror angle is also adjusted so that the maximum number of reflections takes place in the cavity. For optimum transmission from the plasma tube, its ends are cut at the *Brewster angle*, which is defined as the incident angle whose tangent equals the refractive index of the mirrors. For instance, glass or quartz with an index of refraction of 1.5 would have a Brewster angle of $\approx 56°$.

Figure 6-18 shows a simplified energy-level diagram for the He-Ne laser. As previously mentioned, the electrons from the glow discharge acquire considerably more kinetic energy than the ions and atoms, and during electron-atom collisions the electron's energy is transferred to the atoms. (Atom-to-atom collisions also contribute when the population density increases.) In the He-Ne laser the pressure of He is ≈ 10 times that of Ne, and the helium merely serves as an energy transfer agent.

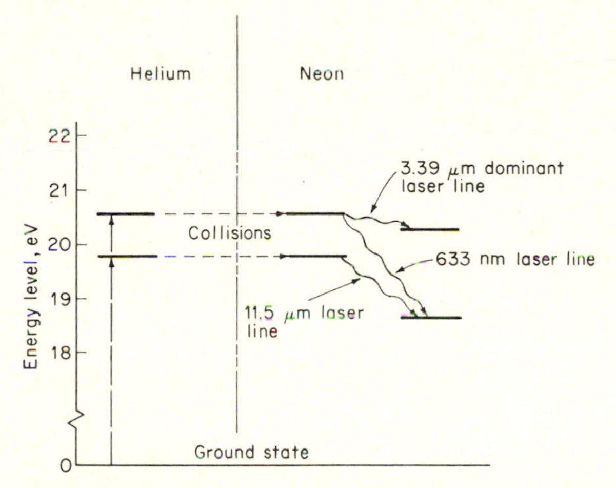

Fig. 6-18 He-Ne laser energy-level diagram, showing the He excitation, energy transfer to Ne by collision, and three main radiative transitions.

Referring to Fig. 6-18, we see that population inversion is accomplished by the following sequence. The energetic electrons excite the He atoms from the ground state to the 19.8- and 20.6-eV levels. These particular helium states have no allowed downward radiative transitions; therefore they are *metastable*, meaning the atom's lifetime in this state is long. These states can therefore become densely populated, which is conducive to stimulated emission. Part of the He excitation energy is transferred to the Ne atoms by collision, and the Ne is thus excited to very similar energy levels as shown. The main Ne radiative transition is the infrared 3.39-μm line, but proper reflector or prism design can cause the 633-nm or 1.15-μm line to dominate (Fig. 6-18). The output power of small laboratory He-Ne lasers is generally less than 5 mW.

Figure 6-19 is an excellent example of the usefulness of inexpensive gas lasers. It also demonstrates the close coupling of light sources, optics, photodetector, and electronics in a typical optoelectronic system. The instrument is a hydrocarbon gas detector designed primarily for the detection of methane or natural gas. It turns out that methane (CH_4) has a strong absorption line almost exactly centered on one of the possible He-Ne laser emission lines (3.394 μm). In other words, any methane in the path of such a laser beam would strongly absorb the light. The very narrow spectral bandwidth of the laser ($\approx 10^{-4}$ nm)

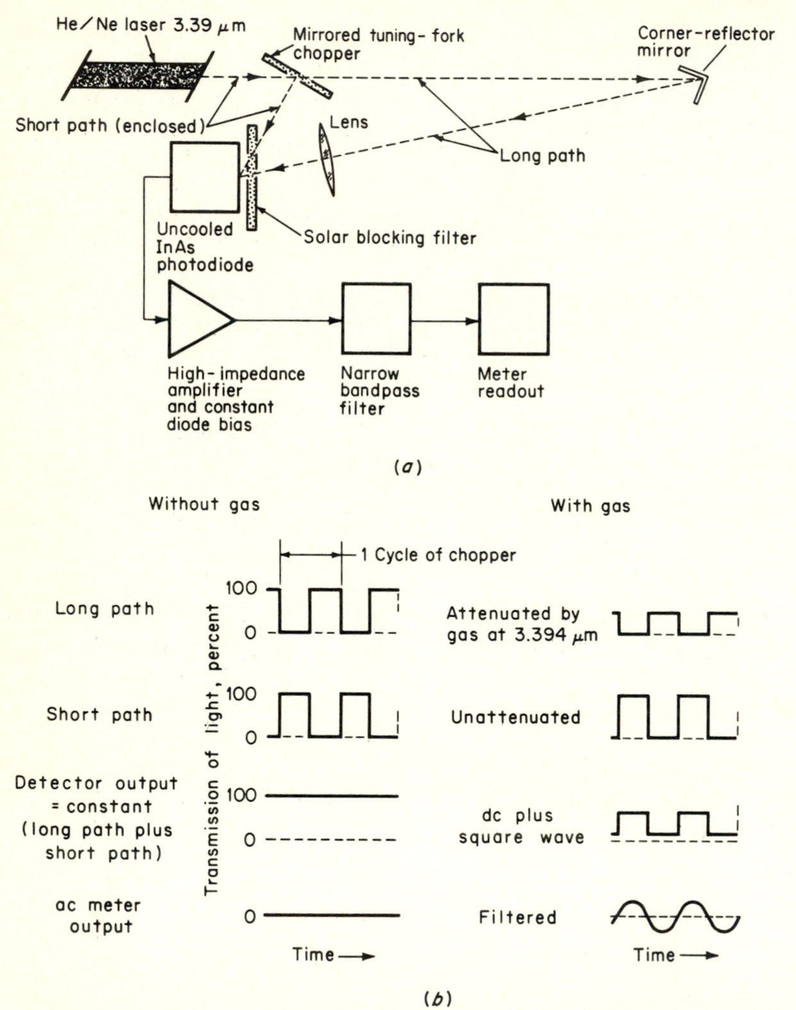

Fig. 6-19 Hydrocarbon gas detector: (*a*) schematic; (*b*) waveforms. *NASA/Ames.*

offers an important advantage compared to other broad-band infrared sources. That is, with a spectrally broad light source the absorption lines of several elements and compounds may be contained within the width of the source emission line. Hence, one could not tell which element was doing the absorbing. However, with a very narrow 3.394-μm laser emission line, most other nonhydrocarbon absorption lines do not overlap into that line. Hence the laser provides better specificity or less interference from other similar absorbing media.

The operation of the hydrocarbon detector is as follows (Ref. 2). There are a single He-Ne 3.39-μm light source, a single infrared photodiode, and two light paths—one short, enclosed reference path and the longer signal path in which the gas sample can enter. The beam is directed to a mirrored tuning-fork chopper which serves the dual function of chopping or converting the beam to ac, and reflecting it to the photodiode when the chopper tines close. When the tines are open, the laser beam traverses the indicated long path and is focused at the same point as the short path. The two beams are thus summed at the photodiode (Fig. 6-19*b*). An indium arsenide photodiode was chosen because its uncooled spectral response peaks near the desired 3.39 μm, whereas the responses of other uncooled junction IR detectors are lower at that wavelength. The amplifier is similar to that of Fig. 7-10*b*, and the electronic bandpass filter, with center frequency set at the chopping frequency, reduces the system noise and provides a better null when no absorbing gas is present. The waveforms of Fig. 6-19*b* illustrate the long- and short-path summing and the fact that an ac signal exists when a hydrocarbon gas is present. The reason for the two-beam arrangement is that, if the source or detector outputs drift with time, temperature, etc., both beams are equally affected. Since the output signal is the difference between the two beams, the output remains constant irrespective of such influences. The conversion to ac also reduces detector and amplifier drift, as discussed in Sec. 7-1-3.

6-2-6 SEMICONDUCTOR INJECTION LASERS

Laser action is also possible in a semiconductor junction, as shown in Fig. 6-20. The main differences between this structure and that of a conventional LED are the precision of initial plane cutting and end polishing, operation at higher

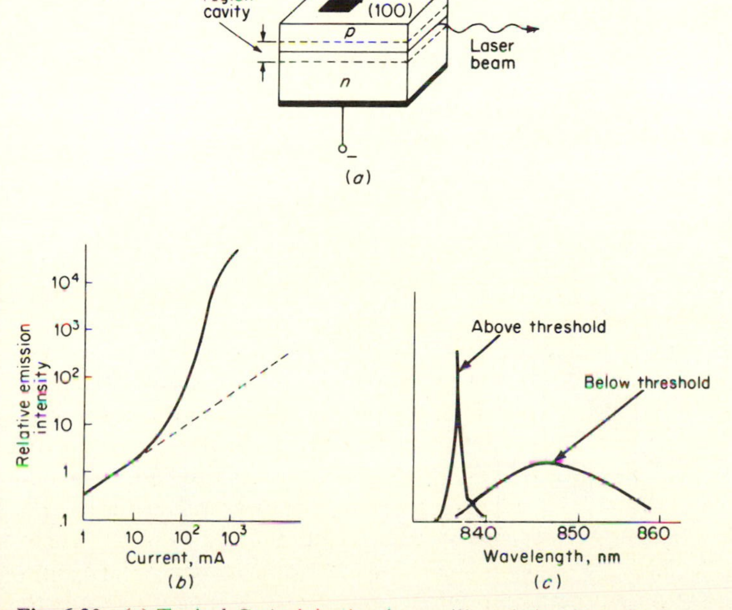

Fig. 6-20 (*a*) Typical GaAs injection laser; (*b*) emission intensity versus current for a GaAs injection laser; (*c*) typical spectral narrowing observed above lasing threshold.

forward currents and voltages, and the fact that coherent light is produced. Such lasers are called semiconductor *injection* lasers. The main applications of semiconductor injection lasers are communications, optical ranging and guidance, and instrumentation.

The optical cavity in the injection laser consists of a parallelepiped defined by the crystal end faces and the depletion region, as shown in Fig. 6-20*a*. Hence the ends serve the same function as the reflector mirrors of the gas laser. The depletion-region boundaries, although neither sharp nor very different in index of refraction, serve the same purpose as the ruby rod of Fig. 6-16. For this type of injection laser the output increases somewhat linearly with current density, similarly to the spontaneous emission of the LED, until a threshold is reached. Then the brightness increases by two or three orders of magnitude, as shown in Fig. 6-20*b*. The increase in brightness is accompanied by a drastic narrowing of the spectral width, as shown by Fig. 6-20*c*. In the injection laser the forward current provides the population inversion or pumping; the same current provides the electrons and holes which recombine and produce adequate photon density for sustained stimulated emission. The directionality is provided by the previously described optical cavity.

Figure 6-21 shows data for a typical GaAs injection laser capable of room-

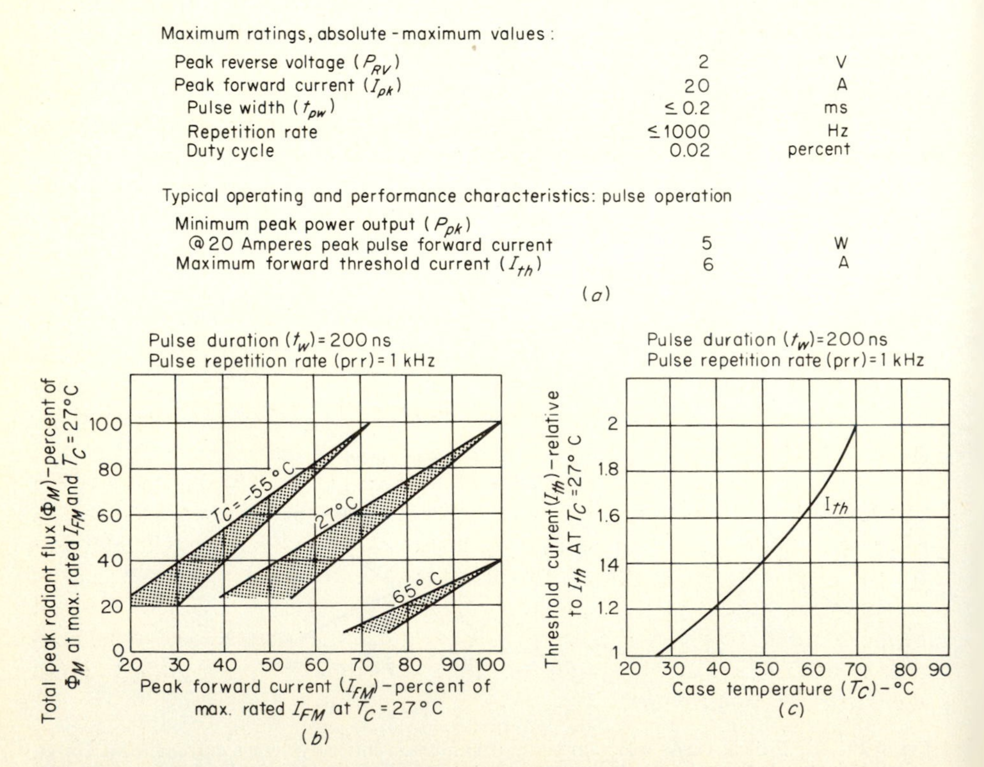

Maximum ratings, absolute-maximum values:		
Peak reverse voltage (P_{RV})	2	V
Peak forward current (I_{pk})	20	A
Pulse width (t_{pw})	≤ 0.2	ms
Repetition rate	≤ 1000	Hz
Duty cycle	0.02	percent
Typical operating and performance characteristics: pulse operation		
Minimum peak power output (P_{pk}) @ 20 Amperes peak pulse forward current	5	W
Maximum forward threshold current (I_{th})	6	A

(*a*)

Fig. 6-21 GaAs injection-laser diode data: (*a*) specifications; (*b*) typical total peak radiant flux versus peak forward current; (*c*) temperature dependence of lasing threshold current. *RCA*.

temperature pulsed operation. Many of the characteristics of semiconductor injection lasers are temperature-dependent, as evidenced by Fig. 6-21*b* and *c*. As in the conventional LED, as temperature increases the energy gap decreases, trapping sites increase, and more transitions become radiationless—thus the output power decrease (Fig. 6-21*b*). The lasing threshold current is also a strong function of junction temperature, as shown in Fig. 6-21*c*. (The emission wavelength also varies with temperature.) For these reasons solid-state injection lasers are often operated at cryogenic temperatures.

6-3 DISPLAYS

Dramatic changes have taken place in the electronic display field in the last few years. First, analog displays are rapidly being replaced by digital readouts in many electronic products for the following reasons: the cost competitiveness of digital displays; the switch from analog to digital circuit techniques (thus making the circuit/display interface simpler with digital displays); the improved accuracy and resolution inherent in digital readouts; and the ruggedness, sales appeal, ease of reading, and versatility. As an example, consider the new digital watch and digital-voltmeter displays described in Sec. 5-6. Second, the Nixie (Burroughs Corp.) stacked-character single-digit gas-discharge display has almost totally disappeared from new products, and in its place we find displays that did not exist commercially a few years ago—that is, light-emitting diodes and liquid crystals. This section is devoted to displays and the process by which digital data are converted to a form that can be comprehended by humans (generally either numeric and/or alphanumeric). It contains subsections covering available display comparisons, characteristics, nomenclature, and specifications; decoding/driving/multiplexing; and LED and liquid-crystal displays (LCDs).

6-3-1 DIGITAL-DISPLAY-TECHNOLOGY COMPARISON

Table 6-4 (Ref. 3) compares the five most prominent digital-display technologies and the older Nixie display. The left-hand column of Table 6-4 lists the characteristics and parameters that are important in display selection. Most of the terms are easily understood; however, a few are peculiar to display technology and are defined below.

Brightness refers to the display intensity. There is no industry standard for measuring brightness. Some manufacturers use point-source luminous intensity (Table 6-2*a*), and others use area source luminance (Table 6-2*b*) which accounts for the display's area and the distance to the observer. Because of this lack of standardization and the fact that contrast is what the eye perceives, the best means of evaluating display brightness is to view actual samples under the conditions of distance, ambient lighting, angle, drive, etc., that are anticipated in the particular application.

Contrast ratio is defined as the assessment of the difference in appearance (brightness) between two parts of a field viewed simultaneously. A high contrast ratio between the lighted portion of the display and its background is desired for

TABLE 6-4 Digital display comparisons*

	LED	Nixie	Gas Discharge	Fluorescent	Incandescent	Dynamic Scattering LCD	Field Effect LCD
Power/digit	10 to 140 mW, depending on color	350 mW	30 to 100 mW	100 mW	250 mW to 1 W	100 μW	1 to 10 μW
Voltage	5 V	175 V	180 V	15 to 25 V	5 V	18 V	3 to 7 V
Temperature range (°C)	−55 to 125	0 to 70	0 to 55	−55 to 100	−55 to 100	0 to 80	0 to 70
Switching speed	1 μs	150 μs	1 ms	1 ms	150 μs	300 ms	100 to 300 ms
Life in hours	100,000+	200,000	30,000	100,000	10,000 to 100,00	10,000	10,000
Colors	red, orange, yellow, green	neon orange	orange, others with filters	blue-green; others with filters	all with filters	depends on illumination	depends on illumination
Brightness	good to excellent	excellent	good	good	excellent	not applicable	not applicable
Contrast ratio	10:1	8:1	20:1	10:1	20:1	20:1	20:1
Appearance	good to excellent	fair	excellent	excellent	excellent	good	good to excellent
Viewing angle	150°	100°	120°	150°	150°	90 to 150°	90 to 120°
Font	7 and 16 segment, 5 × 7 dot matrix	individual characters	7 segment	7 segment	7 and 16 segment shaped characters	7 and 16 segment	7 and 16 segment
Vertical size	0.1 to 0.6 in.	0.3 to 2 in.	0.2 to 0.7 in.	0.5 to 0.75 in.	up to 1 in.	0.2 to 8 in.	0.2 to 2 in.
MOS compatibility	small yes, large no	no	yes	yes	no	yes	yes
Cost per digit (0.3 to 0.6 in.) 10 k pieces	$1.50	$2.35	60¢	$1	$1 to $2	$1	$1.50
Ruggedness	excellent	poor	fair	poor	fair	good	good
Ease of mounting	excellent	good	fair	good	good	poor	poor

* *Electronic Design*, vol. 26, p. 60, Dec. 29, 1974.

maximum readability. Since reflected ambient light from the display background reduces the contrast ratio, a black background and/or filter are frequently employed.

The *viewing angle* refers to the off-axis angle at which the display's luminous intensity falls off a given amount, such as 3 dB. Again, there is no industry standard for this specification, and spectral-radiation polar plots such as shown in Fig. 6-10*f* are the most meaningful means of describing viewing angle. Viewing angle decreases with magnification. For liquid crystal displays, the viewing angle depends on the contrast ratio and is generally much greater in the vertical dimension than in the horizontal plane.

Life or *life expectancy* is not standardized, but frequently refers to the time at which the display brightness decreases to 50 percent of its original value.

Lastly, *font* refers to the arrangement of the various elements of the display. Figure 6-22 shows the two most popular display fonts. Notice that the seven-segment font has, as the name implies, seven segments which can be ON in any combination to display a particular numeral or letter. The seven-segment display is limited to displaying the numbers 0 through 9 and a few letters, because the capital letters B and D would not be distinguishable from the numbers 8 and 0, respectively. The 5- by-7 dot-matrix array of Fig. 6-22*b* provides the best display of all alphanumerics, and the reader should go through the alphabetical and numerical possibilities to become convinced of the validity of this statement. Figure 6-22 also demonstrates the potential electrical decoding/driving complexity for the various fonts; the seven-segment being the simplest since it has the least number of display elements per character.

The digital-display comparisons of Table 6-4 show that:

- Liquid-crystal displays (LCD) consume far less power than the others.
- The Nixie and gas-discharge displays require high voltage drive.
- The operating-temperature range is lower for the LCD, the Nixie, and the gas discharge displays.
- LEDs are by far the fastest display, and LCDs the slowest.
- LCDs have the shortest life.
- Brightness is not given for LCDs, since they are not light sources (Sec. 6-3-4).
- The LCD, gas-discharge, and incandescent displays have the best contrast ratios.
- LCDs are available in the largest size.
- Costs are comparable, but the Nixie is the most expensive.

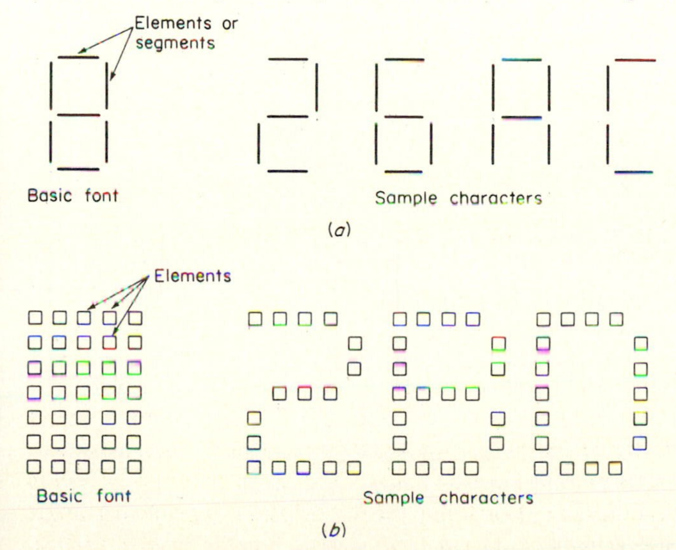

Fig. 6-22 Popular digital display fonts: (*a*) seven-segment; (*b*) 5-by-7 dot matrix.

The selection of the best display for a given application is frequently complicated, because there is no one perfect display with all its characteristics optimized. For example, battery-operated digital-wristwatch applications require a display with very low power consumption, small size, and low drive voltage. These requirements eliminate all the displays of Table 6-4 except the LED and LCD, and neither of these is perfect. The high LED power consumption requires a "push-to-see" button so that the display draws power only when the owner desires to see the time. The LED display is also susceptible to "washout" in high ambient lighting conditions. The LCD is ideal from the power-consumption standpoint; however, it requires a built-in light source for reading under low ambient lighting, and hence also requires a push-to-see button. The limited lifetime and limited operating-temperature range of the LCD are also problems.

6-3-2 DIGITAL-DISPLAY DECODING/DRIVING AND MULTIPLEXING

The visible portion of a display is only part of the total electronic system required to convert digital data into a form readily comprehended by a human viewer. Figure 6-23 shows a general block diagram of a typical digital-display electronic system for a single-character display. Although the details of such a system vary somewhat with the type of display, all the functions of Fig. 6-23 are required for any of the displays of Table 6-4. The user can purchase all the functions of Fig. 6-23 in a single package, or each of the blocks in a separate package (Sec. 5-5-5).

The operation of the display decoder/driver system of Fig. 6-23 is as follows. The four-bit bistable latch provides temporary storage for the binary-coded decimal (BCD) data. Such temporary storage is required because the BCD data source is typically an IC decade counter (Fig. 5-31) which does not incor-

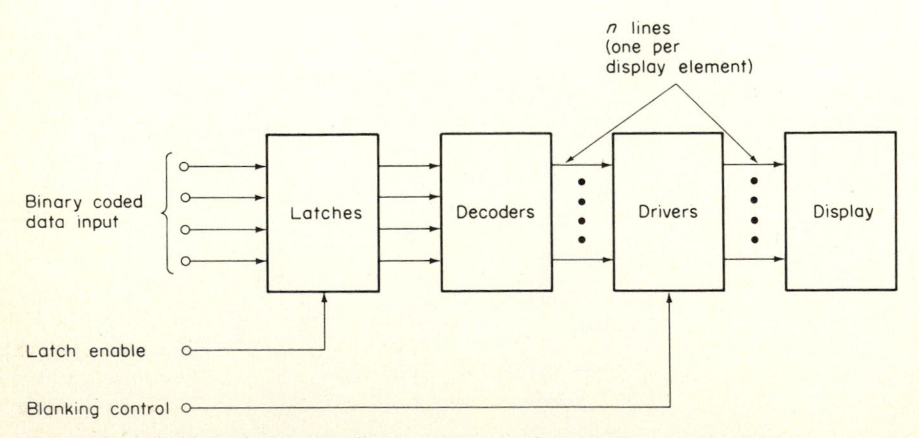

Fig. 6-23 Digital display decoder/driver electronic block diagram (single digit).

porate any storage. Unless the display is operated at exactly the same data rate as the counter, improper sample timing and loss or confusion of data may occur without the latch. When the latch enable is low, the input data are transferred to the latch outputs, and these outputs follow the input data. When the latch enable goes high, the data inputs at that time are stored in the latch outputs until the enable goes low again. The decoder circuit decodes (converts) the BCD data to the appropriate display code (seven-segment, for example), so that the required display elements or segments are turned ON via the display drivers, which provide the appropriate constant current or voltage for each display element. For example, for a BCD input of six (0110), the decoding logic enables the proper drivers and display elements so that the number 6 appears on the display. The blanking control is provided to modulate the display elements if a continuous dc drive is not desired. This control is at the driver circuit rather than at the latch so that its operation does not alter the input data or the data stored in the latches.

Figure 6-23 shows the nonmultiplexed decoder/driver method of display driving, whereby each character (or digit) of the display is driven by its own decoder/driver circuit. When the number of display characters or digits exceeds four, it becomes economical to *multiplex* or time-share a single decoder/driver for all the display digits. This reduces the number of required decoder/drivers and interconnections. Multiplexed operation requires additional timing and scan-decoding logic to steer the information to the proper digit locations; however, most MOS display-related ICs contain this additional circuitry. Figure 6-24 shows that in multiplexed operation the characters arrive sequentially or serially (although all the bits for a given character arrive simultaneously), compared to the nonmultiplex case in which all the information is presented to each digit of the display simultaneously. Multiplexing is possible because the human eye is slow and will perceive a multiplexed display as a continuous flicker-free phenomenon as long as the display *refresh* rate exceeds $\approx$70 times per second. [The fraction-of-a-second switching speeds of LCD displays (Table 6-4) are not adequate for multiplexing.]

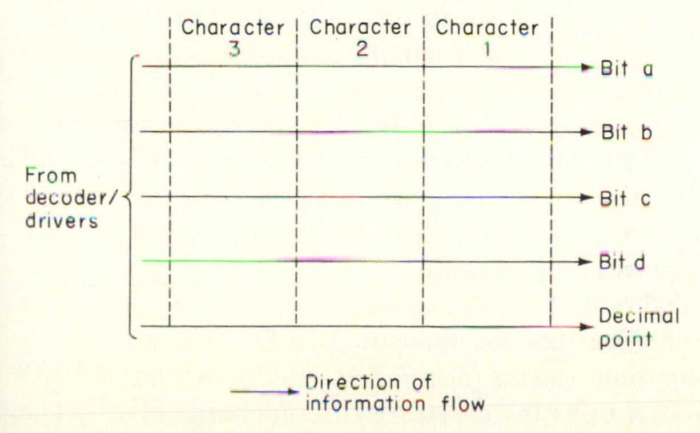

Fig. 6-24 Character-serial, bit-parallel data flow in a multiplexed display.

6-3-3 LED DISPLAYS

Visible light-emitting-diode (LED) displays are presently the most popular form of digital display because of their solid-state reliability, long life, low cost, low voltage drive requirements, wide operating-temperature range, high speed, and appearance. The main LED display disadvantages include their washout under high ambient light conditions, high current requirements, and high cost for large digits (>0.5 in [12.7 mm]).

The reader is referred to Sec. 6-2-4 for the basic theory of discrete IR and visible LEDs, and Sec. 5-6 for other examples of digital LED displays.

Figure 6-25 shows the construction of typical large and small seven-segment LED displays. For small LED displays, such as those for hand-held calculators

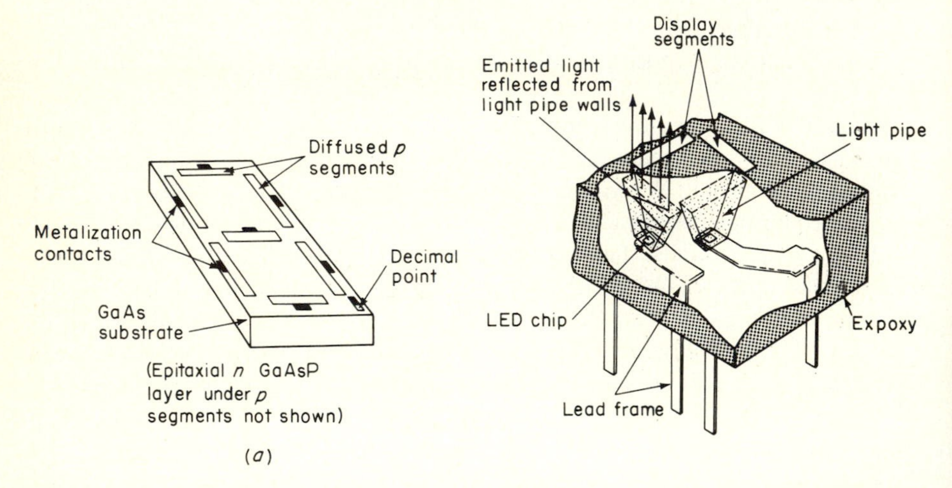

Fig. 6-25 LED seven-segment-display construction: (*a*) small (<0.15 in [3.8 mm]) monolithic LED display construction; (*b*) large (>0.15 in [3.8 mm]) light-pipe LED display construction. *Hewlett-Packard.*

with character heights less than approximately 0.15 in [3.8 mm], standard monolithic construction techniques such as shown in Fig. 6-25*a* are used (Sec. 3-2). For larger LED displays monolithic construction would require too much expensive GaAsP material, and hence manufacturers have developed the light-pipe fabrication techniques shown in Fig. 6-25*b*. In this case a reflecting light-pipe cavity is placed over the LED chip; the emitted LED light is reflected off the reflecting-cavity walls and the suspended glass particles and emerges as a significantly larger lighted surface area at the face of the display (Ref. 4). Figure 6-26 shows a typical seven-segment LED display.

Figure 6-27*a* is a block diagram of a typical single-digit seven-segment LED display and decoder/driver. Notice that the individual LEDs have a common anode, an external current-limiting resistor (Sec. 6-2-4), and an individual LED driver or transistor switch which sinks the necessary LED forward current. The clockwise arrangement of the seven segments A through G is standard, and the

Fig. 6-26 Seven-segment LED display. *Hewlett-Packard.*

RBO and RBI decoder/driver terminals provide blanking input/output. The truth table of Fig. 6-27*b* shows that the appropriate LED display segments are grounded or provided with a completed circuit path for the normal BCD 0-through-9 count.

LED displays are presently available in 0.1- to 1.0-in character heights; single- and multiple-character displays; the colors red, green, and yellow; 7- and 16-segment, and *hexadecimal* (partially alphanumeric) or complete alphanumeric (5-by-7 dot-matrix) fonts; with separate or integral decoder/driver electronics.

LED displays are a natural for multiplexing. The LED luminous efficiency (light output per unit input current) increases as the LED peak forward-current level increases (Sec. 6-2-4); hence the use of higher peak LED currents and lower duty cycles results in increased light output. Multiplex operation thus consumes less average power than dc drive for the same intensity.

6-3-4 LIQUID-CRYSTAL DISPLAYS

Liquid-crystal displays (LCDs) have become commercially available only in the last few years; however, they are an important contender in the digital-display field. Table 6-4 shows that the LCD's main advantages are its microwatt power requirements, low cost (not as size-dependent as others), and good contrast. LCDs suffer from lack of established reliability, limited temperature range, poor visibility in low ambient lighting, slow speed, and the need for an ac drive.

Liquid-crystal displays do not emit or generate light, but rather alter externally generated illumination. Figure 6-28 shows the construction of a typical LC display (Ref. 5). As shown, a thin layer of liquid-crystal fluid is sandwiched between two glass plates. The plates are coated with a transparent conductive film in the form of the desired alphanumeric image, and a ground plane or reflective coating is added (depending on the type of LCD).

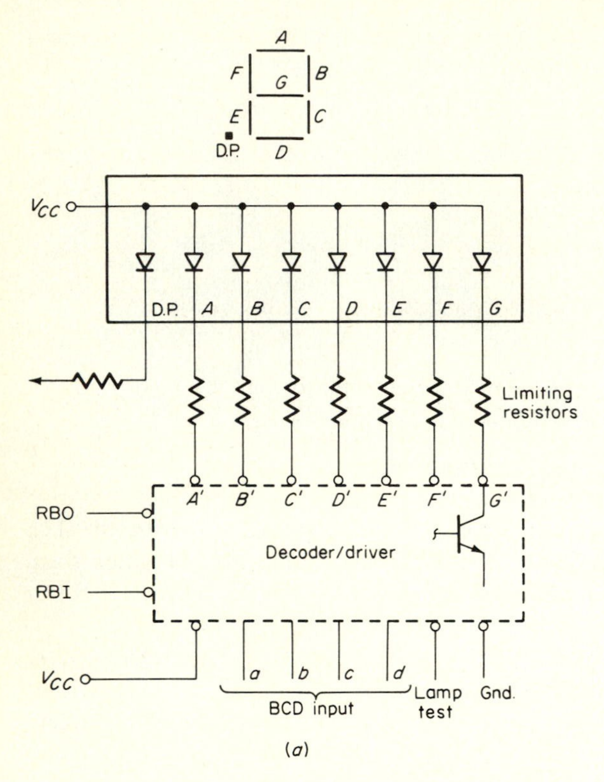

BCD input				Output state							Display
d	*c*	*b*	*a*	*A'*	*B'*	*C'*	*D'*	*E'*	*F'*	*G'*	
0	0	0	0	0	0	0	0	0	0	1	0
0	0	0	1	1	0	0	1	1	1	1	1
0	0	1	0	0	0	1	0	0	1	0	2
0	0	1	1	0	0	0	0	1	1	0	3
0	1	0	0	1	0	0	1	1	0	0	4
0	1	0	1	0	1	0	0	1	0	0	5
0	1	1	0	1	1	0	0	0	0	0	6
0	1	1	1	0	0	0	1	1	1	1	7
1	0	0	0	0	0	0	0	0	0	0	8
1	0	0	1	0	0	0	1	1	0	0	9

(*b*)

Fig. 6-27 Typical seven-segment LED display and decoder/driver: (*a*) block diagram; (*b*) truth table.

Figure 6-29 shows the theory of operation of the two general types of LC displays—dynamic scattering and field effect (Ref. 5). As shown in Fig. 6-29*a*, the liquid-crystal fluid in a dynamic-scattering LCD is normally clear. Under the application of an electric field, ion activity causes turbulence which scatters the external light, yielding a lighted display segment on a dark background, as shown in Fig. 6-30*a*.

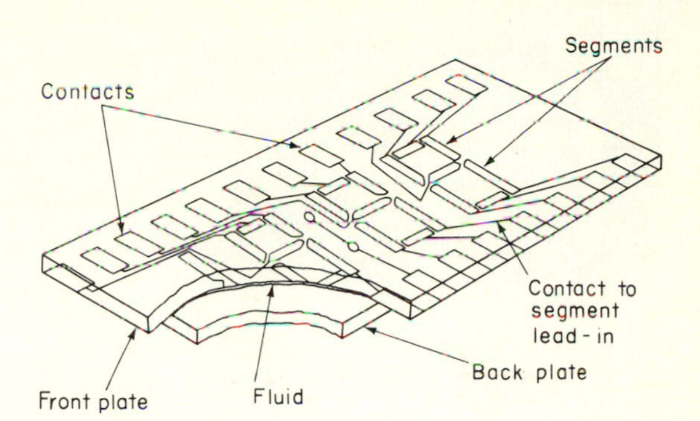

Fig. 6-28 Liquid-crystal-display (LCD) construction.

Figure 6-29*b* shows that the field-effect LCD contains front and back polarizers at right angles to each other. Without electrical excitation, the light coming through the front polarizer is rotated 90° in the fluid, passes through the rear polarizer, and is reflected by the mirror. When an electrostatic field is applied, the LC-fluid molecules rotate 90° so that the light is not rotated 90° and is therefore absorbed by the rear polarizer. This causes the appearance of dark digits on a light background, as shown in Fig. 6-30*b*.

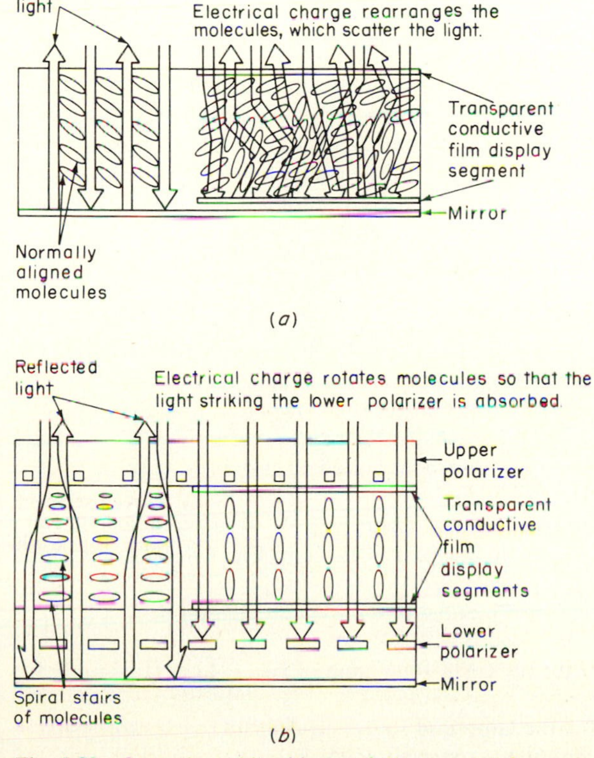

Fig. 6-29 Operation of liquid-crystal displays: (*a*) dynamic scattering; (*b*) field effect. *Hamlin Inc.*

Fig. 6-30 Typical liquid-crystal displays: (*above*) dynamic scattering; (*below*) field effect. *Hamlin Inc.*

Most LCDs are prone to premature failure if operated with dc potentials; hence an ac square-wave drive is usually employed. Reference 6 is an example of an ac MOS LCD driver similar to Fig. 6-23.

PROBLEMS

6-1 Construct a simple graph demonstrating how rapidly the available irradiance (H of Table 6-2a) varies with source-to-detector separation for a point source. Assume a radiant intensity I_R of 1 W, and graph H for 1-cm, 2-cm, 4-cm, and 8-cm separations.
6-2 State the key condition under which a light source can be considered a point source.
6-3 What is the main reason for using radiometric terminology and units in all optoelectronic source/detector applications other than those involving the human eye?
6-4 Calculate the energy, in eV, associated with light of the following wavelengths: 900 nm (IR LED emission wavelength), 550 nm (peak of the eye's response), 3.39 μm (IR line from the He-Ne laser).
6-5 With the aid of an atomic electron shell diagram, briefly describe the basic physical principle and theory of operation of most solid-state photodetectors and sources.
6-6 What is the solar constant at the earth's surface?
6-7 Assuming a solar-cell power-conversion efficiency of 10 percent, how many square centimeters of solar-cell surface area would be required to provide 1.0 W of electric power on earth?
6-8 Briefly explain the operation of an incandescent lamp.
6-9 Calculate the rerated MSCP and lifetime of a 1175 lamp (Fig. 6-4) if the operating voltage is increased to 7 V.
6-10 Calculate the rerated MSCP and lifetime of a CM8-1096 lamp operated at 12 V rather than 5 V (Fig. 6-4).
6-11 Calculate the irradiance H for the CM8-1096 lamp of Fig. 6-4 operated at design levels with a 5-cm source-to-detector separation.
6-12 Calculate the irradiance for the lamp and source-to-detector separation given in Prob. 6-11, but with the filament voltage reduced to 3 V.

6-13 (*a*) Calculate the lamp current for an A2B neon lamp with a 100-kΩ series resistance at 115 V line operation (see Fig. 6-7*b*).

(*b*) Choose the best neon lamp from Fig. 6-7*c* for turn-ON above 60 V and extinction below the same voltage.

6-14 Calculate the appxoximate irradiance for a 10-W incandescent lamp operated at normal temperature with a source-to-detector separation of 5 in [127 mm].

6-15 List two advantages and two disadvantages of miniature incandescent lamps.

6-16 Briefly explain the operation of an LED.

6-17 Assuming transitions approximately equal to the energy gap, at what wavelength would GaAs emit (Table 6-3)?

6-18 Calculate the minimum power-conversion efficiency of the GaAs LED of Fig. 6-10*b*.

6-19 List two advantages and two disadvantages of LED sources, compared to incandescent lamps.

6-20 Calculate the actual radiant output of the GaAs LED of Fig. 6-10 operated at an I_F of 20 mA dc and 40°C ambient temperature. Take into account the LED's junction temperature under these conditions (see Fig. 6-10 and Eq. 8-1).

6-21 Calculate the irradiance H available from an LED with an actual P_{out} of 1.0 mW, 30° divergence angle, and a source-to-detector separation of 5 cm [127 mm]. Assume a point source.

6-22 Calculate the irradiance available for pulsed LED operation with the LED of Fig. 6-10, with a 200-mA I_F pulse, 10^{-3} duty cycle, 1.0-in separation, and 25°C ambient temperature.

6-23 List two differences between GaAs infrared LEDs and GaAsP visible LEDs.

6-24 Calculate the value of the LED series limiting resistor for a 12-V dc supply and an LED with I_F and V_F maximum continuous ratings of 100 mA and 1.3 V, respectively.

6-25 For the He-Ne laser of Fig. 6-17, explain the function of the end mirrors, high voltage, filament, Brewster angle, capillary tube, and He gas.

6-26 Calculate the Brewster angle for glass with an index of refraction of 2.4.

6-27 Explain the difference between a spontaneous LED and a semiconductor injection laser.

6-28 What is the most popular type of digital display?

6-29 List two general advantages of digital displays over conventional analog displays.

6-30 List two advantages and two disadvantages of each of the following displays: LED, liquid crystal, incandescent.

6-31 Sketch the seven elements of a seven-segment display, and show which are ON for each of the numbers 0 through 9.

6-32 Which digital-display technology would you select for:

(*a*) Lowest power consumption

(*b*) Highest speed

(*c*) Lowest drive-voltage requirement

(*d*) Largest available physical size

6-33 (*a*) List two advantages of multiplexing digital displays.

(*b*) Why can't liquid-crystal displays be multiplexed?

(*c*) Why is the LED display a natural choice for multiplexing?

6-34 Briefly explain the operation of the dynamic-scattering and field-effect liquid-crystal displays.

REFERENCES

1. Burrous, C., and G. Deboo: "Temperature Compensation of LEDs and Photodiodes," Electro-Optics Systems Design Conference, New York, Sept. 14–16, 1971.
2. Beam, B., C. Burrous, and D. Jaynes: "Hydrocarbon Gas Detection Using a HeNe Laser," ISA Conference, Las Vegas, Nev., Nov. 1969.
3. Gilder, J.: "Focus on Displays," *Electron. Des.*, Vol. 26, p. 60, Dec. 29, 1974 (prepared by Symon Chang of AMI).
4. Hewlett Packard: "Optoelectronic Applications Seminar Handbook," p. 118, May 1974.
5. Hamlin Inc.: "LCD Application Manual," 1974.
6. RCA: CD 4054-6 Data Sheet, File no. 634, April 1973.

chapter 7

Photodetectors and Source-Detector Combinations

7-1 PHOTODETECTORS

Virtually all solid-state photodetectors are covered in this section, including bulk photodetectors (photoresistors), photodiodes, solar cells, phototransistors, photofield effect transistors, and photothyristors. There are also subsections on basic photodetector parameters and infrared, ultraviolet, and photoposition detectors. Several examples involve the various light sources of Chap. 6.

The various forms of source-detector combinations or opto-isolators are discussed in Sec. 7-2, with particular emphasis on the LED/phototransistor combination.

7-1-1 BASIC PHOTODETECTOR PARAMETERS

The photodetector is the heart of most optoelectronic systems; hence the selection of the best detector for the particular application is extremely important. This is not meant to deprecate the role of the light source in such systems, but often the source is a natural type such as the sun so that no choice is involved.

Photodetectors can be divided into two rather all-encompassing categories:

1. *Thermal detectors*, in which the radiation is absorbed and transformed into heat, and the detector responds to the change in temperature (energy).
2. *Quantum detectors*, which respond directly to the incident photons. This latter group is traditionally divided into the following subgroups:
   (*a*) *Photoemissive*, in which incident photons free electrons from the detector's surface. This phenomenon generally occurs in a vacuum, as in a vacuum photodiode or multiplier phototube.
   (*b*) *Photoconductive*, in which the conductivity of the photosensor changes as a function of incident light.
      (*i*) *Bulk photoconductors*, such as photoresistors, which are generally intrinsic or undoped (Sec. 7-1-2).
      (*ii*) *Doped photoconductors*, such as photodiodes, phototransistors, etc.
   (*c*) *Photovoltaic*, in which a voltage is self-generated as light strikes the device, without any external bias. Solar cells are the best example of this type, and photodiodes can be operated in either the photoconductive or photovoltaic mode.

There are four key photodetector characteristics that are pertinent to any optimization selection. They include the sensor's (Ref. 1):

1. *Responsivity*, which refers to the detector's sensitivity or output per unit of input light, such as output amperes/input radiant flux in watts.
2. *Spectral response*, which indicates the ability of the detector to respond to radiation of different wavelengths. For instance, if a source's spectral output ranges from 400 to 800 nm with a peak approximately midway, the detector should have a somewhat similar spectral response, or at least there should be appreciable overlap (Table 6-1).
3. *Frequency response*, or the speed with which the detector can respond to modulated radiation.
4. *Noise*, which refers to random fluctuations of the detector's output current or voltage. (The term *dark current* is used to denote the dc photodetector output current that exists under dark conditions, whereas the *noise current* is the random ac output.) Noise is directly proportional to the square root of detector area.

The signal-to-noise ratio of the detector is obtained by dividing I_{signal} by I_{noise} for a given input and operating condition.

A figure of merit commonly quoted as a means of specifying photodetector performance is *noise equivalent power* (NEP), which is defined as the power necessary to produce a signal-to-noise ratio of 1 for a noise bandwidth of 1 Hz. This is a measure of the minimum detectable signal level:

$$\text{NEP for a photodiode} = \frac{\text{noise current (A/Hz}^{1/2})}{\text{current responsivity (A/W)}} \tag{7-1}$$

the final units being watts per root hertz (W/Hz$^{1/2}$). The lower the NEP, the lower the detection limit. At a bandwidth of 1 Hz and a detector input of 10^{-13} W, an NEP of 10^{-13} W/Hz$^{1/2}$ would mean that the detector's output signal would not quite be discernible from the noise.

7-1-2 BULK PHOTOCONDUCTORS (PHOTORESISTORS)

As pointed out in Sec. 6-1, when enough energy in any form is added to a particular material, valence electrons can escape from their parent atoms and become free current carriers. By definition a hole is the absence of an electron; therefore a free hole is created for each free electron. The energy necessary to accomplish this ranges from 0.2 to 3 eV, depending on the particular material (Table 6-3); therefore radiation of wavelengths between 400 and 6000 nm would be adequate [Eq. (6-2)]. This phenomenon is the basis of operation for all practical *bulk photoconductors*, as portrayed by Fig. 7-1. Notice that there is no junction necessary for the operation of these devices, just a layer of photoconductive material whose resistance decreases (or conductance increases) proportionally to the

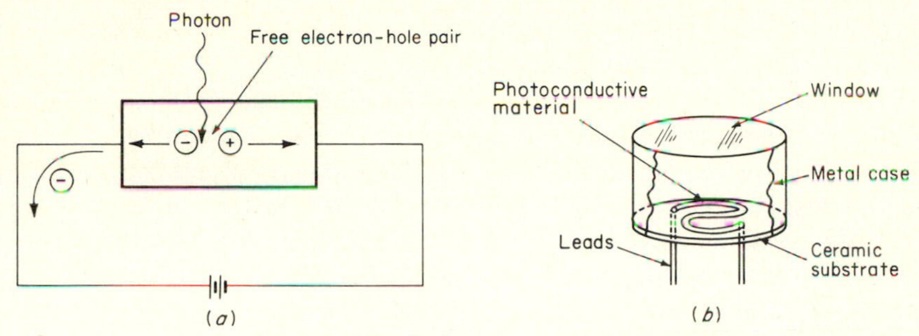

Fig. 7-1 Bulk photoconductor (or photoresistor): (*a*) schematic; (*b*) cutaway view.

intensity of light. The applied field is necessary to make the electrons flow through the detector and external circuit and finally recombine with the available holes at the negative photoconductor terminal. Actually, the basic idea of any bulk or junction photodetector is to convert light to an electrical signal, or get the incident photons into the detector with as little reflection as possible and efficiently extract the resulting free electrons. To do this, the phototransducer should be made of a material which is somewhat transparent at the wavelengths of interest, with an energy gap less than the photon energy [Eq. (6-2)]. An external (or internal self-generated) potential is necessary in order to extract the light-induced current.

A typical bulk photoconductor (or photoresistor) cutaway view is shown in Fig. 7-1*b*. The device is manufactured by vapor depositing or sintering the photoconductive material onto a ceramic substrate. Metallic electrodes and a suitable enclosure and window are then added to complete the device.

The variation in photoconductor resistance with irradiance is fairly linear as will be shown later, and ratios of dark to light resistance are often greater than 1000:1.

The spectral responses of several common intrinsic photoconductor materials are given in Fig. 7-2. (For comparison, a typical incandescent lamp

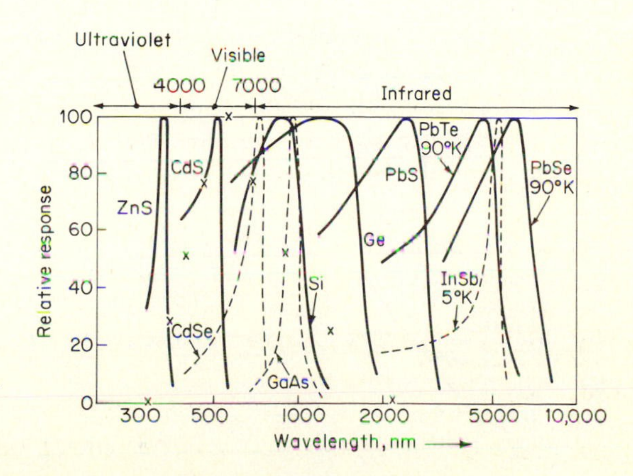

Fig. 7-2 Spectral responses of several common intrinsic photoconductors. *Hewlett-Packard.*

output spectrum is indicated by the *x* datum points.) The response for any particular material drops off at longer wavelengths because the photon's energy becomes less than the material's band gap; therefore the energy is inadequate for excitation of the material's valence electrons to the conduction band (Sec. 6-1). On the short-wavelength (ultraviolet) side the response falls off because the incident photons create electron-hole pairs very near the detector's surface with a higher probability of these free current carriers recombining before they reach a sufficient electrostatic field strength to be separated and hence contribute to the photocurrent. Detectors with response at wavelengths longer than 2000 nm will be covered in Sec. 7-1-8.

Figure 7-3 shows the rise and fall times of a typical bulk photoconductor. Remember that bulk-photoconductor resistance decreases with light, hence the

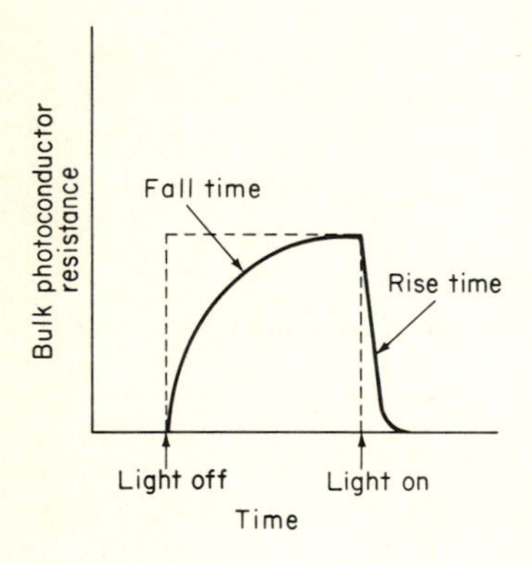

Fig. 7-3 Typical bulk-photoconductor speed of response.

apparent reversal of rise and fall times. Fall times are considerably longer because it takes longer for the electrons to "drain" back to the valence band, owing to trapping in imperfections.

Resistive photoconductors exhibit temperature coefficients which vary considerably with light level and material, as shown in Fig. 7-4. Notice that, for this case, the cadmium sulfide (CdS) type is superior to cadmium selenide (CdSe) as far as temperature stability is concerned, and operation at high intensities is desirable for maximum stability.

Photoresistors also exhibit a light history or memory effect; that is, their specific resistance depends on the intensity and duration of previous light exposure, and the length of time since such exposure.

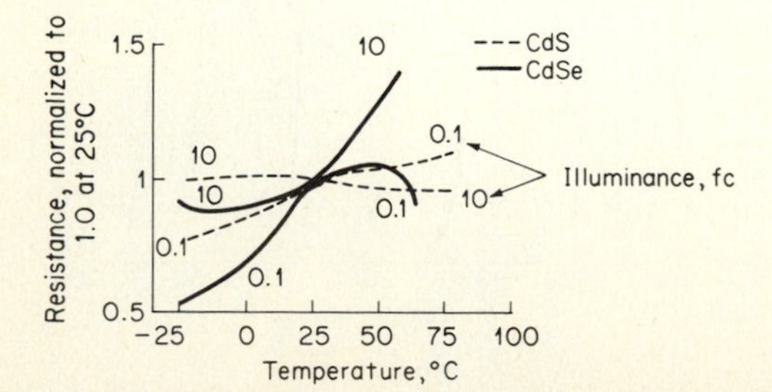

Fig. 7-4 Temperature dependence of typical CdS and CdSe photoconductors.

Figure 7-5 gives excerpts from typical CdS and CdSe photoconductor data sheets. Notice the difference in spectral response, speed, and temperature coefficient.

Resistive photoconductors possess the following advantages:

1. High sensitivity (due to large area)
2. Ease of use (simple light-sensitive resistor)
3. Low cost
4. No junction potential

Their disadvantages include:

1. Narrow spectral response
2. Light-history effects
3. Generally poorer temperature stability for the faster materials
4. Generally slower response for the stable materials

Recent advances in bulk-photoconductor technology have resulted in detectors which exhibit excellent speed and temperature stability in the same device; however, disadvantages 3 and 4 above still generally hold true.

The main areas of application of photoresistors are: inexpensive low-accuracy light measurement and alarm/relay controls.

Figure 7-6 shows several simple photoresistor applications. The lightmeter of Fig. 7-6*a* can be built to accommodate a wide range of light levels by proper selection of the battery, photoresistor, meter, and shunts. For example, assume a 5-V battery, a CL905 photoresistor (type 5 in Fig. 7-5*c*), and a 0-to-100-μA meter. Assuming that the meter resistance is zero, the illuminance that would cause the meter to read (1) full scale and (2) one-tenth of full scale can be calculated as follows:

1. Full scale = 100 μA, so $R = 5\ \text{V}/(100 \times 10^{-6})\ \text{A} = 50\ \text{k}\Omega$, which corresponds to 10 fc (foot candles) from Fig. 7-5*c*.
2. One-tenth of full scale = 10 μA, so $R = 5\ \text{V}/(10 \times 10^{-6})\ \text{A} = 500\ \text{k}\Omega$, which corresponds to 0.6 fc from Fig. 7-5*c*.

(Notice from Fig. 7-5 that photoresistor data are typically expressed in photometric terms. Radiometric units will be used for all other photodetectors.)

Replacing one leg of a simple voltage divider with a photoresistor yields a light-dependent voltage, as shown by Fig. 7-6*b*. As an example of the sensitivity of such photoresistors, consider the left-hand circuit of Fig. 7-6*b* with $R_1 = 10$ kΩ, a CL905L photoresistor (Fig. 7-5*b*), and light conditions varying from dark to 2 fc. Figure 7-5*b* shows that the CL905L photoresistor varies from 670 kΩ at dark to 10 kΩ at 2 fc. Hence, $V_{\text{out}} \approx 10$ V under dark conditions and, at 2 fc,

$$V_{\text{out}} = \frac{V_{\text{in}} R_2}{R_1 + R_2} = \frac{(10\ \text{V})(10\ \text{k}\Omega)}{10\ \text{k}\Omega + 10\ \text{k}\Omega} = 5\ \text{V}$$

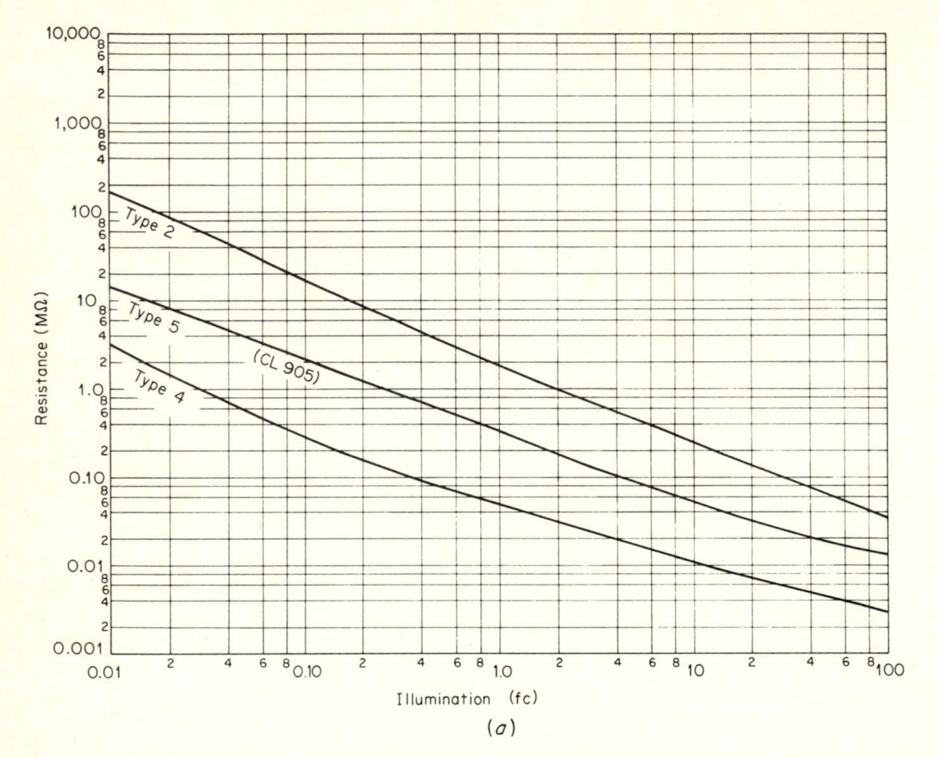

Fig. 7-5 CdSe and CdS photoconductor data: (*a*) resistance versus illumination for 900 Series; (*b*) spectral response, temperature coefficient, and speed; (*c*) typical specifications. *Clairex Corp.*

As an example of the incandescent-lamp rating required to produce an illuminance of ≈ 2 fc in a photoresistor application such as the above, consider a 25-W household incandescent lamp, which is typically rated at ≈ 220 lm (lumens) or MSCP. Assuming a point-source geometry, the luminous intensity can be calculated from the photometric equation of Table 6-2*a*:

$$I_L = \frac{F(\text{lm})}{4\pi\ (\text{steradians})} = \frac{220}{12.56} = 17.5\ \text{cd(candles)}$$

From Table 6-2*a*, the illuminance (flux at the photodetector) for a 3-ft source-to-detector separation is

$$E(\text{fc}) = \frac{I_L(\text{cd})}{d^2(\text{ft}^2)} = \frac{17.5}{9} = 1.94\ \text{fc}$$

(Subminiature lamps such as discussed in Fig. 6-4*a* cannot provide this level of illuminance because of their low MSCP rating.)

TYPE 4 CdSe, peak spectral response 6900 angstroms, lowest resistance photocell available. Can be used for "on-off" applications when low resistance is desired For use with incandescent or neon lamps.

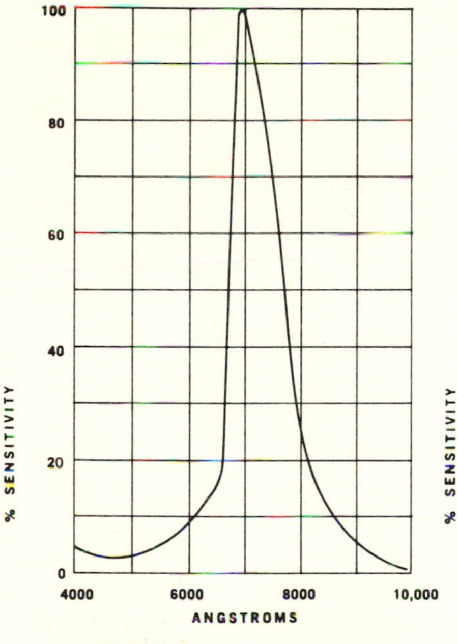

Variation of Conductance With Temperature and Light

Foot Candles	.01	0.1	1.0	10	100
Temperature	% Conductance				
−25°C	220	138	114	97	95
0°C	166	118	110	100	99
25°C	100	100	100	100	100
50°C	53	70	83	95	96
75°C	23	30	57	80	85

Response Time Versus Light

Foot Candles	.01	0.1	1.0	10	100
Rise (Seconds)*	1.780	.430	.088	.023	.005
Decay (Seconds)*	.160	.047	.030	.015	.008

*Time to (1 - 1/e) of final reading after 5 seconds Dark adaptation.

TYPE 5 CdS, peak spectral response 5500 angstroms (closely matches the human eye), most stable, lowest memory photocell available. Can be used in light measuring applications and precision low speed switching. For use with incandescent, fluorescent or neon lamps.

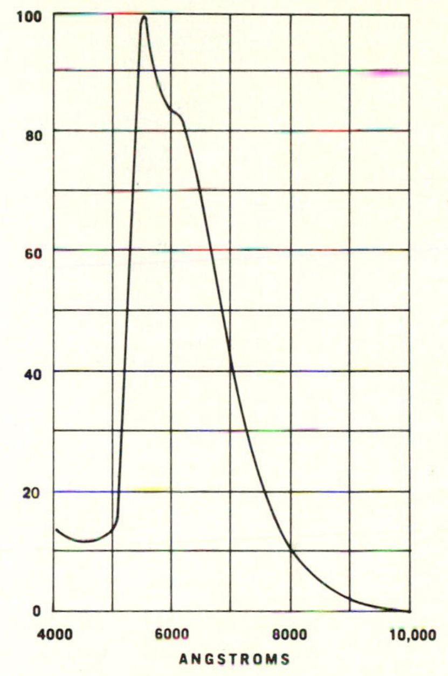

Variation of Conductance With Temperature and Light

Foot Candles	.01	0.1	1.0	10	100
Temperature	% Conductance				
−25°C	117	108	98	95	96
0°C	105	102	96	95	97
25°C	100	100	100	100	100
50°C	96	95	101	106	104
75°C	82	81	98	111	110

Response Time Versus Light

Foot Candles	.01	0.1	1.0	10	100
Rise (Seconds)*	5.80	.82	.140	.035	.010
Decay (Seconds)*	2.96	.56	.110	.043	.014

*Time to (1 - 1/e) of final reading after 5 seconds Dark adaptation.

(b)

TYPE	Sensitive Material	Peak Spectral Response (Angstroms)	Resistance @ 2 ft-c (Ohms)	Min. Dark Resistance 5 sec. After 2 ft-c	Maximum Voltage Rating (Peak A.C.)	Measurement Voltage
CL902	Type 2 CdS	5150	1 Meg	66.7 Meg	250V	60V
CL902N			500K	33.3 Meg	100V	30V
CL902L			67K	4.5 Meg	100V	45V
CL904	Type 4 CdSe	6900	30K	20 Meg	250V	20V
CL904N			15K	10 Meg	100V	10V
CL904L			2K	1.3 Meg	100V	10V
CL905	Type 5 CdS	5500	166K	11 Meg	250V	100V
CL905N			83K	5.5 Meg	100V	50V
CL905L			10K	670K	100V	20V

(c)

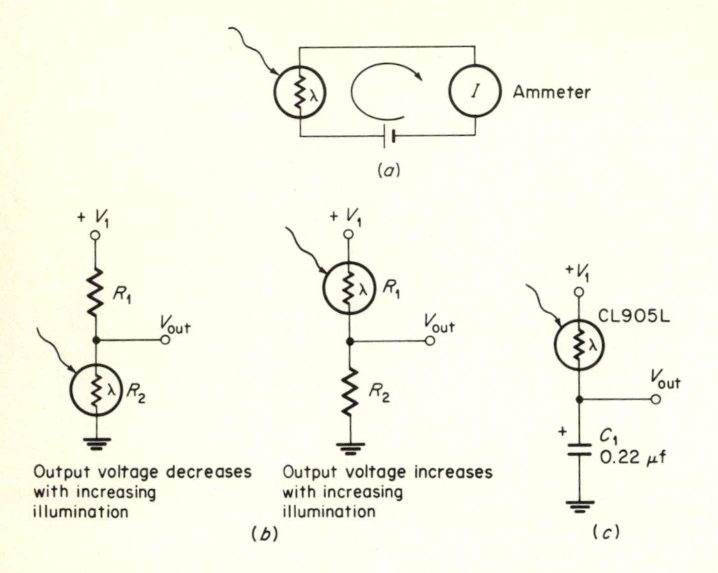

Fig. 7-6 Photoresistor applications: (*a*) simple lightmeter circuit; (*b*) photosensitive voltage-divider circuits; (*c*) light-to-time conversion circuit.

If a capacitor is substituted for the fixed resistor as shown in Fig. 7-6*c*, its rate of charge will be proportional to the incident radiation. For example, if the same photoresistor and light conditions of the previous example are assumed, along with a 0.22-μF capacitor, the *RC* time constant will vary from $\approx$147 ms at dark to 2.2 ms at 2 fc. (See Sec. 9-5 for an application of the circuit of Fig. 7-6*c* to UJT relaxation oscillators.)

Figure 7-7 shows a more sophisticated light-measuring circuit consisting of a balanced bridge for measuring the attenuation of light at detector R_2 due to some attenuating medium in the active cell, with R_4 serving as a reference. If the two photoresistors are matched, their magnitudes will track up and down together, tending to keep the bridge balanced under varying environmental conditions. This circuit reduces the effects of power-supply shifts, light history and temperature drift of the photoresistors, and variations in lamp output because both photodetectors (or bridge halves) will experience similar changes, and therefore the output $E_1 - E_2$ will tend to remain constant. To analyze the circuit of Fig. 7-7, neglect R_p and consider each output as a separate voltage divider. Thus,

$$E_1 = \left(\frac{R_2}{R_1 + R_2}\right)(+\text{V}) \qquad \text{and} \qquad E_2 = \left(\frac{R_4}{R_3 + R_4}\right)(+\text{V})$$

For balance, $E_1 = E_2$ or

$$\frac{R_2}{R_1 + R_2} = \frac{R_4}{R_3 + R_4}$$

Hence,

$$R_2 R_3 = R_4 R_1 \qquad \text{for balance} \tag{7-2}$$

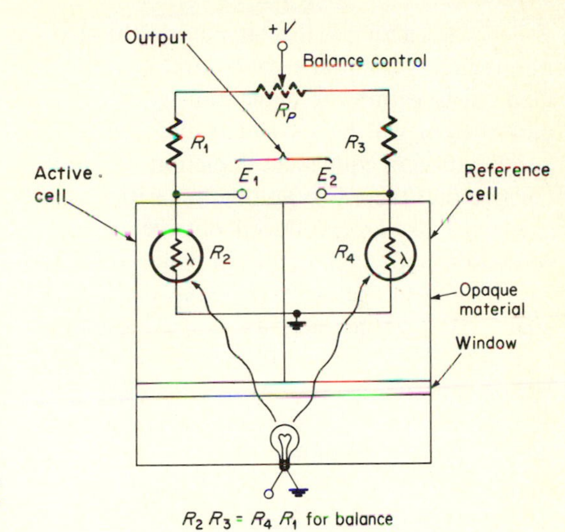

Fig. 7-7 Photoresistor bridge light-measuring circuit for improved stability.

7-1-3 PHOTODIODES

For almost every type of junction semiconductor device there exists an optical equivalent which responds to light instead of (or in conjunction with) an electrical signal. The first observation that semiconductor diodes were light-sensitive probably took place in much the same manner as you might observe a considerable increase in a conventional junction diode's leakage current when it is exposed to light.

Figure 7-8 shows the basic structure and operation of an extrinsic *pn*-junction silicon photodiode. When photons of energy greater than that of the Si energy gap are absorbed in the device, hole-electron pairs are generated. The pair generation will occur at various depths, depending on the energy of the photons and the nature and thickness of the materials. In essence the device looks like a capacitor, with the depletion region serving as the dielectric and the

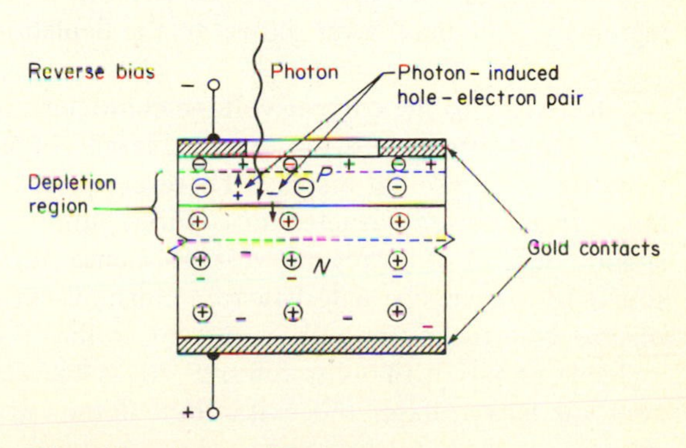

Fig. 7-8 Cross section of a typical *pn*-junction photodiode.

doped areas acting as the capacitor's charged plates. The region within the depletion region is essentially void of free current carriers. Hence the main electrostatic field exists within the depletion region, and any photon-induced hole-electron pair therein will be separated and drawn out (in opposite directions) by the combination of equivalent depletion-region voltage and the applied reverse bias. (Notice that these two voltage polarities are in the same direction.) If the hole-electron pairs are produced outside the depletion region, they have a higher probability of recombining, in which case no photocurrent would be available. It is therefore necessary to make the top *p* layer as thin as possible and the effective depletion region as wide as possible, in order to maximize the quantum efficiency. The depletion region can be widened by increasing the diode's reverse bias. Increasing the reverse bias also decreases the transit time for the hole and electron to reach the extremities. The advantages and disadvantages of reverse bias will be covered after some of the basic characteristics of photodiodes have been discussed.

Portions of a *pin* photodiode data sheet are given in Fig. 7-9. These data will be referred to throughout this section to illustrate various photodiode characteristics. In the particular structure of Fig. 7-9*a*, an *I* or an undoped *intrinsic* layer exists between the *n* and *p* ends. Since a depletion region extends farther into a lightly doped area, a wider depletion region is provided by the *pin* structure. The increased width of the *I* layer can be thought of as an effective increase in capacitor plate separation; therefore the junction capacity is reduced because capacity is inversely proportional to separation. Thus a *pin* diode is much faster than a conventional *pn* diode. This structure also exhibits extremely low noise and dark current and greater efficiency at the longer wavelengths.

Figure 7-9*b* shows the spectral response of the *pin* photodiode. The long-wavelength cutoff is explained by the fact that the Si energy gap is 1.1 eV [Eq. (6-2)]. Hence we see that the longer-wavelength photons are so low in energy that they can pass through the entire photodiode without producing an electron-hole pair. The response falls off on the short-wavelength (high-energy) side of the peak response because the higher-energy ultraviolet radiation tends to create pairs close to the surface of the top *p* region, and even with a thin *p* region the pair may never diffuse to the depletion region before recombination occurs.

Referring to the current/voltage characteristics of Fig. 7-9*c*, remember that these characteristics are the reverse-biased characteristics turned upside down from the conventional display. There are several important observations to be made from these characteristics. First, notice that the photocurrent is only slightly dependent on reverse voltage; hence the device exhibits a high dynamic source resistance, characteristic of a constant-current source. Also, at a constant reverse bias the output photocurrent is linear for more than 10 decades of incident radiation (This is referred to as the *current mode* of operation, i.e., constant bias voltage and extraction of the current as a measure of incident radiation.)

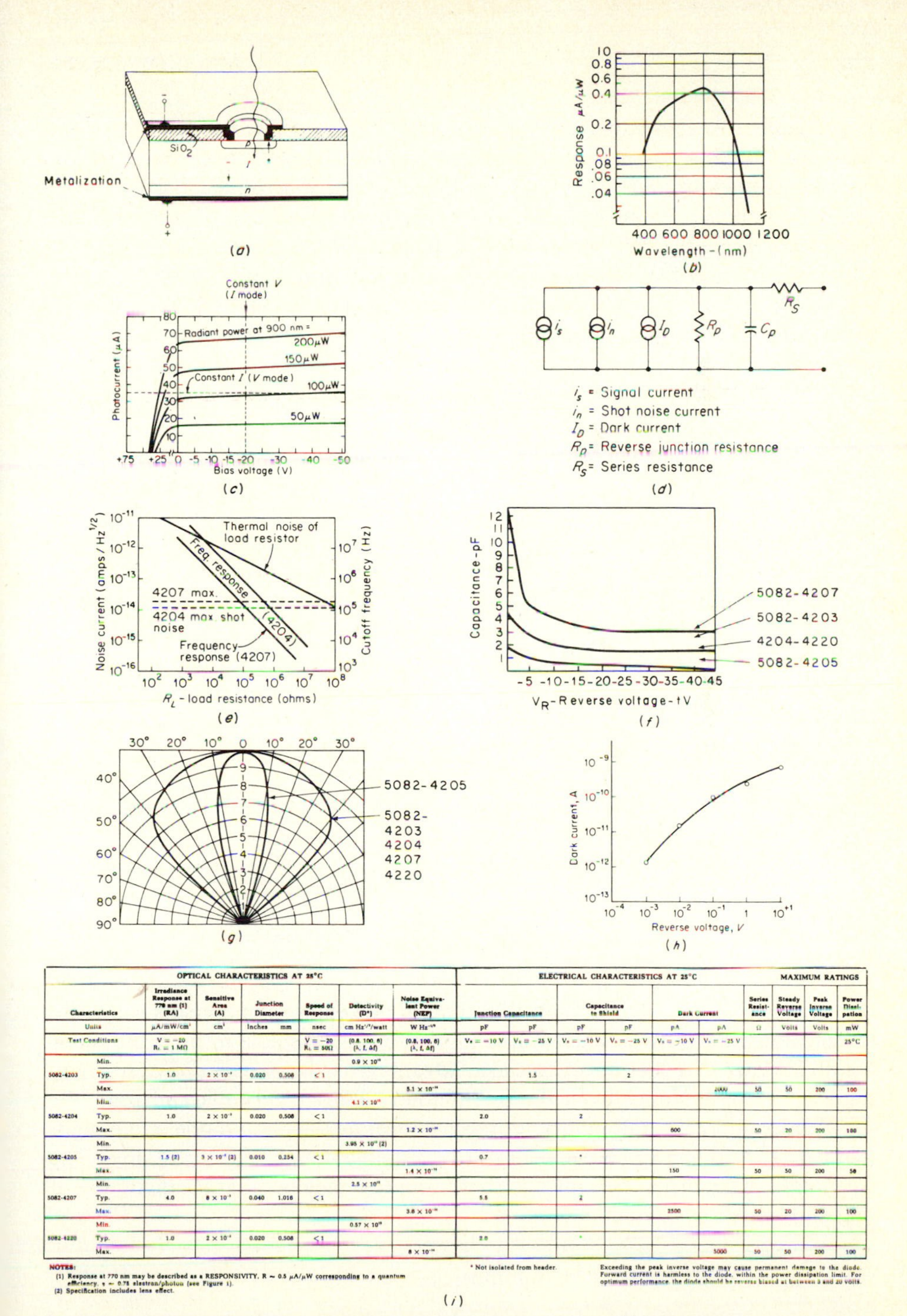

OPTICAL CHARACTERISTICS AT 25°C									ELECTRICAL CHARACTERISTICS AT 25°C						MAXIMUM RATINGS			
Characteristics		Irradiance Response at 770 nm (1) (RA)	Sensitive Area (A)	Junction Diameter		Speed of Response	Detectivity (D*)	Noise Equivalent Power (NEP)	Junction Capacitance		Capacitance to Shield		Dark Current		Series Resistance	Steady Reverse Voltage	Peak Inverse Voltage	Power Dissipation
Units		µA/mW/cm²	cm²	Inches	mm	nsec	cm Hz$^{1/2}$/watt	W Hz$^{-1/2}$	pF	pF	pF	pF	pA	pA	Ω	Volts	Volts	mW
Test Conditions		V = −20 R$_L$ = 1 MΩ				V = −20 R$_L$ = 50Ω	(0.8, 100, 6) (λ, f, Δf)	(0.8, 100, 6) (λ, f, Δf)	V$_R$ = −10 V	V$_R$ = −25 V	V$_R$ = −10 V	V$_R$ = −25 V	V$_R$ = −10 V	V$_R$ = −25 V				25°C
5082-4203	Min.						0.9 × 10^{12}											
	Typ.	1.0	2 × 10^{-3}	0.020	0.508	<1				1.5		2						
	Max.							5.1 × 10^{-14}						2000	50	50	200	100
5082-4204	Min.						4.1 × 10^{12}											
	Typ.	1.0	2 × 10^{-3}	0.020	0.508	<1			2.0		2							
	Max.							1.2 × 10^{-14}					600		50	20	200	100
5082-4205	Min.						3.95 × 10^{12} (2)											
	Typ.	1.5 (2)	3 × 10^{-4} (2)	0.010	0.254	<1			0.7		*							
	Max.							1.4 × 10^{-14}					150		50	50	200	50
5082-4207	Min.						2.5 × 10^{12}											
	Typ.	4.0	8 × 10^{-3}	0.040	1.016	<1			5.5		2							
	Max.							3.6 × 10^{-14}					2500		50	20	200	100
5082-4220	Min.						0.57 × 10^{12}											
	Typ.	1.0	2 × 10^{-3}	0.020	0.508	<1			2.0		*							
	Max.							8 × 10^{-14}						5000	50	50	200	100

NOTES:

(1) Response at 770 nm may be described as a RESPONSIVITY, R ~ 0.5 µA/µW corresponding to a quantum efficiency, η ~ 0.75 electron/photon (see Figure 1).

(2) Specification includes lens effect.

* Not isolated from header.

Exceeding the peak inverse voltage may cause permanent damage to the diode. Forward current is harmless to the diode, within the power dissipation limit. For optimum performance, the diode should be reverse biased at between 5 and 20 volts.

(i)

Fig. 7-9 *pin*-photodiode data-sheet excerpts: (*a*) cross section; (*b*) spectral response; (*c*) current/voltage characteristics; (*d*) equivalent circuit; (*e*) noise current and cutoff frequency versus load resistance at −10 V bias; (*f*) junction capacitance versus reverse bias; (*g*) relative directional sensitivity; (*h*) dark current versus reverse bias (*Ref. 3*); (*i*) specifications. (*All but h courtesy Hewlett-Packard.*)

Figure 7-9*c* shows that the device could also be operated with no external applied bias, that is, open circuited on a horizontal constant-current line. The output open-circuit voltage would then serve as a logarithmic indication of incident radiation. We shall refer to this as the *voltage mode*, i.e., constant current bias and sensing the voltage change with light.

Operation in the current mode produces less distortion and nonlinearity, bias shift, sensitivity to environment, etc., as pointed out in Ref. 2. Obviously, operation is not limited to horizontal or vertical load lines, and in fact the load line can have any slope.

Figure 7-9*d* is a photodiode equivalent circuit. Notice the signal-current source along with the undesirable sources of noise current and dc *dark current* (I_D), or the dc leakage current that flows without any light. The values of I_D and i_n, of course, determine the lower limit of device sensitivity. Dark current doubles approximately every 8°C for Si diodes, and there are two components involved. One is leakage current, which varies with reverse bias and active area, and the other is the conventional reverse saturation current, which increases with temperature. The components of the noise current are

1. $I_{\text{shot}} = \sqrt{2qI_D \, \Delta f}$ (7-3)
2. $I_{1/f}$, which increases as the frequency of operation decreases
3. $I_{\text{thermal}} = \sqrt{\dfrac{4KT \, \Delta f}{R_T}}$ (7-4)

where q = charge on an electron (1.602×10^{-19} coulomb)
I_D = dark current
K = Boltzmann's constant (8.62×10^{-5} eV/K or 1.38×10^{-23} joule/K)
T = temperature, K (K = 273 + °C)
$R_T = R_{\text{source}} + R_{\text{load}}$ in series, Ω
Δf = bandwidth, Hz

The preceding formulas show that there are two noise components which are proportional to electrical bandwidth; therefore it is desirable to limit the bandwidth in order to reduce noise. It is also advantageous to operate above 20 Hz to reduce $1/f$ noise. Notice from the thermal-noise-current formula and Fig. 7-9*e* that for low-noise operation it is desirable to operate the photodiode into a high-resistance load. This of course may not be possible if high speed is desired. The key to obtaining the proper units in Eq. (7-3) is to remember that the definition of 1 A is 1 coulomb/s. One thus obtains

$$I_{\text{shot}} = \sqrt{(2)(1.6 \times 10^{-19} \text{ coulomb})[I_D \text{ (coulomb/s)}][\Delta f \, (\text{s}^{-1})]}$$
$$= \sqrt{K(\text{coulomb}^2/\text{s}^2)} = \text{coulomb/sec} = \text{A}$$

As an example of a typical photodiode (current) signal-to-noise-ratio calculation, consider the 4204 photodiode of Fig. 7-9*i* operated at room temperature (25°C) with 10 V reverse bias. Assume 10^{-7} W radiant input at 770 nm, a 10-MΩ load resistor, and 10 Hz bandwidth. From Fig. 7-9*i* and Eq. (7-3),

$$I_{\text{shot}} = [(2)(1.6 \times 10^{-19})(600 \times 10^{-12})(10)]^{1/2} = 4.38 \times 10^{-14} \text{ A}$$

For a 10-MΩ load resistance Eq. (7-4) gives

$$I_{\text{thermal}} = \left[\frac{(4)(1.38 \times 10^{-23})(298)(10)}{10^{+7}}\right]^{1/2} = 1.28 \times 10^{-13} \text{ A}$$

The units $[\text{joules}/(\text{ohms})(\text{seconds})]^{1/2}$ are equivalent to amperes because

$$\text{Power} = \frac{\text{energy}}{\text{time}} = I^2R = \frac{\text{joules}}{\text{time}}$$

and solving for I gives

$$I = \left[\frac{\text{joules}}{(\text{ohms})(\text{time})}\right]^{1/2}$$

The total noise (excluding $1/f$ noise) is equal to the sum of the squares of the individual noise components; i.e.,

$$\begin{aligned} I_{\text{total noise}} &= [(I_{\text{thermal}})^2 + (I_{\text{shot}})^2]^{1/2} \\ &= [(1.28 \times 10^{-13})^2 + (0.438 \times 10^{-13})^2]^{1/2} \\ &= 1.35 \times 10^{-13} \text{ A} \end{aligned}$$

The signal current can be estimated from the 0.5 A/W conversion-efficiency note on the bottom of Fig. 7-9*i* or graphically from Fig. 7-9*b*. That is,

$$I_{\text{signal}} = \frac{0.5 \text{ A}}{\text{W}}(10^{-7} \text{ W}) = 5.0 \times 10^{-8} \text{ A}$$

The current signal-to-noise ratio is thus

$$\frac{I_{\text{signal}}}{I_{\text{noise}}} = \frac{5 \times 10^{-8} \text{ A}}{1.35 \times 10^{-13} \text{ A}} = 3.7 \times 10^{+5}$$

At room temperature the $I_{\text{signal}}/I_{\text{dark}}$ ratio is $\approx 5 \times 10^{-8}/600 \times 10^{-12}$ or 83:1.

Figure 7-9*f* shows the reduction of diode junction capacity that may be obtained by increasing the reverse bias. (The active area of the upper trace diode is four times that of the lower.)

Typical directional sensitivities obtained with lenses and flat windows are shown in Fig. 7-9*g*. For applications requiring a reduction of look angle, the narrow directional lobe produced by a lens would obviously be preferable. The addition of a lens does not appreciably effect the responsivity as shown by the 4205 specifications of Fig. 7-9*i*. However, the lens gathers light over a larger area and focuses it on the photodiode, which effectively increases the active area.

Comparing junction photodiodes to bulk photoconductors, we see that the photodiode possesses considerably better frequency response, linearity, spectral response, and lower noise. The photodiode's disadvantages include: small active area, rapid increase in dark current with temperature, offset voltage, and necessity of amplification at low irradiances.

As previously mentioned, the current mode, or operation at constant voltage while sensing the diode photocurrent, is generally the best mode of photodiode operation. The photodiode current in that case is linear for several decades and is equal to

$$I_{\text{photo}} = HS_p\sigma \tag{7-5a}$$

or

$$I_{\text{photo}} = HS_eA \tag{7-5b}$$

where I_{photo} = photodiode output current, μA for Eq. (7-5*a*) and mA for Eq. (7-5*b*)
H = irradiance striking the photodiode, mW/cm^2
S_p = photodiode sensitivity at the *peak* of its spectral response, $\mu\text{A}/(\text{mW/cm}^2)$ (Fig. 7-9*i*)
σ = relative photodiode response at the actual source emission wavelength
S_e = photodiode response at the *actual* source emission wavelength, mA/mW as given by Fig. 7-9*b*
A = detector sensitive area, cm^2

The output current of the 4207 photodiode (Fig. 7-9) can thus be calculated as follows. Assume the photodiode is irradiated by the GaAs infrared (900-nm) LED of Figs. 6-10 and 6-11 with a 1.0-in [2.54-cm] source-to-detector separation. For this case the LED irradiance at the detector was previously calculated to be 0.32 mW/cm^2. Figure 6-10*c* shows that the GaAs infrared LED's spectral output peaks at $\approx$900 nm and is near zero at $\approx$800 and 1000 nm. Comparing this LED output spectrum to the Si photodiode spectral response (Fig. 7-9*b*) shows that all of the LED's output spectrum is within the photodiode spectral response, and the photodiode sensitivity to the LED spectral peak at 900 nm is $\approx$0.35 $\mu\text{A}/\mu\text{W}$ or mA/mW, and is $\approx$0.17 and 0.45 mA/mW at 1000 nm and 800 nm, respectively. An accurate photodiode sensitivity figure can be calculated by graphical integration; however, a reasonable estimate of the average photodiode sensitivity to the total IR LED spectrum is $\approx 0.3\ \mu A/\mu W$ or mA/mW (Figs. 6-10*c* and 7-9*b*). Substituting these quantities into Eq. (7-5*b*) yields

$$\begin{aligned} I_{\text{photo}} &= HS_e A \\ &= (0.32\ \text{mW/cm}^2)(0.3\ \text{mA/mW})(8 \times 10^{-3}\ \text{cm}^2) \\ &= 7.7 \times 10^{-4}\ \text{mA or } 0.77\ \mu\text{A} \end{aligned}$$

Figure 7-10*a* shows two methods of amplifying the low (microampere) photodiode output current to obtain a voltage linearly proportional to the incident irradiance. The simplest (Fig. 7-10*a*) involves reverse biasing the photodiode as indicated and sensing or amplifying the light-induced $I_{\text{photo}} R_1$ voltage. (Note that I_{photo} is a reverse current.) The only disadvantage of this circuit is that the bias on the photodiode will vary as the light level varies. That is, as the irradiance increases, I_{photo} becomes larger, and the $I_{\text{photo}} R_1$ voltage also increases. This voltage drop actually reduces the photodiode reverse bias; hence the photodiode bias is not constant and in fact changes owing to the very phenomenon we wish to detect. However, Fig. 7-9*c* shows that the photocurrent from a *pin* photodiode is relatively independent of reverse-bias voltage; hence the error will be small. A configuration for achieving constant reverse bias is shown in Fig. 7-10*b*. The circuit consists of an inverting operational amplifier with a high input impedance plus low noise and drift. (A high-impedance input such as a FET or MOS stage is required for most applications.) An explanation of this type of operational-

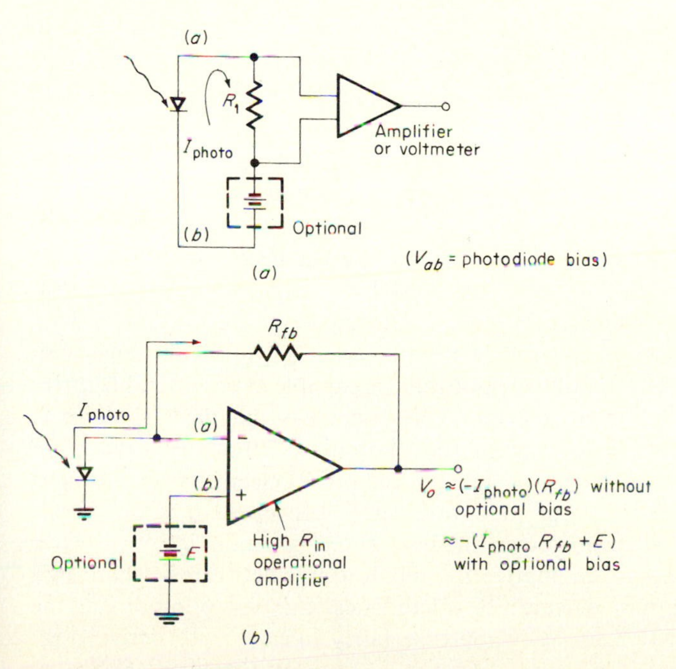

Fig. 7-10 Photodiode circuits: (*a*) simple circuit; (*b*) inverting operational-amplifier circuit that provides constant photodiode bias voltage.

amplifier configuration is given in Chap. 4, but the operation can be sufficiently explained by quoting the two summing-point restraints that apply to an ideal operational amplifier. (1) The amplifier is assumed to have infinite input impedance and hence draws no current. This allows all the photodiode output current to flow into the feedback element (R_{fb}) as shown in Fig. 7-10*b*. (2) Owing to the high open-loop gain, there is also no voltage difference between the two amplifier input terminals. This means that whatever negative voltage appears at terminal *b* as a consequence of the battery also appears at the anode *a*; and since V_{ab} is small (≈ 1 mV) and fairly constant for a good-quality operational amplifier, the photodiode bias ($V_a \stackrel{+}{\equiv}$) remains essentially constant. Since the input is connected to the negative or inverting terminal, the output will be inverted. Thus the output voltage, without optional bias E, approaches the ideal R_{fb} times the diode output current:

$$V_{out} = -I_{photo} R_{fb} \tag{7-6}$$

As an example, if the photodiode output current is 0.77 μA from the previous example, and the feedback resistance is 1.0 MΩ, the output voltage would be -0.77 V. A common figure of merit for such amplifiers is transresistance or output voltage per unit of input current ($\Delta E_{out}/\Delta I_{in}$). The transresistance for the above example would be 1 V/μA. Since the photodiode output current (under proper conditions) is a linear function of the incident radiation, the output of the circuit in Fig. 7-10*b* is also linear with incident radiation. Section 4-6-6 gives a similar circuit with a logarithmic element in the feedback loop. In that case the output voltage is proportional to the logarithm of input irradiance. Such logarithmic amplifiers offer a much broader dynamic range, and their percentage of reading error is fixed regardless of the particular input radiation level. This is not the case for a linear amplifier, in which the percentage of reading accuracy decreases drastically as one approaches the sensitivity limit (Ref. 3).

In practice the capacitance across R_{fb} of Fig. 7-10*b* limits the frequency response. Equation (7-6) and Fig. 7-9*e* demonstrate the necessary compromise between sensitivity and speed for the circuit of Fig. 7-10*b*. Notice that as R_{fb} increases, the sensitivity increases but the frequency response decreases. This points out the reason why photomultipliers are superior when high sensitivity *and* speed are required. The multiplier phototube is capable of considerable internal current gain with minimum thermal noise, whereas full thermal noise is generated in the feedback resistance in the photodiode circuit of Fig. 7-10*b* [Eq. (7-4)]. Also, neither the photomultiplier speed nor its electron transit time is seriously affected by the value of load resistance; hence, there is no speed-sensitivity compromise as evidenced in the case of the photodiode.

At this point we should summarize the relationships between detector area and some of the important detector characteristics. For the photodiode, the signal current and junction capacity increase fairly linearly with active area. Noise current is proportional to $\sqrt{\text{area}}$; however, dark current often increases more rapidly than area because of the greater probability of including wafer imperfections. To summarize: Increasing the detector area (1) increases the current S/N ratio by a factor of $\Delta\text{area}/\sqrt{\Delta\text{area}}$; (2) may actually decrease the $I_{\text{signal}}/I_{\text{dark}}$ ratio; and (3) increases the detector junction capacitance on approximately a 1:1 basis.

Although rarely mentioned in applications other than solar-energy conversion and infrared detection, circuits with photodiodes such as the *pin* types can be operated at 0 V reverse bias. This would eliminate the battery in Fig. 7-10*b*, and the positive amplifier terminal would be grounded. The operational-amplifier action would still maintain constant near-zero photodiode bias voltage (V_{ab}). The advantages of reverse biasing have already been mentioned, the chief among these being the possibility of increasing the speed by a factor of 5 or so, owing to the decreased junction capacity and transit time (Fig. 7-9*a* and *f*). Figure 7-9*c* also shows that a slight increase in quantum efficiency can be realized. There are, however, some disadvantages of high reverse bias. The dark current increases rapidly as shown in Fig. 7-9*h*, and this poses two problems: lower signal-to-dark-current ratios and increasing difficulty of temperature compensation owing to the higher initial room-temperature dark current. {The shot noise is also increased with increasing reverse bias because of its dark-current dependence [Eq. (7-3)].} There are two related phenomena that allow the efficiency of *pin* photodiodes to remain high even at very low reverse biases. First, the self-induced depletion region will extend much farther into the pure *I* layer than in a conventional highly doped *pn*-junction device; therefore a large depleted volume is available for pair separation. [Even at 0 V reverse bias the *pin*-photodiode depletion region extends approximately halfway through the *I* layer (Fig. 7-9*a*).] Second, the long carrier lifetimes available in high-purity silicon allow some of the photon-induced hole-electron pairs which are created outside the depleted *I* layer to diffuse to the depletion region prior to recombining.

Another type of *pin* photodiode is the surface or Schottky barrier diode, which is similar to the *pin* diode except that its *p* and *n* regions are formed by metallization instead of diffusion. Recent surface-barrier photodiode speeds, noise-equivalent powers, etc., are comparable to those of silicon-diffused *pin* photodiodes, and they have the advantage of simpler, lower-temperature fabrication.

Although we have discussed only individual photodiodes, several manufacturers offer arrays of two to several hundred photodiodes on a single silicon chip. Dimensions of the devices themselves and the spacing between units approach the limit of microminiature lithography (a fraction of 0.001 in—see Sec. 3-2). Such arrays find application in character-recognition page readers, star tracking, and even replacement of the traditional vidicon image devices.

7-1-4 SOLAR CELLS

The advent of the space age brought about a tremendous effort in the area of direct conversion of solar energy to electric energy. The photovoltaic solar cell has become the workhorse for producing spacecraft power, and advances in solar-cell technology have found their way into many earth-based applications, including heating homes.

In order to effectively utilize the sun as a source of power, any converter should have the following characteristics:

1. Its spectral response should somewhat correspond to the sun's spectral output, which peaks at approximately 500 nm with a long infrared tail (Fig. 6-3). At the earth, the sun's irradiance is approximately 100 mW/cm^2.
2. The converter should possess a high power-conversion efficiency.

Figure 7-11*a* shows a typical Si solar-cell cross section. Its structure, fabrication, and operation are similar to those of a silicon photodiode, but there are several

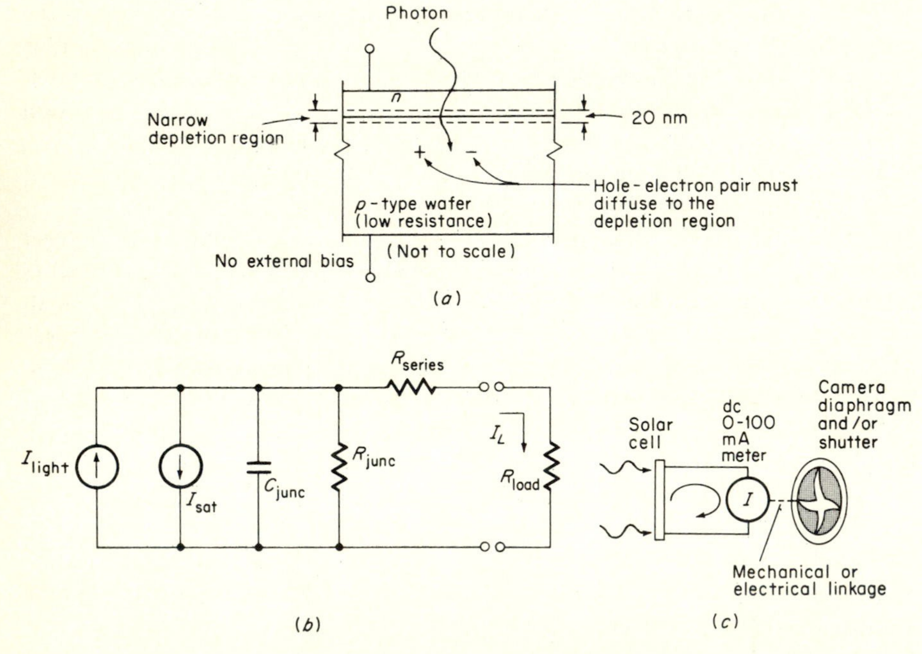

Fig. 7-11 (*a*) Typical solar-cell cross section; (*b*) solar-cell equivalent circuit; (*c*) solar-cell photographic exposure control.

important differences necessary to optimize the device for solar-energy conversion. The differences are:

1. The series resistance R_s shown in the solar-cell equivalent circuit of Fig. 7-11*b* must be very low, of the order of 1 Ω. Maximum power transfer would not occur if there were an appreciable drop across R_s. (R_s is composed of two resistances—the contact resistance and the sheet resistance.)
2. The doping of the solar cell is several orders of magnitude higher. Since the doping level approaches that of a conductor, the series resistance is signifi-

cantly reduced, the depletion region becomes very narrow, and the open-circuit voltage increases.

3. The solar-cell area is several orders of magnitude higher, so that the maximum radiant flux can be intercepted.
4. The high doping and antireflection coatings cut down reflection losses.
5. The top layer is generally as thin as possible in order to extend the ultraviolet response to match the solar spectrum.
6. Solar cells are always operated unbiased (photovoltaic mode).
7. The speed of the solar cell is considerably lower than that of the photodiode.

A solar cell operates in somewhat the same manner as other junction photodetectors in that a built-in depletion region is generated without an applied reverse bias, and photons of adequate energy create hole-electron pairs. In the solar cell, as shown in Fig. 7-11*a*, the pair must generally diffuse a considerable distance to reach the narrow depletion region to be drawn out as useful current. Hence, there is a higher probability of recombination, with resulting low quantum efficiency. The current generated by the separated pairs increases the depletion-region voltage (the photovoltaic effect), and, when a load is connected across the cell, the potential causes the photocurrent to flow through the load.

Figure 7-11*c* shows how solar cells are used in photographic exposure meters and automatic exposure controls. For this application the solar cell is superior to most other photodetectors because (1) it is a large-area detector with milliampere output currents compared to microamperes from photodiodes; hence the meter cost is reduced; (2) owing to its photovoltaic operation, a battery is not required as in the bulk-photoconductor circuit of Fig. 7-6*a*; and (3) its output current is linear with incident radiation for several decades, as with the photodiodes of Sec. 7-1-3.

Figure 7-12 shows typical silicon solar-cell data. The rectangular-shaped voltage/current characteristics of Fig. 7-12*a* are similar to the photodiode characteristics of Fig. 7-9*c*, but the voltage axis is inverted. The axis intersections are the maximum short-circuit current (at $R_L = 0$) and open-circuit voltage (at $R_L = \infty$), and the locus of maximum power (voltage times current) points is shown.

Solar-cell conversion efficiency is defined as the ratio of power out to power in, or

$$\text{Conversion efficiency (\%)} = \frac{\text{electrical output (mW)} \times 100}{\text{total radiation input (mW/cm}^2) \times \text{active area (cm}^2)} \qquad (7\text{-}7)$$

Notice in Fig. 7-12 that the optimum load line would pass through the graph's origin and the intersection of the maximum power line at a particular irradiance curve. Under such conditions, present Si solar cells possess a conversion efficiency

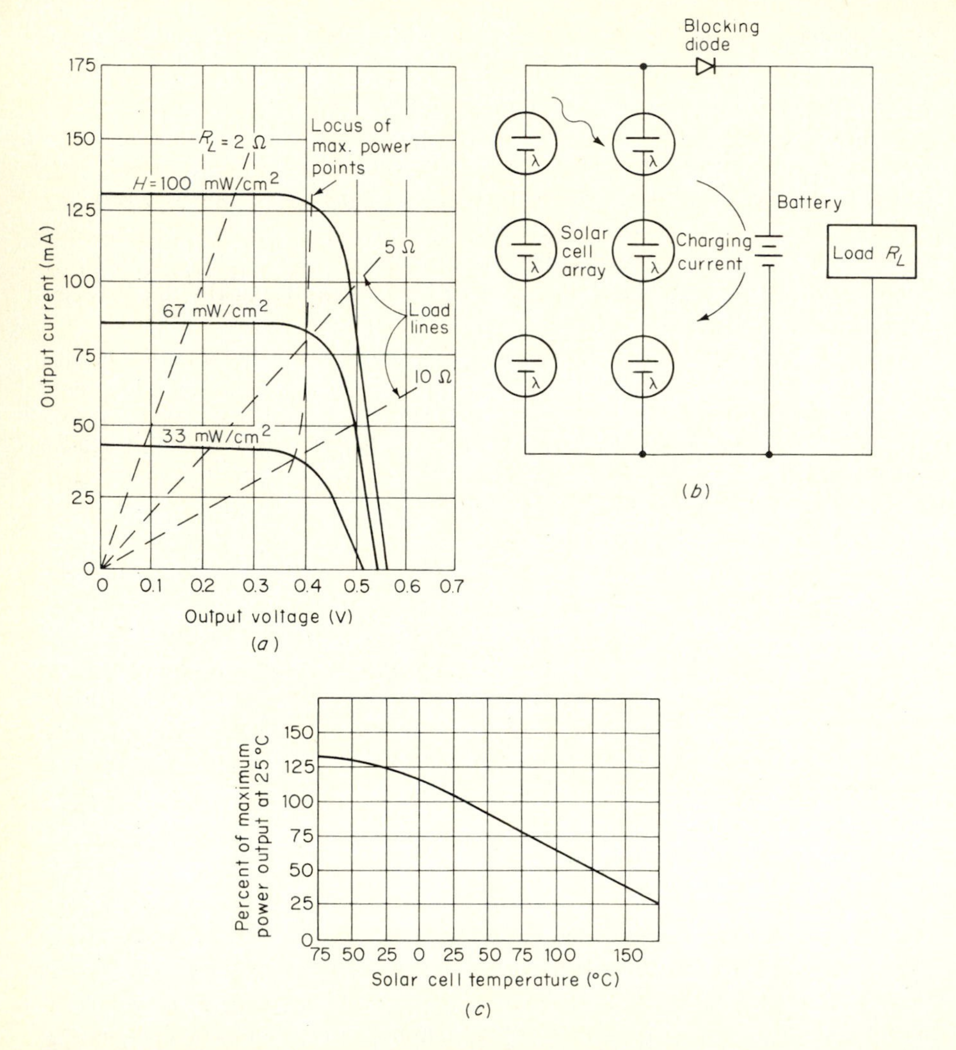

Fig. 7-12 Typical solar cell: (*a*) current/voltage characteristics at various irradiances; (*b*) remote battery-charging-array application; (*c*) temperature dependence. *Silicon Sensors, Inc.*

of 10 to 12 percent and are capable of producing 15 W/lb. (Recently, experimental GaAs solar cells have been fabricated with conversion efficiencies in the 20-percent range.)

As an example of solar-cell calculations, assume that we desire to calculate the power delivered by the solar cell of Fig. 7-12*a* to a 5-Ω load for an earth application. In Sec. 6-2-1 we noted that the solar irradiance on the earth's surface was $\approx 100\ \text{mW/cm}^2$ on a clear day. The intersection of the 5-Ω load line and the $H = 100\ \text{mW/cm}^2$ IV characteristic line of Fig. 7-12*a* gives $V_{out} \approx 0.48$ V and $I_{out} \approx 96$ mA, or $P_{out} = (0.48\ \text{V})(0.96\ \text{mA}) = 0.46$ mW. To find the optimum load resistance for this same irradiance, the intersection should be at the maximum power point ($V_{out} \times I_{out}$) on the 100-mW/cm^2 line of Fig. 7-12*a*. This point corresponds to a load resistance of $V_{out}/I_{out} \approx 0.4$ V/0.128 A = 3.1 Ω.

Individual solar cells are frequently arrayed in series and/or parallel combinations to obtain a higher output voltage and/or current. As an example of such an array, consider the remote battery-charging application of Fig. 7-12*b*. Assume an N91 6.25-V (1.2 A capacity) NiCd rechargeable battery for which the manufacturer gives a preferred charging voltage of 7.0 V at a trickle charge of 24 to 40 mA. Also assume a 100-mA continuous load. Solar-cell manufacturers warn potential users that for battery-charging applications:

1. Applying a trickle charge current as designated by the battery supplier may not provide adequate charging. The main consideration should be the replacement of the ampere-hours that have been depleted from the battery by the load.
2. Do not use the load-line method for the analysis of solar-cell battery-charging applications. Also, do not use the battery series-resistance values quoted by the battery supplier, because series resistance varies drastically with the state of the battery's charge, as evidenced by the large variation of unlimited charging current which a battery consumes during a typical charging cycle.
3. Most users do not provide enough solar-cell array capacity because they neglect reductions in solar-cell output due to cloudly days, the reduced number of daylight hours during the winter, the daily variation of the sun's angle of incidence on a stationary solar panel (solar-cell ratings are given for an optimum angle of incidence of 90°), and the reduction in solar-cell efficiency at higher temperatures (Fig. 7-12*c*).

To provide the 7.0 V of charging voltage plus the 0.6 V of blocking-diode forward voltage drop in the example of Fig. 7-12*b* requires ≈7.6 V. The 100 mW/cm^2 solar constant should be reduced to about one-third to account for the various degradations listed above. Assuming ≈0.35 V and 33 mA from each cell (from Fig. 7-12*a* at 33 mW/cm^2), we calculate that ≈7.6 V/0.35 V, or 22 series cells, and 100 mA/33 mA, or three parallel strings, are required to guarantee battery charging under all conditions.

We have not mentioned selenium (Se) photovoltaic cells, which are also used for solar-energy conversion. These detectors are made by depositing a layer of selenium, CdSe, and metal in that order on a metal base. The spectral response of selenium cells peaks at ≈570 nm or much closer to the peak of the eye's response or the solar spectrum than Si at 850 nm. The output current of Se cells is somewhat logarithmic with illumination but suffers from light history, slow response, and temperature effects (both positive and negative temperature coefficients), much as do the bulk photoconductors of Sec. 7-1-2. With loads below a few kΩ the conversion efficiency of Se cells is ≈20 times less than that of Si. For these reasons Si solar cells have become much more popular.

7-1-5 PHOTOTRANSISTORS

Phototransistors combine the ability to detect light and to provide gain within a single device. Their construction is similar to that of conventional transistors, except that the top surface is exposed via a window or lens as shown in Fig. 7-13.

The incident photons generate electron-hole pairs in the vicinity of the large collector-base (*CB*) junction. The reverse-biased *CB* junction voltages draw the induced holes to the base area and the electrons toward the collector region. The forward base-emitter bias causes these holes to flow from base to emitter, whereas the electrons flow from emitter to base. At this point conventional transistor action takes over, with the emitter-injected electrons crossing the narrow base region and being drawn to the more positive collector. This flow of electrons constitutes a light-induced collector current. The photon-induced hole-electron pairs thus contribute to the base current, and, if the phototransistor is connected in the common-emitter configuration, the light-induced base current is multiplied by the transistor's β or h_{fe} and appears as the collector current.

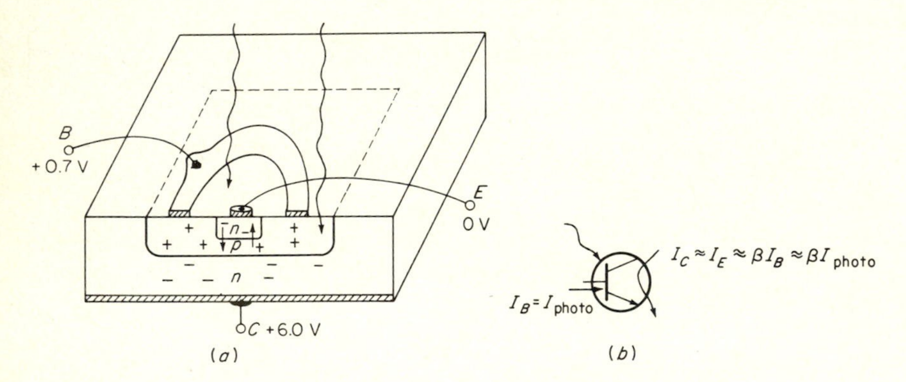

Fig. 7-13 Phototransistor: (*a*) cross section; (*b*) symbol and main currents.

Figure 7-14 shows typical phototransistor data. The collector-to-emitter radiation sensitivity (S_{RCEO}) of Fig. 7-14*a* is defined as the output collector current per unit of incident irradiance with the base open; hence the S_{RCEO} definition includes the phototransistor's β multiplier. From the definition of S_{RCEO} we see that

$$I_c = S_{RCEO} H_{\text{effective}} \tag{7-8}$$

where I_c = phototransistor collector current, mA
S_{RCEO} = phototransistor collector-emitter radiation sensitivity, mA/(mW/cm^2)
$H_{\text{effective}}$ = portion of the source spectral irradiance which is effective in producing I_C, mW/cm^2

(Notice from Fig. 7-14*c* that S_{RCEO} varies significantly with irradiance, and the S_{RCEO} given in Fig. 7-14*a* is the minimum value at $H = 1$ mW/cm^2.)

STATIC ELECTRICAL CHARACTERISTICS (T_A = 25°C unless otherwise noted)

Characteristic	Symbol	Min	Typ	Max	Unit
Collector Dark Current	I_{CEO}				μA
(V_{CC} = 20 V, R_L = 100 ohms) T_A = 25°C		—	—	0.050	
T_A = 100°C		—	10	—	
Collector-Emitter Breakdown Voltage (I_C = 100 μA)	BV_{CEO}	35	50	—	V

OPTICAL CHARACTERISTICS (T_A = 25°C unless otherwise noted)

Characteristic	Symbol	Min	Typ	Max	Unit
Collector-Emitter Radiation Sensitivity (V_{CC} = 20 V, R_L = 100 ohms)	S_{RCEO}	0.2	-	—	mA/mW/cm^2
Photo Current Saturated Rise Time	t_r (sat)	—	2.0	—	μs
Photo Current Saturated Fall Time	t_f (sat)	—	2.5	—	μs

(a)

Relative response %
100
80
60
40
20
0
2300 2500 2700 2900
Color temperature (°K)

(b)

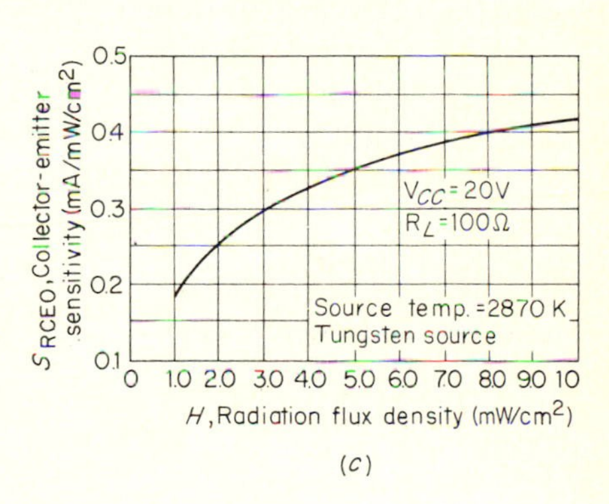

(c)

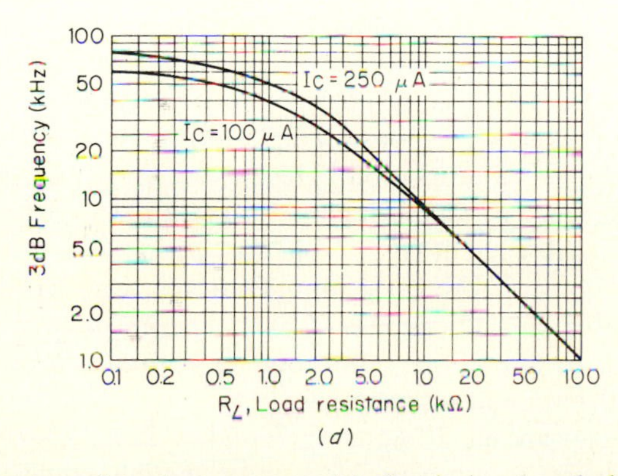

(d)

Fig. 7-14 Phototransistor data-sheet excerpts: (*a*) electrical and optical characteristics; (*b*) relative response versus incandescent-lamp color temperature; (*c*) S_{RCEO}, collector-emitter radiation sensitivity versus radiation flux density; (*d*) frequency response versus load resistance. *Motorola*.

Figure 7-14*b* is useful in determining the phototransistor's response to an incandescent lamp operated at other than the rated color temperature (Sec. 6-2-2).

As an example of the use of the phototransistor data of Fig. 7-14, consider the 7152 subminiature lamp of Fig. 6-4 operated at the rated filament current and voltage with a lamp-to-phototransistor separation of 1.25 in [31.75 mm] (adequate to assume point-source conditions). Figure 6-4 and Eq. (6-3) of Sec. 6-2-2 can be used to calculate the 7152 color temperature under these conditions. The lamp's luminous efficiency = MSCP/(input watts) = 0.147 MSCP/(0.115 A)/(5.0 V) = 0.26 MSCP/W. Entering Fig. 6-4*c* at this value, we see that the lamp's color temperature is ≈2200 K (vacuum lamp). This color temperature is lower than the 2870 K standard lamp color temperature for which the phototransistor of Fig. 7-14 is specified, and an extrapolation of Fig. 7-14*b* shows that the phototransistor's actual response at 2200 K will be ≈35 percent of the specified response. The example of Eq. (6-3) and Fig. 6-4 showed that the irradiance at the detector for these conditions (and 2870 K) is 3.63 mW/cm^2. Since the effective phototransistor irradiance is ≈35 percent of this value, $H_{\text{effective}} = (3.63 \text{ mW/cm}^2)(0.35) = 1.27 \text{ mW/cm}^2$. Figure 7-14*c* gives S_{RCEO} at this irradiance as ≈0.2 mA/(mW/cm^2). Finally, Eq. 7-8 gives the phototransistor collector current as

$$\begin{aligned} I_c &= S_{RCEO} H_{\text{effective}} \\ &= [0.2 \text{ mA}/(\text{mW/cm}^2)](1.27 \text{ mW/cm}^2) \\ &= 0.25 \text{ mA} \end{aligned}$$

One might think that combining the light-detection and amplifying functions into one semiconductor device would provide the ultimate light sensor; however, this is not true for many analog applications. First, it is important to realize that the phototransistor dark current I_{CBO} is also multiplied by β, just as its base photocurrent; hence there is no basic improvement in the signal-to-dark-current ratio. The frequency response of a phototransistor is also less than that of a photodiode-amplifier combination. This is due to the large phototransistor base-to-collector capacity which is highly charged and can only be discharged by the relatively low dark current (Fig. 7-14*d*).

The region of linear phototransistor operation is several orders of magnitude less than that of the photodiode (Sec. 7-1-3). The linearity problem and many other phototransistor limitations stem from the well-known variation of transistor β with current level and temperature.

Phototransistors are most frequently found in digital ON-OFF applications in which the nonlinearity is not a problem and their inherent gain eliminates the need for further amplification. In fact the largest phototransistor market is for computer punched-card and paper-tape readers, where phototransistors offer higher speeds than bulk photoconductors and more gain (and lower cost) than photodiodes, so that the need for additional amplification is often eliminated. In this application one is only concerned with whether or not there is a hole in the card or paper tape (a digital phenomenon); hence the nonlinearities and other disadvantages can be ignored. Owing to the size of the market for punched-card and paper-tape readers, semiconductor manufacturers have expended considerable effort to produce inexpensive phototransistors including linear arrays of 6 to 12 devices. There are several sources of epoxy phototransistors with small quantity prices under $1.00 each.

Figure 7-15 shows phototransistors in their most general application—that of a light-sensitive switch. In such applications the base terminal is not connected and in fact many phototransistors do not have an external base lead. As shown in Fig. 7-15*a*, the phototransistor is OFF prior to the application of adequate irradiance; hence no emitter current flows through R_E, and the output is ≈ 0 V. With adequate illumination the phototransistor switches ON, and the high

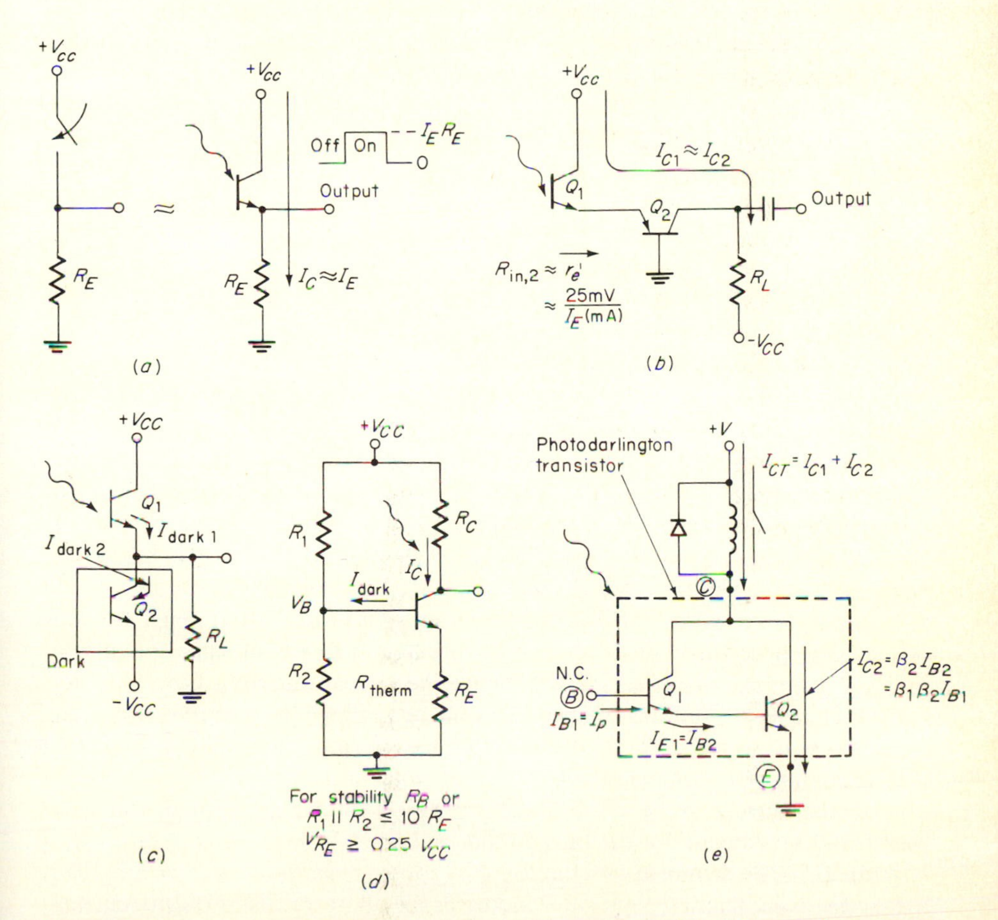

Fig. 7-15 Phototransistor circuits: (*a*) simple phototransistor switch and equivalent circuit; (*b*) high-frequency circuit; (*c*) and (*d*) temperature compensation; (*e*) high-current photodarlington configuration driving a sensative relay.

emitter current develops an output equal to $I_E R_E$. The output increases as R_E is increased; however, the frequency response decreases owing to the $C_{CB} R_L$ time constant, as shown by Fig. 7-14*d*.

As an example of the use of the phototransistor data of Fig. 7-14, assume that V_{CC} of Fig. 7-15*a* equals +20 V, $R_E = 1$ kΩ, and an output of 5 V is desired when the phototransistor switches. Since $V_{out} = I_E R_E$, $I_E = 5$ V/1 kΩ = 5 mA ($\approx I_C$); from Eq. 7-8 and Fig. 7-14*a*, an $H_{effective}$ of 25 mW/cm^2 is required. Figure 7-14*d* shows that the frequency response would be down to ≈50 kHz at an R_E of 1 kΩ. One method of avoiding this reduction in speed while retaining approximately the same gain is demonstrated in Fig. 7-15*b*. In this case R_L is paralleled (reduced) by the low input resistance of the common-base (Q_2) stage. Hence the frequency response becomes relatively load-independent. Since the current gain of a common-base stage is approximately unity, the current through R_L is approximately the same as if Q_2 were not present. Assuming the same conditions as the example of Fig. 7-15*a*, the frequency response of Fig. 7-15*b* would be increased by the ratio of the effective load-resistance reduction:

$$\frac{R_L}{r'_e} = \frac{1000\ \Omega}{5\Omega} \qquad \text{at } I_C = 5 \text{ mA}$$

or 200:1. As with any transistor switch, the recovery time of the phototransistor increases if it is saturated. Thus, the optical input should be limited to that value which produces satisfactory I_C without saturations.

Phototransistors are subject to the same temperature variations as conventional bipolar transistors. That is, their dark current I_{CEO} doubles approximately every 8 to 10°C of temperature rise, and at elevated temperatures this thermal error current may be significant compared with the photocurrent. Figure 7-15*c* shows how two-lead matched phototransistors can be operated to provide temperature compensation. Q_1 serves as the photosensor, and Q_2 is kept dark. Since the two phototransistors are matched, the Q_1 and Q_2 dark currents remain approximately the same with temperature and cancel each other; hence no dark current flows into the load R_L. When Q_1 is exposed to a light signal, its photocurrent (I_C or I_E) exceeds $I_{dark\ 2}$, and hence the Q_1 photocurrent flows into R_L.

A three-lead phototransistor has one advantage in common with any three-terminal device; that is, the third or base terminal provides a means of electrical control or compensation. A degree of stabilization of collector current with temperature can be maintained by employing conventional transistor stabilization techniques in the bias circuit, as shown in Fig. 7-15*d*.

Figure 7-15*e* demonstrates the use of a *photodarlington transistor* amplifier. In this case the integral second stage increases the phototransistor output current. As shown in Fig. 7-15*e*, the initial photoinduced Q_1 base current is multiplied by $\beta_1\beta_2$, providing a total collector current sufficient to energize a sensitive relay.

7-1-6 PHOTOFIELD EFFECT TRANSISTORS

The photofet is similar to a conventional junction FET (Chap. 1), with the exception of a lens for focusing light onto the gate junction. As such, the photofet provides a useful combination of a photosensitive *pn* junction with a high-impedance, low-noise amplifier in one device. The operation of a photofet is demonstrated with the aid of Fig. 7-16. The photons enter the gate area and

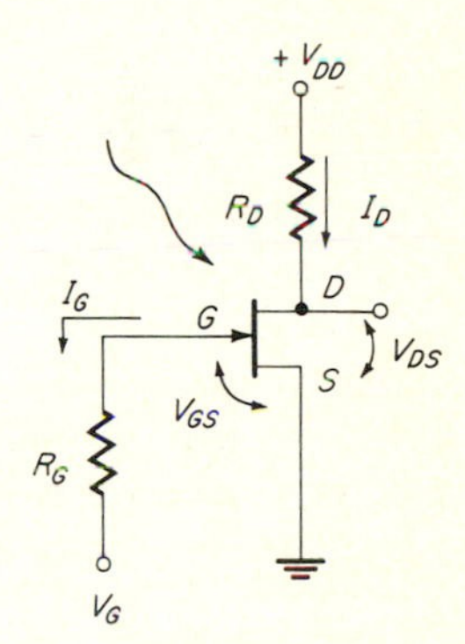

Fig. 7-16 Photofield effect transistor schematic.

excite valence electrons into the conduction band. The photon-excited current carriers cause a small change ΔI_G in the nominal gate current. ΔI_G flows through the gate resistor and thus develops a change in the gate-to-source voltage ΔV_{GS}. The ΔV_{GS} is multiplied by the photofet's transconductance g_{fs} (Sec. 1-3) and a ΔI_D is produced. The ΔI_D changes the voltage drop across R_D and thus alters the drain-to-source voltage. In summary,

$$\Delta V_{DS} = \Delta I_G R_G g_{fs} R_D \tag{7-9}$$

The voltage developed across R_G by the induced photocurrent is thus amplified by FET action.

The main advantages of the photofet as compared to other semiconductor photodetectors include: combination of photosensitive junction and low-noise, high-impedance amplification; low offset voltage, allowing superior switching performance (Sec. 1-6); power gain greater than that of many phototransistors; and frequency response superior to that of phototransistors.

Photofet disadvantages include: small effective sensitive area; low sensitivity; increased power consumption inherent at the low gate-to-source voltages necessary for high g_{fs} and low dark current (Sec. 1-2); and slow response and large thermal noise if a large gate resistor is required for sensitivity.

As an example of a simple photofet design, refer to Fig. 7-16 and assume the following: A typical photofet sensitivity and g_{fs} of 3 μA/(mW/cm^2) and 2000 μmho, respectively,

$$V_{DD} = +12 \text{ V} \qquad R_G = 1 \text{ M}\Omega \qquad R_D = 2.7 \text{ k}\Omega$$

and an incident irradiance of 0.1 mW/cm^2. Then

$$I_{\text{photo}} \approx [3\ \mu\text{A}/(\text{mW/cm}^2)](0.1\ \text{mW/cm}^2) = 0.3\ \mu\text{A}$$
$$\Delta V_{GS} \approx R_G I_{\text{photo}} \approx 0.3\ \text{V}$$
$$\Delta I_D = \Delta V_{GS}\, g_{fs} \approx (0.3)(2 \times 10^{-3}) = 0.6\ \text{mA}$$
$$\Delta V_D = \Delta I_D R_D = (0.6\text{mA})(2.7\text{k}\Omega) = 1.62\text{V}$$

The ratio of ΔI_G to I_{GSS} (signal-to-dark current ratio) at $V_{GS} = 2$ V is probably greater than 10^4.

Matched pairs of photofets are employed in a dual source-follower, differential-amplifier configuration. In such an application one of the two photofets is kept dark, and, since the differential amplifier responds only to a difference in input voltages, a degree of temperature compensation is achieved through the close thermal tracking of matched photofet gate leakage currents. The source-follower configuration produces a very high input impedance, allowing all the photocurrent to flow into the feedback resistor (Sec. 2-5-2).

7-1-7 PHOTOTHYRISTORS

The operation of conventional thyristors is covered in Chap. 8. Thyristors are four-layer *pnpn* structures which are generally regarded as semiconductor switches. The oldest member of the thyristor family is the *silicon-controlled rectifier* (SCR). The SCR is like a latching relay which changes from high to low resistance when properly triggered at its control or gate terminal. The SCR does not fall out of conduction until the anode current decreases below some minimum holding current. The reader is referred to Chap. 8 for such thyristor details as theory of operation, characteristics, triggering, and turn-OFF methods. (Figure 8-7 shows a conventional low-current sensitive-gate SCR whose electrical characteristics are very similar to those of typical photo-SCRs.)

Photothyristors are like phototransistors or photofets in that they are very similar to their conventional counterpart except for the addition of a window or lens to focus light onto an appropriate area. They also have three leads, and therefore their optical triggering threshold can be varied electronically. The photothyristor's main advantage is that it is an excellent switch with power-handling capacities far beyond those of other photodetectors.

The operation of a typical photo-SCR is as follows. With the proper bias, entering photons create electron-hole pairs in the vicinity of the second junction, and these free current carriers are swept across all junctions, producing a current from anode to cathode. At a certain radiation level the net current gain of the device exceeds unity, and the anode-to-cathode current becomes limited only by external impedances (Sec. 8-2). At this point the photo-SCR has changed from a near open circuit to a near short circuit, as depicted by Fig. 8-3. The output of a photo-SCR is *not* proportional to the incident irradiance as is the case for the previous photodetectors. The photo-SCR is OFF (low anode current) prior to adequate triggering irradiance (typically 2 mW/cm^2) and ON any time the optical threshold is exceeded. The anode current does not vary significantly with light level. Since photothyristors closely approximate switches, their main applications include optical logic systems such as counters, sorters, and relay functions.

Sensitive photo-SCRs require an external resistance between the gate and cathode, which effectively shunts part of the photocurrent in the *npn* portion of the device. The gate-to-cathode resistance R_{GK} determines the light sensitivity which, in turn, influences the temperature effects, frequency response, and dv/dt (Sec. 8-2). Increasing R_{GK} increases the sensitivity to light and temperature but decreases the response time. As might be expected, the photo-SCR triggering level depends on the junction temperature, applied voltage, etc.

7-1-8 INFRARED DETECTORS

The materials and methods used to detect radiation in the infrared region are sufficiently different to warrant a separate section devoted to that portion of the spectrum. Some of the differences between infrared detectors and those previously discussed include: the use of cooling equipment for many detectors, the atmospheric absorption and background problems, the emergence of various lead salts, etc., as basic detector materials (Table 6-3), and the elimination of photomultipliers because of their lack of response past 1 μm.

There are, however, many similarities to the photodetectors previously discussed, such as the existence of bulk, junction, and thermal detectors; the amplifier techniques and electrical filtering (bandwidth and noise reduction) discussed in Sec. 7-1-3 are still appropriate, and often a compromise must be reached between speed and sensitivity.

Many infrared applications are military in nature and as such have been shrouded in secrecy. However, with the passage of time a great deal of the basic material has become unclassified, and commercial users have benefited from this research. Examples of military operational infrared systems include:

1. The Sidewinder infrared guided missile
2. Infrared haze penetration
3. Infrared photography and reconnaissance (such as the detection of rocket launches)

Some of the commercial infrared applications include:

1. Infrared spectroscopy for chemical analysis, mainly by the measurement of infrared absorption spectra
2. Weather observation via meteorological satellites
3. Optical pyrometers for remote measurement of high temperatures
4. Automatic shutdown for industrial boilers by sensing the infrared radiation from the flame, and flame detection through haze or smoke (used by the Forest Service in fire mapping)

5. Crop forecasting and disease spotting (Ref. 4)
6. Detection of cardiovascular circulation problems

Prior to discussing infrared systems in detail, it is necessary to expand somewhat on the physics-of-light material of Sec. 6-1 to include some aspects of the infrared portion of the electromagnetic spectrum. All objects above absolute zero temperature emit electromagnetic radiation in the infrared region. To explain this fact without delving too deeply into the physics of the matter, let us refer to the blackbody concept, wherein a *blackbody* is defined as a substance which absorbs all radiation of any wavelength (and a good absorber is also a good radiator).

Figure 7-17 gives the spectral radiant emittance of blackbodies at various temperatures as a function of wavelength. These curves explain the variation of color at different temperatures (see Fig. 6-1*b* for color-wavelength conversions). Many actual materials have characteristics similar to those of the ideal balckbody, with their actual emittances depending on their temperature (Fig. 7-17) and *emissivity*, or ratio of radiant output to that of the ideal blackbody at the same temperature. Notice that at room temperature (300 K) all the radiation is in the infrared region, and, as the temperature increases, the curves shift toward the ultraviolet. The walls of a room and people emit infrared radiation near the

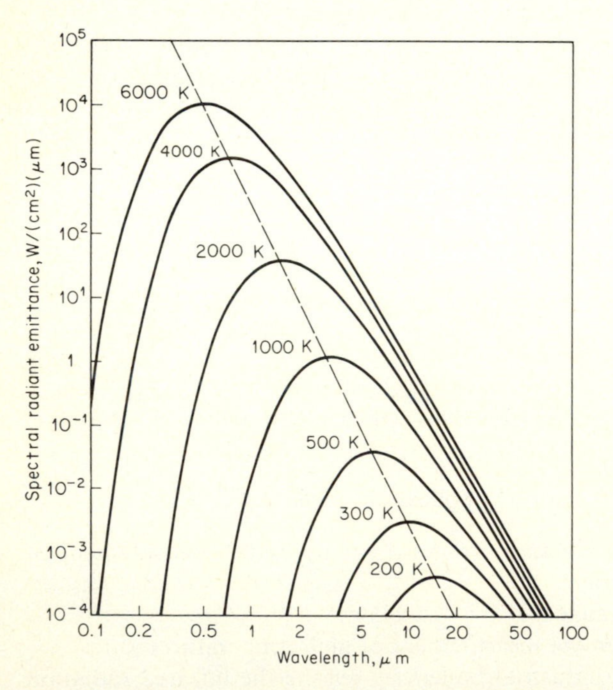

Fig. 7-17 Spectral radiant emittance of a blackbody versus wavelength at various temperatures. *From J. Janieson, et al., "Infrared Physics and Engineering," p. 20, McGraw-Hill Book Company, New York, 1963.*

10-μm peak suggested by the 300-K curve. This brings up the problem of background in infrared detection systems. Such systems must discriminate between the background and the object to be detected. Typical infrared systems discriminate between heated objects such as transformers, engines, factories, etc., and normal ground phenomena. The hot metal and exhaust of aircraft or ships also emit infrared radiation which can be discriminated from the background sky or ocean.

The infrared spectrum has been divided into three areas, according to the type of detector required for each area. The near-infrared region can be handled by conventional Si and Ge detectors, as previously mentioned (Fig. 7-2). The intermediate range requires special infrared detectors. In the far infrared (20 to 1000 μm), the separation between energy bands becomes so small that thermal detectors must be employed in lieu of quantum detectors. Infrared detectors are divided into two categories: (1) thermal detectors such as thermistors (*bolometers*) and thermocouples, which respond to the heating produced by absorption of radiation, and (2) quantum detectors, similar to the intrinsic and doped types previously discussed.

If is often necessary to cool infrared detectors and housings to reduce the intermediate infrared radiation emitted from the housing at room temperature. Regardless of the type of infrared detector, this radiation will appear as a background that will decrease the possibility of detecting an infrared target which radiates near the background wavelengths. There are also various detector noise sources whose magnitude is strongly temperature-dependent. For these reasons, the means of maintaining infrared-detector elements at temperatures far below room temperature have become an integral part of infrared detection. The most common means of achieving this cooling is with a dewar flask, which is similar in concept to the common vacuum thermos bottle.

It is often necessary to optically chop the incident radiation in infrared-detection applications to eliminate the excessive detector thermal drift associated with dc operations because such thermal drift makes it impossible to separate the signal from the background. Light chopping is usually accomplished by a motor-driven disk containing a pattern of alternate opaque and clear areas, or a tuning fork whose tines alternately open and close in the light path. In either case a time-varying periodic signal is superimposed onto the normal signal. By Fourier analysis it may be shown that, if this multiplexed signal is averaged, the dc term and all the even harmonics become zero. Hence, a major portion of the background noise and dc drift are eliminated.

Figure 7-18 shows a simple passive method of optical chopping for applications with a moving IR source. The operation of the passive IR intrusion detector is as follows. The detector mask establishes areas in the detector's field of view which are alternately opaque and transparent. As intruders enter this field of view and move across the transparent/opaque pattern, their bodies effectively emit a "pulse" of infrared radiation (in the 9- to 11-m wavelength region) as they pass a transparent region. The detector responds to this emitted radiation

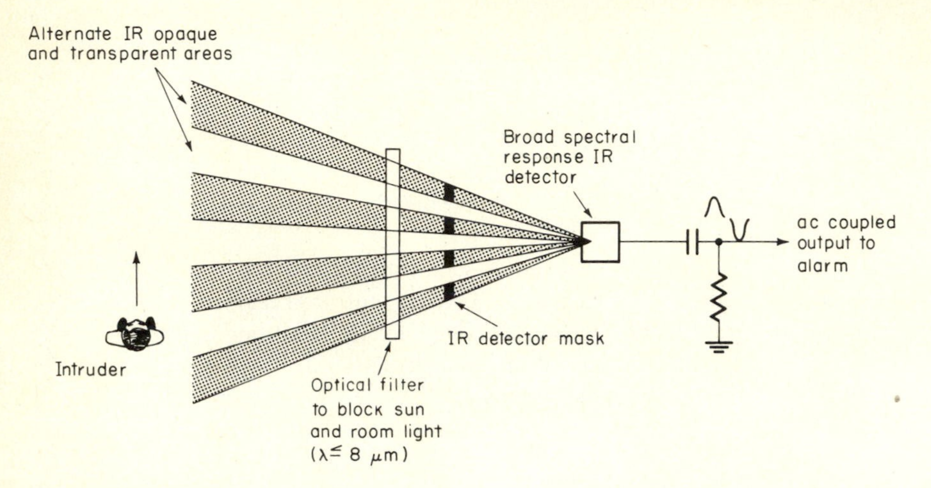

Fig. 7-18 Passive infrared intrusion detector.

and produces either a positive or negative pulse at the ac-coupled output. The alarm is designed to be triggered by such pulses and not by lower-frequency or dc signals. (If the intruders stand still, the alarm will not trigger, but they obviously are not intruding if they are not moving.) This arrangement is not susceptible to dc or slowly varying thermal-emission gradients (such as walls whose IR emission changes as the sun rises and sets), because the ac coupling prevents responses to dc or slowly varying IR radiation.

Figure 7-2 shows the spectral response of several bulk intrinsic infrared detectors (cooling generally increases the response and shifts it further into the infrared).

One of the earliest and most versatile infrared photoconductors is lead sulfide (PbS). The detectors are fabricated with a thin film of PbS on a quartz substrate, much as the visible bulk photoconductors of Sec. 7-1-2. PbS detectors respond to infrared from ≈ 1 to 4 μm, depending on the operating temperature, and PbS has high sensitivity even at room temperature. A bias voltage is generally applied to these detectors, and the optimum value is a few volts where the noise starts to increase more rapidly than the signal. Thin-film lead selenide (PbSe) detectors are generally faster and have spectral response farther in the IR than PbS detectors, but their response is over an order of magnitude less.

Indium antimonide (InSb) has one of the lowest energy gaps, as shown in Table 6-3, and responds up to 5.6 μm, as evidenced by Fig. 7-2. InSb is used to make bulk or *pn*-junction detectors. Intrinsic Si and Ge photosensors have only limited near-infrared response, with Ge providing slightly more, as shown in Fig. 7-2.

In contrast to the cost decline of most semiconductors, one finds that bulk or junction infrared detectors are quite expensive, especially when compared to similar photodetectors for the visible region. Thermistors, however, are less expensive but may require additional optical filtering to eliminate response from unwanted wavelengths.

Thermal detectors for the far infrared are generally of the bolometer or thermocouple type. The sensor for the first type is usually a flake semiconductor thermistor with a few percent per degree Celsius negative temperature coefficient. For infrared detection the thermistors are usually blackened to produce uniform absorption over a wide spectral range. The thermistors are also generally operated in a matched-bridge configuration to reduce temperature shifts (Fig. 7-7). Radiation thermocouples are non-solid-state devices used in the far infrared and throughout the spectrum because of their stability and flat spectral response. Nearly all commercial spectrometers employ thermocouples. A radiation thermocouple consists of a suitably blackened metal plate in contact with a two-metal wire junction which develops an electromotive force (emf) proportional to the temperature difference. Thermocouples are characteristically very-slow-response detectors and are generally not used in applications with greater than 10-Hz chopping rates. Another disadvantage of thermocouples is that their sensitivity is limited to approximately 10^{-8} W/cm^2.

7-1-9 ULTRAVIOLET DETECTORS

The number of semiconductor detectors capable of ultraviolet detection is very small compared to the variety of solid-state photodetectors for the visible and infrared portions of the spectrum. This might seem peculiar because the energy of ultraviolet radiation is higher and one would thus expect ultraviolet detection to be easier. However, there are some basic problems encountered in the detection of ultraviolet. First, conventional glass windows cut off below approximately 300 nm, and even quartz and ultraviolet-grade sapphire become opaque below 180 nm. Below $\approx$100 nm there is no suitable ultraviolet window material at all. Second, the transmission of silicon and most other materials falls off drastically in the ultraviolet. This ultraviolet absorption very near the detector's surface often results in loss of the photon-induced pair, as in the case of a photodiode, where the pair has a high probability of never reaching the depletion region and thus never being separated or contributing to the photocurrent.

Traditionally, detection in the intermediate- and vacuum-ultraviolet areas has been accomplished by multiplier phototubes with photocathodes made of one of the metals plus telluride, cesium, or a fluorescent coating of sodium salicylate. Many of these photocathodes are "solar blind" or unresponsive to wavelengths greater than $\approx$300 nm. This characteristic is desirable if false responses from sunlight are to be eliminated. There are also CdSe bulk photoconductors available with spectral-response peaks at $\approx$250 and 400 nm, and zinc sulfide detectors with a peak at $\approx$370 nm and negligible response past 400 nm.

In the junction-photodiode and solar-cell area, manufacturers have decreased the thickness of the top layer in order that more ultraviolet radiation may penetrate that layer and create hole-electron pairs within the depletion region; however, a thickness of 0.3 μm is still longer than the characteristic wavelength of intermediate-ultraviolet radiation.

It has been shown that the area outside the normal active area of a *pin* photodiode structure (as shown by Fig. 7-9*a*) is capable of excellent ultraviolet response to less than 200 nm (Fig. 7-19) (Ref. 5). As shown in Fig. 7-9*a*, if the intrinsic layer is depleted, ultraviolet radiation will pass through the SiO_2 (quartz) passivation layer, and the induced carriers will be immediately separated by the electrostatic field in the depleted *i* layer.

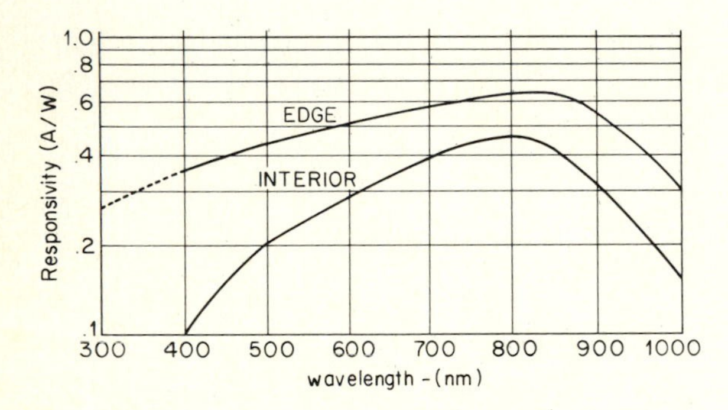

Fig. 7-19 Ultraviolet spectral response for the *pin* photodiode of Fig. 7-9*a*. Edge and interior refer to outside and within the *p* region of Fig. 7-9*a*, respectively. *Hewlett Packard.*

Most common photodetector windows are made of glass because of its low cost and ease of sealing; therefore the quoted ultraviolet response of such detectors reflects the glass-window transmission cutoff. Considerably extended ultraviolet response may be available merely by replacing the original glass window with quartz or sapphire.

7-1-10 PHOTOPOSITION DETECTORS

Figure 7-20 shows two types of special photodetectors used for light-beam-position detection. They provide an output signal proportional to the location of the light beam. The older four-quadrant detector of Fig. 7-20*a* consists of a circular CdS detector divided into four isolated quadrants, each with its own output lead. The illustrated spot produces the relative analog output levels shown in Fig. 7-20*a*. Such detectors have the disadvantages of poor thermal stability, slow speed, dead areas, and lack of resolution if the spot is entirely within one quadrant.

An improved version of the four-quadrant position detector is shown in Fig. 7-20*b*. This device is a silicon *pin* photodiode with a single large active area

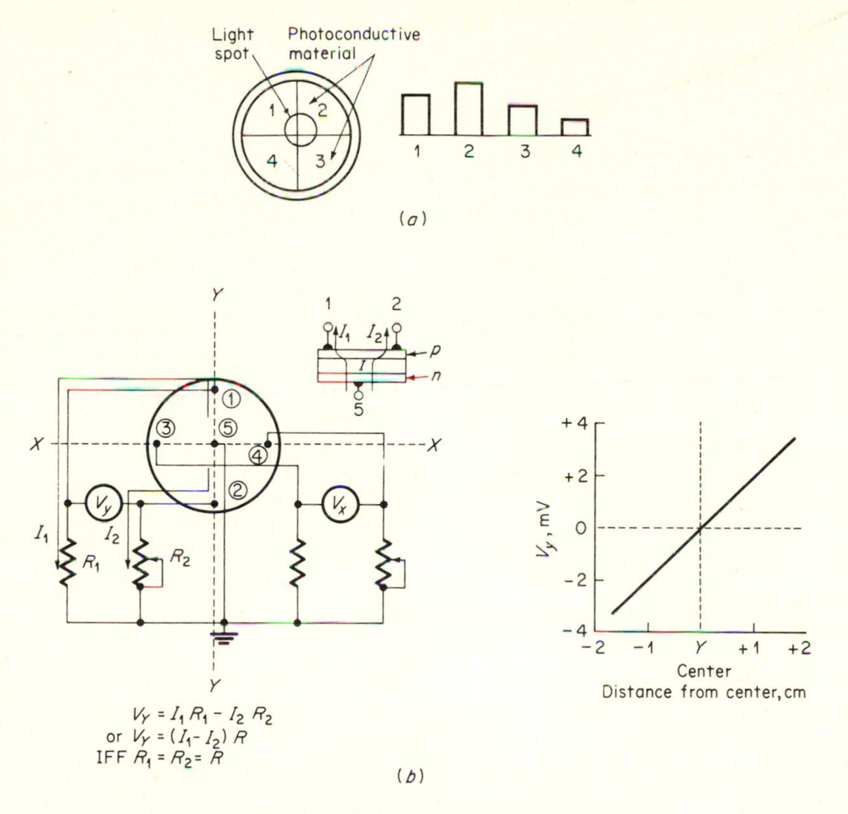

Fig. 7-20 Position detectors: (*a*) four-quadrant position detector; (*b*) *pin* continuous position detector showing typical *X*- and *Y*-axis readout and cross section. *Adapted from United Detector Technology pin-Spot 10 data sheet.*

and symmetrical connections as shown. The area is not divided into quadrants in this case, and the magnitudes of the various indicated photocurrents depend only on the relative closeness of the light spot to the particular lead, assuming a fixed spot size and intensity. Notice that the Y axis is defined by leads 1, 5, and 2, and the reading of the V_y voltmeter corresponds to the difference between I_1 and I_2, assuming $R_1 = R_2$. This difference is then proportional to the light-spot location; i.e., if the device were properly nulled by adjusting R_2, I_1 would equal I_2 when the spot was centered in the Y axis, and V_y would be 0. If the spot were on lead 1, I_2 would approach zero, and V_y would be high (as indicated by the linear relationship between V_y and the distance from the Y-axis center in Fig. 7-20*b*). A battery is not required in the center (5) lead unless high speed is required (Sec. 7-1-3). The speed and temperature stability of this detector are superior to that of the four-quadrant detector as long as the spot irradiance is high enough to produce photocurrents well above the dark-current level. Its output is relatively independent of spot size, and there are no dead spots or lack of position knowledge if the spot is within one quadrant.

7-2 SOURCE-DETECTOR COMBINATIONS (OPTO-ISOLATORS)

The combination of a miniature light source and a photodetector in the same package has led to a very useful family of devices commonly referred to as *opto-isolators* (or *optocouplers*). As Fig. 7-21 shows, opto-isolators are available with virtually every possible combination of light source and photodetector discussed in Secs. 6-2 and 7-1, as well as combinations with an integrated-circuit amplifier in the detector output. The main application of opto-isolators involves electronic-circuit isolation, where the isolator eliminates common-ground connections (provides signal transfer between isolated grounds), which prevents ground loops and substantially reduces common-mode noise (Fig. 7-24*a*).

The coupling in all opto-isolators is via photons instead of charged particles, and because of this a number of important advantages occur:

1. Electrical isolation can be superior to that of transformer isolation or RF coupling. This is because the chargeless photons are not influenced by electrostatic or magnetic fields; hence nearly infinite common-mode rejection is attainable with photon coupling.
2. In switching or chopping applications the inherent mechanical problems such as contact bounce, noise, intermittency, maintenance, reliability, transients, and lifetime are all eliminated by contactless opto-isolators.
3. The signal transfer in opto-isolators is unilateral; therefore changing load conditions cannot affect the input.
4. Opto-isolators are much faster than isolation transformers and relays.

Figure 7-21 shows the key opto-isolator characteristics, two of which are new and are therefore defined below. *Current-transfer ratio* (*CTR*) is defined as the ratio of the opto-isolator output current to its input current in percent $[(I_{out}/I_{in})(100)]$. Notice from Fig. 7-21 that this quantity varies from less than 1 percent for the LED/photodiode combination to 600 percent for the LED/photodiode-amplifier/gate IC. (The LED/photodiode's low CTR, and the resultant need for additional amplification, is the reason for its lack of popularity.) The CTR varies with operating temperature and forward current, as shown in Fig. 7-22*c*, and hence the designer must account for these variations. (Since most opto-isolator applications are digital in nature, one need only ensure that the opto-isolator provides the minimum required output under the worse-case conditions of high temperature and low input drive.) The second new characteristic is the *insulation voltage*, which is the voltage between input (source) and output (detector) below which the opto-isolator will not break down. This value is determined by the coupling capacitance and leakage resistance between the opto-isolator source and detector. Most opto-isolators are designed with the source and detector in rather close proximity, separated only

Opto-isolator type / Characteristic	Discrete						Integrated circuit	
	Lamp/photoresistor	LED/photoresistor	LED/photodiode	LED/phototransistor	LED/photodarlington	LED/photo-SCR	LED/photodiode trans. amplifier	LED/photodiode amplifier / gate
	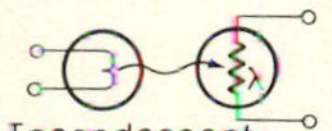 Incandescent or neon	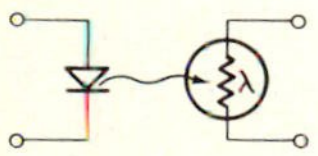	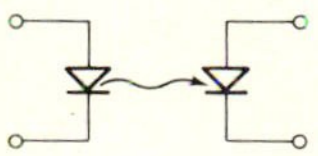	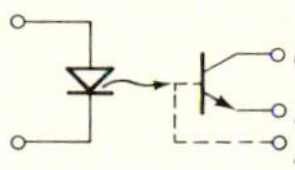 C E B		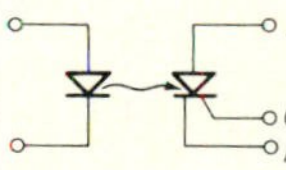 A G K	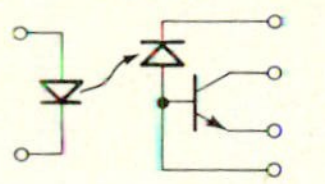	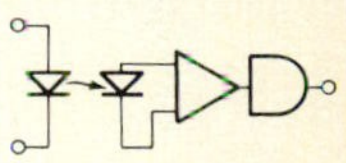
Current transfer ratio (CTR) I_{in}/I_{out} (100)(%) (*$\Delta R_o/\Delta$ source input for photoresistors)	*$\Delta R_{out}/\Delta V_{in}$ = 100 Ω to 10 kΩ/V	*$\Delta R_{out}/\Delta I_{in}$ = 10 kΩ/A	0.2 %	10 % to 200 %	600 % typ.	I_{in} to trigger = 10-30 mA	20 % typ.	600 % typ.
Data rate (MHz)	10^{-5} to 10^{-4} MHz	10^{-4} MHz	4 MHz	0.2 MHz	0.02 MHz conventional 0.3 MHz split darlington		1.0 MHz	10 MHz
Insulation Voltage (kV)	0.5 to 1.6 kV	0.5 to 2.0 kV	0.2 to 50 kV	0.5 to 10 kV	1.5 to 4.0 kV	2.0 kV	2.5 kV	2.5 kV
Input Current (mA) Required	1.0 to 40 mA		2 to 80 mA	15 to 60 mA	0.5 to 20 mA		15 mA	5 mA
Input Current (mA) Max		40 mA	60 to 100 mA	60 to 150 mA	60 mA			

Fig. 7-21 Opto-isolators.

by a transparent layer of silicon dioxide (Fig. 7-22*a*), with a typical isolation voltage of 1.0 to 2.0 kV. Opto-isolators which are specifically designed for high-voltage applications utilize a light pipe to couple the source and detector in order to decrease the coupling capacity. Such high-voltage opto-isolators are available with isolation-voltage ratings as high as 50 kV.

Figure 7-21 shows that the maximum data rate of opto-isolators varies from a few hertz for photoresistor units to 10 MHz for the IC version. Also notice from Fig. 7-21 that data rate is generally sacrificed for CTR in the discrete LED/phototransistor and photodarlington opto-isolators, whereas the integrated-circuit isolators offer both high CTR and excellent speed. (Some digital logic families discussed in Sec. 5-3 have data rates even higher than these opto-isolators.) Since most opto-isolators utilize a LED for the source, the input-current requirements are similar; however, the higher-CTR isolators and the IC versions generally require less LED drive. (LED-input opto-isolators require the same magnitude of input current as many incandescent-lamp isolators; however, the LED input-voltage requirement is typically an order of magnitude less than lamp-filament voltages.) The LED protection techniques discussed in Sec. 6-2-4 are also applicable to LED-source opto-isolators. That is, the input LED must generally be current-limited by a series resistor and protected from reverse voltages which exceed its rating.

A wide variety of opto-isolators is available; the following subsections contain data and applications for only the most popular versions.

7-2-1 BULK-PHOTOCONDUCTOR OPTO-ISOLATORS

The first column of Fig. 7-21 gives the characteristics for a typical lamp/CdSe photoconductor or photoresistor opto-isolator. The lower data rate is representative of an incandescent-lamp source, whereas the higher speed can be obtained with a neon gas-discharge-lamp source. The speed of these isolators is limited by the millisecond fall times of photoresistor detectors and the thermal time constant of the lamp sources. The main advantages of photoresistor opto-isolators are their high sensitivity ($\Delta R_{out}/\Delta V_{in}$), the simplicity of a resistor whose magnitude is controlled by light, and their bilateral characteristics with no offset voltage as is found in semiconductor photodetectors. (Lamp/photoresistor opto-isolator suffer from large temperature coefficients.) The main photoresistor isolator applications include analog circuits such as audio receivers, where the bilateral feature of the photoresistor is used to adjust AGC, and high-voltage monitoring or trigger circuits.

7-2-2 LED/PHOTOTRANSISTOR OPTO-ISOLATORS

One of the most popular opto-isolators is the LED/phototransistor (or the related LED/photodiode-transistor amplifier) of Fig. 7-21. Figure 7-22 gives excerpts from a typical LED/phototransistor data sheet. Notice the packaging details and the maximum LED ratings of Fig. 7-22*b*. (The LED *IV* characteristics are similar to those of the GaAs LED of Sec. 6-2-4.) Figure 7-22*c* gives the most

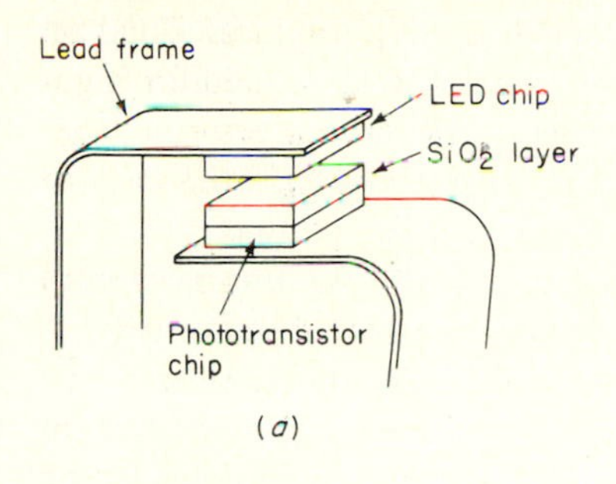

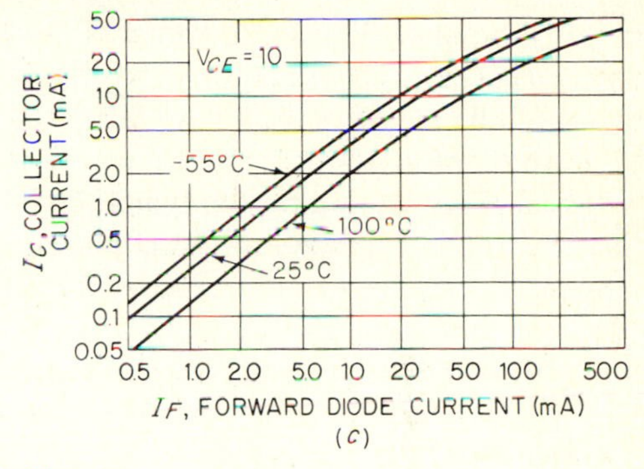

INFRARED EMITTING DIODE MAXIMUM RATINGS

Forward Current - Continuous	I_F	80	mA
Forward Current - Peak Pulse Width = 300 μs, 2.0% Duty Cycle	I_F	3.0	A

OPTO-INSULATOR CHARACTERISTICS
LED CHARACTERISTICS (T_A = 25°C unless otherwise noted)

Characteristic	Symbol	Min	Typ	Max	Unit
Reverse Leakage Current (V_R = 3.0 V, R_L = 1.0 MΩ)	I_R	–	0.05	100	μA
Forward Voltage (I_F = 10 mA)	V_F	–	1.2	1.5	V

COUPLED CHARACTERISTICS (T_A = 25°C unless otherwise noted)

Characteristic	Symbol	Min	Typ	Max	Unit
Collector Output Current (1) 4N25, A, 4N26 (V_{CE} = 10 V, I_F = 10 mA, I_B = 0)	I_C	2.0	3.5	–	mA
Isolation Voltage 4N25, A (60 Hz Peak) (60 Hz RMS For 1 Second)	V_{ISO}	2500	–	–	V

SWITCHING CHARACTERISTICS

		Symbol	Min	Typ	Max	Unit
Delay Time	(I_C = 10 mA, V_{CC} = 10 V) 4N25, A, 4N26	t_d	–	0.07	–	μs
Rise Time		t_r	–	0.8	–	μs

(b)

Fig. 7-22 LED/phototransistor opto-isolator data-sheet excerpts: (*a*) construction; (*b*) LED input and opto-isolator characteristics; (*c*) opto-isolator phototransistor output collector current versus LED input forward current. *Motorola*.

important opto-isolator characteristic; that is, how the phototransistor output current varies with LED input current I_F and temperature. To illustrate the use of this figure, consider an application which requires a minimum of 5 mA of phototransistor collector current over a -55 to $+100$°C temperature range. Entering Fig. 7-22*c* at an I_c of 5 mA shows that, in order to ensure a minimum I_c of this magnitude, the LED I_F must be approximately 25 mA to account for the decrease in I_c at the worst-case $+100$°C temperature. [To allow for the typically large variations in this parameter from one opto-isolator to another, an even higher I_F is required, as shown by the minimum/typical I_c ratings of Fig. 7-22*b*. If the phototransistor isolator's minimum guaranteed output current (or CTR) is not adequate, the photodarlington isolator may be required, assuming its reduced speed is acceptable.] Notice that the opto-isolator of Fig. 7-22*b* has a typical CTR $[(I_{out}/I_{in})(100)]$ of $(3.5/10)(100) \approx 35$ percent.

Most opto-isolator applications are in digital systems; hence the ease of interfacing opto-isolators with typical digital logic families is a prime consideration. Figure 7-23*a* shows the LED/phototransistor of Fig. 7-22 interfaced with typical TTL 7400 series digital logic gates. First, notice that the LED source is interfaced with the TTL logic in the *current sinking mode* (Q_2 effectively grounds the LED drive circuit). The *current-source* active LED drive illustrated in Fig. 7-23*b* is less desirable because of the large variation in TTL output high current ($I_{OH} = 18$ to 55 mA) from one gate to another, due to the large variation in the totem-pole diffused collector resistance R_C. Since the radiant output of a LED is very sensitive to its forward current I_F, as shown in Fig. 6-10*e*, and $I_F = I_{OH}$ for the current-source drive of Fig. 7-23*b*, the current-source drive is not recommended. For the current-sinking interface shown in Fig. 7-23*a*, R_C is not involved since the LED is only ON when Q_1 is OFF and Q_2 is ON.

The operation of Fig. 7-23*a* is as follows. When the logic output is low, Q_2 saturates and its V_{CE} is reduced to 0.4 V or less, leaving $\approx$4.6 V across R_{limit} and the LED. Assuming a 1.2-V LED forward voltage drop, the LED $I_F = 3.4$ V/220 Ω or 15.5 mA. When the logic output drive is high, the LED will not have adequate forward voltage or current to turn ON. Checking polarities through Fig. 7-23*a* shows that if the logic output is high, the LED is OFF; therefore the phototransistor is also OFF and its output (collector) is thus high. Hence, the circuit of Fig. 7-23*a* is noninverting. (Some MOS logic families can drive the isolator LED directly without a limiting resistor, owing to their inherently low current drive capability.)

Figure 7-23*a* also shows the interface for the opto-isolator driving a TTL input. The pull-up collector resistor is added to allow the isolator output to be pulled up to $\approx V_{CC}$ when the LED is OFF, thus providing an adequately positive input (2.0 V minimum) to the TTL gate. When the opto-isolator output goes low (a few tenths of a volt), the phototransistor sinks $\approx$1.6 mA from the TTL gate input (maximum guaranteed) plus 0.33 mA of collector current.

The seventh column of Fig. 7-21 shows a variation of the LED/phototransistor opto-isolator which offers a significant improvement in data rate while maintaining an equivalent CTR. In this case the LED illuminates a photodiode and its output current drives the base of a high-speed transistor (not a phototransistor). Such an arrangement eliminates the large base-collector capacitance inherent in the large base area of a phototransistor, and hence offers higher speed.

Figure 7-24 shows two opto-isolator applications. The circuit of Fig. 7-24*a* provides a high-speed line-voltage monitor/status indicator. The split photodarlington isolator is used in this case to reduce the LED drive requirements and hence lower the power consumption of the monitor. For ac line monitoring, the diode is necessary to keep from exceeding the LED reverse voltage rating, and

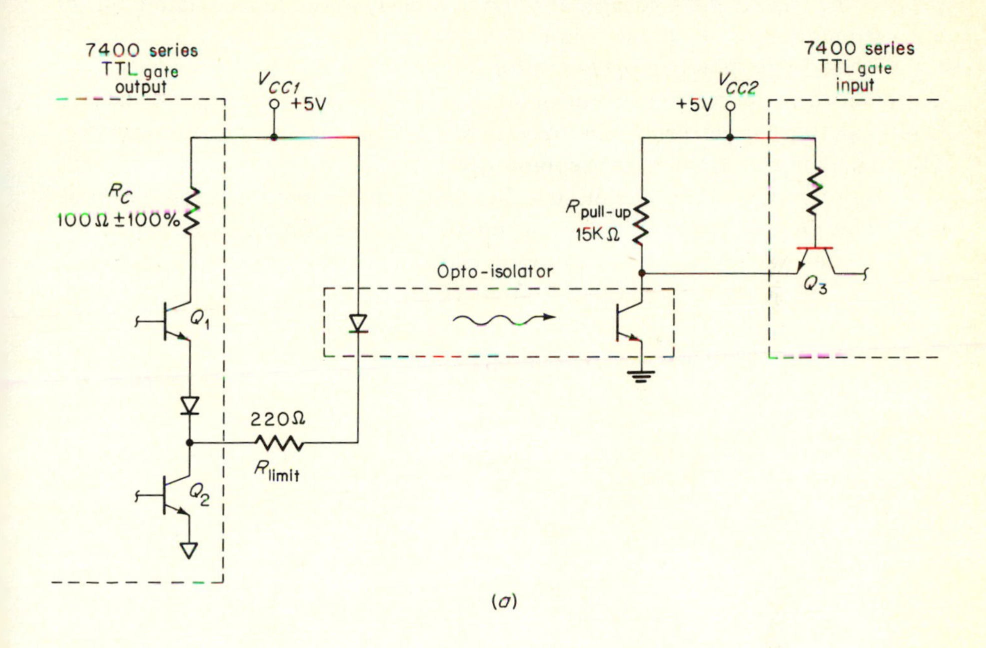

(*a*)

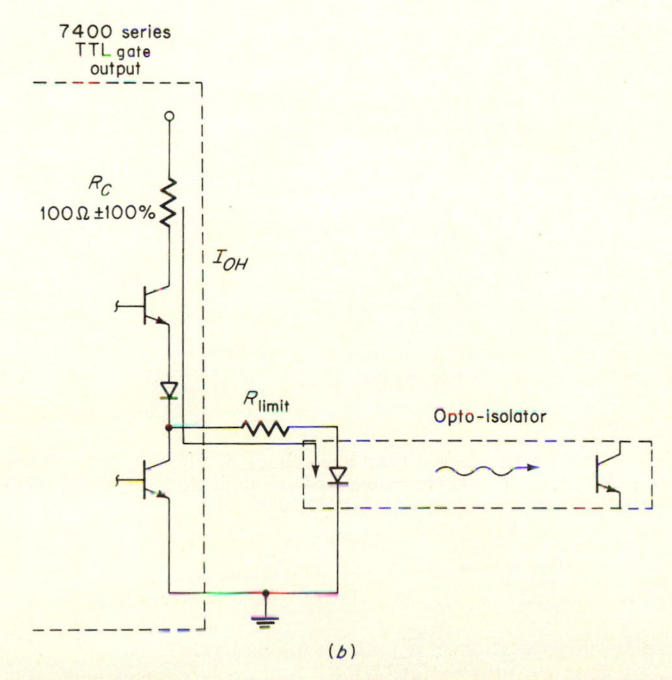

(*b*)

Fig. 7-23 LED/phototransistor opto-isolator interface with 7400 series TTL: (*a*) recommended current-sinking drive interface and output interface; (*b*) nonrecommended current-source drive interface.

the limiting resistor must be large enough to limit I_F to the desired value. For dc voltage monitoring, the diode is not required.

Figure 7-24*b* shows an opto-isolator in a line-receiver application typical of computer-to-peripheral signal transmission. Opto-isolators are superior to conventional transformer or IC line receivers for such applications because they offer less complexity and higher common-mode rejection.

In order to maintain the electrical isolation and common-mode reduction offered by opto-isolators, it is important to lay out the circuit elements in a manner that will minimize the coupling capacitance between the input circuit (including opto-isolator LED) and output circuit.

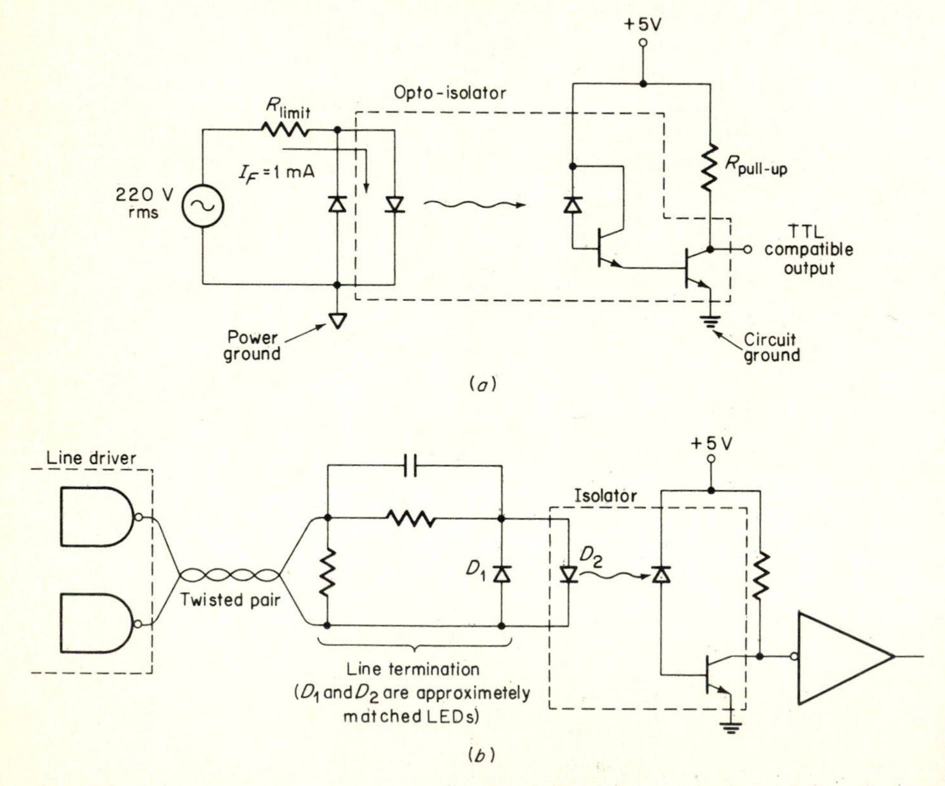

Fig. 7-24 Opto-isolator applications: (*a*) ac line-voltage monitor utilizing LED/split-photo-darlington opto-isolator; (*b*) line receiver using LED/photodiode transistor amplifier. *Adapted from Hewlett Packard Application Note 951-1, p. 3, June, 1975.*

PROBLEMS

7-1 Briefly explain the operation of a bulk photoconductor.

7-2 If the energy gap of CdS is 2.4 eV, what is the longest wavelength of light that it would theoretically respond to? Repeat the calculation for CdSe, whose energy gap is 1.7 eV.

7-3 For a typical neon lamp whose output peaks at ≈ 650 and 850 nm (Sec. 6-2-3), would a CdS or CdSe photo resistor offer the best spectral match?

7-4 What is Δconductance/ΔT for the two bulk photoconductors of Fig. 7-5*a* from room temperature (25°C) to +50°C at 1.0 fc?

7-5 List three advantages and three disadvantages of conventional bulk photoconductors.

7-6 Design a lightmeter with a full-scale sensitivity of 10 fc. Utilize the photoresistor circuit of Fig. 7-6*a* with a 0- to 1.0-mA ammeter and the type-4 photoresistor of Fig. 7-5*c*.

7-7 Design a simple lightmeter using the circuit of Fig. 7-6*a*. Assume that the maximum illuminance is 2 fc and that a 1.25-V battery and 0- to 150-μA ammeter are available. Select a photoconductor from the list of Fig. 7-5*b*.

7-8 Design one of the two circuits of Fig. 7-6*b* to provide ≈ +0.5 V in the dark and +10 V with 2 fc. Use the CL905L of Fig. 7-5*b*.

7-9 Refer to Fig. 7-7. If it were experimentally found that the nominal ON resistance of the photoconductors was ≈ 100 kΩ, select the resistors for the other half of the bridge (R_1 and R_3) for initial balance. Assume a 20-kΩ pot to allow for zeroing with cell mismatches of +10 percent.

7-10 (*a*) Design the light-sensitive relay-activating circuit of Fig. Prob. 7-10. The relay should close when an illuminance of ≈2 fc is present, and remain open under dark conditions. The relay requires 15 mA of pull-in current for closure. Assume $V_E = 5$ V, and see Fig. 7-5.

(*b*) What photodetector would activate the relay directly?

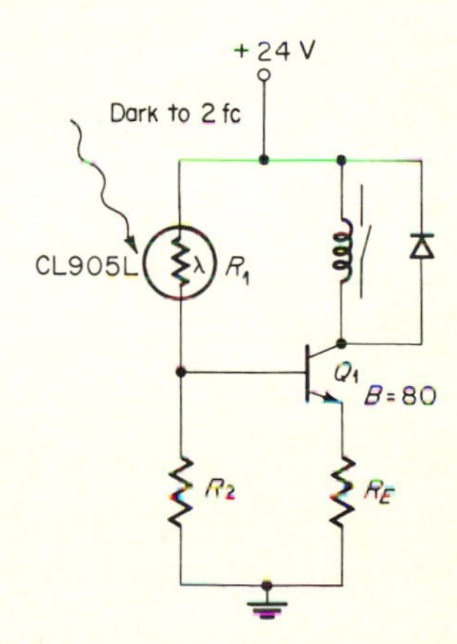

Fig. Prob. 7-10

7-11 Briefly describe the operation of a typical photodiode, and list three of its advantages.

7-12 List two advantages of a *pin* photodiode compared to a *pn* photodiode.

7-13 Calculate the approximate photodiode dynamic source impedance from Fig. 7-9*c* at 100 μW and reverse bias.

7-14 Plot output photocurrent versus input radiant flux from Fig. 7-9*c* at constant 5 V and 0 V reverse bias under open-circuit conditions. Check for linearity.

7-15 (*a*) Define dark current. (*b*) What is the equivalent current in a *pn* junction which is not a photodetector?

7-16 Sketch the spectral response of a Si photodetector and the output spectrum of a GaAs LED together. Comment on the spectral compatibility of this source/detector combination.

7-17 Calculate the shot-noise current for a 4204 photodiode at 10 V reverse bias at 1-, 10-, and 100-Hz bandwidths. (See Fig. 7-9*i*.)

7-18 Compute the current signal-to-noise ratio for a 4204 photodiode at 10 V reverse bias operating at room temperature, into a 1-MΩ load, at 100 Hz bandwidth, with 10^{-10} W of total radiant flux at 770 nm. Consider all sources of noise except that operation is out of the high $1/f$ noise region. Note the 0.5-A/W (at 770 nm) response in the note of Fig. 7-9*i*. Repeat the calculation for +85°C.

7-19 Construct a table for the 4204 and 4207 photodiodes comparing the manufacturer's quoted active areas, responses, junction capacities, and dark currents. Remark on the correspondence to theory.

7-20 What is the high-frequency 3-dB roll-off point for a 4204 photodiode at 10 V reverse bias with a 1-kΩ load?

7-21 (*a*) Calculate the 4207 photodiode output current for a 1.0 mW/cm² irradiance at the photodiode's spectral-response peak of ≈ 770 nm.

(*b*) Using the photocurrent calculated in part *a*, compute V_{out} for the circuit of Fig. 7-10*b* if $R_{fb} = 1.0$ MΩ.

(*c*) What output voltage is generated by the nominal 4207 photodiode dark current with −10 V reverse bias and the same 1.0-MΩ R_{fb}? (See Fig. 7-9*h*.)

7-22 Calculate the value of feedback resistance required to give an output of 1.0 V from the circuit of Fig. Prob. 7-22 for a GaAs IR LED (with peak emission at ≈900 nm) with an irradiance of 1.0 mW/cm².

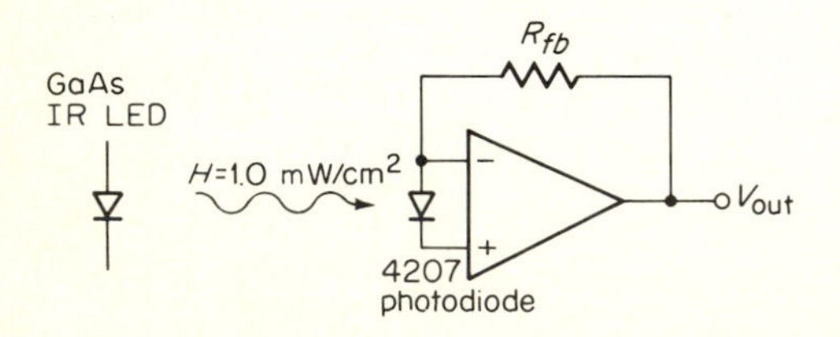

Fig. Prob. 7-22

7-23 In Fig. Prob. 7-23 calculate I_C for a diode flux of 200 μW at peak response.

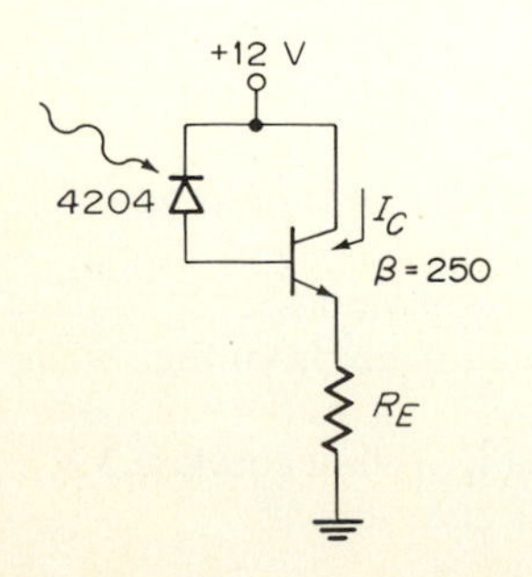

Fig. Prob. 7-23

7-24 Calculate the output voltage for the circuit of Fig. Prob. 7-24. (See Figs. 6-10 and 7-9.) Neglect the LED junction-temperature increase due to the pulsed operation.

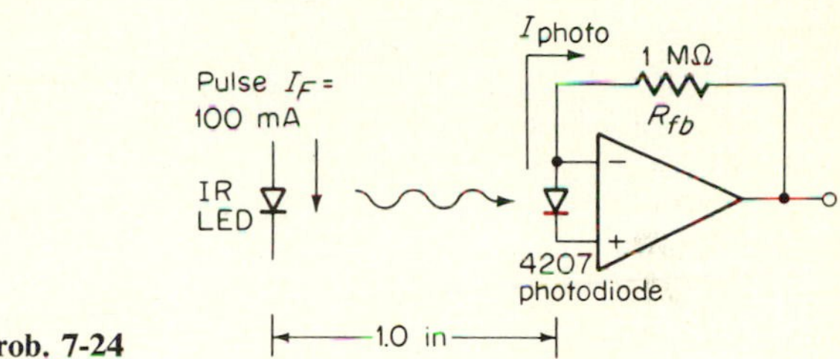

Fig. Prob. 7-24

7-25 (*a*) Approximately what operational-amplifier input resistance and leakage current are allowable for the circuit of Fig. 7-10*b* with a photodiode current of 1.0 μA and a 1.0-MΩ feedback resistance.

7-26 Name three differences between conventional photodiodes and solar cells.

7-27 Assuming a nominal solar-cell power-conversion efficiency, how much solar-cell area is required to produce 5 W of electric power on earth?

7-28 (*a*) Calculate the output power from the solar cell of Fig. 7-12*a*, assuming a 10-Ω load and nominal earth solar constant.

(*b*) What load resistance would give the optimum P_{out}?

7-29 (*a*) What is the purpose of the blocking diode of Fig. 7-12*b*?

(*b*) List three factors which tend to reduce solar-cell outputs in applications such as earth-based remote battery charging.

(*c*) Assuming an available solar-cell output of 0.4 V and 50 mA, design a series/parallel combination to produce $\approx$12 V at 2.5 A.

7-30 Design the circuit of Fig. Prob. 7-30 so that $V_{out} = 10$ V with an effective irradiance of 7 mW/cm². Assume $S_{RCEO} = 0.4$ mA/(mW/cm²).

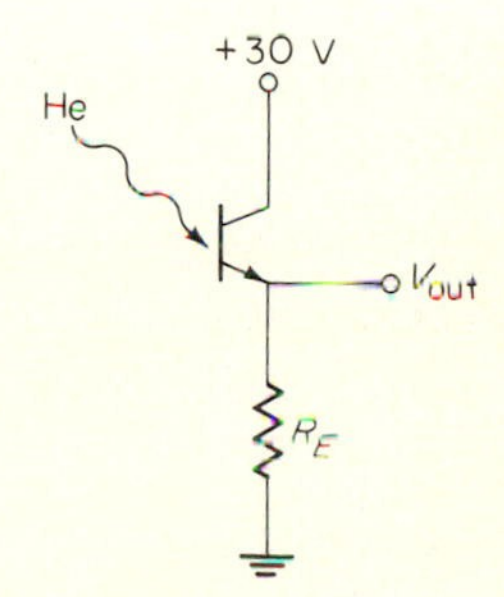

Fig. Prob. 7-30

7-31 Calculate the collector photocurrent for the phototransistor of Fig. 7-14 with an incandescent lamp with an irradiance of 5 mW/cm² operated at 2400 K.

7-32 Sketch a circuit to temperature-compensate a three-lead phototransistor.

7-33 Explain the operation of a photofet.

7-34 Name two advantages and two disadvantages of (*a*) photofets and (*b*) phototransistors.

7-35 In Fig. Prob. 7-35 calculate the value of R_G necessary to obtain a change in V_{DS} of 5 V with 0.1 mW/cm² input. Assume $g_{fs} = 2700$ μmho and the gate-current sensitivity = 3 μA/(mW/cm²).

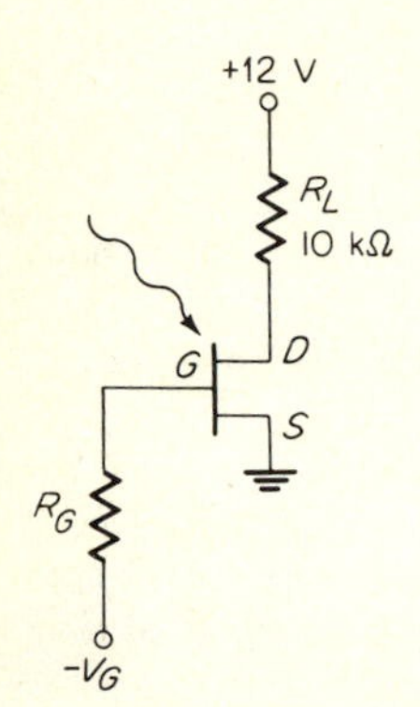

Fig. Prob. 7-35

7-36 Why is cooling necessary for many infrared detector applications?

7-37 Why are thermal detectors employed in the far infrared region?

7-38 Design the circuit of Fig. Prob. 7-38 to give −50 mV bias with at least ±50 percent adjustment to an extrinsic InSb detector.

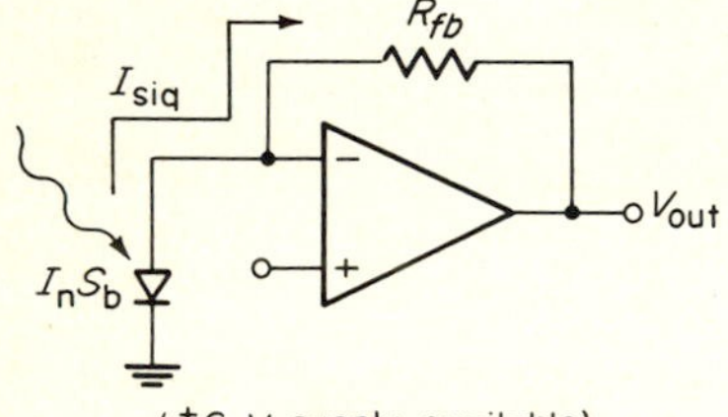

Fig. Prob. 7-38

7-39 For the bridge circuit of Fig. Prob. 7-39, choose an R_2 to ensure that the bridge can be balanced under dark conditions if the detector $R_{dark} = 2$ MΩ ± 20 per cent. Assume $R_1 = R_3 = 2$ MΩ and that the R_2 wiper is adjusted to the far right. If the detector resistance decreases to 1 MΩ when illuminated, what is the output with an amplifier voltage gain of 5?

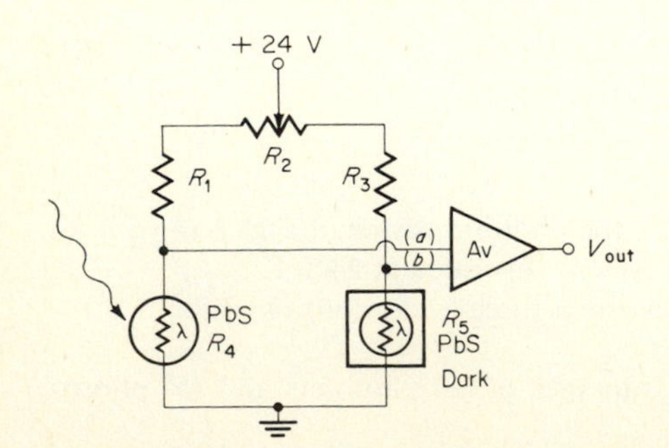

Fig. Prob. 7-39

7-40 List three advantages of opto-isolators that are a result of their photon-coupled mode of operation.

7-41 What type of opto-isolator would you use for:

(*a*) Maximum current gain

(*b*) Maximum speed

(*c*) Lowest input-current requirement

(*d*) Best combination of the above three characteristics

7-42 Why does the photodiode-transistor-amplifier opto-isolator have a higher data rate than the phototransistor opto-isolator?

7-43 (*a*) What is the current-transfer ratio (CTR) of the opto-isolator of Fig. 7-22 at room temperature, with $I_F = 5$ mA? With $I_F = 10$ mA? With $I_F = 50$ mA?

(*b*) What are the CTRs for the same values of I_F at 100°C?

7-44 What limits the data rate of photoresistor opto-isolators?

7-45 What value of input current I_F is required for the LED/phototransistor opto-isolator of Fig. 7-22 if a minimum of 10 mA of output current I_C is required over a -55 to $+100$°C operating-temperature range?

7-46 Explain why opto-isolators should be driven by TTL digital logic in the current-sink versus source mode.

7-47 (*a*) Calculate the value of R_{limit} of Fig. 7-23*a* so that $I_F \approx 20$ mA. Assume the opto-isolator of Fig. 7-22.

(*b*) Assuming the opto-isolator of Fig. 7-22, what value of I_C is anticipated for the conditions of (*a*) above at $+25$°C? At $+100$°C?

7-48 Explain how the opto-isolator circuit of Fig. 7-23 maintains the original polarity (noninverting)?

REFERENCES

1. Merriam, J., and W. Eiseman: "Interpretation of Photodetector Parameters," NOLC Report No. 558, 1962.
2. Perry and Skvarna: "Current-Mode Amplification of Signal and Noise Currents Generated by InSb Photovoltaic IR Detectors," IRIS Detector Specialists Group Meeting, 1964.
3. Burrous, C., *et al.*: "A New Logarithmic Radiometer," Electro-Optical Systems Design Conference, New York, Sept. 16–18, 1969.
4. Weaver, K.: "Remote Sensing—New Eyes to See the World," *Natl. Geograph.* pp. 47–73, Jan. 1969.
5. Whiting, E., C. Burrous, and H. Sorensen: "Improved UV Response of a PIN photodiode," *Appl. Opt.*, p. 2141, Oct. 1968.

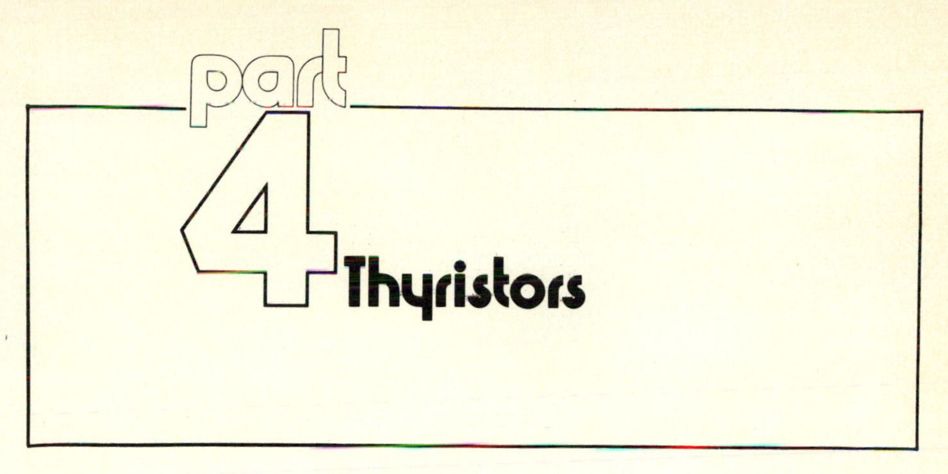

part 4 Thyristors

chapter 8

Thyristors

8-1 INTRODUCTION—THE THYRISTOR FAMILY

Thyristor applies to a family of multilayer semiconductor devices that exhibit bistable (ON-OFF) switching action due to their inherent regenerative feedback. The main advantage of thyristors is their ability to control large amounts of power with a very minimal expenditure of control power. The closest vacuum-tube thyristor equivalent is the gas-filled thyratron, which fires at the proper grid potential and remains ON without any grid control until the plate current approaches zero. In this respect, the thyristor is also similar to the latching relay.

The major thyristor applications involve the control or conversion of electric power. They include:

1. Power interfaces requiring rectification to dc; dc-to-ac inversion; or regulation for power supplies, battery charging, welding, electrochemical processing, and remote power generation
2. Replacing electromechanical devices such as relays, lamp dimmers and flashers, automobile and gas-appliance ignition systems, overvoltage/current protectors and bimetal-strip temperature controls
3. Motor speed controls for consumer and industrial applications, ranging from small appliances such as hand drills to large traction-drive vehicles
4. Logic functions such as power multivibrators and shift registers, timers, level or threshold detection, counters, alarm circuits, and emergency lighting control
5. Miscellaneous power control including modulation (such as radar), laser pulsers, and ultrasonic generators

The power levels involved in these applications range all the way from milliwatts for the low-level logic circuits to over 100 MW for industrial controls employing arrays of individual thyristors capable of continuous operation at more than 250 kW.

Although the integrated circuit has captured the lead in semiconductor sales, we find that thyristors and other solid-state power devices are running a close second. The reason is that in many applications the low-level signals must be converted to high power. In this respect we see the use of thyristors in consumer products increasing at a rapid rate. As an example, many manufacturers of power hand tools and light dimmers have switched to thyristors.

This does not mean that thyristors have replaced power or high-voltage transistors. On the contrary, power transistors are linear devices even better suited to some linear moderate-power applications such as automobile alternator-regulator systems, audio- and high-frequency power amplifiers, and servo amplifiers. When sufficiently saturated, power transistors also make excellent switches within their ratings. However, a disadvantage of power transistors for switching applications is that they require a high and continuous base current to be held in saturation. On the other hand, the regenerative action of the thyristor (Sec. 8-2) allows it to be held in saturation by a very-short-duration, low-power trigger pulse. For example, a 50-A transistor may require 1.0 A of base current at $\approx$0.7 V base-to-emitter voltage, or 0.7 W of *continuous* power, to hold it in saturation. A similarly rated SCR thyristor may be triggered into conduction by a *momentary* 1.5-V, 50-mA gate trigger pulse.

By comparison to the previous 250 kW thyristor power rating, few power transistors are capable of operation above 500 W, and those with high voltage capability rarely handle much current. Although major strides have been made to increase transistor voltage ratings, the transistor is at a distinct disadvantage owing to its thin base region necessary for high β, and the susceptibility of such thin regions to voltage breakdown. This problem does not exist in thyristors because of their low β requirement as discussed in Sec. 8-2-1. (The overcurrent capability of the thyristor is also superior to that of the power transistor.)

Although thyristors are basically switches, they can perform linear functions when one considers the concept of *averaging* as shown in Fig. 8-1. All the load-voltage waveforms indicated in Fig. 8-1 are possible with the appropriate thyristor and triggering. Hence, the average load voltage or power can be varied from approximately zero to full input by varying the time of effective switch closure. Control of many types of motors, heaters, and lighting fixtures can thus be effected by such a circuit, because these implements respond to the average power regardless of whether the waveshape is dc, pulse, or sinusoidal.

The ac load-voltage waveforms of Fig. 8-1 illustrate a method of power control used extensively in thyristor circuits. This technique is referred to as *phase control* and involves switching the thyristor to connect the supply to the load for a portion of the input waveform. This can be accomplished by phasing the thyristor triggering with respect to its anode-to-cathode voltage. The angle θ_F of Fig. 8-1 is defined as the *firing angle*, or the number of degrees the input sine wave has gone through prior to thyristor triggering. θ_C is the *conduction angle*, or the number of degrees during which the thyristor is ON and power is applied to the load. For complete control the phase-control circuit must allow θ_C to vary from 0 to 180°.

In Fig. 8-1 the thyristor has been represented as an ideal *switch*. An ideal switch would have no voltage drop or internal resistance when closed, and infinite resistance and zero leakage current when open. For such power-control

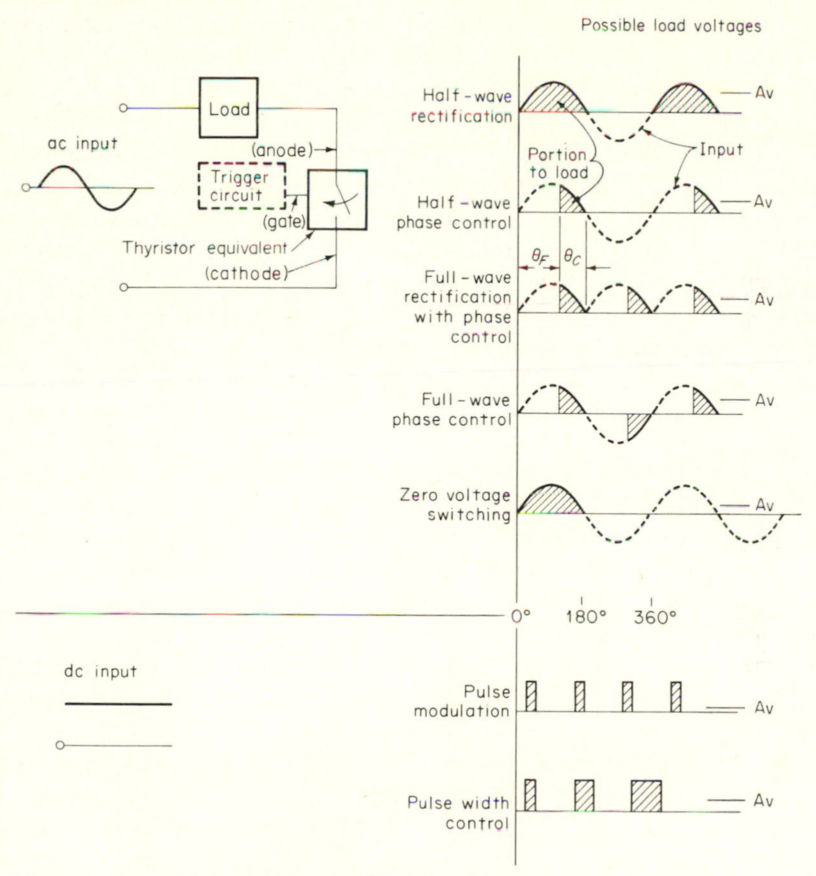

Fig. 8-1 Typical load-voltage waveforms for thyristor circuits with ac or dc inputs.

applications the device would also have to be capable of handling the applied power. No thyristor (or any semiconductor switch) meets the requirements of an ideal switch; however, their microamp to milliamp OFF leakage currents and 1- to 2-V ON voltage drops make them approach the ideal for solid-state switching. The rectification shown in most of the waveforms of Fig. 8-1 is inherent to many thyristors—that is, they *conduct* with one polarity of applied voltage and *block* with the other. This is due to their *pn* junctions which rectify like any solid-state diode. However, one type of thyristor is capable of conduction with either polarity of applied voltage (Sec. 8-3).

Previous to 1960 the number of available thyristors was limited to the four-layer diode and silicon-controlled rectifier (SCR). However, Fig. 8-2 shows that the number of devices within the thyristor family has grown significantly since that time. Figure 8-2 is an important figure because it summarizes all the commercially available thyristors—their names, symbols, cross sections, current-voltage characteristics, equivalent circuits, triggering means, maximum ratings, and

Type	Number of leads	Official name	Common name	Symbol	Cross section	Main trigger means	Maximum ratings available		Major applications	Principal current – voltage characteristics	Equivalent circuit
Unidirectional (reverse blocking)	2 (diode)	Reverse blocking diode thyristor	Four-layer (Shockley) diode		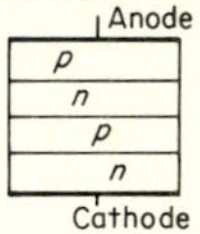	Exceeding anode breakover voltage	400 V	300 A peak pulse	SCR Trigger timing circuits, pulse generators	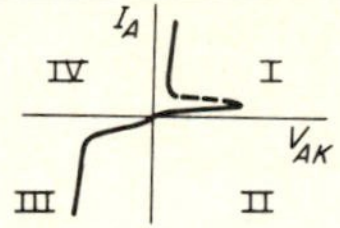	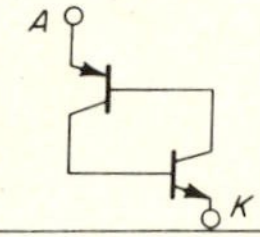
	3 (triode)	Reverse blocking triode thyristor	Silicon controlled rectifier (SCR)		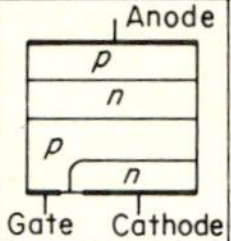	Gate signal	1800 V	550 A av	Power conversion replace electro-mechanical device motor speed control, switching phase control	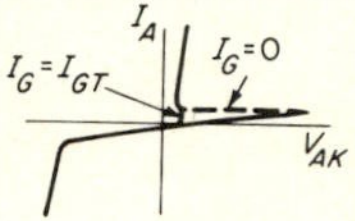	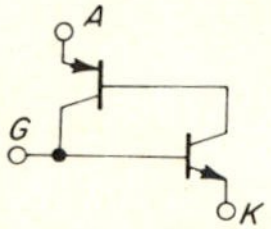
		—	Amplifying gate SCR		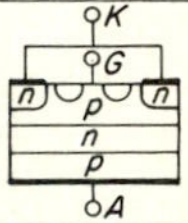	Gate signal	1200 V	110 A	Inverters and choppers	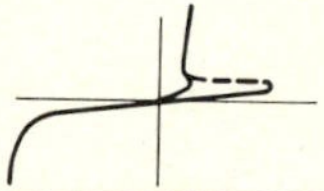	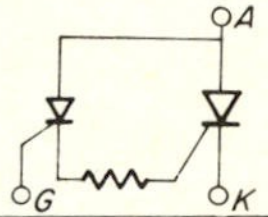
		Reverse blocking triode thyristor	Light-activated SCR (LASCR)		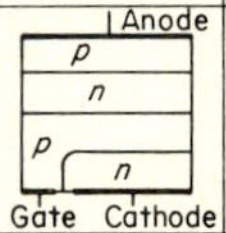	Gate signal and / or radiation	200 V	1 A av	Position monitors, static switches, limit switches, trigger circuits, photoelectric controls	Same as SCR	
		Turnoff thyristor	Gate – turnoff (GTO)		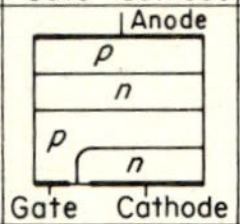	+ Gate signal turns GTO ON and – turns OFF	500 V	10 A	Dc switches inverters, choppers, logic	Same as SCR	

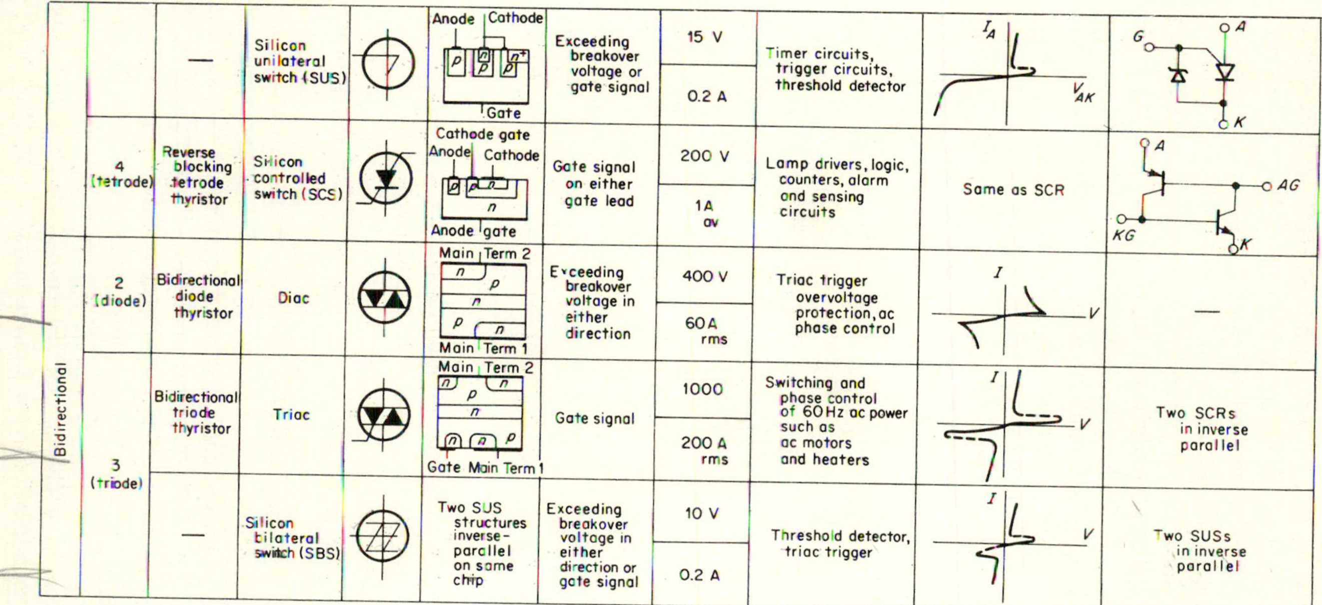

		—	Silicon unilateral switch (SUS)		Anode Cathode p n p n^+ p Gate	Exceeding breakover voltage or gate signal	15 V / 0.2 A	Timer circuits, trigger circuits, threshold detector	I_A V_{AK}	G A K
	4 (tetrode)	Reverse blocking tetrode thyristor	Silicon controlled switch (SCS)		Cathode gate Anode Cathode p n p n Anode gate	Gate signal on either gate lead	200 V / 1 A av	Lamp drivers, logic, counters, alarm and sensing circuits	Same as SCR	A AG KG K
Bidirectional	2 (diode)	Bidirectional diode thyristor	Diac		Main Term 2 n p n p n Main Term 1	Exceeding breakover voltage in either direction	400 V / 60 A rms	Triac trigger overvoltage protection, ac phase control	I V	—
	3 (triode)	Bidirectional triode thyristor	Triac		Main Term 2 n n p n n n p Gate Main Term 1	Gate signal	1000 / 200 A rms	Switching and phase control of 60 Hz ac power such as ac motors and heaters	I V	Two SCRs in inverse parallel
		—	Silicon bilateral switch (SBS)		Two SUS structures inverse-parallel on same chip	Exceeding breakover voltage in either direction or gate signal	10 V / 0.2 A	Threshold detector, triac trigger	I V	Two SUSs in inverse parallel

Fig. 8-2 The thyristor family. *Adapted from F. Gutzwiller, Thyristors and Rectifier Diodes—The Semiconductor Workhorses, IEEE Spectrum, p. 107, August, 1976.*

major applications. The details of Fig. 8-2 will not be discussed until later, but it is important to make a few general comments at this point. First, notice that thyristors are divided into *unidirectional* and *bidirectional* types. This refers to the passage of current through the device; the bidirectional thyristors are capable of conduction in either direction, as symbolized by the two-way arrows in their symbols. Within these two general classifications we find that thyristors range from simple two-lead diodes to four-lead tetrodes. All the thyristors contain four (or more) layers of alternate p and n diffusions, with an anode and cathode terminal contacting the external regions, plus access to one or more of the internal layers on some devices. The addition of the third lead has a great many control advantages similar to those offered by the vacuum-tube triode or bipolar transistor, as compared to their two-lead diode counterparts.

Several similarities among the various thyristors of Fig. 8-2 should be noted. Several of the unidirectional triode structures, symbols, characteristics, and equivalent circuits are essentially the same, with only minor variations in the accessible leads or triggering. Referring to the *principal* (anode-to-cathode) current-voltage characteristics of Fig. 8-2, we see that the unidirectional devices are capable of being switched from the OFF state (essentially no current flow) to the ON state (high conduction) with positive anode-to-cathode voltages (V_{AK}).

When V_{AK} becomes negative (third quadrant), these devices only block current, breakdown generally being avoided. Hence the terminology, *unidirectional reverse-blocking thyristor*. The bidirectional thyristors can be triggered into the high-conduction state in either the first or third quadrant, with a different current path existing for each quadrant. This results in a more complex structure for the bidirectional devices. (Notice the symmetry of the bidirectional-device characteristics.)

One of the problems facing thyristor users is understanding the differences between the various thyristors so that the optimum choice can be made for each application. The most popular types are the silicon-controlled rectifier (SCR) and the triac, the first being available in much higher power and frequency ratings and the latter being superior for full-wave ac control applications (Sec. 8-3).

The order of the thyristor listing in Fig. 8-2 is fairly chronological, and as such we see that many of the newer devices are more versatile than the original SCR. The main changes are improvements in turn-OFF control (Sec. 8-4) and the addition of the bidirectional devices. The latter demonstrates the attempt to more closely match thyristors to the commercial ac power system. More progress along this line, and the trend to integrate more within a single package, can be expected in the future.

This chapter will proceed in much the same sequence as the thyristor listing of Fig. 8-2. Each thyristor section will include a discussion of the device's theory of operation, characteristics, ratings, triggering and turn-OFF methods, and typical circuits. Many of the basic concepts common to all thyristors are introduced in the SCR section (Sec. 8-2); hence this material is a necessary prelude to the other thyristors. Many of the low-power thyristors that function well as trigger devices are discussed in the triggering subsections (Secs. 8-2-3 and 8-3-3).

Throughout this chapter the various thyristors will be referred to by their common name or abbreviation. These titles are not as descriptive or precise as the more formal nomenclature; however, they are briefer and much more commonly used.

8-2 SILICON-CONTROLLED RECTIFIERS

The *silicon-controlled rectifier* (SCR) is a unidirectional, three-terminal, reverse-blocking thyristor as shown in Fig. 8-2. It is one of the oldest of the thyristor family and by far the most used and highest-power thyristor.

8-2-1 SCR THEORY OF OPERATION

The cross section and symbol of the SCR are repeated in Fig. 8-3*a* and *b*. Although the SCR symbol does not portray all the structural details, at least the proximity of gate and cathode areas and the fact that the anode is a *p* region and the cathode an *n* region are shown. The SCR symbol suggests the inherent rectifying action, and the gate terminal hints at why the device is called a silicon-*controlled* rectifier instead of just a silicon rectifier. Figure 8-3*b* also shows that the SCR is roughly equivalent to a switch, as mentioned in the introduction. Although the details as to how the device achieves this equivalence are more complicated, the reader is reminded that the SCR is just that—*a semiconductor switch.*

The SCR is converted to a two-transistor equivalent circuit in Fig. 8-3*c* and *d*. Any four-layer device can be considered as an interconnected *npn* and *pnp* transistor if we hypothetically separate the two center regions and then reconnect them as demonstrated by Fig. 8-3*c*. If transistor symbols are substituted, the SCR two-transistor equivalent circuit of Fig. 8-3*d* is obtained. The reader should verify to himself that the interconnections and external leads of Fig. 8-3*d* are equivalent to the SCR itself. As mentioned in Sec. 8-1, internal thyristor feedback occurs because the collector of Q_1 drives the base of Q_2, and vice versa. That this internal feedback is regenerative (positive) is demonstrated by the circled polarity arrows of Fig. 8-3*d*. Assuming the normal transistor common-emitter phase reversal and proceeding from Q_2 to Q_1, we see that a positive-going voltage at the base of Q_2 is inverted at its collector, which is common with Q_1's base. Q_1 also inverts, and the signal at its collector is then in phase with the original input voltage.

The overall current gain of Fig. 8-3*d* equals $\beta_1 \times \beta_2$, where β_1 and β_2 are the common-emitter current gains of Q_1 and Q_2, respectively. Figure 8-3*d* shows that the total anode-to-cathode current is

$$I_A = I_{C1} + I_{C2}$$

(since $I_E \approx I_C$ for the bipolar transistor).

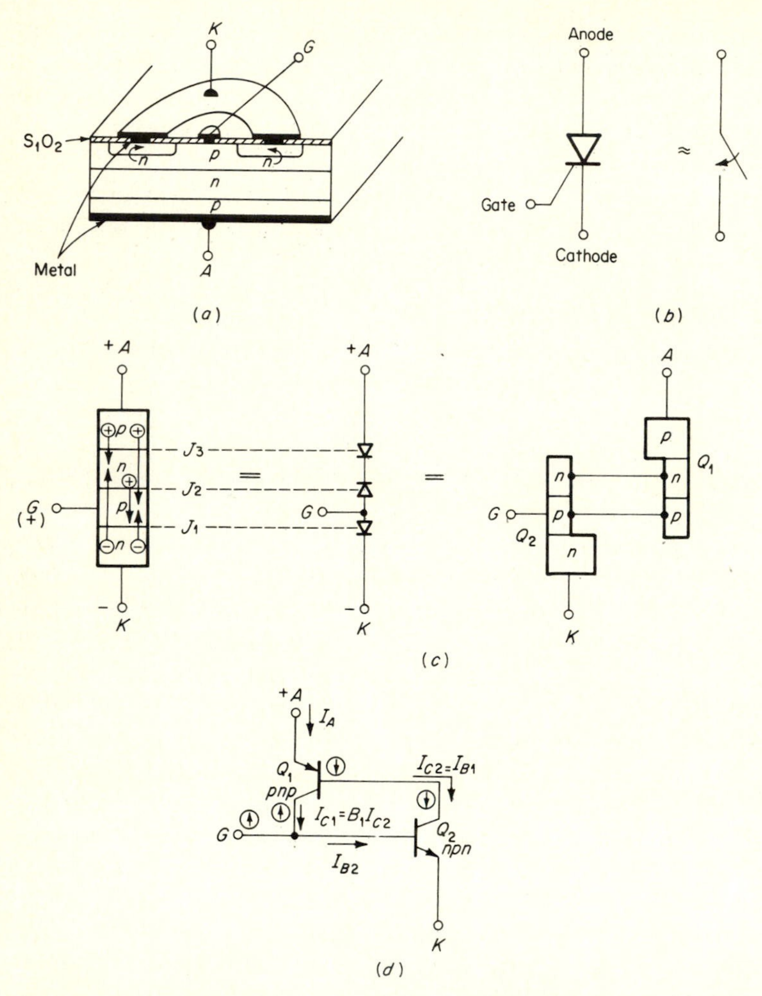

Fig. 8-3 Silicon-controlled rectifier: (*a*) cross section; (*b*) symbol; (*c*) and (*d*) equivalent circuits.

We now summarize the ON-OFF action in terms of the applied gate voltage.

1. If V_{GK} *is zero or negative*, the *npn* transistor (Q_2) is biased OFF and therefore Q_1 is never turned ON. Thus I_A equals the sum of the leakage currents of Q_1 and Q_2, the resistance from anode to cathode is high, the anode-to-cathode voltage drop is high, and *the switch is open*.
2. *If* V_{GK} *is adequately positive* ($\approx +1.0$ V), Q_2 is forward-biased, I_{C2} increases, and Q_1 is turned ON. Thus the circuit becomes regenerative; both transistors saturate, reducing the forward resistance and voltage drop drastically, and *the switch is closed.* In fact, once the SCR is ON there is no need for gate current because I_{C1} (Fig. 8-3*d*) is adequate to maintain I_{B2} and keep Q_2 ON.

> The only way to turn the SCR OFF is to reduce I_A below some minimum holding current. This is sometimes a disadvantage with the SCR, as discussed in Sec. 8-2-3. In this sense the SCR is very much like the vacuum-tube thyratron or latching relay, since both stay ON as long as primary current flows.

Owing to this ON-OFF action, the SCR can be considered a switch from anode to cathode, which can be triggered from the OPEN or blocking state to the CLOSED or low-resistance–high-conduction state (Fig. 8-3*b*). This switch analogy will be used throughout this chapter to facilitate the analysis of the various thyristor circuits.

Figure 8-3*c* can also be used to illustrate the actual SCR turn-ON mechanism. Notice that with the indicated anode-to-cathode potential, junctions J_1 and J_3 are automatically forward-biased. However, junction J_2 is reverse-biased. Upon application of an adequately positive gate voltage J_1 becomes highly forward-biased, causing a high concentration of electrons to drift from the cathode area toward the gate region and appear at J_2. These electrons experience the positive anode potential, and some are drawn into the *n* region between J_2 and J_3. Once in this region the electrons recombine with holes from the top *p* region, and the hole vacancies leave the way open for electrons to flow from the anode to cathode, thus completing the current flow path. A similar situation exists for the holes. In essence we see that J_2 is reverse-biased and thus limits the anode-to-cathode current prior to the application of positive gate potential or complete breakdown. When the gate is forward-biased, electrons are swept across J_2 and encounter a forward-biased anode area to complete the current flow path.

The β of either thyristor-equivalent transistor need not (and cannot) be high, since their product need only exceed unity for regeneration to occur. β varies with a number of transistor parameters, and a combination of these are utilized to increase $\beta_1\beta_2$ to > 1.0 in order to turn the SCR ON. Specifically, V_{BE} and I_C for both equivalent transistors (and hence $\beta_1\beta_2$) increase as the gate signal increases. β also increases with temperature, but this is more a problem than an asset because it causes premature SCR triggering at high temperatures and possible failure to trigger at low temperatures. This temperature effect on β and triggering is common to all thyristors.

8-2-2 SCR CHARACTERISTICS AND RATINGS

A typical SCR current voltage plot is shown in Fig. 8-4*b*. This is a plot of anode-to-cathode current I_A versus anode-to-cathode voltage V_{AK} with the *gate open*. When the anode voltage is negative, the SCR is essentially composed of two reverse-biased *pn* junctions (Fig. 8-3*c*), and a small leakage current flows. When $-V_{AK}$ exceeds the reverse breakdown voltage, $-I_A$ increases rapidly. The region from $V_{AK} = 0$ to V_{RSOM} is referred to as the *reverse-blocking* region. (The SCR terminology will be discussed shortly.) A similar *forward-blocking* region exists with positive anode potentials until the forward breakover voltage $V_{(BO)O}$ is

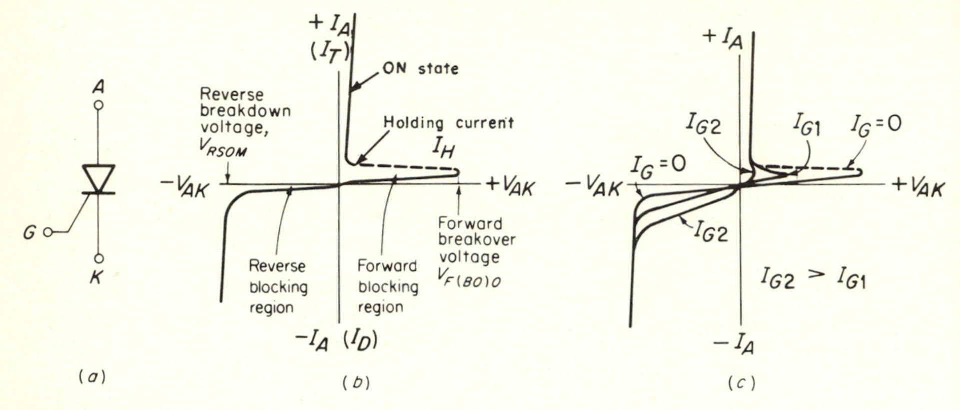

Fig. 8-4 SCR principal current/voltage characteristics: (*a*) symbol; (*b*) current/voltage characteristics with the gate open; (*c*) current/voltage characteristics with various levels of gate current.

reached, whereupon the SCR changes from the nonconducting to the high-conduction state. It is this dashed transition region of the SCR characteristic that exhibits regeneration; i.e., V_{AK} decreases as I_A rises. At $V_{(BO)O}$, the J_2 junction of Fig. 8-3*c* breaks down, allowing the anode current to increase rapidly. To turn the SCR OFF, or bring it out of the high-conduction region, I_A must be reduced below some minimum value called the *holding current* I_H.

The reader is referred to Table 8-1 for a complete listing of SCR terminology.

Figure 8-4*c* shows the effect of gate current on the SCR current voltage characteristics. Referring to the first quadrant, we see that increasing the gate current drastically reduces the forward breakover voltage. In essence this means that the SCR can be triggered into conduction at a relatively low forward voltage by inserting a small current into the gate terminal. In most applications SCRs are not triggered by V_{AK} and an SCR is chosen with a $V_{(BO)O}$ much greater than any likely circuit voltage. It is then triggered by a moderate amount of gate trigger power, thus greatly increasing the overall reliability and ease of design. Once the SCR is triggered, the gate loses control, and the only method of turning it off is to reduce I_A below I_H.

Figure 8-4*c* demonstrates two general SCR design principles. First, if gate triggering is to be employed, one must choose an SCR with forward and reverse voltage ratings sufficiently above the maximum applied voltage. For example, since the standard 115-V rms ac line voltage has ± 163-V peaks ($V_{\text{rms}} = V_{\text{peak}}/\sqrt{2}$), one should choose an SCR which will not breakover at less than $\approx +163$ V (V_{DSOM}) nor break down at less than -163 V (V_{RSOM}). A more practical value would be $\approx$200 V, to allow for fluctuations and a safety factor. Second, an undesirable increase in reverse leakage current results if gate current is applied when V_{AK} is negative. In practice, this problem is eliminated by making sure that the gate trigger signal is removed whenever V_{AK} goes negative.

Since most thyristor applications involve the control of power, considerable emphasis is placed on their power ratings. This is the main concern of the

TABLE 8-1 SCR terminology*

Maximum Ratings	
V_{RSOM}: Nonrepetitive Peak Reverse Voltage	The nonrepetitive peak reverse voltage rating is the maximum negative instantaneous voltage that may be applied to the main terminals (anode, cathode) of an SCR. This rating applies for an open-gate condition.
V_{RROM}: Repetitive Peak Reverse Voltage	The repetitive peak reverse voltage rating is the maximum peak reverse voltage that may be continuously applied to the main terminals of an SCR. This rating is specified for an open-gate condition.
V_{DSOM}: Nonrepetitive Peak Forward (Off-State) Voltage	The nonrepetitive peak forward (off-state) voltage rating is the maximum instantaneous positive non-repetitive voltage that may be applied to the main terminals of an SCR. The rating applies for an open-gate condition.
V_{DROM}: Repetitive Peak Forward (Off-State) Voltage	The repetitive peak forward (off-state) voltage rating is the maximum peak forward voltage which may be continuously applied to the main terminals of an SCR. This rating is specified for an open-gate condition. The repetitive peak forward (off-state) voltage rating applies for case temperatures up to the maximum operating temperature.
I_T: Current Rating, SCR	The limiting factor for rms and average currents is usually restricted so that the power dissipated during the on-state and as a result of the junction-to-case thermal resistance will not produce a junction temperature in excess of the maximum operating junction temperature.
I_{TSM}: Peak Surge (Nonrepetitive) On-State Current	The peak surge current is the maximum peak current (normally specified as sinusoidal at 50 or 60-Hz) that may be applied to the device for one full cycle of conduction without damaging the device.
I_{TRM}: Peak Repetitive On-State Current	This rating defines the maximum peak current that may be applied to the device during short, repetitive pulses. Blocking capability is maintained when the device operates under these rated conditions. The minimum pulse duration and shape are defined; these pulse characteristics control the applied di/dt.
di/dt: Rate of Change of Forward Current	This rating defines the maximum rate of rise of current through the SCR during turn-on. The conditions under which this rating applies include the value of anode voltage prior to turn-on and the magnitude and rise time of the gate trigger waveform during turn-on.
P_{GM}: Gate Power Dissipation	The gate power-dissipation rating defines both the peak forward and reverse (P_{GM}, P_{GRM}, respectively), and the average power ($P_{G(AV)}$) that may be applied to the gate. These ratings must be observed to avoid damage to the gate. Since the peak power allowed is a function of time, the width of the applied gate pulses must be considered in determining the voltage and current allowed.

TABLE 8-1 (Cont.)

Characteristics	
$V_{(BO)O}$: Instantaneous Forward Breakover Voltage	The breakover voltage is the voltage at which a device would turn on (switch to the on-state by voltage breakover) if the junction temperature were at maximum. This value applies for open-gate or negative-bias conditions.
I_{DROM}: Peak Forward Off-State (Blocking) Current	The peak forward off-state (blocking) current is the maximum leakage current permitted through the SCR when the device is operated with rated positive voltage on the anode (dc or instantaneous) at rated junction temperature and with the gate open.
V_T: Instantaneous On-State Voltage (Forward Drop)	The instantaneous on-state voltage (forward drop) is the anode-to-cathode voltage for a specified instantaneous current and case temperature when the SCR is in the conducting state.
I_{GT}: DC Gate Trigger Current	The dc gate trigger current is the minimum dc gate current required to cause the SCR to switch from the nonconducting to the conducting state for a specified anode voltage and case temperature.
V_{GT}: DC Gate Trigger Voltage	The dc gate trigger voltage is the dc gate-cathode voltage that exists just prior to triggering when the gate current equals the dc gate trigger current.
I_L: Latching Current	The latching current is the dc anode current above which the gate signal can be removed and the device will stay on. It is related to, has the same temperature dependence as, and is somewhat greater than the dc gate trigger current.
I_H: Holding Current	The holding current is the dc anode current below which the device will not stay in the on state if the gate is open. This current is slightly lower in value than the latching current and is related to, and has the same temperature dependence as the dc gate trigger current.
dv/dt, Static: Critical Rate of Rise of Off-State Voltage	The critical rate of rise of off-state voltage (static dv/dt) is the exponential rate of rise of off-state voltage that a minimum device will support, gate open, without turning on.

Subscripts

	First Subscript	Later Subscript
R	Reverse	Repetitive
D	Forward (off-state)	
T	On-state	Trigger
O		Gate terminal open circuited
M		Maximum
G	Gate	
S		Steady state (non-repetitive)
X		With gate resistance (R_{GK})
A	Anode	
H	Holding	
K	Cathode	
L	Latching	

* From RCA App. Note AN-6328

thyristor manufacturer, as evidenced by the fact that a large percentage of the data on a typical thyristor data sheet involve maximum power ratings (Figs. 8-6 and 8-7). Such ratings are important to the user, who desires maximum reliability but does not want to sacrifice economy by specifying a higher rating and more expensive thyristor than needed.

Before delving into the details of thyristor ratings, we should briefly review two general concepts which are used extensively in the derivation of semiconductor power ratings. The first involves the relationship of peak, rms, and average current, voltage, or power. One type of rating, such as rms voltage, cannot be used universally because some loads and thyristor characteristics are sensitive to average or peak voltage. Figure 8-5 can be used to convert from one of these units to another for sinusoidal waveforms. All the curves are plotted as a function of conduction angle, since phase control (Fig. 8-5*a*) is the most common method of power control.

Figures 8-5*b* and *c* are normalized to the rms value of applied voltage for half wave and full wave, respectively. To obtain the absolute peak or average voltage at a particular conduction angle, multiply the normalized value by the rms line voltage; i.e., at a 140° conduction angle, a half-wave 115-V rms voltage would have a (0.4)(115 V) = 46.0-V average value and a (1.42)(115 V) = 163-V peak level (Fig. 8-5*b*). Figure 8-5*b* and *c* are also valid for current in a resistive load and for power normalized to full conduction into a constant-impedance load. As an example of the use of Fig. 8-5, consider the task of selecting an SCR with adequate repetitive peak OFF-state voltage V_{DROM} with a 115-V rms source, a half-wave application, and a 60° maximum conduction angle. Entering Fig. 8-5*b* at $\theta_c = 60°$, we see that $V_{\text{peak}} = (1.23)(115\text{ V}) = 141\text{ V}$. V_{DROM} should thus be greater than 141 V.

The slopes of the power curves of Fig. 8-5*b* and *c* show that it is pointless to strive for conduction-angle control at less than $\approx 40°$ or over $\approx 140°$. Conduction angles outside this range have very little effect on the percentage of applied power. Figure 8-5 will be referred to many times in this chapter.

The second rating consideration common to all power devices involves the maximum operating temperature. Every semiconductor device has some maximum operating junction temperature (Figs. 8-6 and 8-7) above which the device is subject to catastrophic failure through overheating. Equation (8-1) expresses the junction temperature as a function of the other related variables:

$$T_J = T_A + \theta_{JA} P_J \tag{8-1a}$$

or

$$T_{J,\max} = T_{A,\max} + \theta_{JA} P_{J,\max} \tag{8-1b}$$

where T_J = junction temperature, °C
T_A = ambient temperature, °C
θ_{JA} = total thermal resistance from the device junction to the ambient, °C/W
P_J = total (or average) power dissipated at the junction, W

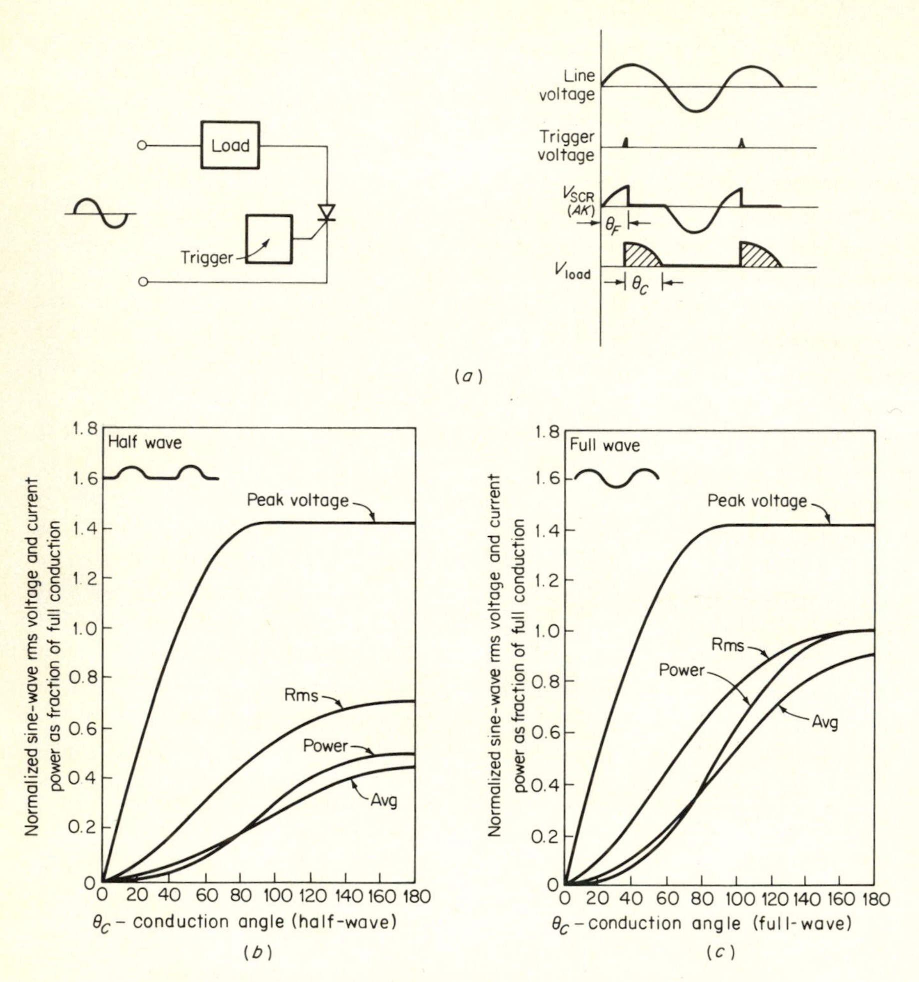

Fig. 8-5 Conversion charts for peak, rms, and average units of voltage, current, and power for resistive loads and sinusoidal waveforms: (*a*) general SCR phase-control circuit and waveforms; (*b*) half-wave chart; (*c*) full-wave chart. *From D. Zinder, "SCR Power Control Fundamentals," Motorola Appl. Note, AN-240, p. 3, 1971.*

As shown in Eq. (8-1*b*), the equation is equally valid for $T_{J,\,\max}$ if all other quantities are maximized. θ_{JA} is the total *thermal resistance* (from junction to ambient) which relates the junction temperature to the dissipated power. In other words, θ_{JA} tells us the number of degrees Celsius the junction temperature will rise per watt of junction power dissipation. (The lower the θ_{JA}, the less the SCR will heat up at a given dissipation.) θ_{JA} is the sum of the thermal resistances in the entire thermal path from junction to air, which includes the junction-to-case resistance θ_{JC} plus the case-to-ambient resistance θ_{CA}. If a heat sink is involved, the sink-to-ambient thermal resistance θ_{SA} must also be added to the total θ_{JA}, so that

$$\theta_{JA} = \theta_{JC} + \theta_{CS} + \theta_{SA} \tag{8–2}$$

Since there are so many junctions involved in thyristors, we shall assume that T_J is actually the temperature of the entire semiconductor chip; which is valid at least under steady-state conditions.

Equation (8-1) is a mathematical statement of the fact that a device's ambient temperature plus the temperature rise due to its internal power dissipation must never exceed $T_{J,\max}$. Generally, the manufacturer specifies $T_{J,\max}$ and θ_{JC}, and the real question becomes one of how high can the ambient temperature or power dissipation become under a given set of cooling conditions before the maximum junction temperature is exceeded. In fact many of the manufacturer's specifications involve graphs of case temperature or power dissipation versus average forward current (Fig. 8-6*e*). Such characteristics are derived from formulas such as Eq. (8-1) and destructive testing. As one would expect, Eq. (8-1*b*) states that as the ambient temperature T_A increases, the power dissipation must be decreased in order to remain under $T_{J,\max}$ (or, the power dissipation could remain constant if a more efficient heat exchanger were employed.)

As an example of the use of Eqs. (8-1) and (8-2), assume an SCR with a $T_{J,\max}$ of 125°C and a θ_{JA} (without heat sink) of 1.5°C/W. If the maximum ambient temperature were estimated to be ≈80°C, the maximum internal power dissipation would be

$$P_{J,\max} = \frac{T_{J,\max} - T_{A,\max}}{\theta_{JA}} = \frac{125°\text{C} - 80°\text{C}}{1.5°\text{C/W}} = 30 \text{ W}$$

On the other hand, suppose at the desired operating-current levels the power dissipation is actually 50 W for the same SCR and environment. This means that a heat sink must be added. Since most manufacturers give θ_{JC}, assume a θ_{JC} of 0.4°C/W and an ideal thermal contact from case to heat sink so that θ_{CS} drops out of Eq. (8-2). Substituting Eq. (8-2) into Eq. (8-1) and solving for the required θ_{SA} gives

$$\theta_{SA} = \frac{T_J - T_A}{P_J} - \theta_{JC} = \frac{125°\text{C} - 80°\text{C}}{50 \text{ W}} - 0.4°\text{C/W} = 0.5°\text{C/W}$$

Since 1 in^2 [6.45 cm^2] of aluminum surface has a θ_{SA} of ≈50°C/W, and θ_{SA} varies inversely with surface area, 100 in^2 [645 cm^2] of surface is required. This is a fair amount of surface area, and if it is not available, a more elaborate sink such as extruded aluminum fins, forced air, or water cooling must be employed.

Excerpts from a typical medium-current SCR data sheet are introduced in Fig. 8-6, and some data for a sensitive-gate, low-current SCR are shown in Fig. 8-7. These figures demonstrate typical SCR nomenclature, ratings, and the use of such data sheets. We shall refer to Fig. 8-6 in our discussion of ratings, and use Figs. 8-6 and 8-7 for several SCR problems.

As previously mentioned, the maximum operating junction temperature $T_{J,\max}$ is the most important thyristor rating, and all other ratings are somehow related to it. [Note that several portions of Fig. 8-6 are in terms of case (in lieu of junction) temperature, because of its accessibility.]

Figure 8-6*b* shows typical voltage ratings for a family of SCRs. The *repetitive peak forward* OFF-*state voltage* (V_{DROM}) refers to the maximum *forward* anode-to-cathode voltage that can be applied without triggering the SCR into the ON or high-conduction state. These voltage ratings are specified at $T_{J,\max}$ with the gate terminal open. These conditions represent the worst-case conditions because any thermally generated leakage current is multiplied by the SCR's regenerative action. If such currents cross the gate-to-cathode junction, the breakover voltage is reduced, just as if gate current were inserted (Figs. 8-3*d* and 8-8*a*). The two reverse-voltage ratings of Fig. 8-6*b* define the maximum repetitive and nonrepetitive *reverse* voltages that can exist before breakdown occurs. V_{DROM}

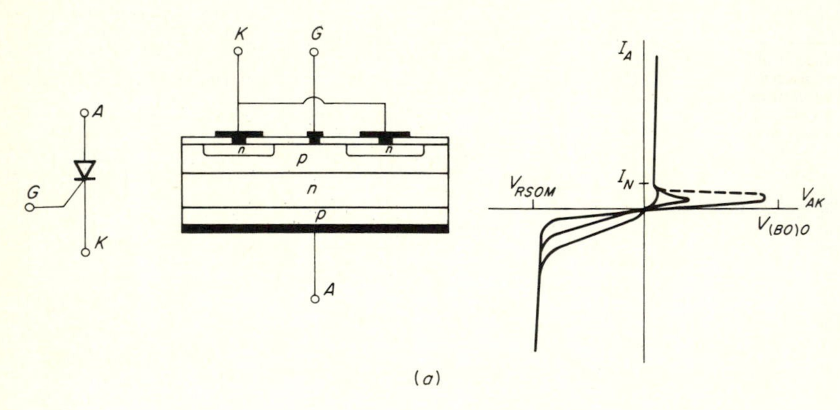

(*a*)

Type	Repetitive peak forward (off-state) voltage V_{DROM} T_C = -65°C to +125°C	Repetitive peak reverse voltage V_{RROM} T_C = -65°C to +125°C	Nonrepetitive peak reverse voltage(<5.0 ms) V_{RSOM} T_C = -65°C to +125°C
C35F(2N682)	50 V	50 V	75 V
C35A(2N683)	100 V	100 V	150 V
C35B(2N685)	200 V	200 V	300 V
C35C(2N687)	300 V	300 V	400 V
C35D(2N688)	400 V	400 V	500 V
C35E(2N689)	500 V	500 V	600 V

(*b*)

RMS forward current, on-state $I_{T(RMS)}$	35 A (all conduction angles)
Average forward current, on-state $I_{T(av)}$	Depends on conduction angle
Peak one-cycle surge forward current, I_{TSM}	150 A
Peak gate power dissipation, P_{GM}	5 W
Average gate power dissipation, $P_{G(av)}$	0.5 W
Peak reverse gate voltage, V_{GM}	5 V
Operating temperature, T_J	-65° C to +125°C

(*c*)

Fig. 8-6 Typical medium-current SCR data: (*a*) symbol, cross section, and current/voltage characteristics; (*b*) maximum voltage ratings; (*c*) other maximum ratings; (*d*) characteristics; (*e*) maximum allowable case temperature as a function of I_A and conduction angle (sinusoidal input). *General Electric.*

Test	Symbol	Min	Max	Units	Test conditions
Peak reverse or forward blocking current					$T_C = -65°C$ to $+125°C$ $V_{RROM} = V_{DROM}$ = Peak
C35F (2N682)	I_{RROM} or I_{DROM}	–	13.0	mA	= 50V
C35A (2N683)		–	13.0		= 100V
C35B (2N685)		–	12.0		= 200V
C35C (2N687)		–	10.0		= 300V
C35D (2N688)		–	8.0		= 400V
C35E (2N689)		–	6.0		
Gate trigger current	I_{GT}	–	40	mAdc	$T_C = +25°C$, $V_A = 12$ Vdc, $R_L = 50$ ohms
		–	80	mAdc	$T_C = -65°C$, $V_A = 12$ Vdc, $R_L = 50$ ohms
Gate trigger voltage	V_{GT}	–	3.0	Vdc	$T_C = -65°C$ to $125°C$, $V_A = 12$ Vdc, $R_L = 50$ ohms
		0.25	–	Vdc	$T_C = +125°C$, V_A = rated V_{DROM} $R_L = 1000$ ohms
Peak on-voltage	V_T	–	2.0	V	$T_C = +25°C$, $I_A = 50$ A peak, 1 millisecond wide pulse
Holding current	I_H	–	100	mAdc	$T_C = +25°C$, Anode supply = 24 Vdc, Gate supply = 10 V, 20 ohms, 45 μsec min. pulse width. Initial forward current pulse = 0.5 A, 0.1 millisecond to 10 milliseconds wide.
Critical rate of rise of forward blocking voltage (Higher values may cause device switching)	dv/dt			V/μs	$T_C = +125°C$ Gate open circuited. V_{DROM} = rated
C35F		10	–		
C35A		20	–		
C35B		20	–		
C35C		25	–		
C35D		25	–		
C35E		25	–		
Circuit commutated turn-off time	t_{off}	–	75	μs	
Effective thermal resistance (dc)	θ_{JC}	–	1.7	°C/W	

(*d*)

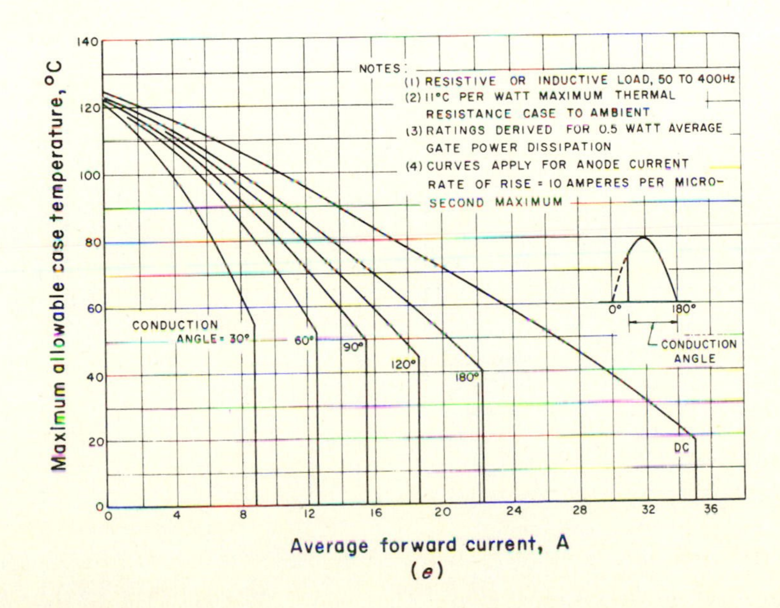

(*e*)

Fig 8-6 (cont.)

MAXIMUM RATINGS, Absolute-Maximum Values:

	Symbol	Q	Y	F	A	B	C	D	E	M	
							Suffix Letter				
NONREPETITIVE PEAK REVERSE VOLTAGE R_{GK} = 1000 Ω, T_C = −40 to 110°C	V_{RSXM}	25	50	75	125	250	400	500	600	700	V
NONREPETITIVE PEAK OFF-STATE VOLTAGE R_{GK} = 1000 Ω, T_C = −40 to 110°C	V_{DSXM}										
REPETITIVE PEAK REVERSE VOLTAGE R_{GK} = 1000 Ω, T_C = −40 to 110°C	V_{RRXM}	15	30	50	100	200	300	400	500	600	V
REPETITIVE PEAK OFF-STATE VOLTAGE R_{GK} = 1000 Ω, T_C = −40 to 110°C	V_{DRXM}										
ON-STATE CURRENT: Conduction angle = 180°, T_C = 85°C											
Average ac value	$I_{T(AV)}$						2.5				A
RMS value	$I_{T(RMS)}$						4				A
DC operation	$I_{T(DC)}$						2.75				A
PEAK SURGE (NONREPETITIVE) ON-STAGE CURRENT: For one cycle of applied principal voltage, T_C = 85°C 60 Hz (sinusoidal)...................................	I_{TSM}						35				A
PEAK GATE CURRENT (t = 10 μsec).....................	I_{GFM}						0.2				A
PEAK GATE REVERSE VOLTAGE	V_{GRM}						6				V
RATE POWER DISSIPATION:											
PEAK FORWARD (for 10 μs max.).....................	P_{GM}						0.5				W
AVERAGE (averaging time = 10 ms max.)	$P_{G(AV)}$						0.1				W
TEMPERATURE RANGE:											
Operating (case)*	T_C						−40 to +110				°C

(*a*)

CHARACTERISTIC	SYMBOL	LIMITS FOR ALL TYPES UNLESS OTHERWISE SPECIFIED			UNITS
		MIN.	TYP.	MAX.	
PEAK OFF-STATE CURRENT: Forward, $V_D = V_{DRXM}$, R_{GK} = 1000 Ω; T_C = 25°C	I_{DRXM}	–	0.1	10	μA
T_C = 110°C		–	10	100	
Reverse, $V_R = V_{RRXM}$, R_{GK} = 1000 Ω; T_C = 25°C	I_{RRXM}	–	0.1	10	
T_C = 100°C		–	10	100	
INSTANTANEOUS ON-STATE VOLTAGE: For i_T = 4 A and T_C = 25°C	v_T	–	1.25	2.2	V
DC GATE TRIGGER CURRENT: V_D = 12 V (dc), R_L = 30 Ω, T_C = 25°C: S2060 Series	I_{GT}	–	–	200	μA
DC GATE TRIGGER VOLTAGE: V_D = 12 V (dc), R_L = 30 Ω, T_C = 25°C	V_{GT}	–	0.5	0.8	V
INSTANTANEOUS HOLDING CURRENT: R_{GK} = 1000 Ω, V_D = 12 V, $I_{T(INITIAL)}$ = 50 mA, T_C = 25°C: S2060 Series	i_H	–	1.7	3	mA
CRITICAL RATE OF RISE OF OFF-STATE VOLTAGE: $V_D = V_{DRXM}$, R_{GK} = 1000 Ω, Exponential rise, T_C = 110°C	dv/dt	5	8	–	V/μs
CIRCUIT COMMUTATED TURN-OFF TIME: $V_D = V_{DRXM}$, i_T = 1 A, R_{GK} = 1000 Ω, Pulse Duration = 50 μs, dv/dt = V/μs, di/dt = −10 A/μs, I_{GT} = 1 mA at turn on, T_C = 110°C	t_q	–	30	100	μs
THERMAL-RESISTANCE: Junction-to-Case	$R_{\theta JC}$	–	–	3.5	°C/W

(*b*)

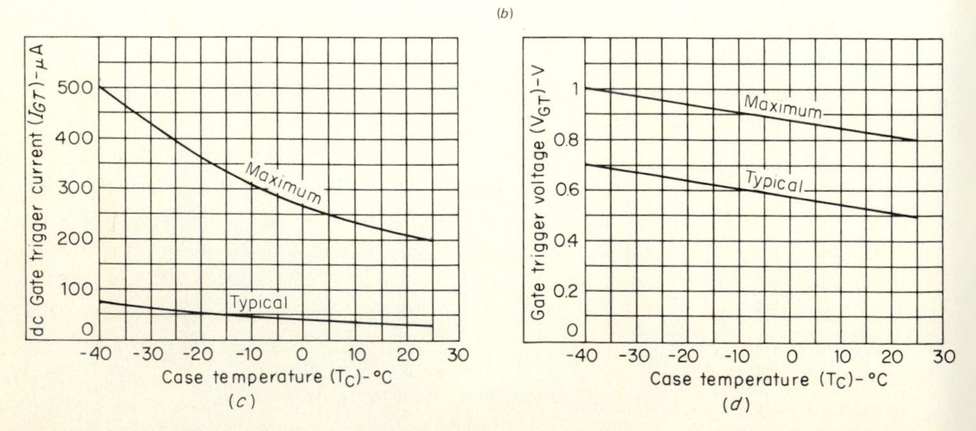

Fig. 8-7 Typical low-current sensitive-gate SCR data: (*a*) maximum ratings; (*b*) characteristics; (*c*) dc gate-trigger-current temperature dependence; (*d*) gate-trigger-voltage temperature dependence. *RCA*.

is replaced by V_{DRXM} for the sensitive-gate SCR of Fig. 8-7*a* because of the requirement for a gate-to-cathode resistor (Table 8-1 subscripts).

SCRs exhibit a *critical rate of rise of forward voltage* (dv/dt) above which the device may undesirably turn ON even though the necessary gate triggering signal does not exist. Figure 8-6*d* gives a typical 10- to 25-V/μsec rating of principal V_{AK} rise; however, higher ratings are available. This "false" triggering is caused by the parasitic capacity that exists between the bases of the two transistors of the SCR equivalent circuit of Fig. 8-3*d*. This capacitance allows rapidly rising anode voltages to couple to the base of the second equivalent transistor, thus biasing it ON even without a gate signal. The simplest method avoiding dv/dt (rate) effects is to utilize the dv/dt limiting ability of a capacitor to apply anode voltage gradually as shown in Fig. 8-9*a*. Gate biasing can also help reduce rate effect.

The last important principal (V_{AK}) voltage is the ON-state voltage (v_T of Fig. 8-7*b*). This is not a maximum allowable value, but rather a specification of the principal *forward voltage drop* when the SCR is ON.

We shall now discuss SCR current ratings. The first current rating to be determined is the maximum ON-state current (Figs. 8-6*c* and 8-7*a*). Notice in Fig. 8-6*e* that $I_{T,\,\mathrm{rms}}$ becomes the upper limit of all steady-state current ratings and is never exceeded at any conduction angle.

Prior to introducing the rest of the SCR current ratings we should refer to Fig. 8-5*a* which shows typical load and SCR voltage waveforms with an ac source. Notice that the SCR experiences a portion of the positive-half-cycle and all of the negative-half-cycle input voltage. In fact, the less the SCR is ON, the greater the voltage impressed across it. Figure 8-6*e* shows the maximum allowable case temperature as a function of $I_{T,\,\mathrm{av}}$ for various conduction angles. The important observations of Fig. 8-6*e* include the following.

1. All the curves terminate on the zero-current axis at $T_{J,\,\mathrm{max}}$ because no power dissipation would be allowed when $T_{J,\,\mathrm{max}} = T_A$ [Eq. (8-1)].
2. Figure 8-5*a* shows that the forward voltage across the SCR decreases as θ_C increases. Hence at a constant $I_{T,\,\mathrm{av}}$ of Fig. 8-6*e* the maximum allowable case temperature is seen to increase with increasing θ_C. Conversely, at a constant junction temperature the allowable $I_{T,\,\mathrm{av}}$ increases with θ_C.

We shall now consider the SCR gate characteristics. The main considerations with this portion of the SCR are that adequate voltage and current are available for reliable triggering; that the peak and average gate power dissipation (P_{GM} and $P_{G,\,\mathrm{av}}$) are not exceeded; and that the desired turn-ON time is realized. The P_{GM} rating is often of no concern because triggering is accomplished with rather low voltage and current levels (a few volts and a few milliamperes are generally sufficient for triggering). However, in many applications the gate triggering signal is derived from the same source which is being controlled, and complicated attenuation is not desirable. The biggest reason for overdriving

SCRs (triggering well above the required gate levels) is that very precise triggering can be achieved (Sec. 8-2-3). Using large gate-current pulses (versus dc or slow ac) for triggering reduces the variations in turn-ON time and the effects of temperature changes. Figure 8-6*c* gives the maximum gate power specification (P_{GM} and $P_{G,\,av}$) for the medium-current SCR. The peak or average gate current times the gate voltage must not exceed these levels. Figure 8-6*d* shows the gate current and voltage required to trigger the medium-current SCR into conduction (I_{GT} and V_{GT}). Notice that I_{GT} and V_{GT} must increase as the SCR's temperature decreases.

The low-current SCR data of Fig. 8-7*b* show that the sensitive-gate SCR is much more sensitive, i.e., the I_{GT} requirement is approximately two orders of magnitude less. Such low triggering requirements point out one of the most important SCR advantages—direct transducer-to-gate operation, thus eliminating additional stages of gain.

Another gate-triggering consideration is the time to complete turn-ON, which decreases as the gate current is increased. For rectangular gate trigger pulses the maximum allowable triggering frequency can be calculated from the relation

$$f_{\max} = \frac{P_{G,\,av}\text{ rating}}{(P_{G,\,peak})(\text{pulse width})} \tag{8-3}$$

Before leaving the subject of SCR gate characteristics, we should consider the effect of adding impedance between the gate and cathode (a requirement for many sensitive-gate SCRs such as shown in Fig. 8-7). As one might expect, adding an impedance from one terminal to another affects several SCR parameters. Let us first consider the consequences of adding a resistance from gate to cathode, as indicated by Fig. 8-8*a*. [Such a resistance is produced internally in many newer SCRs by metallizing the cathode contact onto part of the gate region (Fig. 8-3*a*). This produces a partial gate-to-cathode short, adding a small

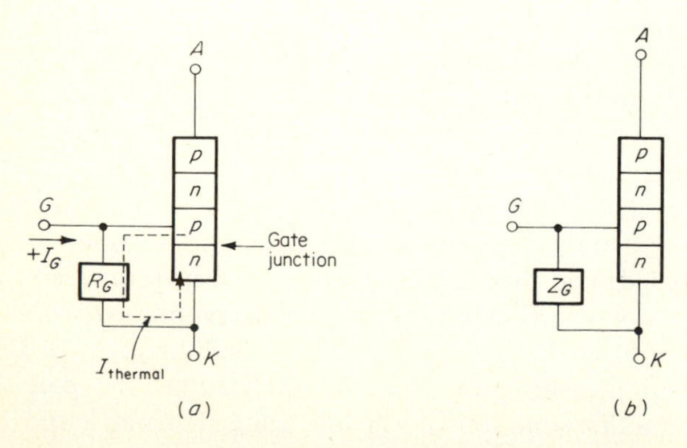

Fig. 8-8 (*a*) Resistance from gate to cathode; (*b*) reactance from gate to cathode.

resistance in parallel with the gate junction.] Regardless of whether such a low shunt resistance is added internally or externally, the resistance provides a lower-impedance path than the gate *pn* junction for the theramlly generated reverse-leakage current. Hence, some of the thermally generated current (and dv/dt induced anode current) is diverted around the gate-cathode junction as shown by Fig. 8-8*a*. Since regeneration is a direct function of the thermal leakage current, the addition of a small R_G causes the SCR to be less sensitive to temperature and dv/dt. Of course, the triggering sensitivity is decreased, but often this is required in sensitive-gate (20 to 100 μA I_{GT}) SCRs in order to prevent false triggering (Fig. 8-7).

Now let us consider the effect of a capacitor or inductor in the gate circuit (Fig. 8-8*b*). A capacitor of appropriate value will decrease the triggering sensitivity to high-frequency noise and dv/dt, because its low X_C will bypass these frequencies. Low-frequency sensitivity will be maintained because these frequencies will not be bypassed. The disadvantages of a gate-to-cathode capacitance are that it causes the rise time di/dt of I_G to increase, and it may cause failure to turn OFF or commutate because the capacitor continues to supply energy to the gate after I_H is reached. An inductance from gate to cathode causes a differentiating effect and thus reduces sensitivity to slow variations while maintaining high sensitivity to sudden changes. The effect on turn-OFF time is just the opposite of a capacitor; that is, t_{off} can be significantly reduced, owing to the reverse (negative) gate current induced in the gate inductance during turn-OFF.

Figure 8-9 shows other precautionary measures that are taken to protect thyristors from false triggering or catastrophic failure. As previously mentioned, such improper operation may occur if the anode voltage or current is applied too rapidly. As an example, closing the power switch in the SCR circuit of Fig. 8-9*a* will probably exceed the dv/dt rating and spontaneously trigger the

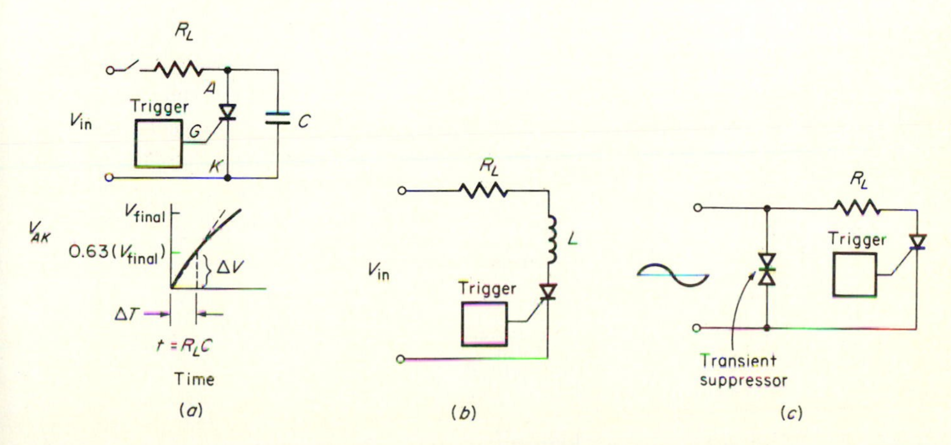

Fig. 8-9 Thyristor protection circuits: (*a*) dv/dt suppression; (*b*) di/dt suppression; (*c*) transient suppression.

SCR prior to normal gate activation. However, since the voltage across a capacitor cannot change instantaneously, adding a parallel capacitor will limit the rate of rise of the anode voltage.

As an example, consider the circuit of Fig. 8-9*a* with a 24-V supply, 12-Ω load R_L, and an SCR dv/dt rating of 20 V/μs. Referring to the *RC*-time-constant information of Fig. 8-9*a*, we see that V_{AK} would rise to (0.63)(24 V) or ≈15 V in one time constant t. Since a rate of rise of 20 V in 1 μs is allowed, the time for a 15-V rise would be

$$\frac{20\text{ V}}{1\ \mu\text{s}} = \frac{15\text{ V}}{X\ \mu\text{s}}$$

$$X = 0.75\ \mu\text{s}$$

Since $t = R_L C$, we can solve for the required C:

$$C = \frac{t}{R_L} = \frac{0.75 \times 10^{-6}}{12} = 0.0625\ \mu\text{F} \tag{8-4}$$

Figure 8-9*b* shows the simplest method of preventing SCR failure due to rapidly rising anode currents (di/dt effect). In this case the natural inductive suppression of rapid current changes slows down the rise of I_A.

Although not shown in any of the following thyristor circuits, a transient suppressor is often added between the ac line and the SCR circuit, as shown in Fig. 8-9*c*. Typical transient suppressors such as neon glow lamps or inverse series breakdown diodes are normally open and then conduct when a transient voltage of either polarity exceeds their breakdown voltage. Hence the device provides a path to ground for overvoltages.

All mechanical or solid-state switches produce considerable *radio-frequency interference* (RFI). The source of this RFI is the rapid (microsecond) switching of current from zero to the full load value. In a typical 60-Hz phase-control circuit this step function of current produces a noise spectrum which causes considerable AM radio interference. The amplitude of such RFI decreases with frequency and hence is not a problem at TV and FM broadcast bands. The type of RFI of most concern is that conducted over the power lines, requiring some sort of filtering between the thyristor and the line, as shown in Fig. 8-10.

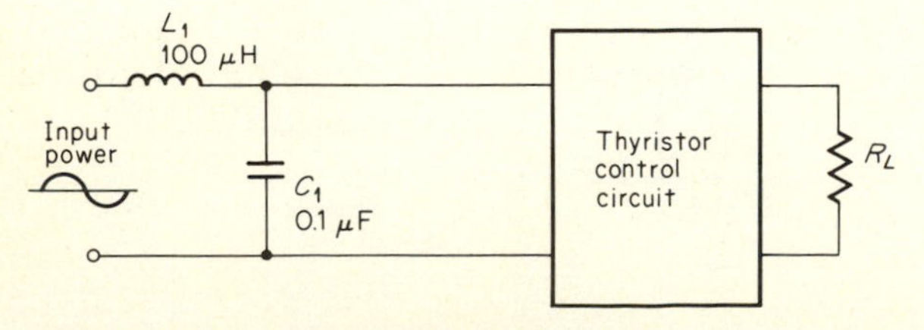

Fig. 8-10 Typical thyristor RFI filter.

Adding an inductor in series with the load limits the rate of rise of load current, and a parallel capacitor bypasses the higher frequencies to ground while passing the 60-Hz power. An LC filter as shown in Fig. 8-10 is often designed with equal LC and LR cutoff frequencies [3-dB down frequencies (f_o)]. Since $X_L = 2\pi fL$ and the 3-dB point occurs when $X_L = R_L$, L_1 can be found as follows:

$$2\pi f_o L_1 = R_L$$

and

$$L_1 = \frac{R_L}{2\pi f_o}$$

The resonant frequency of a series RLC circuit is $1/(2\pi\sqrt{LC})$, and, since the LC and LR frequencies are equal,

$$\frac{1}{2\pi\sqrt{L_1 C_1}} = f_{oL1} = \frac{R_L}{2\pi L_1}$$

Squaring the right and left sides and solving for C_1 gives

$$C_1 = \frac{L_1}{{R_L}^2}$$

With R_L and the desired cutoff frequency specified, L_1 and C_1 are determined from

$$L_1 = \frac{R_L}{2\pi f_o} \qquad \text{and} \qquad C_1 = \frac{L_1}{{R_L}^2} \tag{8-5}$$

The disadvantages of such a filter are its cost, size, power consumption of the series element, and the induced power-line reflections. (Sections 8-2-3 and 8-3-4 include discussions of a technique which reduces the RFI so that a filter is not needed.)

8-2-3 SCR TRIGGERING AND TURN-OFF METHODS

The general SCR triggering requirements were covered in the previous subsection. Recall from Fig. 8-6*d* that only a few milliamps of gate current and 1 to 2 V are required to trigger a medium-current SCR. In this section we shall discuss the various means of producing an adequate combination of current and voltage at the SCR gate, along with the various methods of turning the SCR OFF (commutating).

There are two general types of SCR-triggering circuits—those in which the triggering signal is derived from the same power source that is applied to the load (Fig. 8-11), and those with a completely separate trigger source (Fig. 8-9*a*). In either case the triggering and/or the controlled power source can be ac or dc. (The most prevalent type is the single ac source). Triggering can be accomplished by a dc or low-frequency gate signal or by a high-frequency pulse. The latter overdriving technique is more popular, owing to the reduction in turn-ON variations, temperature effects, switching time, and SCR selection as covered in Sec. 8-2-2. Within this pulse-triggering category, simple *RC* triggering-device circuits employing a variety of switching devices are used to produce triggering pulses. The particular application generally determines whether a simple (and inexpensive) or more sophisticated triggering circuit is employed.

The simplest type of SCR triggering is shown in Fig. 8-11. This is an example of deriving the trigger signal from a common power source. (At first

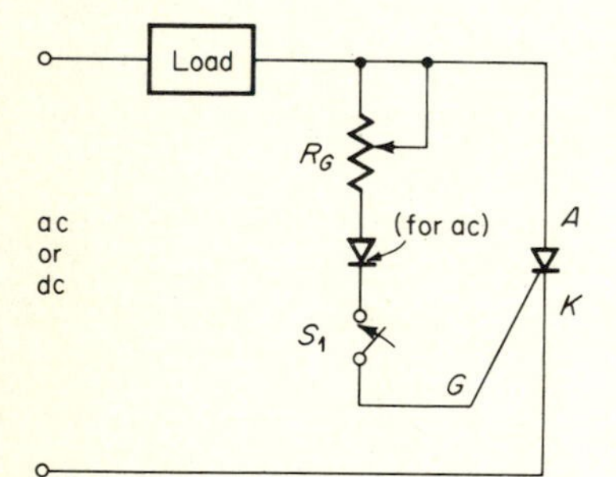

Fig. 8-11 Simple SCR triggering circuit.

glance it may appear that the load is continuously powered. However, when the SCR is OFF, it is an open circuit from A to K, and only a small current flows through R_G and the gate.) When the gate switch is closed, adequate gate voltage is applied, and R_G limits the gate current to the desired level. When the SCR switches ON, V_{AK} falls to a couple of volts (V_T), and the same voltage (V_T) is impressed across the gate trigger circuit since it parallels the SCR. (This automatic reduction of derived gate voltage is an important point to remember in all such circuits.) If R_G were low, a significant portion of the total current might flow through the gate; hence R_G is made large enough to limit I_G. (R_G generally consists of a fixed and a variable resistance to ensure I_G limiting if the variable component is inadvertantly reduced to 0 Ω.)

The operation of Fig. 8-11 with a dc source is as follows. With R_G large enough the required gate trigger current I_{GT} is never reached. Hence the SCR remains OFF with no voltage impressed on the load, owing to the lack of a completed circuit. As R_G is reduced I_{GT} is reached, and the SCR triggers. At that point its anode-to-cathode voltage drops to V_T; hence virtually all the dc input voltage is impressed across the load.

The ac operation of Fig. 8-11 is similar, except that the gate diode is added to limit the gate current when the source goes negative. As R_G is decreased, I_G increases, and the SCR triggers earlier in the cycle, thus transferring more of the source voltage to the load (Fig. 8-5*a*). The simple circuit of Fig. 8-11 can be turned ON only between ≈0 and 90° and cannot be turned OFF between 90 and 180°. Hence only ≈50 percent of the positive half cycle is controllable, because the gate and anode currents and voltages are all in phase in such a simple resistive circuit.

We shall not dwell on triggering circuits such as shown in Fig. 8-11, because they have a major disadvantage. That is, the gate voltage rises *slowly* to V_{GT}. This produces very unrepeatable SCR triggering because of the V_{GT} temperature dependence (Figs. 8-6 and 8-7) and the V_{GT} (and I_{GT}) spread within a batch of SCRs.

The disadvantages of the previous triggering circuit are best reduced by employing a gas-filled or semiconductor *triggering device* to provide a pulse to trigger the SCR. The devices most frequently used are the neon gas-discharge lamp (Fig. 6-7), unijunction transistor (UJT of Chap. 9), and many of the thyristors shown in Fig. 8-2, including the Shockley diode, SUS, SCS, SBS, and the diac; the bilateral devices being more applicable to the bidirectional triac. As shown in Fig. 8-12*a*, an *RC* network is used, and the triggering device is inserted in series with the gate. The triggering device acts like an ideal normally open switch, "closing" when the capacitor charges to the trigger device's firing voltage. At that point the capacitor rapidly discharges through the trigger device

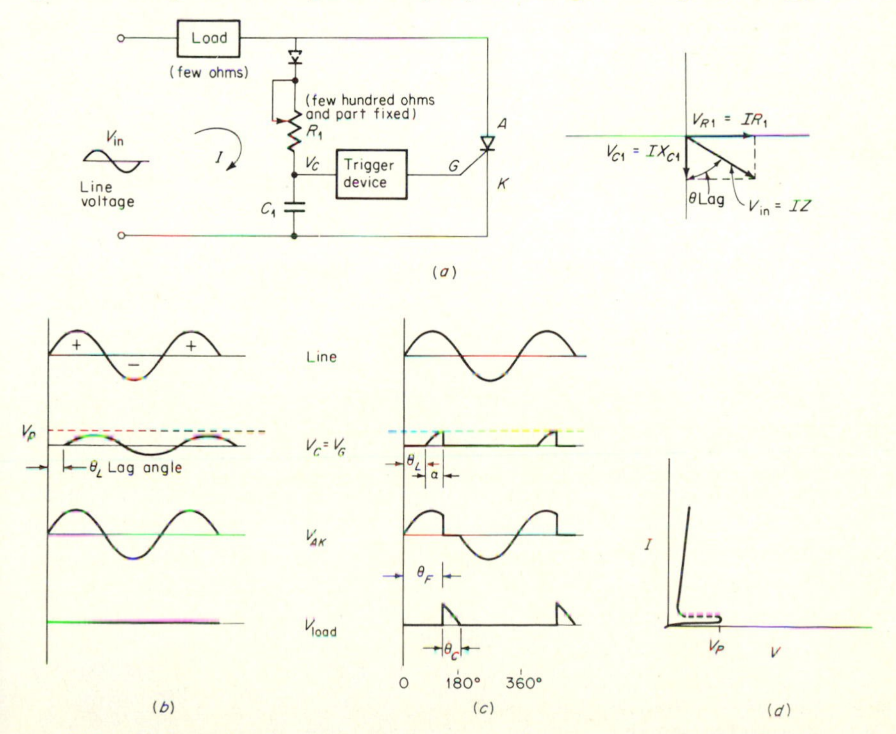

Fig. 8-12 (*a*) General pulse-trigger circuit; (*b*) waveforms before V_C reaches the trigger-device firing voltage V_P; (*c*) waveforms for $V_C > V_P$; (*d*) typical current/voltage characteristic for thyristor triggering devices.

and provides a step function of gate voltage to the SCR gate, instead of a slowly rising voltage.

Figure 8-12*d* shows a typical triggering device current voltage characteristic. Notice that it is normally OFF (low current flow) and enters a high-conduction state only after a certain critical voltage is reached. Many exhibit a negative-resistance characteristic which allows their voltage drop to decrease as the current increases. This means that very little of V_{C1} will be dropped across the trigger device when it fires. These devices exhibit a very small triggering-voltage spread.

Figure 8-12*a* is a typical SCR phase-control circuit with V_C lagging the line voltage because of the reactive element (C_1). [To briefly clarify the lag between V_G and the line (or SCR) voltage, remember that the current through the resistive load, R_1, and C_1 is in phase with the line current because this is a simple series circuit with a common current. Since the voltage across a capacitor always lags its current, V_G lags the line (or SCR) voltage; thus the lagging gate voltage.]

R_1 is the triggering control, and when it is large enough to limit V_C to less than the triggering device's critical voltage V_p, the waveforms of Fig. 8-12*b* are obtained. In this case neither the triggering device nor the SCR switches to the high-conduction state, and thus the load is never supplied with a low-impedance return path. Hence the load voltage remains near zero. When R_1 is decreased enough so that $V_C \geq V_p$, the triggering device breaks down and C_1 rapidly discharges through the low ON impedance of the triggering device. At this point the gate pulse switches the SCR to the ON (low-impedance) state, and most of the line voltage is impressed across the load as shown in Fig. 8-12*c*. The SCR does not conduct when the line voltage goes negative; thus that half cycle is impressed across the SCR and not the load. When the SCR is in the ON state, its forward voltage drop from anode to cathode falls to ≈ 2.0 V. Since the gate triggering circuit is in parallel with the SCR, V_C is also reduced when the SCR is ON.

Simple *RC* triggering circuits such as that shown in Fig. 8-12*a* are limited to a maximum conduction angle θ_C of substantially less than 180°. The main reasons for this are the initial V_C lag (θ_L) due to C_1 and R_1 and the attenuation of V_C by the $R_1 - C_1$ divider, as shown in Fig. 8-12*c*. One way to analyze such circuits is by vector analysis, as shown in Fig. 8-12*a*. Recall that V_{C1} lags V_{R1} by 90° and the lag angle between the input voltage V_{in} and V_{C1} can be calculated from

$$\tan \theta_L = \frac{R_1}{X_{C1}} \tag{8-6}$$

where $X_{C1} = 1/2\pi f C_1$ and θ_L is shown in Fig. 8-12*a*.

As an example, consider that $C_1 = 0.5\ \mu\text{F}$ and $V_{\text{in}} = 115$ V rms (60 Hz), and we wish to calculate both the value of R_1 which just allows triggering of a 40-V trigger device and the resulting conduction angle. We have

$$V_{\text{peak}} = (V_{\text{rms}})(\sqrt{2}) = 163 \text{ V}$$

The voltage across R_1 at triggering is

$$V_{R1} = (V_{\text{in}}{}^2 - V_{C1}{}^2)^{1/2} \tag{8-7}$$

and since $V_{C1\,\text{peak}} = 40$ V at triggering,

$$V_{R1} = (163^2 - 40^2)^{1/2} = 158 \text{ V}$$

$X_{C1} = 1/(2\pi f C_1) = 5.31$ kΩ. Since I is the series current through R_1 and C_1,

$$I = \frac{V_{C1}}{X_{C1}} = \frac{40 \text{ V}}{5.31 \text{ k}\Omega} = \frac{V_{R1}}{R_1} = \frac{158 \text{ V}}{R_1} \tag{8-8}$$

or

$$R_1 = \frac{(158 \text{ V})(5.31 \text{ k}\Omega)}{40 \text{ V}} = 20.9 \text{ k}\Omega$$

The firing angle (θ_F of Fig. 8-12*c*) can be calculated by adding the V_{C1} lag angle θ_L to the additional delay α required before V_{C1} rises to the triggering-device firing voltage V_P. This additional delay is calculated using the sine-wave formula

$$v = V_{\text{max}} \sin \omega t$$

In this case

$$\sin \alpha = \frac{V_P}{V_{\text{in}}} = \frac{40 \text{ V}}{163 \text{ V}} = 0.247 \quad \text{or} \quad \alpha = 14.2^\circ \tag{8-9}$$

and

$$\tan \theta_L = \frac{R_1}{X_{C1}} = \frac{20.9 \text{ k}\Omega}{5.31 \text{ k}\Omega} = 3.94 \quad \text{or} \quad \theta_L = 75.8^\circ \tag{8-10}$$

Therefore,

$$\theta_F = \theta_L + \alpha = 75.8^\circ + 14.2^\circ = 90^\circ \tag{8-11}$$

and conduction is limited to

$$\theta_C = 180^\circ - \theta_F = 180^\circ - 90^\circ = 90^\circ \tag{8-12}$$

[Firing angles (θ_F) less than 90° can be obtained by reducing R_1.]

One of the simplest, least expensive, and most reliable triggering devices is the neon (gas-discharge) lamp (Fig. 6-7). These lamps exhibit a few thousand ohms of internal resistance prior to triggering and break down at 60 to 90 V, depending on the particular lamp. When the discharge is initiated, the current through the lamp is limited only by the external resistance, and the device exhibits a negative resistance. The neon lamp makes an excellent trigger for circuits operated from 115 V (or greater) line voltage. However, it is not without its disadvantages. For example, it exhibits >20 V spread of firing voltages for a nonselected batch; is among the slowest of the triggering devices; requires a very high breakdown voltage; and its breakdown is sensitive to background illumination. The high breakdown voltage of the neon lamp causes a reduction in available power to the load. The reason for this is that a high breakdown voltage results in a higher $\theta_{F,\min}$ (Fig. 8-12*c*). This in turn reduces the available conduction angle θ_C.

The use of the unijunction transistor (UJT) as a thyristor triggering device is covered in detail in Sec. 9-5-5. The UJT makes an excellent triggering device with an all-dc system, but since it requires a dc power supply, additional complexity results with ac line operation.

Two of the unidirectional thyristors of Fig. 8-2 are also used for SCR triggering. These are the four-layer diode and the silicon unilateral switch (Sec. 8-4). Such thyristors switch from the blocking to the conducting state at a fairly low (≈8 V minimum) breakover voltage in much the same manner as the SCR without gate current.

One can also duplicate the regenerative action of the thyristor with two appropriately connected transistors (similar to the SCR equivalent circuit in (Fig. 8-3*d*), except that the threshold can be controlled. Like the UJT, the two-transistor switch must have a dc supply.

Sensitive-gate SCRs (and triacs) can be triggered directly by many digital logic families (Sec. 5-3). That is, the digital IC output drive is generally sufficient to trigger the gate of a sensitive-gate thyristor even under worst-case conditions.

Figure 8-13 demonstrates another form of SCR triggering, referred to as *zero-crossing switching*. This technique allows the SCR to be triggered very near the zero-voltage crossing of the ac line voltage, and thus significantly reduces the RFI generation which occurs when thyristors are switched at higher V_{AK} voltages (Fig. 8-10). With such triggering, *complete half cycles* of the ac line voltage are applied to the load, as compared to switching *during* every positive half cycle. (For triac or inverse-parallel SCR full-wave control circuits, *complete cycles* of the line voltage are applied to the load. This important application is discussed further in connection with Fig. 8-29.)

To achieve a significant reduction in RFI, switching must take place very close to the zero crossing of the applied voltage. This would require extreme precision and minimal drifts; therefore pulse-triggering circuits are seldom used. The most popular means of achieving true zero-crossing triggering is to apply a positive gate signal prior to (and somewhat past) the anode-voltage zero crossing.

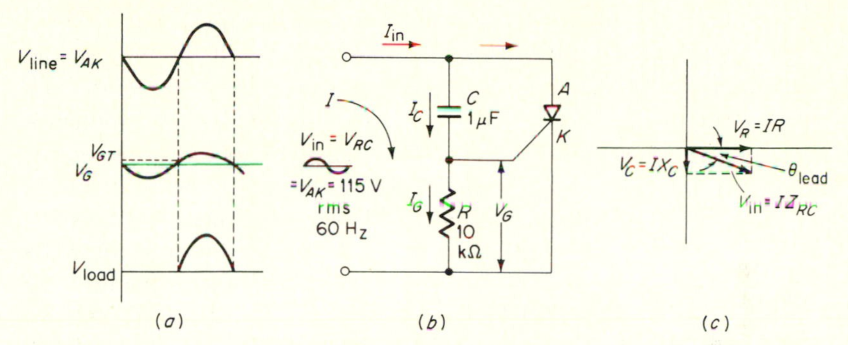

Fig. 8-13 SCR zero-crossing switching using a capacitor to obtain a leading gate signal to ensure zero-point switching: (*a*) waveforms; (*b*) circuit; (*c*) vector diagram.

Thus the SCR will switch ON very shortly after the anode voltage crosses zero in a positive direction. One of the simplest methods of obtaining such an "anticipating" gate drive is to derive a leading gate signal from a capacitor, as shown in the vector diagram of Fig. 8-13*c*.

[Since both the magnitude and phase of V_G and V_{AK} are important in thyristor ac phase-control circuits, a brief review of such calculations for Fig. 8-13 is offered here. Since

$$R = 10\ \text{k}\Omega \qquad C = 1\ \mu\text{F and } f = 60\ \text{Hz}$$

we have

$$X_C = \frac{1}{2\pi f C} = 2.65\ \text{k}\Omega$$

and the impedance of C and R in series is $Z_{RC} = R - jX_C$. Assuming that the voltage vector plot of Fig. 8-13*c* is just an impedance plot (dropping all I's), it can be seen that $\tan\theta = X_C/R = 0.265$; therefore the V_G lead angle $\theta = 14.8°$. Also, $Z_{RC} = X_C/\sin\theta = 10.4\ \text{k}\Omega$. The series current I is then equal to V_{RC}/Z_{RC} $= 115\ \text{V}/10.4\ \text{k}\Omega = 11.1\ \text{mA rms}$. Finally, the voltages are

$$V_G = V_R = IR = (11.1\ \text{mA})(10\ \text{k}\Omega) = 111\ \text{V rms}$$
$$V_C = IX_C = (11.1\ \text{mA})(2.65\ \text{k}\Omega) = 29.4\ \text{V rms}$$

and V_{RC} checks as equaling the input voltage:

$$V_{RC} = \sqrt{(111)^2 + (29.4)^2} \approx 115\ \text{V rms}$$

Note that the peak magnitude of V_G increases as C increases (less X_C) and as the lead angle decreases.]

The next important characteristic is SCR turn-OFF or *commutation*. There are two reasons for concern over SCR turn-OFF. First, some definite time interval must elapse, after the SCR's anode current passes through zero, before it is capable of blocking forward voltage without turning ON again. This problem is part of the maximum-frequency limitation. Second, many applications require the anode-current flow to be stopped at other than the natural ac zero-current crossing. Since the gate has no control over SCR turn-OFF, some external commutation means must be employed. The fact that the SCR is not naturally commutated in dc circuits is one of its major disadvantages. [The gate turn-OFF (GTO) of Sec. 8-4 can be turned OFF by the application of negative gate signals.]

To better understand the problem of SCR commutation, let us first define *SCR turn-OFF time* (t_{off}) as the total time required from zero anode-current crossing to when the SCR is able to block reapplied forward voltage. Typical SCR t_{off} times range from 10 to 200 μs.

There are two basic SCR commutation methods. One involves anode-current interruption, as demonstrated by the simplified schematic of Fig. 8-14*a*. There are only a few applications for which such a mechanical switch is suitable,

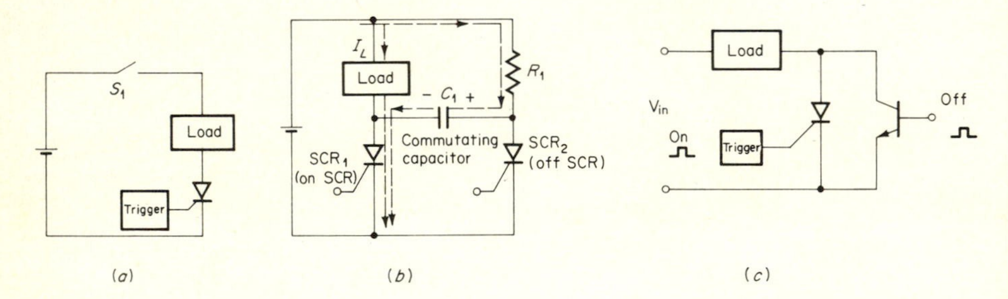

Fig. 8-14 SCR turn-OFF via: (*a*) anode-current interruption; (*b*) commutating-capacitor technique; (*c*) momentary-shorting technique.

but the same function can be accomplished by various solid-state switches. The disadvantages of such a scheme are that a second high-power solid-state switch is required, and the switching action produces high dv/dt, which may undesirably retrigger the SCR.

The other type of SCR commutation is forced commutation, depicted by Fig. 8-14*b* and *c*. In Fig. 8-14*b* the capacitor performs the commutation. Assuming that the SCRs are switches with SCR_1 ON and SCR_2 OFF, current flows through the load and C_1 as shown. When SCR_2 is triggered ON, C_1 is effectively paralleled across SCR_1. The charge on C_1 is then opposite to SCR_1's forward voltage, SCR_1 is thus turned OFF, and the current is transferred to the R_1–SCR_2 path. The rating of SCR_2 can be lower than that of SCR_1, since it must be ON only for the required SCR_1 t_{off} time. R_1 should reduce the SCR_2 anode current to less than one-tenth the minimum forward leakage current of SCR_2 (I_{DROM} at specified voltage from the data sheet). The value of the commutating capacitor (for a resistive load) must be sufficient to supply reverse voltage to SCR_1 until it turns OFF. C_1 is determined by the techniques of Ref. 1:

$$C_1 \geq \frac{1.5 t_{off} I_{L,\max}}{E_{B,\min}}. \tag{8-13}$$

where t_{off} = minimum SCR t_{off}
$I_{L,\max}$ = maximum load current at commutation
$E_{B,\min}$ = minimum dc supply voltage

As an example, assume that SCR_2 is the C35F SCR of Fig. 8-6, the dc supply is 12 V, and the load is resistive and $\approx 12.0\ \Omega$. From Fig. 8-6*d*, R_1 would equal

$$R_{1,\max} = \frac{1}{10}\begin{pmatrix}\text{minimum forward}\\ \text{leakage resistance}\end{pmatrix} = \frac{1}{10}\frac{V_{DROM}}{I_{DROM}} = \frac{1}{10}\frac{50\ \text{V}}{13\ \text{mA}} \approx 385\ \Omega$$

Using the manufacturer's t_{off} and Eq. (8-13), C equals

$$\frac{(1.5)(75 \times 10^{-6})(1)}{12} \approx 9.4\ \mu\text{F}$$

The disadvantage of this technique is that a large and expensive unpolarized capacitor is required because of the bidirectional current flow through C_1.

Figure 8-14*c* demonstrates a turn-OFF technique which momentarily deprives the SCR of its forward current and voltage. In this case the saturated bipolar transistor provides a momentary short, yet it need not be a high-power device because a few microseconds duration is all that is required for the SCR to fall out of latch-up.

There are many other methods of commutating SCRs, but all require additional circuitry because of the loss of gate control once the device is ON.

8-2-4 SCR POWER-CONVERSION CIRCUITS

In the next four subsections many practical SCR circuits are discussed. The organization is approximately the same as the original SCR application listing of Sec. 8-1. That is, the SCR circuits are separated into

1. Power-conversion circuits
2. SCR replacements for electromechanical devices
3. Motor speed control
4. Logic circuits

This breakdown does not include a separate section devoted to ac phase control or static switching circuits. The general concepts of ac phase control are first introduced in Sec. 8-2-5, and SCR static switching circuits are covered mainly in Secs. 8-2-5 and 8-2-7. The circuit functions that are much better suited for other devices, such as the triac, are dealt with in Sec. 8-3 and 8-4. Also, it is beyond the scope of this text to cover the vast multitude of SCR applications, especially the very specialized high-power or complex circuits.

The SCR is used in a great many power applications which require interfacing with the common 60-Hz ac power system and conversion to dc. Storage-battery charging is one of the best examples of such conversion, and Fig. 8-15 shows a very simple but effective circuit for such purposes (Ref. 2). Operation

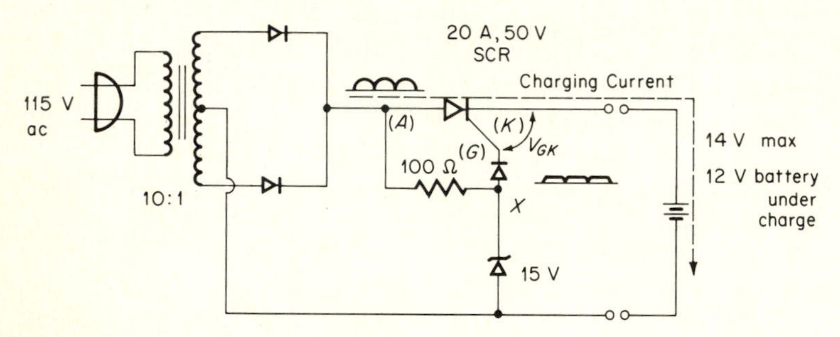

Fig. 8-15 Simple SCR battery charger.

of the circuit is as follows. The 115-V ac is full-wave-rectified by the center-tapped transformer and diode circuit, and the 15-V zener clamps point X to a maximum of +15 V. When the battery becomes discharged, its voltage drops below 12 V, and the SCR's gate thus becomes positive enough compared to the cathode to turn the SCR ON. That is, V_{GK} is (15 V − 0.7 V) − 12 V = +2.3 V, which is greater than the required gate trigger voltage V_{GT}. With the SCR in the high-conduction state the battery is charged through the SCR as shown. When the battery becomes fully charged, V_K approaches 14 V, allowing the SCR to turn OFF. In this condition the battery cannot discharge through the circuit because the diodes and SCR block such a discharge current. Hence the circuit can be permanently connected to the battery, charging it whenever power is applied and $V_{GK} = V_{GT}$. This general technique of using V_{GT} as a sensing element is quite common. It is accomplished by applying a reference voltage to either the gate or cathode terminal, and a voltage proportional to the desired control function on the other. Whenever $V_{GK} > V_{GT}$ the SCR conducts, bringing the battery charge, motor speed, etc., back to normal.

SCRs are also used extensively in dc-to-ac inverters (or dc-to-dc converters if the output ac is rectified). There are several classes of SCR inverter circuits with the classification by method of turn-OFF (Sec. 8-2-3). The reader is referred to Refs. 3 and 4 for examples of SCR inverters.

The SCR is also used in switching regulator circuits; however, linear regulation is better suited to the power transistor because of its inherent linear operation. In a typical linear transistor voltage regulator the output voltage is sampled, compared to a stable reference voltage, and fed back to control the base current (and hence the conduction) of the regulating transistor. This action is comparatively smooth and does not produce the RFI common to switching regulators. The operation of the switching regulator is similar to the linear version except that the ON-OFF duty cycle of the regulating device is varied. Such regulators are very efficient and ideally suited to the SCR, but require filtering to smooth the dc output.

8-2-5 SCR REPLACEMENTS FOR ELECTROMECHANICAL DEVICES

Solid-state devices are rapidly replacing mechanical devices while providing superior performance, reliability, and lower cost. SCRs (and other thyristors) have been used to replace power relays; conventional automobile ignition systems; rheostats or variacs for lighting and motor speed control; bimetal-strip thermostat temperature controls and interrupters for flashers; and mechanical timers. Some functions, such as low-frequency full-wave applications, are better adapted to the bidirectional triac than the SCR; hence they are covered in Sec. 8-3.

There is no exact solid-state equivalent to the mechanical relay, because no semiconductor switch exhibits the mechanical-switch characteristics of an almost ideal short circuit when closed and an open circuit when open. However, several semiconductor devices come rather close to duplicating the role of a mechanical switch with ON impedances ranging from 1 Ω to 1 kΩ and OFF impedances from 10 kΩ to 10^{+10} Ω. Hence the figure of merit (R_{off}/R_{on}) for many semiconductor switches is excellent. For instance, the SCR switch of Fig. 8-16 is excellent in

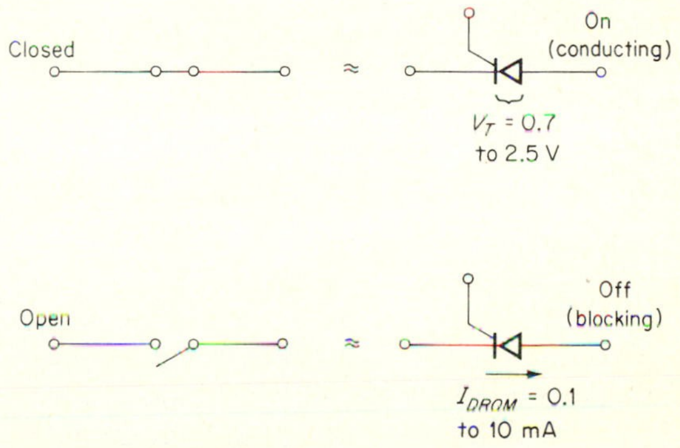

Closed ≈ On (conducting)

V_T = 0.7 to 2.5 V

Open ≈ Off (blocking)

I_{DROM} = 0.1 to 10 mA

Fig. 8-16 The SCR as a static switch.

applications where 0.7 to 2.5 V of forward voltage drop can be tolerated when the SCR is conducting, and 0.1 to 10 mA of reverse leakage current can be tolerated when it is OFF (depending on the SCR). In a 115-V power application, a 1-V SCR loss and a few milliamps of leakage current are certainly negligible (and the SCR *latching* characteristic is frequently desired).

Many of the circuits in this section fall under the classification of *static switches*—that is, they open or close a circuit completely, applying all or none of the power to the load. (This is in contrast to SCR phase-control circuits, in which it is desirable to control the *average* power at any level from zero to full power. Static-switching applications are covered in Secs. 8-2-3, 8-2-7, 8-3-3, and 8-4.) SCR static switches can operate with ac or dc supplies and in a virtual normally open or normally closed state, just as the relay.

We shall now consider thyristor *phase-control* circuits as mentioned in the introduction (Fig. 8-1). This technique involves transferring a controlled fraction of an ac input voltage to loads such as motors, lamps, and heaters. The number of electrical degrees of the line or input waveform transferred to the load is defined as the conduction angle (θ_C). The firing angle (θ_F) is then defined as the number of electrical degrees prior to switching to the load Several thyristor phase-control circuits are shown in Fig. 8-17. In all the configurations of Fig. 8-17 the loads are resistive and the triggering circuits are represented as boxes capable of initiating a triggering pulse at any predetermined time. (Figure 8-17*a* shows the load connected to the SCR anode *or* cathode; however, the rest of the figures show only one of the two possible load locations.)

Although the details are not shown, we assume that the triggering pulses can be derived and thus synchronized to the input ac waveform. Note that the single SCR is adequate for half-wave phase control, but a pair of SCRs can produce full-wave control. The two-SCR configuration of Fig. 8-17*c* is referred to as the *inverse parallel* connection, wherein each SCR passes one phase of the input waveform. The reader is encouraged to study Fig. 8-17 to become convinced that the indicated output waveforms are indeed obtained.

As demonstrated in Fig. 8-17*d*, *full-wave* control can be effected with far fewer parts utilizing bidirectional *triacs* rather than SCRs, and hence such applications are discussed in Sec. 8-3-4.

The automobile ignition system is an excellent example of an electromechanical system which is being replaced by a semiconductor equivalent. The requirements for an automobile ignition are as follows:

1. Depending on the spark-plug gap, condition of the plugs, mixture, and cylinder pressure, ≈ 20 kV is needed to arc across the gap.
2. At engine speeds of 5000 rpm (or ≈ 83 rps) a complete 360° revolution takes ≈ 12 ms, and this sets an upper limit on ignition-voltage rise time. Since only 30° (dwell angle) is allowed in an eight-cylinder engine, the storage system must become recharged in $(30/360)(12 \text{ ms})$ or ≈ 1 ms at 5000 rpm.
3. The spark duration (after ignition) should be between 50 and 200 μs.
4. The current is generally limited to <5.0 A to prevent damage to the points. (Many newer systems are "contactless.")
5. The minimum energy is ≈ 0.01 watt-seconds.

There are two major deficiencies in conventional ignition systems. First, the maximum voltage (and energy) is limited. Then, as the engine speed is increased, the coil current cannot reach its maximum, owing to the charging LR time constant; therefore the primary current and secondary voltage drop off at high rpm. Some of the difficulties associated with conventional ignition systems have been reduced by replacing the points with a magnetic pick-off.

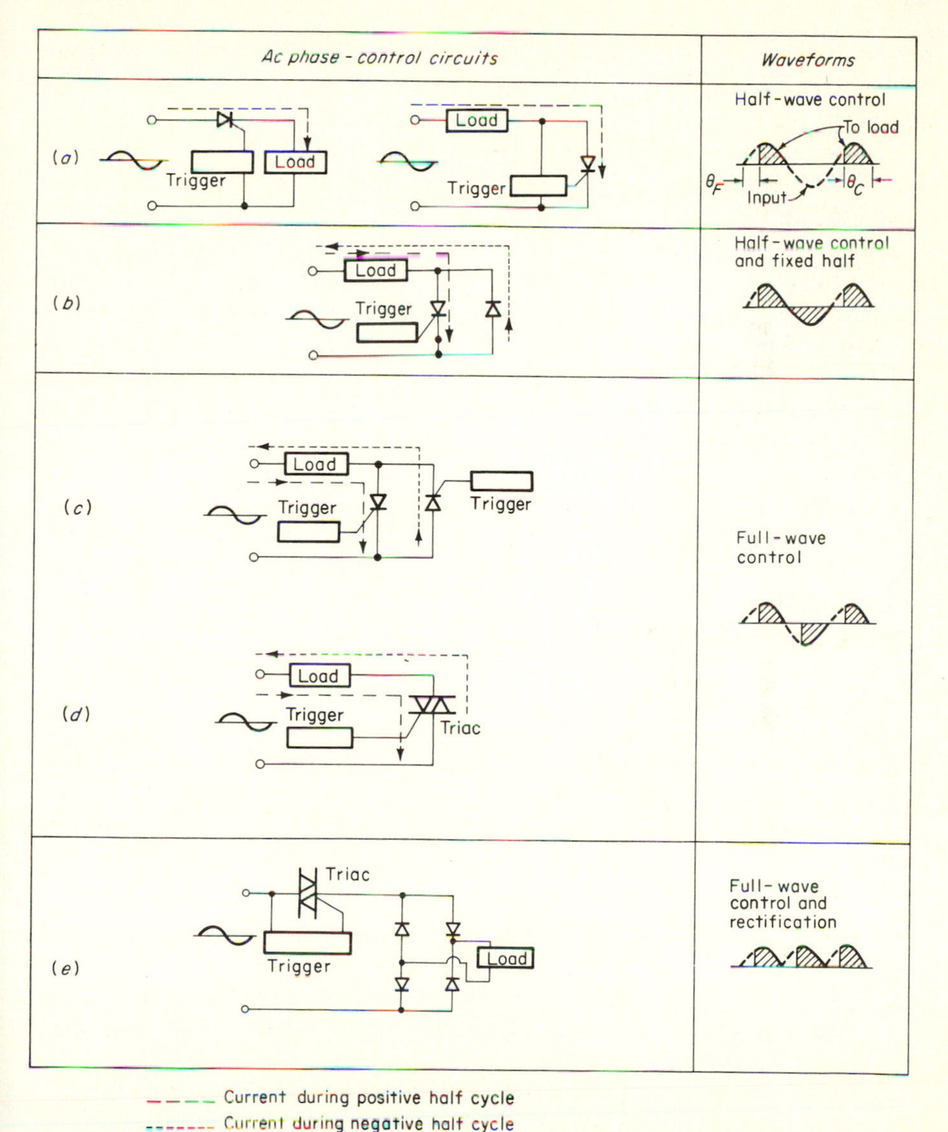

Fig. 8-17 Various types of thyristor phase-control circuits and associated waveforms.

The latest solid-state ignition system is the SCR capacitive-discharge system. In a capacitive-discharge system the storage element is a capacitor, as shown in the simplified schematic of Fig. 8-18. A transformer (or the same coil) is still used, but only as a pulse transformer and not for energy storage. A converter is needed to increase the battery voltage to a higher voltage for the storage element, and the points are still necessary, but only to key the SCR triggering device at very low current. (In some systems the points are replaced with a magnetic or optoelectronic system, making an all-electronic, contactless ignition system.)

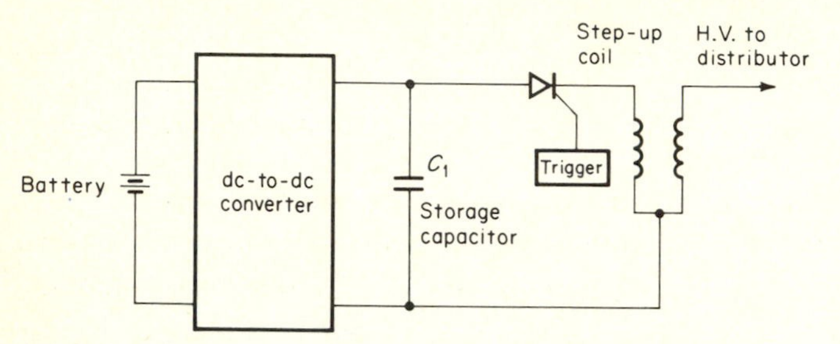

Fig. 8-18 SCR capacitive-discharge ignition-system block diagram.

The operational sequence of Fig. 8-18 involves the battery voltage being stepped up to ≈400 V by the transistor dc-to-dc converter. This voltage charges the capacitor, and at the appropriate time the SCR is turned ON, allowing C_1 to discharge through the transformer. The transformer steps up this voltage by the turns ratio, and the energy is fed to the appropriate spark plug by the distributor.

The main advantages of the capacitive-discharge system are

1. The recharging time constant is not determined by a coil but only by the time required to charge C_1. Hence high-speed operation is better.
2. Faster rise times are possible because this time is determined by the small leakage inductance of the coil rather than its magnetizing inductance as is the case when the coil is used as the storage element. (The shorter rise time significantly improves the system's ability to fire fouled plugs.)
3. The power actually increases as required at higher speeds, whereas just the opposite is true for the inductive-discharge system. The system is a demand type, storing power in the capacitor until needed.
4. The battery and generator-alternator drain are reduced, owing to the improved high-speed efficiency, the fact that input power can be terminated as soon as C_1 is charged, and the demand feature.
5. When the battery is low, the engine starts easily because full output voltage is still developed through the regulation of the converter.

The chief disadvantage of capacitive-discharge systems is the fact that they are more complicated and expensive, owing to the added cost of the converter and the low-leakage, high-voltage capacitor. Also, the resulting RFI generation often means that shielding of the spark-plug cables and the electronics package is necessary for AM radio reception.

Another interesting SCR capacitive-discharge applications involves igniters for gas appliances, such as ranges and clothes dryers. In such applications the SCR ignition system replaces pilot lights or lighting with a match; hence the name "electronic match." The circuit is similar to that of the SCR capacitive-discharge automobile ignition system.

Although most of the thyristor temperature-control applications (and the related zero-crossing triggering technique) involve the bidirectional triac, an interesting SCR heater-control application is the electronic clothes dryer dryness sensor of Fig. 8-19. In contrast to conventional dryers whose heat cycle is merely a function of a set time, this control actually "feels" or measures the amount of moisture in the clothing and terminates the heat cycle when a preset dryness level is reached. As shown in Fig. 8-19*a*, the resistance of the clothing is monitored within the tumbler when the clothing falls across the sensor. This measurement is brought out of the tumbler, and the clothing resistance becomes part of

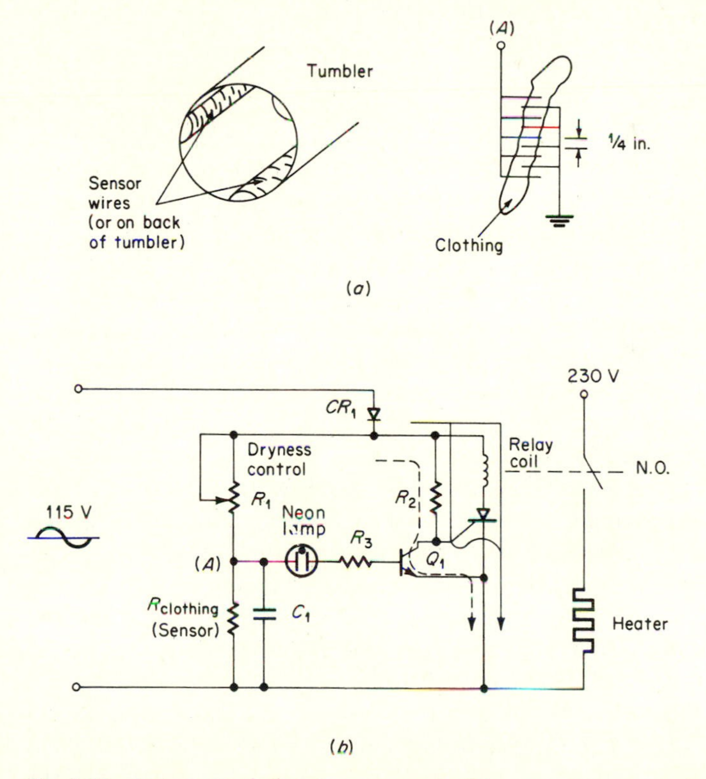

Fig. 8-19 Automatic clothes-dryness sensor: (*a*) sensor measures resistance of clothing which is related to dampness; (*b*) simplified schematic.

a voltage divider (R_1 and R_{clothing} of Fig. 8-19*b*). In normal operation current flows through R_2 and the SCR gate plus the relay coil, and the relay contacts are closed to apply power to the heater element. As the moisture content falls, the clothing resistance increases, C_1 charges for several positive half-cycles until the neon lamp ignites, and Q_1 is saturated. When Q_1 saturates, it acts as a closed switch, shorting the gate-to-cathode junction during positive half-cycles, thereby reducing V_{GK} to $< V_{GT}$ and thus turning the SCR OFF and removing the relay-coil current. At that point the relay contact reverts back to the normally open position, opening the heater circuit. The slow response is desirable to eliminate false triggering due to clothes-dryness variations, poor contact, etc.

As an example, let us calculate the approximate component values for Fig. 8-19. Assume a 60-V neon-lamp trigger voltage and a clothing resistance between elements of 200 kΩ when wet and 200 MΩ at the desired dryness. We wish to design for a 10-s response (integration time) assuming no C_1 discharging during negative half cycles and a total of 10 elements in the sensor. Ten parallel 200-MΩ resistors are equivalent to 20 MΩ, and at 60 Hz the number of positive half cycles in 10 s is 10/0.0167 = 600. Therefore V_{C1} must charge to 60 V in 600 cycles, or 0.1 V/cycle. For a point A to equal to 60 V,

$$163\ V_{\text{peak}} \frac{R_{\text{cloth}}}{R_{\text{cloth}} + R_1} \geq 60\ \text{V}$$

or $R_{1,\max} = 34.3$ MΩ at an R_{cloth} of 20 MΩ. For a 10-s response $(R_1 \| R_{\text{cloth}})C_1$ must equal 10 s, and solving for C_1 gives 0.8 μF. However, charging occurs only during one-half of the cycle, so C_1 must actually equal 0.4 μF.

The incandescent lamp flasher is another example of bimetallic-strip replacement. The SCR offers greater reliability than the electromechanical equivalent. Compared to the power transistor, the SCR provides lower power consumption and better capability of handling the high lamp inrush currents. Typical lamp-flasher applications include: dc-operated automobile turn indicators, remote warning lights for barricades and traffic signals, and line-operated flashers for aircraft beacons and advertisement signs.

A UJT-SCR combination is ideal for lamp flashers in that the SCR is well suited for dc power switching, and the UJT relaxation oscillator (Sec. 9-5-4) provides the necessary timing or chain of triggering pulses. Figure 8-20 is an example of a UJT-SCR lamp flasher. This circuit is used in automobiles to provide sequential turn signals. That is, assuming the three lamps are on the right fender, the normal light sequence would be: inboard lamp ON, inboard and center lamps ON, and finally all three ON, followed by extinction of all three and a repeat of the sequence.

The operation of the circuit is as follows. When the turn-signal switch is closed, the inboard lamp lights immediately. At the same time the first UJT emitter starts charging toward the supply voltage at a rate determined by R_1C_1 (Sec. 9-5-4). When the UJT emitter firing point (V_p) is reached, C_1 discharges through R_2, providing a positive gate trigger pulse to the gate of SCR_1.

The R_1C_1 time constant is $\approx$0.5 s, so SCR_1 and the center lamp will trigger 0.5 s after the inboard lamp has turned ON. At the same time that SCR_1 provides power to the center lamp, it also starts the second UJT timer. The R_3C_2 time constant is the same as before, so SCR_2 and the outboard lamp are turned ON after an additional 0.5-s delay. The other two lamps have remained ON so at this time all three are ON. They are all extinguished when the thermal cut-out switch becomes hot enough (from lamp current) to open, causing the sequence to restart. Hence, we see that the desired delay between lamp triggering is provided by the UJT relaxation oscillators, and the SCRs act as switches to power the lamps and

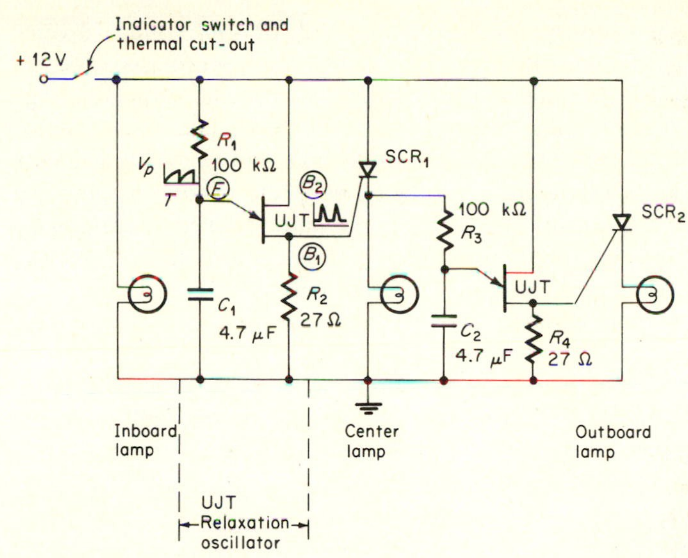

Fig. 8-20 SCR sequential flasher used for automobile turn signals.

initiate each timing cycle. Reference 5 gives a simple one-lamp UJT-SCR lamp flasher used for barricade flashers. This circuit uses a UJT relaxation oscillator for timing plus capacity commutation (Sec. 8-2-3) to extinguish the lamp.

SCRs are also used in overcurrent and overvoltage applications when the slow-speed of conventional circuit breakers is inadequate (a few milliseconds). Figure 8-21 shows a typical SCR overvoltage protection circuit. Under normal supply-voltage conditions, Q_3 and Q_4 are biased ON with the

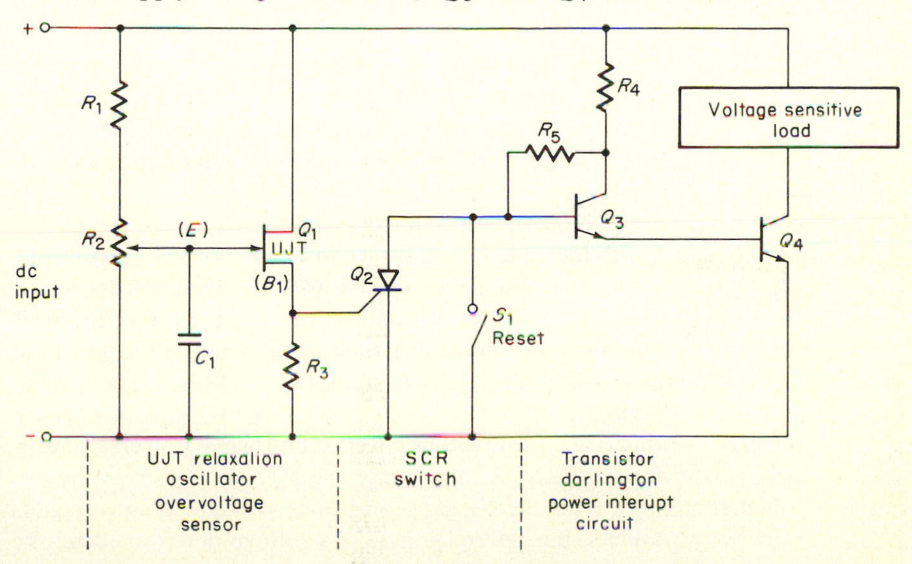

Fig. 8-21 SCR overvoltage-protection circuit (electronic circuit breaker).

saturated Q_4 providing a return path for the load. The R_1-R_2 voltage divider is adjusted so that the UJT's emitter voltage E is less than its firing voltage V_p, and hence Q_1 and Q_2 are OFF under normal conditions. If the dc supply voltage increases to an overvoltage condition, $V_E \geq V_p$ and the UJT relaxation oscillator is started (Sec. 9-5-4). C_1 discharges through the UJT and provides a trigger pulse to the SCR's gate. With the SCR ON, the Q_3 base bias is reduced below the level required to keep it ON; hence the Q_4 base connection is opened, turning Q_4 OFF. When Q_4 switches OFF, the power return path is opened, thus protecting the load from the overvoltage. The circuit is reset by momentarily closing S_1 to reduce the SCR's anode-to-cathode voltage below that required to maintain conduction.

Such protection circuits will open the supply line in a few microseconds, compared to a few milliseconds for conventional electromechanical circuit breakers. The same technique can also be utilized to provide overcurrent protection (Ref. 6).

8-2-6 SCR MOTOR SPEED CONTROL

Thyristors are being applied to motor speed control in a variety of household and shop appliances. These circuits provide a range of constant motor speed under changing load conditions for hand drills, grinders, saber saws, mixers, etc. Often the tool trigger is a potentiometer which provides a continuous adjustment of motor speed, and at each setting the speed is regulated by a simple feedback circuit. The popularity of such speed-control circuits stems from the convenience of ac power, the simplicity and low cost of SCR motor speed-control circuits, and the desire for constant speed in many applications.

The type of motor most often found in these appliances is a *universal* (ac or dc) *motor* of very low horsepower but high torque. These motors were developed long before the speed-control circuits, so the circuits were obviously "molded" to fit the existing motors. Luckily, the concept of controlling the average power (Sec. 8-1) by phase control (Sec. 8-2-5) works well with universal motors. That is, they will operate with portions of a full or half sine wave of applied voltage.

The need for universal motor speed control is evidenced by the typical reduction of motor speed with increasing load shown in Fig. 8-22*a*. As an example, recall the decrease in speed experienced as a hand drill is pushed harder into a board. In order to control the applied power and thus the motor speed, an SCR phase-control circuit requires some sort of feedback signal which is proportional to speed. Such a characteristic is the *back electromotive force* (bemf) developed by the motor (Fig. 8-22*b*).

Figure 8-22*c* demonstrates a simple but effective half-wave universal motor speed control. The firing of the SCR, and hence the average power applied to the motor, is determined by the SCR's gate-to-cathode voltage. This voltage is a function of two variables—the reference gate (G) voltage determined by the R_2 pot setting, and the cathode (K) voltage, which is equal to the back emf and is proportional to the motor speed. Whenever V_{GK} exceeds the SCR gate trigger voltage ($V_{GT} \approx +1$ V) the SCR will conduct and apply power to the motor. This of course occurs only during positive line half cycles, when V_{AK} is positive and CR_1 and CR_2 are forward-biased. R_2 provides the speed adjustment causing the SCR to fire sooner and thus transfer a greater portion of the line voltage to the motor as V_G is increased (Fig. 8-22*c*).

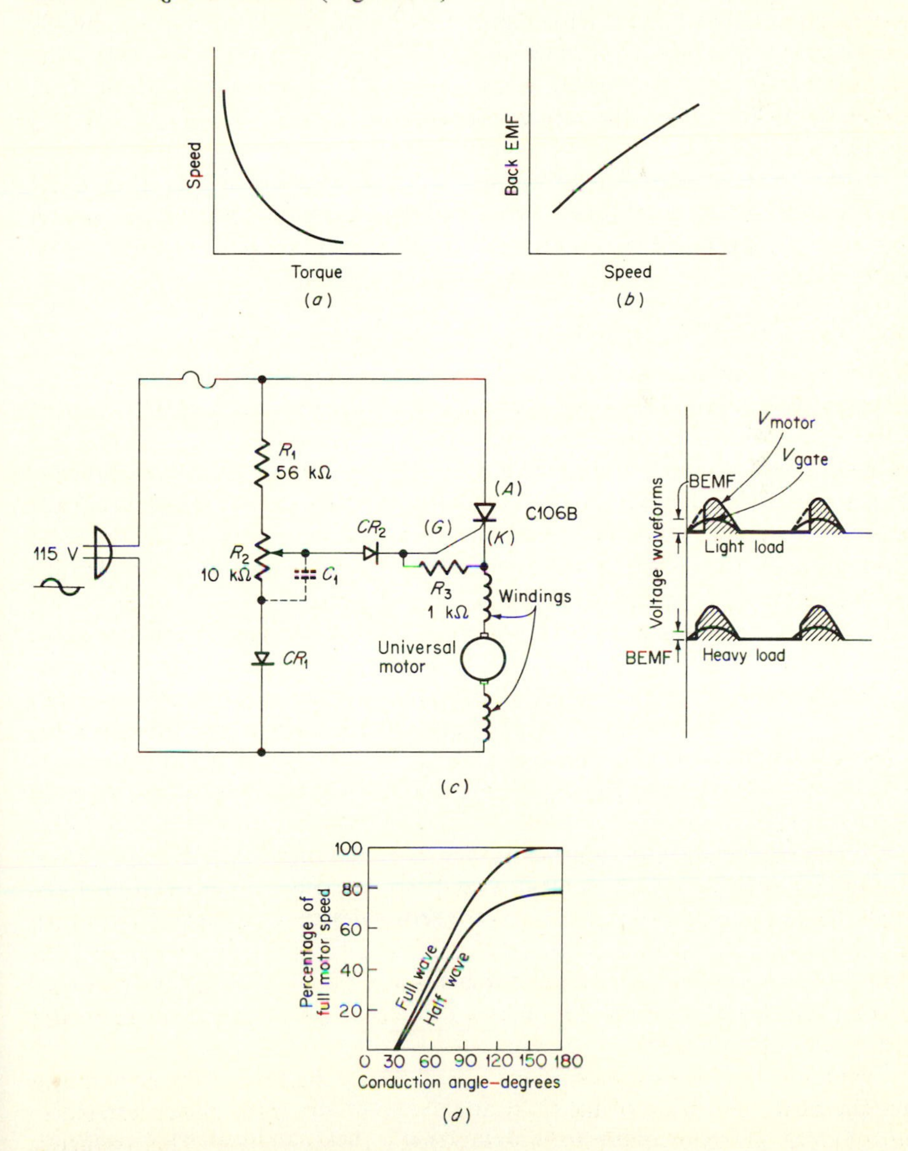

Fig. 8-22 Universal-motor speed control: (*a*) speed versus torque; (*b*) back emf versus speed; (*c*) simple half-wave speed-control circuit and waveforms; (*d*) comparison of full- and half-wave control of universal motors. *RCA*.

As an example of the feedback action, if the load is increased, the motor speed decreases, causing the back emf to decrease. This will cause the SCR to fire sooner in the cycle, thus applying more voltage to the motor armature to increase its speed. It is interesting to note that, owing to the dc characteristics of a universal motor, half-wave operation offers nearly as much speed control as full wave (Fig. 8-22*d*). In fact, full speed can be obtained by merely providing a path around the SCR in order to apply full line voltage to the motor. With a purely resistive gate circuit, the gate voltage of Fig. 8-22*c* is always in phase with V_{AK}, and firing angles are limited to less than 90°. The addition of a capacitor around R_2 can provide a lagging gate drive and thus delay triggering past the 90° point. There are many other possible improvements to the basic scheme of Fig. 8-22*c*, and the reader is referred to Refs. 7 and 8 for such details.

The speed of a dc motor is varied by altering its armature voltage. This used to be done by inserting different amounts of resistance in series with the motor. However, this method is not only inefficient due to the wasted power in the series resistance, but also requires a large variable power resistor. The speed-torque curve of the series dc motor is like that of Fig. 8-22*a*, and its speed-control circuit is very similar to that of Fig. 8-22*c* except that a full-wave bridge rectifier is required.

The control of an ac motor is more complicated. Voltage control is often unsatisfactory, since ac motors are more sensitive to frequency than voltage, and a wide-range variable-frequency control is rather expensive. However, triacs are often used in induction, reversing, and furnace-blower motors. These applications are covered in Sec. 8-3-4.

8-2-7 SCR LOGIC CIRCUITS

We have already discussed a few SCR circuits which fall in the general logic category; including the SCR battery charger (Fig. 8-15), motor speed control (Fig. 8-22), latching relay equivalent (Fig. 8-3), and protection circuits (Fig. 8-21). Some of these applications involve static switching, and others provide continuous control of average power. In both cases advantage was taken of the SCR's gate triggering characteristics, either by comparing a reference gate voltage with the cathode (load) voltage or vice versa. The other thyristors frequently used in logic circuits are threshold and triggering devices such as the four-layer diode, GTO, SCS, and SUS (Fig. 8-2).

A series of light-activated SCR circuits is also possible using simple photoresistors in place of fixed resistors in the gate circuit, or by utilizing light-activated SCRs directly (LASCR of Fig. 8-2). Chapter 7 includes many of these applications.

The SCR is ideal for alarm-type circuits, especially where simplicity is desired. Typical of such circuits are burglar, smoke, heat, and water-level alarms; proximity detectors; power-failure indicators; and emergency lighting. Figure 8-23 is an example of such a circuit. The sensor for such applications can be any

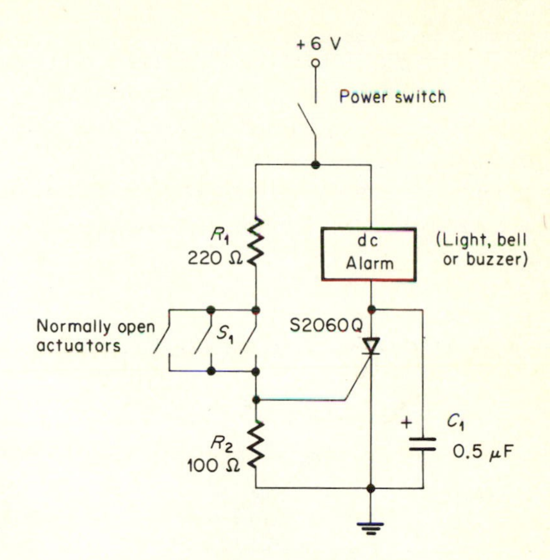

Fig. 8-23 SCR alarm circuit.

electrical or mechanical device, such as a normally open microswitch that closes when the desired phenomenon occurs. This action triggers the SCR, which in turn applies power to a series warning device. Once ON the SCR latches, and the power switch must be opened to stop the alarm. Such latching opration is desirable in applications such as intrusion alarms. For example, the alarm should continue even after an intruder recloses the door containing the activating switch. The presence of such a power switch generally requires some form of dv/dt suppression (C_1) to ensure that the SCR does not spontaneously trigger when the power switch is initially activated. Also, if the actuators are a considerable distance from the circuit, filtering may be required to eliminate false triggering due to wiring-noise pickup. The circuit of Fig. 8-23 has the advantage of no standby power drain because S_1 and the SCR are normally open.

As an example, let us check the component values of Fig. 8-23. Figure 8-7 gives the specifications for this particular sensitive-gate SCR. Figure 8-7*a* shows that the anode current and voltage ratings are 2.75 A and 25 V dc respectively, which should be adequate for most alarm loads. Checking the Fig. 8-7*c* and *d* worst-case (−40°C) gate trigger requirements shows that $I_{GT,\max} = 0.5$ mA and $V_{GT,\max} = 1.0$ V. The R_1-R_2 voltage divider of Fig. 8-23 would provide a V_G of 1.88 V and would not limit I_G to $<I_{GT}$. Therefore adequate triggering is provided even under worst-case conditions, and the low value of R_2 is appropriate for shunting the sensitive gate-to-cathode junction (Sec. 8-2-2). (The gate power rating has a safety factor of >6.0.) To check the adequacy of C_1 to suppress dv/dt effects, recall that the $R_L C_1$ time constant must be such that the rate of rise of V_{AK} does not exceed the dv/dt rating of 5 V/μs in this case. Hence in one time constant $V_{AK,\max} = (0.63)(6\text{ V}) = 3.78$ V or

$$\frac{3.78\text{ V}}{RC} = \frac{5\text{ V}}{1\ \mu\text{s}}$$

Therefore, $RC \geq 0.756\ \mu s$. Even if the alarm resistance were only 2.0 Ω, $R_L C_1$ would be 1.0 μs, and the rate of rise of V_{AK} would be sufficiently slowed.

SCRs are often used in power logic circuits such as multivibrators, ring counters, time delays, lamp flashers (Fig. 8-20), and Nixie drivers. The SCR logic circuits perform the same function as the bipolar and MOS logic of Chap. 5 but are more applicable to higher-power loads.

8-3 THE TRIAC (BIDIRECTIONAL TRIODE THYRISTOR)

The triac is an excellent example of the trend to more closely match the thyristor to the general ac power system. The *triac* is equivalent to (and often replaces) an inverse parallel connection of two unidirectional SCRs (Fig. 8-17*b* and *c*) and as such is capable of conducting (or blocking) with either polarity of applied primary voltage. The triac is a three-terminal thyristor which can also be triggered ON with *either* polarity of gate signal. Owing to its versatility and increasing power-handling capability, the triac has become a very popular semiconductor switch for full-wave control applications such as ac and dc motor speed control (and starting), light dimming, ac static switching, and heater control. Except for the differences in directionality, there are many similarities between the SCR and the triac, including most nomenclature, triggering methods, applications, first-quadrant characteristics, ratings, and manufacturing techniques. Because of these similarities, many of the background details of Sec. 8-2 apply to the triac, and the reader is referred to that section. It is impossible to include all of the triac applications here, so we shall concentrate on the most important and reference others. Although abbreviated, the sequence will be essentially the same as that of the SCR section.

8-3-1 TRIAC THEORY OF OPERATION

Figure 8-24 gives the triac symbol, cross section, and principal current/voltage characteristics. The triac symbol of Fig. 8-24*a* depicts the bilateral current flow with two opposing arrows. The main terminals are labeled MT_2 and MT_1 since the conventional anode and cathode terminology is useless with such bidirectional characteristics.

A glance at the triac structure of Fig. 8-24*b* shows that it is more complicated than the SCR, and the use of the shorted-emitter technique of Sec. 8-2-1 is also apparent. All four of the possible triac operating modes are also demonstrated by Fig. 8-24*b*. Notice the separate paths available for the two different main-terminal current directions. The top half of Fig. 8-24*b* demonstrates the possibility of negative *or* positive gate triggering when MT_2 is positive. (We shall always use MT_1 as a reference for MT_2 and gate signals.) With positive gate

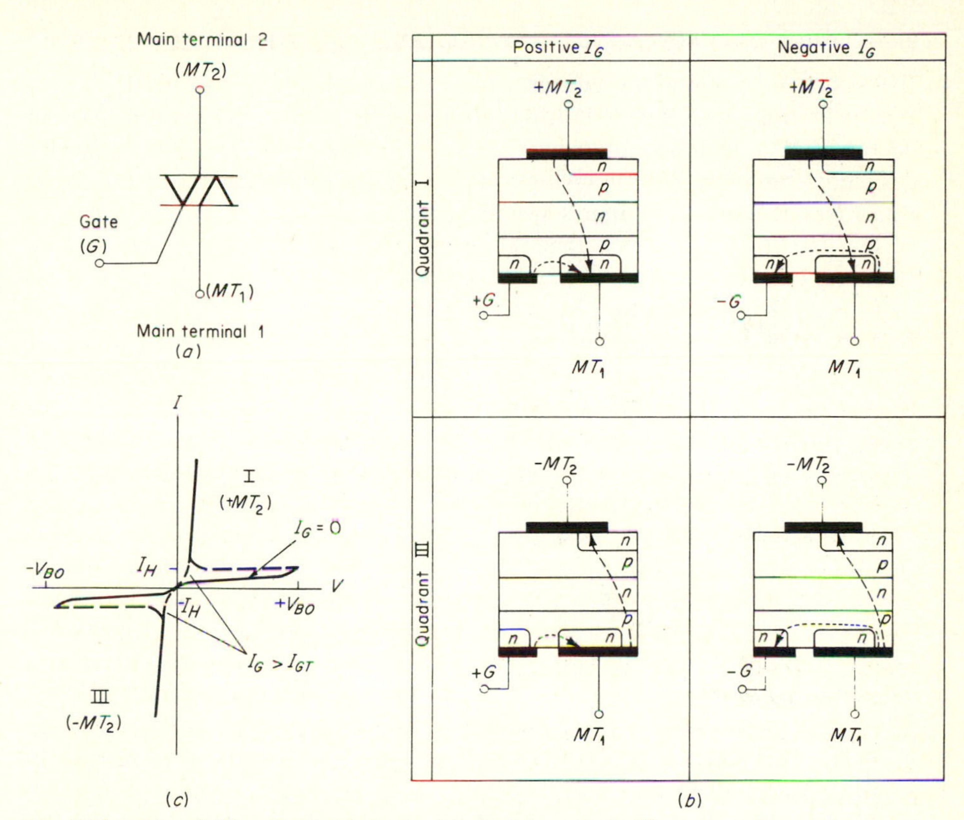

Fig. 8-24 Triac (bidirectional triode thyristor): (a) symbol; (b) cross section and current flow under various operating conditions; (c) principal current/voltage characteristics.

voltage, we see gate current flowing from G to MT_1 via the indicated forward-biased *pn* junction. Although we do not show a two-transistor equivalent circuit for the triac, it possesses the same regenerative capability as the SCR, and the initiation of a small gate current can start the multiplication process and cause the triac to conduct heavily from MT_2 to MT_1.

Negative gate signals can also trigger the triac, the only difference being the path of gate-current flow and some differences in the required level of I_G due to nonsymmetry and effects of principal current.

As shown by Fig. 8-24*c*, the first-quadrant operation is the same as that of an SCR (except for the negative gate-triggering capability). That is, the device can revert from the nonconducting or OFF state to the ON state by exceeding the breakover voltage or by gate triggering. Once ON, the gate also loses control until the principal current is reduced below the holding-current level. Typical triggering levels and ratings are also similar to that of an SCR, i.e., a few tens of milliamperes and 1 or 2 V of gate trigger signal; 1 or 2 V of ON voltage V_T and a wide range of blocking voltages up to several hundred volts. Again we shall consider the triac as a switch just as in the case of the SCR.

8-3-2 TRIAC CHARACTERISTICS, RATINGS, AND NOMENCLATURE

Excerpts from a typical low-current triac are included in Fig. 8-25. Many of the triac definitions, symbols, and ratings are the same as those of the SCR. (Nearly all the SCR terminology of Table 8-1 is applicable to the triac; however, the definitions are more general, owing to the increased number of modes of operation.) Notice that all the triac's specification are for a maximum frequency of 60 Hz. The reason for this is that the SCR has an entire negative half cycle to turn OFF, but the bidirectional triac conducts during both positive and negative half cycles and therefore may have to turn OFF in a very brief interval when the principal voltage passes through zero. This problem will be discussed further in Sec. 8-3-3. The roman-numeral notations with polarity symbols of Fig. 8-25*b* refer to the primary operating quadrant and polarity of gate voltage; i.e., III^- means third-quadrant operation with negative gate triggering.

There are many possible triggering modes and current-flow directions for the triac, and different triggering levels exist for each mode (Fig. 8-25*b*). Some of the triac maximum-rating definitions are new, such as peak gate trigger current I_{GTM}. One must be careful not to confuse the *peak* gate quantities with the lower-value gate trigger levels. The first refers to the maximum allowable gate dissipation, and the latter to the minimum level required for triggering (Fig. 8-25*a* and *b*). As in the case of the SCR, typical triac data include graphs relating maximum temperature to ON-state current, plus parameter variations with temperature. (For the sake of brevity Fig. 8-25 does not include all the triac data; however temperature affects most of the triac parameters in the same manner as it affects the SCR).

Figure 8-26 shows a simple full-wave triac circuit, including the load and the triac voltage waveforms. (The bidirectional operation is obvious when one compares this figure to the waveforms for the single SCR of Fig. 8-17*a*).

Figure 8-26 points out many of the considerations involved in developing the triac ratings of Fig. 8-25. The sinusoidal *full-wave* formulas must be used to develop average, rms, and peak triac current and voltage ratings. Hence we refer to the full-wave conversion graph of Fig. 8-5*c*.

As in all power semiconductor devices, temperature and the means of heat removal greatly influence the allowable operating power. The thermal equations [Eqs. (8-1) and (8-2)] apply here, and some form of heat sinking is obviously needed to operate these low-level triacs (Fig. 8-25*c*) at high ambient temperatures or currents.

8-3-3 TRIAC TRIGGERING AND TURN-OFF METHODS

Since the triac can be triggered by a low-power positive or negative gate signal, a wide variety of triggering devices and techniques can be utilized. Most of the triggering techniques discussed in Sec. 8-2-3 are applicable to the triac, including dc, ac, resistance, *RC*, and the capacitance-discharge pulse techniques using a neon lamp, UJT, pulse transformer, and thyristor trigger devices. Of particular interest are the triggering devices which exhibit bilateral characteristics. This is

		T2800B T2802B	T2800C T2802C	T2800D T2802D	T2800E T2802E	T2800M T2802M	
REPETITIVE PEAK OFF-STATE VOLTAGE:							
Gate open, $T_J = -65$ to $110°C$	V_{DROM}	200	300	400	500	600	V
RMS ON-STATE CURRENT (Conduction angle = 360°):	$I_{T(RMS)}$						
Case temperature							
$T_C = 80°C$		200	300	8	500	600	A
PEAK SURGE (NONREPETITIVE) ON-STATE CURRENT:	I_{TSM}						
For one cycle of applied principal voltage							
60 Hz (sinusoidal), $T_C = 80°C$		200	300	100	500	600	A
PEAK GATE-TRIGGER CURRENT:							
For 1 μs max., ...	I_{GTM}	200	300	4	500	600	A
GATE POWER DISSIPATION:							
Peak (For 1 μs max., $I_{GTM} \leqslant 4$ A, ...	P_{GM}	200	300	16	500	600	W
AVERAGE	$P_{G(AV)}$	200	300	0.35	500	600	W
TEMPERATURE RANGE:							
Operating (Case)	T_C	200	300	−65 to 100	500	600	°C

(*a*)

CHARACTERISTIC	SYMBOL	LIMITS For All Types Except as Specified			UNITS
		MIN.	TYP.	MAX.	
Peak Off-State Current: Gate open, $T_J = 100°C$, V_{DROM} = Max. rated value	I_{DROM}	–	0.1	2	mA
Maximum On-Stage Voltage: For $i_T = 30$ A (peak), $T_C = 25°C$	v_{TM}	–	1.7	2	V
DC Holding Current: Gate open, Initial principal current = 150 mA (dc) $v_D = 12$ V, $T_C = 25°C$, T2800 series	I_{HO}	–	15	30	mA
T2802 series		–	20	60	
Critical Rate-of-Rise of Commutation Voltage: For $v_D = V_{DROM}$, $I_{T(RMS)} = 8$ A, commutating di/dt = 4.3A/ms, gate unenergized, $T_C = 80°C$	dv/dt	4	10	–	V/μs
Critical Rate-of-Rise of Off-Stage Voltage: For $v_D = V_{DROM}$, exponential voltage rise, gate open, $T_C = 100°C$: T2800B, T2802B	dv/dt	100	300	–	V/μs
DC Gate-Trigger Current: For $v_D = 12$ V (dc), $R_L = 12\ \Omega$, $T_C = 25°C$ — Mode I⁺, V_{MT2} positive, V_G positive . .T2800 series.	I_{GT}	–	10	25	mA
. .T2802 series.			25	50	
Mode III⁺ negative negative .T2800 series.		–	15	25	
. .T2802 series.		–	25	50	
Mode I⁻ positive negative .T2800 series only		–	20	60	
Mode III⁻ negative positive .T2800 series only		–	30	60	
DC Gate-Trigger Voltage: For $v_D = 12$ V (dc), $R_L = 12\ \Omega$, $T_C = 25°C$	V_{GT}	–	1.25	2.5	V
For $v_D = V_{DROM}$, $R_L = 125\ \Omega$, $T_C = 100°C$		0.2	–	–	
Thermal Resistance: Junction-to-Case	$R_{\theta JC}$	–	–	2.2	°C/W
Junction-to-Ambient	$R_{\theta JA}$	–	–	60	

(*b*)

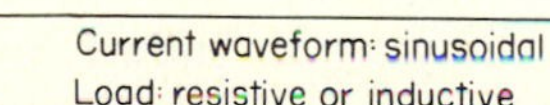
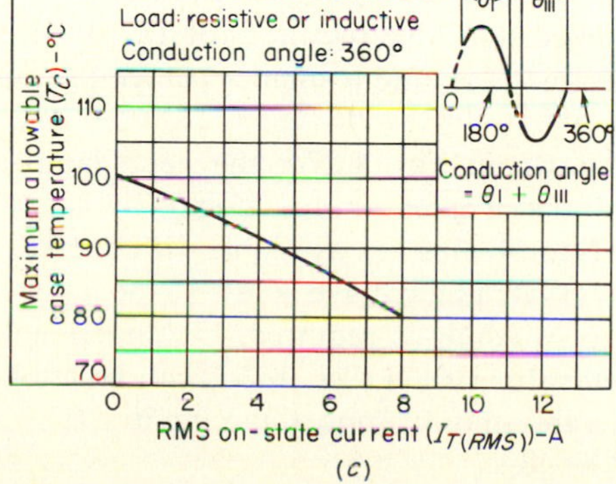

(*c*)

Fig. 8-25 Low-current triac data: (*a*) maximum ratings; (*b*) characteristics; (*c*) maximum allowable case temperature versus ON-state current. *RCA.*

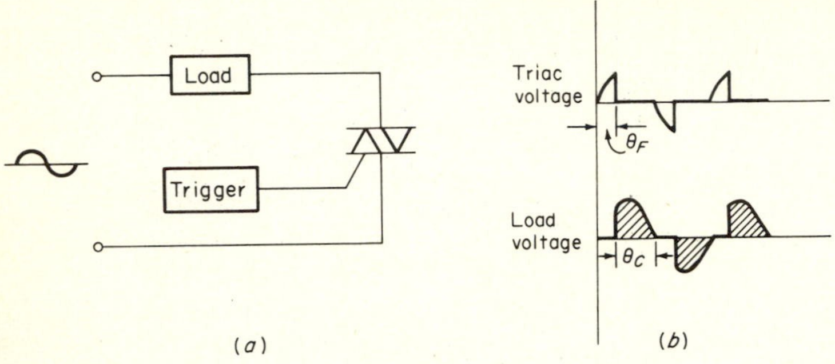

Fig. 8-26 Simple triac circuit: (*a*) circuit; (*b*) waveforms.

important when the triggering-device signal is derived from the ac line, and it is desirable for the trigger device to break down or conduct at some specified *positive and negative* input voltage. Such a device would thus provide the required triggering pulses to the triac gate during both half cycles of the ac line voltage. There are three devices ideally suited for such bidirectional triggering requirements—the neon lamp, the diac, and the silicon bilateral switch thyristor of Fig. 8-2. The latter two have not yet been discussed in detail, so we shall elaborate here.

Figure 8-27*a* shows a diac triggering circuit which is similar to the SCR circuit of Fig. 8-12. The diac current/voltage characteristics are included in Fig. 8-27*c*. This figure shows that the *diac* (or symmetrical trigger diode) breaks over in the forward or reverse direction at $\approx$32 V with fair symmetry. It also exhibits a slight negative-resistance characteristic and low triggering-current requirements. Therefore the diac of Fig. 8-27*a* will break over when C_1 charges to $\approx \pm 32$ V, that being a reasonable level for 115- or 230-V line operation. At negative or positive breakover, C_1 partially discharges through the diac, providing a negative or positive trigger pulse to the triac gate, either of which will trigger the triac. The line voltage is thus transferred to the load for the remainder of the particular half-cycle, as shown by the waveforms of Fig. 8-27*a*. If R_1 is increased, the time required for C_1 to charge to the diac breakover voltage will also increase, and a smaller percentage of the line voltage will be applied to the load.

Although not shown in Fig. 8-27*a*, the neon lamp and silicon bilateral switch (SBS of Sec. 8-4) also exhibit bidirectional characteristics and can be used for triac triggering. Neon lamps are available with breakdown voltages from 50 to 100 V, and the SBS switching voltage is $\approx$8 V, the latter being preferred when precision low-voltage switching is required. Unidirectional devices such as the UJT and the two-transistor switch (Sec. 8-2-3) can be used to trigger triacs, but they require considerably more supporting circuitry than the previously mentioned bidirectional devices.

Commutation of triacs must occur in the rather short time interval when their current passes through zero. This problem is apparent in the 4- to 10-V/μs

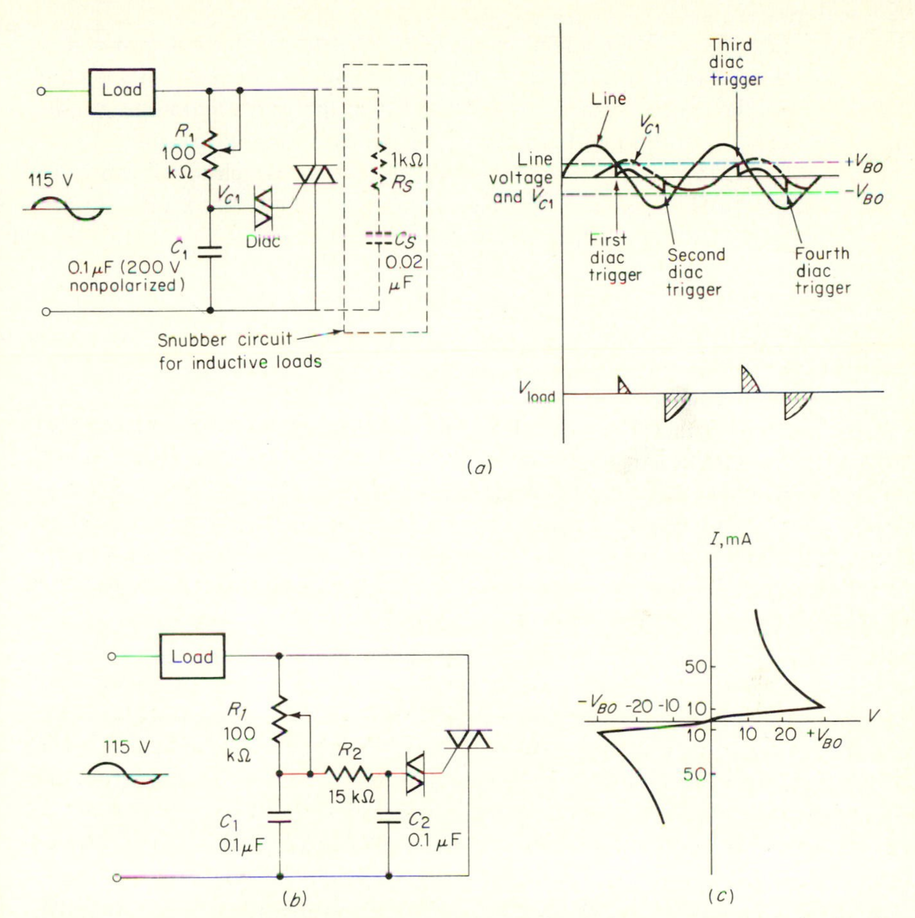

Fig. 8-27 (*a*) Diac-triggered triac phase-control-circuit, single time constant (a fixed resistor is generally added in series with R_1); (*b*) double-time-constant trigger circuit to reduce hysteresis; (*c*) diac current/voltage characteristics.

"critical rate of OFF-state commutation voltage" specification of Fig. 8-25*b*. The problem with triac turn-OFF is that the device can conduct current in either direction; therefore a rapid reversal of applied voltage causes a recovery current which may reinitiate turn-ON in the opposite direction. To eliminate this problem, the triac principal current must be reduced below I_H, and the principal voltage must *not* be reapplied for a time interval roughly equivalent to the rate of change of voltage that exists near the zero-current crossing of 60-Hz ac power. (Four-hundred-hertz triacs are also available.) This amount of time is sufficient for the stored carriers to recombine, so that the triac is ready to conduct again.

With an inductive load the voltage across the inductance can change instantaneously, and the voltage across the triac will rise immediately after its current falls to zero. The rate of rise of dv/dt is limited only by the triac capacitance, since the current in the inductive load is also zero. In such cases an *RC* "snubber" network must be added in parallel with the triac, as shown in Fig. 8-27*a*, to ensure reliable turn-OFF. The analysis of such snubber circuits is complicated; hence the reader is referred to Ref. 9 for simplified design nomographs. The triac heater application of Fig. 8-29 shows a thyristor triggering technique, generally referred to as *zero-voltage-crossing triggering*, which significantly reduces the thyristor RFI generation because the triggering is done very close to the zero-voltage crossing of the ac line voltage.

8-3-4 TRIAC CIRCUITS

In this section we discuss a few of the many triac applications, with emphasis on those which are greatly simplified or better performed by the triac than the SCR. The various triac applications fall in roughly the same categories as those of the SCR; however, they are all combined within this subsection. The triac can significantly reduce the number of components required in full-wave applications, as shown in Fig. 8-17. The reduction in parts is possible because of the triac's ability to conduct with either polarity of primary voltage or gate trigger signal.

The basic diac-triggered triac phase-control circuit of Fig. 8-27*a* is employed in several applications, including lamp dimming, motor speed control, and heater control. The only changes necessary are the addition of a fixed protection resistor in series with potentiometer R_1, dv/dt suppression for inductive loads (dashed), RFI suppression (Figs. 8-10 and 8-29), and a mechanical main power switch. Because of the wide usage of this type of triac circuit, we shall discuss its general characteristics and then briefly mention the alterations required for individual applications.

The capacitor-voltage waveform of Fig. 8-27*a* demonstrates the undesirable effect of the diac on capacitor voltage. That is, at the first diac triggering the capacitor voltage decreases to the level of the diac forward voltage. Thus, C_1 starts to charge from a lower voltage than before, and the breakover voltage occurs sooner in the next half cycle. This decrease in V_{C1} causes an undesirable hysteresis effect; i.e., the R_1 setting required to "start" and "stop" the load will not be the same. In a lamp dimmer this would mean that the lamp turn-ON and extinguish settings would not coincide by as much as 40 percent, thus limiting the actual range of control to 60 percent of the potentiometer range.

The addition of a second *RC* phase-shift network as shown in Fig. 8-27*b* significantly reduces hysteresis and increases range. This improved performance results from the partial restoration of the charge on C_2 from C_1 after the diac has triggered. This allows the triac to be triggered at lower conduction angles and reduces the effect of the diac characteristic on the charging capacitor. The double or single *RC* circuit of Fig. 8-27 is used to control the brightness of incandescent lamps. However, recall from Sec. 6-2-2 that the radiant output of an incandescent lamp is not a linear function of applied power, but rather a third- or fourth-power function. Hence, the illumination will change rather rapidly with the R_1 setting. One of the problems peculiar to lamp controls is the high starting or inrush current caused by the extremely low cold-filament resistance. This resistance is several times lower than the filament resistance at normal operating temperature; however, normal resistance is generally reached within a few hundred milliseconds, so abnormally high current will flow for only a couple of cycles.

The circuit of Fig. 8-27 is also used for full-wave speed control of ac and dc motors, including universal, series-wound, and shaded-pole or permanent split-phase induction motors.

In addition to controlling motor speed, triacs also perform motor static switching functions such as the control of reversing motors for garage door openers; replacement of centrifugal switches in induction-motor starters, and furnace blower-motor control (Ref. 10).

Figure 8-28 shows several triac motor-control circuits. A simplified schematic of an induction-motor speed control is shown in Fig. 8-28*a*. This circuit replaces three-speed motors in appliances such as washing machines and offers the additional advantage of continuous speed control. In this circuit, the triac is in series with the run winding and hence controls the power to this winding and the motor speed. The control and regulation of such a motor are often provided by a UJT pulse circuit (Sec. 9-5) with the UJT supply being derived from a small tachometer. The tachometer coil is built into the end of the motor and

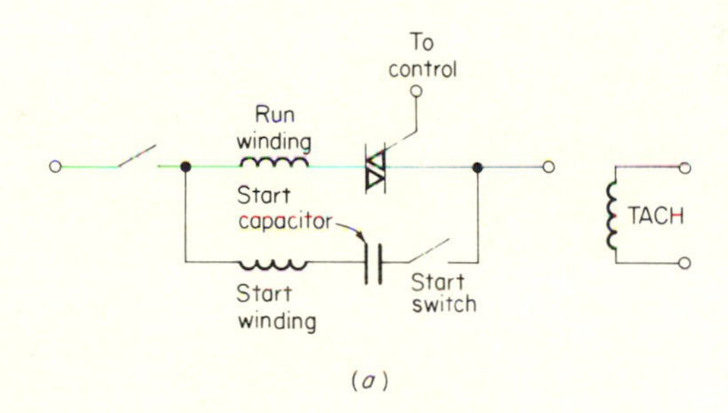

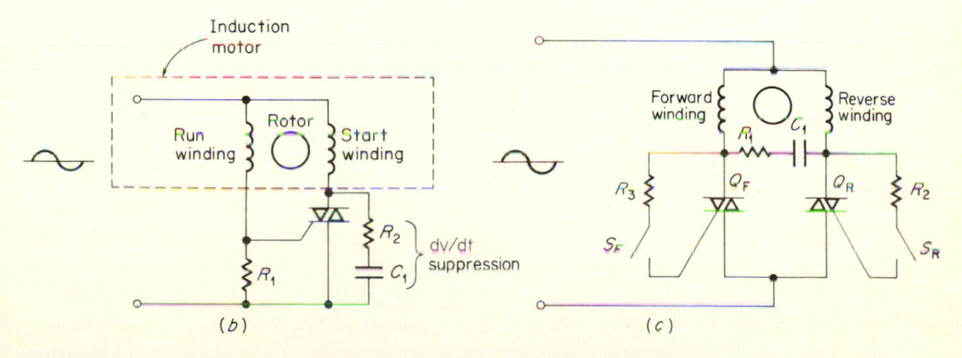

Fig. 8-28 Triac motor-control circuits: (*a*) induction-motor speed control; (*b*) triac replacement for centrifugal-switch induction-motor starter; (*c*) triac reversing motor control.

hence couples to the armature shaft. Its ac output is then proportional to the motor speed and is fed back to maintain the speed at the desired setting.

Another requirement of induction motors is some means of (1) energizing the start winding to bring the motor up to speed and (2) disconnecting it at that point. This task is often performed by an electromechanical device such as a centrifugal switch. In this case power is allowed to flow through the start winding during initial turn-ON and when the electromechanical unit senses that the motor has built up sufficient speed, it disconnects the start winding. (The centrifugal switch senses the increase in centrifugal force as the motor speed increases.) As shown in Fig. 8-28*b*, a triac can be used to replace the centrifugal starting switch. When ac power is applied, the main-winding inrush current is several times the normal running current and the R_1 voltage drop exceeds the triac gate trigger voltage (V_{GT}), thus turning the triac ON and energizing the start winding. The triac turns OFF as the motor and input current approach normal values and the voltage across R_1 falls below V_{GT}.

Another excellent triac application is the control of reversing motors, as portrayed by the simplified schematic of Fig. 8-28*c*. In such applications the triacs perform as static switches, energizing either the forward or reverse winding. The control switches can be any type of mechanical or solid-state switch triggered normally or remotely via a radio-frequency or light link. A limit switch is generally incorporated to stop the motor at the desired limits. The commutating capacitor reverses the triac main-terminal voltages similarly to the SCR circuits of Sec. 8-2-3. R_1 is required to limit the capacitor current in the event that one of the triacs is still conducting when the other is triggered.

The circuit of Fig. 8-27*a* will also control the power applied to a resistance heater if one of the gate-circuit resistances is a thermistor or some such temperature-sensitive component. However, since heaters may be very high power, the RFI would be excessive and an adequate suppression network would be quite large and expensive. For this reason a zero-voltage switching circuit as discussed in Sec. 8-2-3 is most frequently employed in heating applications. These circuits apply complete half- or full-wave cycles to the load, with switching taking place very near the zero-voltage crossing as shown in Fig. 8-29*a*. Since resistance heaters have thermal time constants much greater than the 60-Hz power system, control is not required during every half cycle. Hence complete cycles can be switched to the load with duty-cycle control of the heat, as shown in Fig. 8-29*a*.

The size of the ac power-control market and the popularity of the zero-voltage-crossing (ZVC) switching technique have led to the production of several integrated-circuit versions of this circuit. Figure 8-29*b* shows a block diagram of such an integrated circuit. The integrated ZVC circuit generally interfaces between a sensor element and a triac (or SCR). It provides a gate trigger pulse to the triac when the ac line voltage crosses zero and the input sensor senses the desired level of temperature, pressure, strain, light, etc. (Fig. 8-29*a*). The input sensor is generally an external resistance bridge, as shown in Fig. 8-29*b*, com-

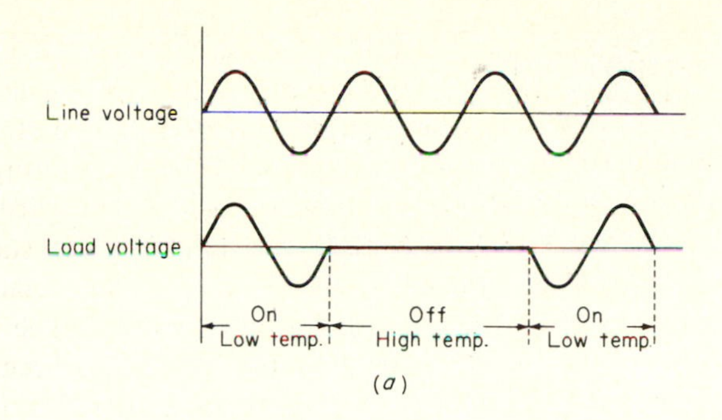

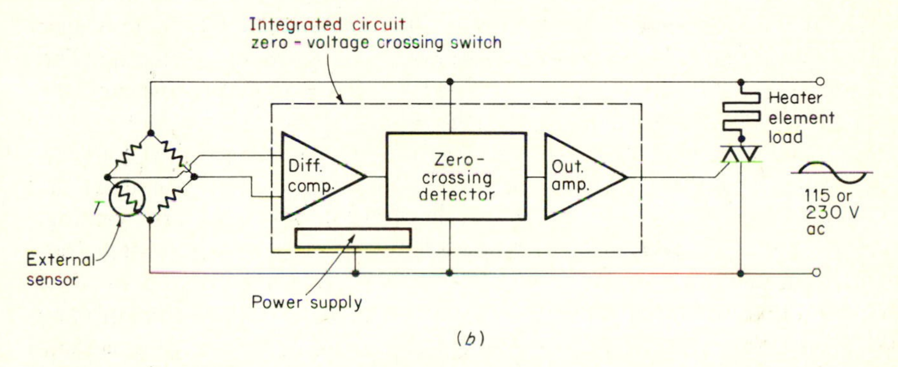

Fig. 8-29 Triac temperature control using an integrated-circuit zero-voltage-crossing trigger: (*a*) waveforms showing application of complete cycles of the line voltage to the load; (*b*) circuit.

posed of such devices as photoresistors (Sec. 7-1-2), resistance strain gauges, or thermisistors. However, undirectional devices such as diodes can also be used.

The ZVC integrated circuit typically contains

1. A power supply and regulator capable of operating from the ac line.
2. An input differential comparator (Sec. 4-3), which saturates whenever one input becomes more positive than the other.
3. A zero-voltage-crossing detector which produces a pulse precisely at the zero crossing. Another gate drives the output amplifier only when the ZVC-detector *and* differential-comparator outputs are high.
4. An output-pulse amplifier which provides the pulse to the triac gate. This pulse triggers the triac and thus connects the load to the line when the proper input is sensed *and* the line ac voltage crosses zero.

The advantage of such an integrated circuit is that one has only to add an appropriate sensor and adequately rated triac to design a variety of zero-crossing controls (Ref. 11).

8-4 OTHER THYRISTORS

We have discussed the SCR, the triac, and the low-power thyristors that are most frequently employed as triggering devices. The unidirectional thyristor trigger devices, including the four-layer diode and the SUS, were covered in the SCR triggering section (Sec. 8-2-3). The bidirectional devices, including the diac and SBS, were dealt with in Sec. 8-3-3.

Of all the thyristors discussed, none have been capable of gate-triggered *turn*-OFF as well as turn-ON. In the previously mentioned thyristors, turn-OFF could be accomplished only by reducing the anode signal to ≈ 0 or by adding a second thyristor or transistor in a commutation arrangement. However, there is one member of the thyristor family which can be turned ON and OFF at its gate terminal. This is the *gate-turn*-OFF SCR (*GTO*) thyristor of Fig. 8-2. The GTO is very similar to the SCR, except that it can be turned OFF by a negative gate signal (and turned ON by a positive gate signal). The application of an adequate negative signal to the GTO gate-to-cathode terminals diverts the principal current to the gate terminal.

The main GTO disadvantage is its inefficient turn-OFF. That is, the required gate turn-OFF voltage is 10 to 20 times its gate turn-ON voltage V_{GT}. Despite the GTO's disadvantages, it is still useful in applications where its bistable nature can be used to reduce the number of components in a given circuit, especially in dc applications where commutation is not automatic.

PROBLEMS

8-1 List two advantages of the thyristor, as compared to the bipolar transistor, for switching applications.

8-2 (*a*) Define the terms "unidirectional" and "bidirectional" as they apply to thyristors.

(*b*) List three similarities among all thyristors.

8-3 Briefly describe the operation of the SCR. Utilize its structure and equivalent circuit to aid in the explanation.

8-4 (*a*) Assuming 230-V half-wave line operation, a resistive load, and gate triggering, select an adequately rated SCR from Fig. 8-6*b*.

(*b*) If conduction were always limited to 20°, which lower-rating C35 series SCR would be adequate for the conditions of part *a*? (See Fig. 8-5*b*.)

8-5 Determine the maximum ambient temperature for a low-power SCR with a $T_{J,\max}$ of 110°C and θ_{JA} of 10°C/W operating at an average level of 0.1 A and 60 V.

8-6 (*a*) Explain the reason for the SCR di/dt and dv/dt limitations. Describe a simple solution to each.

(*b*) Calculate the value of a parallel capacitor to limit dv/dt to 10 V/μs. Assume a 100-V supply, a 12-Ω load resistance, and an exponential waveform. (See Fig. 8-9*a*.)

8-7 Using the SCR of Fig. 8-6*c* and *d*, calculate $P_{J,\max}$ for a room-temperature environment.

8-8 Using the data of Fig. 8-6*d*, find the maximum V_G and I_G to trigger the medium-current SCR at -65°C. Ensure that the maximum gate ratings are not exceeded.

8-9 Compare the room-temperature I_{GT} value for the medium-current and sensitive-gate SCR of Figs. 8-6 and 8-7, respectively.

8-10 Why is a gate-to-cathode resistance R_{GK} required for sensitive-gate SCRs?

8-11 Referring to Fig. 8-12, calculate the minimum and maximum conduction angles θ_c if R_1 is replaced with a fixed 1.0-kΩ resistance in series with a 0- to 10-kΩ potentiometer. Assume a 30-V triggering voltage, 115-V rms line operation, and $C_1 = 0.22\ \mu F$.

8-12 Design the circuit of Fig. 8-12 to obtain a range of conduction angles θ_C of $\approx 90°$ to 150°. Assume 60 Hz, 115-V rms (163-V peak) line operation, and a trigger-device firing voltage of 40 V.

8-13 What is the theoretical minimum conduction angle for a circuit such as Fig. 8-12?

8-14 Design an *LC* RFI filter with a 3-dB cutoff frequency at 50 kHz with a 10-Ω load.

8-15 Referring to Fig. 8-12, calculate the value of R_1 necessary to turn the SCR OFF ($\theta_C = 0°$). Assume the triggering-device firing voltage is 20 V. Operation is from a 115-V rms, 60-Hz line, and $C_1 = 1.0\ \mu F$.

8-16 What is the main advantage of low-voltage triggering devices compared with high-voltage versions such as the neon lamp?

8-17 Referring to the dc static switch of Fig. 8-14*b*, calculate the value of R_1 and C_1, assuming the S2060Y SCR of Fig. 8-7, a 100-Ω load, and a 6-V supply.

8-18 List two advantages of the capacitive-discharge ignition system, as compared to the conventional inductive-discharge system.

8-19 Choose a value for *R* and C of Fig. 8-13 to allow V_G to lead V_{AK} by $\approx 75°$. Assume $C = 1.0\ \mu F$.

8-20 Calculate the values of R_1 and C_1 in Fig. 8-19 for a 90-V neon, 1-min response and a clothing resistance between elements of 100 kΩ when the clothes are wet and 100 MΩ at the desired dryness. Assume 10 elements in the sensor and no C_1 discharge during negative half cycles. Remark on the values of R_2 and R_3 if the coil resistance is 100 Ω and $\beta_{Q1} = 100$, $I_{CQ1,\max} = 1.0$ A, and $I_{G,\max} = 150$ mA.

8-21 Which thyristor rating is in danger of being exceeded when a load is inductive, and what is the conventional limiting technique?

8-22 Design the sequential taillight flasher of Fig. 8-20 to produce the sequence of inboard lamp ON, inboard and center ON, and finally all three ON, with 0.1 s between steps. Assume the lamps operate at 12 V and 0.5 A; choose the SCRs from Fig. 8-7.

8-23 Explain the operation of an SCR universal-motor speed-control circuit.

8-24 Design an SCR alarm circuit similar to that of Fig. 8-23 to accommodate three *normally closed* switches and a 12-Ω, 6-V dc bell. Use an SCR from Fig. 8-7.

8-25 Briefly describe the operation of the triac. Refer to its symbol, cross section, characteristics, and waveforms in your explanation.

8-26 (*a*) List two advantages and two disadvantages of the triac as compared to the SCR.

(*b*) Name two bidirectional trigger devices applicable to triac triggering.

8-27 What is the maximum allowable rms primary current for the triac of Fig. 8-25, assuming an ac waveform and (*a*) 80°C, (*b*) 90°C, and (*c*) 100°C case temperature?

8-28 (*a*) Calculate the values of R_1 of Fig. 8-27 required to produce triggering at $\approx 90°$ (with respect to the line voltage) and to provide complete turn-OFF. Assume 230-V rms line operation, $C_1 = 0.1\ \mu F$, and a diac with a $V_{(BO)O}$ of ± 32 V.

(*b*) By how many degrees would θ_C increase if the trigger device were an 8-V SBS?

8-29 List four triac applications.

8-30 List the major difference between the SCR and each of the other thyristors of Fig. 8-2.
8-31 Referring to Fig. 8-21, calculate the maximum allowable R_3 assuming the sensitive-gate SCR of Fig. 8-7, a UJT interbase current (I_{BB} of Sec. 9-5) of 5 mA, and an SCR case temperature of +20°C.
8-32 Utilizing the low-current SCR of Fig. 8-7, design a simple and inexpensive burglar alarm to give an audio and visual indication of a door opening. Design for battery operation and low (or zero) standby power.

REFERENCES

1. Grafham, D.: *General Electric SCR Manual*, 5th ed., pp. 351–399, 1972.
2. Private communication with Dr. W. Kerwin, NASA/Ames, Moffett Field, Calif.
3. Grafham, D.: *loc. cit.*
4. Solid-State Power Circuits, "RCA Designer's Handbook," pp. 137–143, 1971.
5. Grafham, D.: *op. cit.*, p. 208.
6. Dale, R.: *Semiconductor Power Circuits Handbook*, pp. 4-17–4-19, Motorola, 1968.
7. Grafham, D.: *op. cit.*, pp. 287–298.
8. Yonushka, J.: "Application of RCA Silicon Controlled Rectifiers to the Control of Universal Motors," *RCA Appl. Note AN*-3469, Nov. 1973.
9. Wojslawowicz, J.: "Analysis and Design of Snubber Networks for dv/dt Suppression in Thyristor Circuits," *RCA Appl. Note AN*-4745, Oct. 1971.
10. Grafham, D.: *op. cit.*, pp. 304–305.
11. Sheng, A.: "Features and Applications of RCA Integrated-Circuit Zero-Voltage Switches," *RCA Appl. Note ICAN*-6182, n.d.

part 5 Miscellaneous Semiconductor Devices

chapter 9

Miscellaneous Semiconductor Devices

9-1 RECTIFIER DIODES

A rectifier is a device which will permit the passage of current in one direction but not in the other. It can therefore be used to convert ac to dc, as in power supplies, or to demodulate an amplitude-modulated RF carrier, as in a radio receiver. Most rectifiers are of the silicon *pn*-junction type and can operate at high temperatures, have high conductance and low leakage, and are unsurpassed in putting high-current-carrying capability into a small package.

Figure 9-1 reviews some of the more commonly used rectifier circuits, all of which are assumed to be fed from a sine-wave generator.

Figure 9-1*a* shows the simplest rectifier, which is a half-wave type because only half of the input waveform appears at the output. If a capacitor C is connected across R_L, the voltage across R_L will be a dc level with a small amount of ripple superimposed upon it. The amount of ripple will depend on the product of R_L and C.

If a center-tapped transformer is available, full-wave rectification may be achieved with the circuit in Fig. 9-1*b*. The center tap on the transformer secondary results in each diode conducting alternatively every other half cycle. Thus the load resistor sees a positive half wave for both positive and negative input cycles.

If a center-tapped transformer is not available, four diodes may be used in a bridge to achieve full-wave rectification, as shown in Fig. 9-1*c*.

A center-tapped transformer and a four-diode bridge may be used to obtain plus and minus rectification, as shown in Fig. 9-1*d*. This scheme is often used in obtaining dual dc voltage supplies (for operational amplifiers, for example).

Fourier analysis of the half-wave rectified waveform in Fig. 9-1*a* shows that its average dc level is E/π, where E is the peak value of the rectified half wave. It follows that for full-wave rectification the average dc level is $2E/\pi$.

In addition to their use as simple rectifiers, diodes are also employed as clamps, clippers, slicers, logic elements, and in many other applications. Figure 9-2 shows three such diode applications. Figure 9-2*a* is a dc restorer or diode clamp. Its function is to take an ac-coupled pulse train and clamp one side of it to some level, often ground. In Fig. 9-2*a* the input pulse train swings from $+10$ to $+15$ V. If the diode were omitted from the circuit, the output voltage would

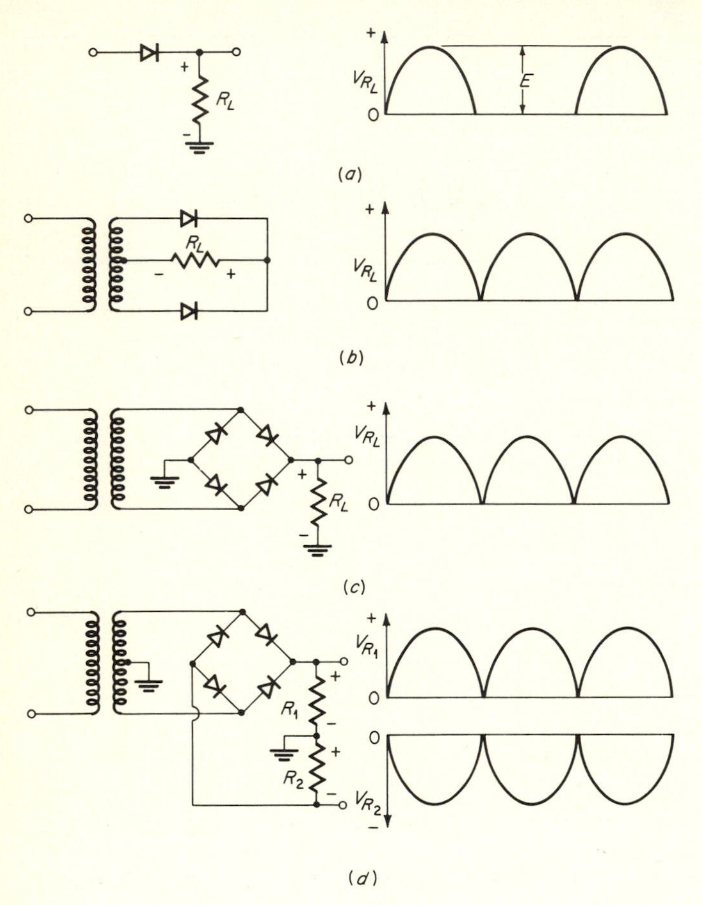

Fig. 9-1 Diode rectifier applications: (*a*) half-wave rectifier; (*b*) full-wave rectifier using tapped secondary winding and two diodes; (*c*) full-wave rectifier using untapped secondary winding and four diodes; (*d*) dual full-wave rectifier using tapped secondary winding and four diodes.

go plus and minus with an average dc level of zero. This is because capacitor C cannot pass any direct current. With D in the circuit the output can never go negative because the impedance of D is essentially a short circuit with the cathode biased negative. Also, when the input swings through 5 V (from +10 to +15 V) the output must also change by 5 V, since the voltage across capacitor C cannot change instantaneously. Thus the output amplitude is 5 V and its negative side is clamped to ground. This results in the output waveform shown in Fig. 9-2*a*. The function of R in the circuit is to stabilize its input resistance and to allow the circuit to recover rapidly from large transients. In a practical circuit the output would swing about 0.6 V negative since this amount of voltage is required before the diode reaches low impedance.

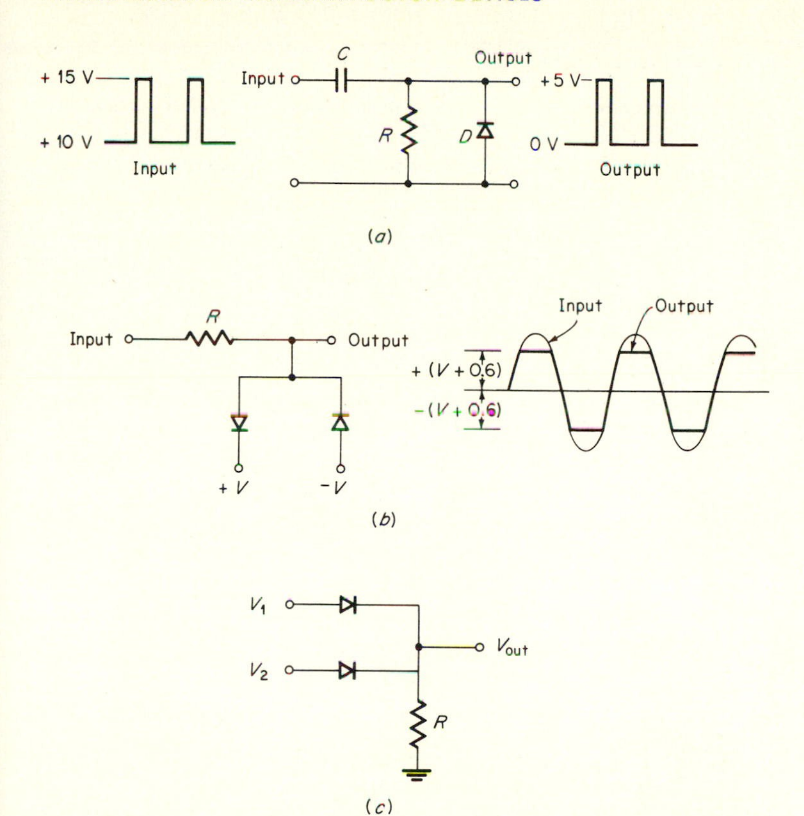

Fig. 9-2 Miscellaneous diode applications: (*a*) dc restorer; (*b*) slicer; (*c*) logic circuit.

Figure 9-2*b* shows a slicer, which is another name for a dual clipper. It slices the top and bottom off a waveform as shown, if the waveform amplitude exceeds $+(V + 0.6)$ and $-(V + 0.6)$ volts. Again, the extra 0.6 V is required to turn the diode ON, so that clamping to low-impedance supplies $+V$ and $-V$ can occur. The value of R is large compared with the diode ON resistance and small compared with the diode OFF resistance. When a diode is ON, the input is attenuated, and clipping occurs. When a diode is OFF, there is no attenuation, and the output waveshape is a duplicate of the input.

Figure 9-2*c* shows how diodes may be used in simple logic circuits. If either V_1 *or* V_2 is positive, the output is positive. Thus, for positive inputs the circuit is an OR gate. If V_1 and V_2 are both normally positive, one of them going to ground will not make the output ground. Both V_1 *and* V_2 must go to ground to make the output ground. In this case the circuit is an AND gate for negative logic. The circuit can therefore be an OR or an AND gate, depending on how the input conditions are defined.

Table 9-1 summarizes the properties of *pn*-junction and other types of diodes.

TABLE 9-1 Summary of diodes

Diode	Structure	Special properties	Typical applications
Rectifier	*pn* junction	High conductance and often high voltages and currents. (Motorola MR1299 has $I_F = 1{,}000$ A, and $PIV = 600$.) Fast recovery pn has 1 μS recovery time Alloyed pn has 5 μS recovery time Diffused pn has 3 μS recovery time	Ac-to-dc conversion
Schottky	Metal-to-semiconductor junction	Almost zero minority carrier storage time Recovery time = typically 0.15 μS	High-frequency mixers, rectifiers, modulators, detectors, waveform generators, and fast-pulse processing
pin	*p* and *n* regions separated by an intrinsic region	Long storage time makes diode an RF resistor.	Matched RF attenuator
Step recovery	*pn* junction designed for long minority carrier lifetime and high-charge storage	Very abrupt cessation of reverse minority carrier current flow	Efficient harmonic generator
Breakdown (Zener)	*pn* junction specially doped to produce controlled breakdown characteristic	Breakdown voltage nearly constant over wide range of reverse currents	Voltage reference source, clamping, clipping, dc coupling, surge protector
Tunnel	*pn* junction with very heavy doping	Has negative resistance region	High-frequency oscillator, amplifier, switch
Backward	Same as tunnel diode but with a low peak current	Has zero reverse-voltage breakdown	Used under reverse bias as a very low voltage rectifier
Voltage-variable capacitance	*pn* diode with graded junction to produce desired capacity vs. reverse-voltage variation	Large-capacity change with reverse-bias-voltage change	Replaces mechanically tuned capacitors

9-2 ZENER DIODES

Practical diodes depart from the ideal in that considerable current can flow if sufficient reverse bias is applied. For many applications reverse-voltage breakdown is a disadvantage, but in zener diodes the phenomenon is used to advantage to produce voltage-regulator diodes. That is to say, it is possible to use a zener diode to provide a relatively constant voltage from a supply voltage which may itself fluctuate considerably.

Zener diodes are more properly called *breakdown diodes*, because what are commonly called zener diodes actually operate by virtue of two distinct breakdown mechanisms. One is the zener effect, named after the man who first proposed a theory for reverse-voltage breakdown. The second is the avalanche effect. Thus, the term *breakdown diode* applies to both kinds of regulator diode now in use, zener and avalanche. However, since Zener's theory came before the discovery of the avalanche mechanism, the name is now well entrenched, and it is common practice for both kinds of breakdown diode to be referred to as zener diodes.

A breakdown diode is a *pn*-junction device which has been doped to produce controlled reverse-bias properties. The voltage level at which breakdown occurs is controlled by doping levels and gradients. Avalanche breakdown is a process similar to that which takes place in gaseous dielectrics. It occurs when holes or electrons are accelerated to very high velocities by external fields. Collisions occur which cause ionization of atoms, which in turn produce hole-electron pairs. These can then be accelerated to produce more ionizations, more hole-electron pairs, and hence more current. The increase in current tends to occur suddenly at a particular voltage and results in the typical zener (really avalanche) characteristic shown in Fig. 9-3*a*. External potentials of about 5 V and above are necessary to produce the energies which cause avalanche breakdown. Thus, breakdown diodes operating at about 5 V or above are avalanche devices. Five-volt avalanche diodes have approximately zero temperature coefficient, which becomes positive for higher-voltage devices.

Zener breakdown is a field-emission process that occurs in diodes with reverse breakdown voltages of 5 V or less. Owing to increased doping, the maximum electric field in the junction increases. It can be made sufficiently large to cause ionization without avalanche multiplication. Again, this results in a rapid increase in current and gives rise to the typical zener characteristic shown in Fig. 9-3*a*. Five-volt zener diodes have approximately zero temperature coefficient, which becomes negative for voltages of less than 5 V.

In Fig. 9-3*a*, notice that the "knee" (the point on the *VI* characteristic where breakdown occurs) is sharper for an avalanche diode than for a zener diode. A sharp knee is often desirable because it means that a lower breakdown impedance can be achieved at lower current (and hence power) levels than with a less-sharp knee. Breakdown impedance is called zener impedance for both types of diode and may be understood by referring to Fig. 9-3*b*.

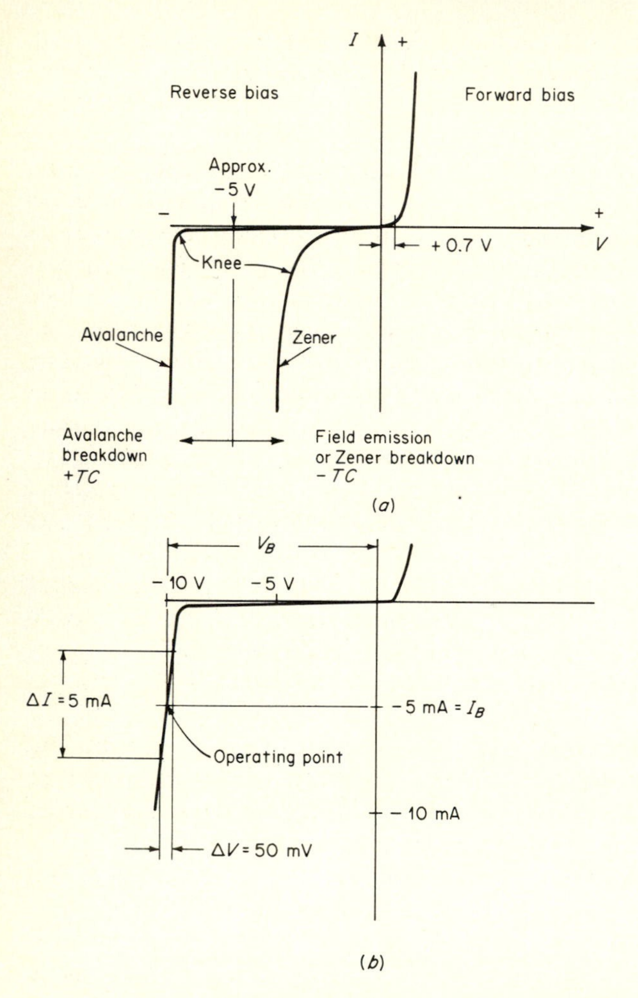

Fig. 9-3 Zener diodes: (*a*) avalanche and zener characteristics; (*b*) zener impedance.

Figure 9-3*b* is a characteristic for a breakdown diode with a breakdown voltage V_B, which is always specified at some test current I_B, such as 5 mA for a low-power (400-mW) breakdown diode. The zener impedance is also specified at this current level and is defined as the slope of the *VI* characteristic at I_B, or

$$\text{Zener impedance} = R_z = \frac{\Delta V}{\Delta I} \tag{9-1}$$

Thus R_z is a dynamic impedance. It is the effective impedance of the diode when operated about an operating point (V_B, I_B). For the *VI* characteristic plotted in Fig. 9-3*b* the zener impedance at $V_B = 10$ V and $I_B = 5$ mA is

$$R_z = \frac{\Delta V}{\Delta I} = \frac{5 \times 10^{-2}}{5 \times 10^{-3}} = 10\ \Omega \tag{9-2}$$

In the breakdown region the slope of the *VI* curve is not quite constant. Because of this, the zener impedance varies slightly with current. For currents above 5 mA, R_z would be less than 10 Ω; for currents below 5 mA, it would be greater than 10 Ω.

Since breakdown diodes are used as voltage references in such precision equipment as digital voltmeters, one of their most important characteristics is their temperature coefficient. This is a measure of how constant the breakdown voltage is as temperature varies. Two methods are used to achieve a low temperature coefficient. One is to use diodes that break down in the region of 5 V, where the temperature coefficient is approximately zero without further compensation. A second method uses two diodes—either a breakdown diode in series with an ordinary forward-biased diode, or two breakdown diodes in series. In either case the pairs of diodes are chosen to have equal and opposite temperature coefficients so that the overall coefficient is as nearly zero as possible. Breakdown diodes are available with temperature coefficients as low as 0.0001 percent/°C. A 10-V diode with this coefficient would vary in voltage only 10 μV for a 1°C change in temperature.

Breakdown voltage varies with time as well as temperature, but in a more erratic and random manner. The best breakdown diodes vary less than ±10 μV over a period of years. Even so, time instability can be a more serious consideration than temperature instability for many applications, especially if a constant-temperature oven is used.

We shall now consider some typical breakdown-diode applications. The first is the simple voltage regulator shown in Fig. 9-4*a*, which also serves to introduce the symbol for a breakdown diode. It is similar to the symbol for an ordinary diode except for the two extra angled lines shown.

Consider the following design problem, which refers to Fig. 9-4*a*. *D* is a 10-V breakdown diode (note the symbol) with a dynamic impedance of 10 Ω. E_{in} is an unregulated supply which can vary from 15 to 20 V. E_{out} is the regulated 10 V connected to a load which draws from 0 to 10 mA of current. Find a suitable value for *R*, and find the maximum variation in E_{out}. *D* requires a minimum current of 5 mA to ensure operation in the breakdown region.

Diode *D* will draw a minimum current when the output current I_3 is 10 mA. Then I_2 must be 5 mA, the minimum diode current. I_1 is the sum of I_2 and I_3, or 15 mA. Knowing I_1 and the voltage drop across *R*, we can calculate *R*. If we calculate *R* for $E_{in} = 20$ V and $I_1 = 15$ mA, we get I_1 less than 15 mA for $E_{in} = 15$ V. Then I_2 will be less than 5 mA, which is not allowed. So we must calculate *R* for $E_{in} = 15$ V and $I_1 = 15$ mA:

$$R = \frac{E_{in} - E_{out}}{I_1} = \frac{15\text{ V} - 10\text{ V}}{15\text{ mA}} = 333\ \Omega \qquad (9\text{-}3)$$

R would therefore be chosen as 330 Ω.

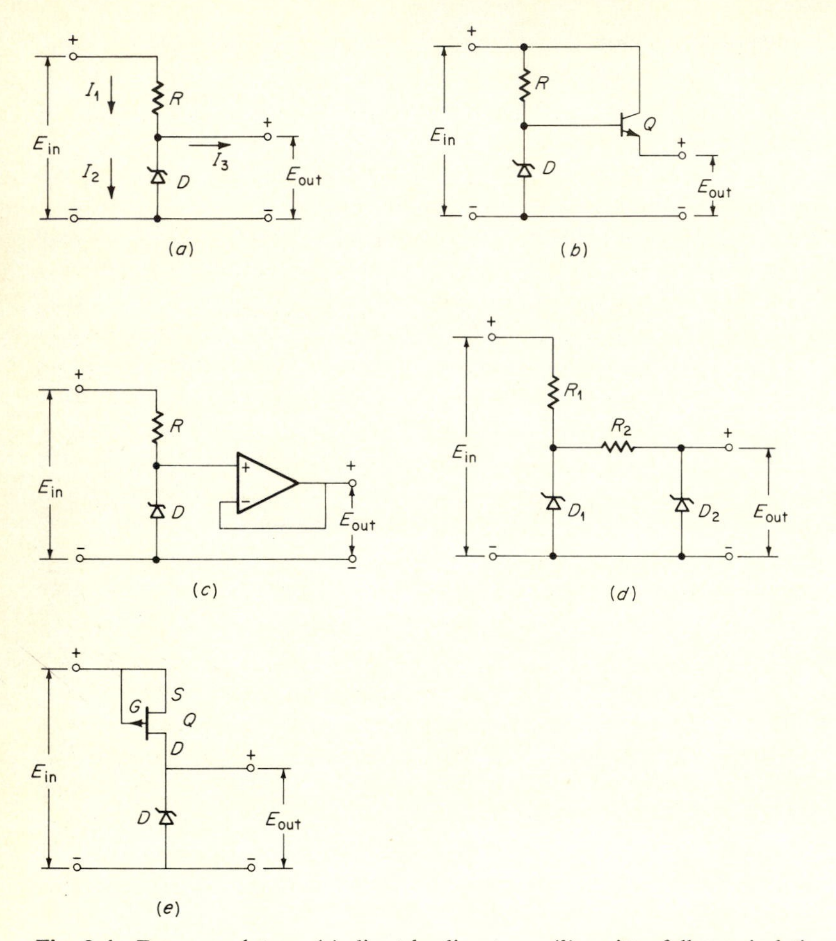

Fig. 9-4 Zener regulators: (*a*) direct-loading type; (*b*) emitter-follower isolation; (*c*) high-gain dc amplifier regulator; (*d*) zener regulator with preregulation; (*e*) constant-current-driven zener regulator.

To find the maximum variation in E_{out} from the nominal 10 V, we find the maximum variation in I_2 and multiply by the dynamic zener impedance. $I_{2,\,min}$, we know, is 5 mA. $I_{2,\,max}$ occurs when $E_{in} = 20$ V and $I_3 = 0$. Then

$$I_2 = I_1 = \frac{E_{in} - E_{out}}{R} = \frac{20 \text{ V} - 10 \text{ V}}{0.33 \text{ k}\Omega} = 30 \text{ mA} \tag{9-4}$$

$$\Delta I_2 = I_{2,\,max} - I_{2,\,min} = 30 - 5 = 25 \text{ mA} \tag{9-5}$$

$$\Delta E_{out} = \Delta I_2 R_z = (25 \text{ mA})(10\ \Omega) = 250 \text{ mV} \tag{9-6}$$

Notice that although the output voltage E_{out} changes up to 250 mV under the worst conditions, it was assumed in Eqs. (9-3), (9-4), and (9-6) that E_{out}

remained constant at 10 V. Usually the errors introduced by this type of assumption are negligible.

The circuit shown in Fig. 9-4*b* can have certain advantages over that in Fig. 9-4*a*. The addition of an emitter follower means that current taken from the divider formed by R and D is a transistor base current. This current will be the load current divided by the current gain of the transistor. For a load current of 15 mA and a current gain of 150, the base current would be only 0.1 mA. This small current flowing in the zener impedance will produce a much smaller voltage variation than would the full load current. However, the circuit does have some disadvantages. The base-emitter voltage of the transistor will vary with temperature, so D and Q have to be chosen to compensate for each other. Also, the output impedance of the emitter follower contains some series resistance which can easily be as large as the dynamic zener impedance R_z. Thus, although the load current does not have to flow through R_z, it must flow through the transistor output impedance, and this might cause large output-voltage fluctuations.

The circuit in Fig. 9-4*c* solves these problems and is the basic block diagram for most types of voltage regulator. It employs a high-gain dc amplifier, which, in Fig. 9-4*c*, is an operational amplifier of the type discussed in Chap. 4. Since it is connected as a voltage follower, the operational amplifier has a high input impedance, a low output impedance, unity voltage gain, and very low voltage offset from input to output. The high input impedance means that the current drain from R and D is negligible. The low output impedance means that the output voltage will change very little with load current. With unity gain and low offset, the output voltage will equal the diode voltage. All these advantages result from the use of large amounts of negative feedback.

If the amplifier in Fig. 9-4*c* is made to have a variable gain, a regulator with a variable output voltage results.

Notice that fluctuations in the diode voltage are caused by two effects. One is current drawn from the divider formed by R and D. The other is variations in E_{in}. The circuit in Fig. 9-4*c* solves the first problem but not the second. As E_{in} varies, the diode voltage will vary slightly, and the change will be fed to the output by follower action. The circuits in Fig. 9-4*d* and *e* solve or reduce the problem of variations in E_{in}.

In Fig. 9-4*d* a double regulator is used. R_1 and D_1 preregulate the voltage fed to R_2 and D_2, which then provide further regulation. Once again the voltage across D_2 can be fed to an operational-amplifier type of circuit which will minimize load-current effects.

In Fig. 9-4*e*, diode D is fed from a constant-current generator, in this case a junction field effect transistor operated with $V_{GS} = 0$. (See Chap. 1.) As E_{in} varies, the diode current remains substantially constant. If the diode is loaded as in Fig. 9-4*c* by a high-impedance isolating amplifier, neither variations in E_{in} nor load current will change the diode current. Thus the diode voltage and hence the output voltage will remain constant.

9-3 VOLTAGE-VARIABLE CAPACITANCE DIODES

In a *pn* junction there exists a depletion region (an insulator), on either side of which are a *p* region and an *n* region (conductors). When two conductors are separated by an insulator, the result is a capacitor. Since the width of the depletion region (and hence the separation of the capacitor "plates") can be varied by varying the external voltage applied to the junction, a *pn* junction can be used as a voltage-variable capacitor (VVC). This is illustrated in Fig. 9-5*a*. Figure 9-5*b* shows the symbol for a VVC.

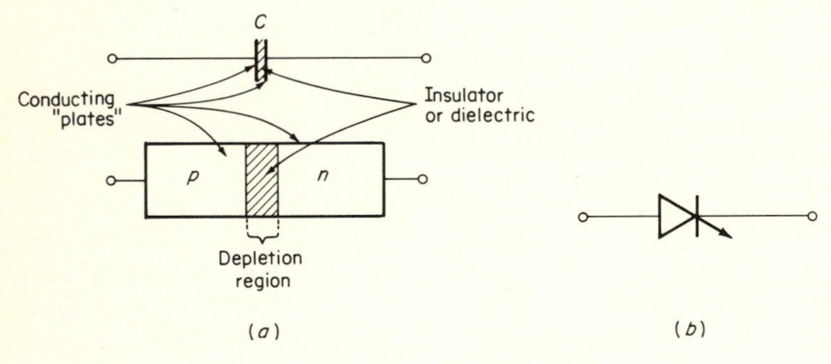

Fig. 9-5 VVC diode: (*a*) construction; (*b*) symbol.

VVCs have significant advantages over other types of variable capacitor, which are usually mechanically tuned. They are small, rugged, and reliable, have no moving parts, operate over a very wide frequency range, have fast response, permit remote tuning, and of course can be tuned electrically rather than mechanically.

The junction capacitance of a reverse-biased diode can be expressed as

$$C_j = \frac{K}{(V_R + V_C)^n} \tag{9-7}$$

where K = constant which is a function of the junction area, the charge on an electron, the dielectric constant, the permittivity of free space, and the doping level in the active region

V_R = externally applied reverse bias

V_C = junction contact potential

n = constant whose value depends on how the junction is graded. For an abrupt junction where the doping changes abruptly from *p* to *n* at the junction, $n \approx \frac{1}{2}$. For a linearly graded junction, $n \approx \frac{1}{3}$. Abrupt and linear junctions are theoretical limits or extremes. n for practical VVCs lies between $\frac{1}{2}$ and $\frac{1}{3}$.

As can be seen from Eq. (9-7), increasing V_R reduces C. This corresponds to the reduction in capacitance obtained by increasing the spacing of ordinary capacitor plates, since increasing V_R increases the depletion width of a *pn* junction.

Figure 9-6 shows how the normalized capacitance of a VVC varies with applied reverse bias. VVCs are available with capacities ranging from a few picofarads up to thousands of picofarads, although the variation in any particular VVC is limited to about 10 to 1.

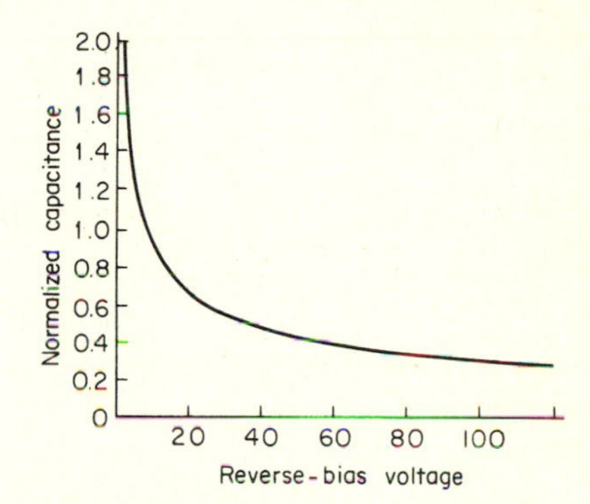

Fig. 9-6 Variation in capacity in a VVC with reverse bias.

The equivalent circuit for a *pn* junction used in VVC applications is given in Fig. 9-7*a*. L and C_s are the package inductance and capacitance, respectively, 5 nH and 0.5 pF being typical values. R_s is the resistance of the bulk semiconductor material and is a few ohms. R_p is the diode leakage resistance, which is responsible for any leakage current and is measured in megohms. For most practical purposes the circuit in Fig. 9-7*a* can be simplified to either of those in Fig. 9-7*b* or *c*.

Since VVCs are often used at RF to tune *LC* circuits, an important parameter is their Q. In this respect they are inferior to the best mechanically variable capacitors, although Q's of over 200 at 50 MHz are possible with VVCs. The Q of a VVC varies inversely with frequency and increases with increased reverse bias. Thus the Q of a VVC depends on how it is used. A figure of merit is desirable

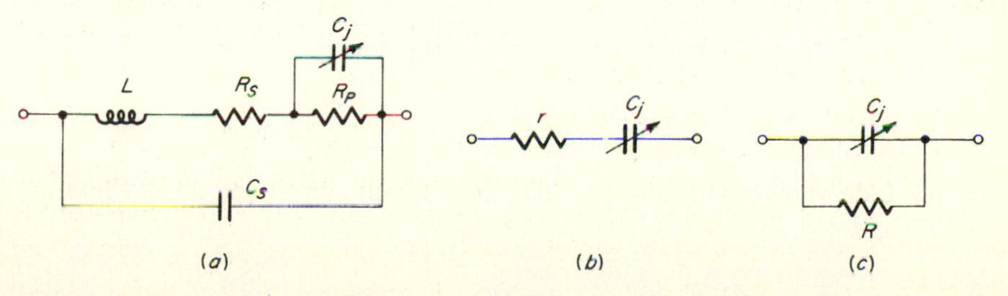

Fig. 9-7 VVC equivalent circuits: (*a*) detailed equivalent circuit; (*b*) series version; (*c*) parallel version.

which gives an indication of the performance of a VVC as a device, independent of external factors. This is done, in part, by using the equivalent circuit of Fig. 9-7*b* and noting that, for a particular value of C_j,

$$Q = \frac{1}{2\pi f r C_j} \tag{9-8}$$

By defining $f = f_c$ as the frequency at which $Q = 1$, we obtain

$$f_c = \frac{1}{2\pi r C_j} \tag{9-9}$$

f_c would be specified for a particular reverse bias, and values up to 10^{10} Hz are attainable.

Since VVCs are used to tune *LC* circuits, the frequency stability of the tuned circuit will depend on the stability of the VVC capacity, especially with temperature. Both the contact potential V_C and K in Eq. (9-7) change with temperature in such a way as to increase C_j as temperature increases. K changes because the dielectric constant of the semiconductor material changes. V_C changes at the usual rate of -2 to -2.5 mV/°C for silicon junctions. If V_R is large compared with V_C, temperature effects due to V_C changes can be swamped [from Eq. (9-7)]. Thus the temperature coefficient of a VVC would be typically 500 ppm at $V_R = -1$ V, and 75 ppm at $V_R = -100$ V. With $V_R = -100$ V, the effects of V_C are negligible, and the 75 ppm would be due almost entirely to temperature variation in the dielectric constant of silicon.

The changes in C_j due to temperature variation of both V_C and K in Eq. (9-7) can be compensated for by using a temperature-sensitive reverse-biased voltage V_R. It is possible to derive design equations for such compensating bias, but these are sufficiently elaborate that in practice compensation, if necessary, is determined experimentally.

A VVC is almost invariably biased to some dc level ($-V_R$) and then varied about this level to produce the desired capacity change. Biasing VVCs therefore involves three things:

1. Applying a reverse-bias voltage V_R, to set the VVC to some initial value of capacity, C_o
2. Making provision for varying the bias voltage about V_R to give a capacity of $C_o \pm \Delta C$
3. Isolating dc and ac paths, so that bias and signal conditions do not interact

Figure 9-8*a* shows a possible biasing method for a VVC used in a tuned circuit, a method which is used to eliminate unreliable mechanically switched turret tuners in TV sets. A variable dc bias voltage is available from the power

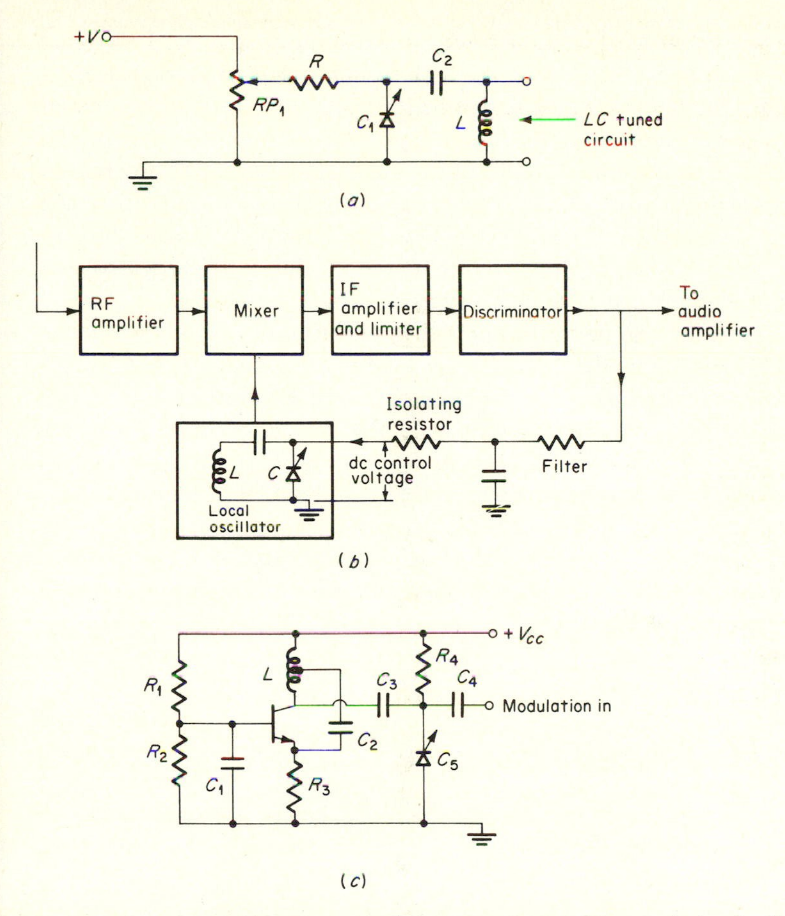

Fig. 9-8 VVC applications: (*a*) biasing a VVC for use in a tuned circuit; (*b*) VVC used for automatic frequency tracking; (*c*) VVC used to produce FM.

supply $+V$ and RP_1. Resistor R isolates the tuned circuit from the shunting effect of RP_1. R should be large for good isolation, but not so large that the reverse leakage current of VVC C_1 could cause the bias voltage on C_1 to change unduly with temperature. R could be replaced with an RF choke that would isolate at RF but be a low impedance to the dc leakage current of C_1. The LC-tuned circuit consists of L and C_1. C_2 is present only to prevent the low dc resistance of L from shorting the bias on C_1 and is much greater in value than C_1. By varying RP_1, C_1 is varied, and the frequency at which the tuned circuit resonates can be changed.

One application of a voltage-tuned LC circuit is in automatic frequency control in FM receivers. How this works is shown in Fig. 9-8*b*. FM signals are received by the antenna and mixed with the local oscillator signal to produce an intermediate frequency (IF). Following IF amplification and limiting, a discriminator converts the FM IF signal to the original audio frequency. This signal is fed to the audio amplifier. The discriminator output also has an average dc level which is linearly related to the frequency of the RF signal received at the antenna. This average dc level can be used to bias a VVC so as to keep the receiver in tune. The VVC is part of the tank circuit of the local oscillator. If for any reason the receiver goes out of tune, the average dc level from the discriminator will change. This dc change is coupled to the VVC via a simple low-pass filter to remove the audio component. The direction of the dc change is chosen so that the local oscillator is retuned in such a way as to retune the receiver.

The circuit in Fig. 9-8*b* points to another application of VVCs, namely the production of frequency modulation. Figure 9-8*c* is a grounded (via C_1) base Hartley oscillator. R_1 and R_2 are bias resistors, and R_3 permits a feedback voltage from a tap in L to be developed via dc blocking capacitor C_2. C_5, the VVC, is biased to $+V_{cc}$ via isolating resistor R_4. C_3 and C_4 isolate the bias on C_5 from dc levels at the transistor collector and the modulation input, respectively. When a modulating signal is applied, the bias on C_5 changes in sympathy so that the capacity of C_5 is modulated. For small changes in C_5 and hence small changes in the oscillator frequency, the FM will be reasonably linearly related to the modulating amplitude.

9-4 SCHOTTKY DIODES

Unlike the conventional *pn* silicon diode, which consists of a *p*-type–to–*n*-type silicon (or semiconductor-to-semiconductor) junction, the Schottky diode consists of a metal-to-semiconductor junction. Metal-to-semiconductor junctions are used in two distinct ways in semiconductor devices and integrated circuits. In the first they are used to attach metal leads to semiconductors; this is done in a manner which results in an *ohmic* or *nonrectifying* connection. The second results in a *Schottky-barrier* diode, which *does have rectifying properties* similar to those of a *pn*-junction device.

Schottky diodes are majority-carrier devices, in contrast to *pn*-junction diodes which are minority-carrier devices. Since minority carriers must be removed from the junction for a *pn* diode to be OFF, there is always a delay due to minority-carrier lifetime when a *pn* diode is turned OFF. Having no minority carriers to be cleared from the junction, a Schottky diode can be turned OFF considerably faster than its *pn*-junction counterpart.

Since the voltage across a forward-biased Schottky diode is only about one-half that of silicon *pn*-junction diodes, the Schottky is a better approximation to the ideal diode.

The two properties—speed and low voltage—are used to advantage in certain digital logic families, as explained in Sec. 5-3-6.

Schottky diodes are also sometimes called *hot carrier diodes.*

9-5 UNIJUNCTION TRANSISTORS

9-5-1 INTRODUCTION

The *unijunction transistor* (UJT) is one of the oldest and simplest semiconductor devices. As the name implies, the UJT (or *double-base diode*, as it was first called) is a three-terminal device with a single *pn* junction. It is primarily a switching device which is different in several ways from other semiconductor devices:

1. Its triggering voltage is approximately a fixed percentage of the power-supply voltage. [Equation (9-12) shows that the frequency of oscillation of UJT oscillators becomes substantially independent of supply voltage because of this characteristic.]
2. It exhibits a stable negative-resistance region which suggests its use in oscillator and trigger circuits.
3. The UJT often reduces the number of components necessary to perform a given function to less than half that required if bipolar transistors are used. The basic simplicity of both the UJT and the resulting circuit increases reliability.
4. The UJT internal resistance in the OFF condition is relatively high (5 to 10 kΩ); hence its idling power is low.
5. The device has very low triggering-current requirements (2 to 10 μA).
6. UJTs have high pulse-current capabilities (2 A).
7. Many UJT structures are very inexpensive.
8. Reasonably high (3 to 5 V) peak output voltages are available for triggering thyristors (Chap. 8 and Sec. 9-5-5).
9. UJT input leakage currents are low (1.0 to 10 nA).
10. Complementary or programmable UJTs (Secs. 9-5-6 and 9-5-7) are capable of even
    (*a*) Higher internal OFF resistances (>30 kΩ)
    (*b*) Lower firing currents (0.1 μA)
    (*c*) Lower operating currents (<1.0 mA)
    (*d*) Higher peak output-voltage pulses (up to 10 V)
    (*e*) Comparable prices

Unijunction transistors are most often used in timing, triggering, sensing, and waveform-generation circuits. These include oscillators; voltage sensing and comparator circuits; frequency dividers; trigger circuits for power-controlling devices such as thyristors; simplified multivibrators; timers; counters; and ramp, triangular, and staircase generators. Most of the UJT literature points out that a reduction in components results if UJTs are employed in the types of circuits just mentioned. The reason for this is that the UJT combines the features of precision voltage sensing, regenerative gain, and very low ON resistance within a single device. These are requirements in many of the circuits mentioned above;

to perform these three functions with conventional bipolar transistors generally requires two or three transistors and some form of feedback. The reduction in components will become more apparent as we proceed with the UJT characteristics and specific applications.

9-5-2 BASIC THEORY OF OPERATION

Figure 9-9 will be used to describe the basic UJT operation. Figure 9-9*a* shows the older "bar" UJT structure, wherein a heavily doped *p*-type emitter (aluminum wire) is alloyed into a lightly doped *n*-type silicon bar. Ohmic contacts, labeled "Base 1" and "Base 2," are connected to both ends of the bar; and the bar is actually a simple resistor whose resistance value depends only on doping and dimensions. Cross sections of the more recent UJT structures will be presented later, but the bar structure helps to simplify the discussion. Figure 9-9*b* is the UJT symbol; the emitter arrow denotes the input *pn* junction and its polarity.

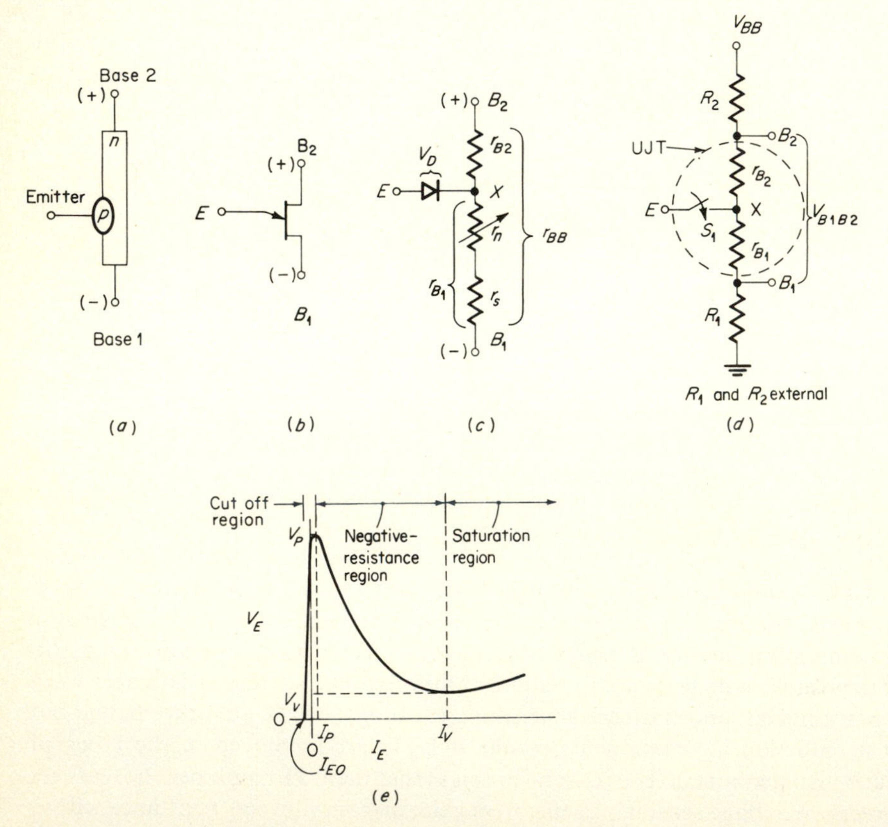

Fig. 9-9 (*a*) Simplified bar-UJT structure; (*b*) UJT symbol; (*c*) UJT equivalent circuit; (*d*) simplified equivalent circuit including external resistors (R_1 and R_2); (*e*) UJT emitter characteristics.

The UJT equivalent circuit and its simplified version with conventional external resistances are given by Fig. 9-9*c* and *d*, respectively. Notice that it takes the form of a simple voltage divider formed by resistances r_{B1} and r_{B2} with the emitter contact between. The voltage at point X is thus a fixed percentage of the applied interbase voltage (V_{B2B1}) (Table 9-2 summarizes UJT nomenclature)—that is, $V_X \approx [r_{B1}/(r_{B1} + r_{B2})](V_{B2B1})$. ($R_1$ and R_2 are generally negligible compared to r_{BB}.) In simplest form, the details from emitter to point X are represented by the ideal switch S_1 of Fig. 9-9*d*. In this ideal representation the switch is open when the emitter diode is reverse-biased, and closed during adequate forward bias. The emitter diode and its inherent voltage drop are shown in series with the emitter lead in Fig. 9-9*c*. This diode exhibits the conventional *pn*-junction reverse leakage current when reverse-biased, and some finite saturation resistance when forward-biased. r_{B1} is composed of a fixed saturation resistance r_s plus a negative resistance component r_n. By negative resistance we mean a resistance which effectively decreases as the current through it is increased. (In a fixed resistor and most semiconductor devices, the voltage drop across the element increases as the current through it increases; however, as Fig. 9-9*e* shows, V_{EB1}, the voltage drop across r_{B1}, decreases with increasing I_E.) As in most equivalent circuits, the relative importance of the various equivalent-circuit components depends on the particular operating conditions. In many cases we are interested in the forward-biased conditions for which the equivalent circuits of Fig. 9-9*c* and *d* are entirely adequate. In the reverse-biased or cutoff region, the only changes in the equivalent circuit of Fig. 9-9*c* are that the emitter diode reverse-biased leakage resistance and the total r_{B1} are more important. The total resistance from B_1 to B_2 is termed the *interbase resistance* (r_{BB}); thus $r_{BB} = r_{B1} + r_{B2}$.

A typical UJT input or emitter characteristic (V_E versus I_E) is shown in Fig. 9-9*e*. Such a characteristic is obtained with constant B_2 voltage, and the negative I_E scale is purposely exaggerated for clarity. One of the most important points to remember is that the characteristic of Fig. 9-9*e* is for the *emitter-to-base-1 current and voltage*. This is easily forgotten, and the result could be misinterpretation.

Normal UJT operation consists of increasing the emitter voltage V_E until it is equal to the firing or *peak-point voltage* V_P of Fig. 9-9*e*. In other words (referring to Fig. 9-9*d*), if $V_E > V_X$ plus the diode voltage drop, S_1 effectively closes. At V_P the UJT enters the negative-resistance portion of its characteristic, and we see a substantial decrease in r_{B1} and consequent increase in current flow from emitter to base 1. At this point the UJT is said to have fired, and an appreciable positive voltage is available at its B_1 terminal. To view this action in an ON-OFF fashion:

1. If $V_E < V_P$ the emitter diode is reverse-biased, resulting in a small reverse *emitter leakage current* I_{EO}, and the UJT is OFF.
2. If $V_E > V_P$ the emitter diode is forward-biased and I_E increases, turning the UJT ON and drastically reducing r_{B1}.

Remember, it is only necessary to adequately forward bias the emitter diode in order to trigger the UJT. Figure 9-9*d* shows that this can be accomplished by holding V_{BB} constant and increasing V_E, or by fixing V_E and decreasing V_{BB}. Hence the peak-point characteristics (V_P and I_P) and their stability are very important because they determine the UJT triggering point. At this stage one might wonder why a discrete diode-and-resistor voltage divider (in the form of the UJT equivalent circuit of Fig. 9-9*c*) wouldn't work just as well as the UJT. The reason, of course, is that a fixed r_{B1} would not exhibit the same negative-resistance characteristic as seen in the UJT.

Referring to Fig. 9-9*a*, let us describe the basic physics involved in UJT action. As V_E is increased, the *pn* junction becomes forward-biased, and holes are injected into the silicon bar. Owing to the polarity of the applied bar voltage, the injected positive holes are attracted toward the negative B_1 terminal. Of course, as holes are injected from the *p* to the *n* region, electrons are simultaneously injected from *n* to *p* to maintain charge neutrality. This increase in free current carriers between the emitter and base-1 region causes a reduction in the resistance of r_{B1}. The decrease in r_{B1} causes a decrease in the voltage at point X and hence an increase in the diode's forward bias (V_{EX} of Fig. 9-9*c*). This in turn allows more holes to be injected from the emitter, and r_{B1} decreases further. This process is thus *regenerative*, and the resistance of r_{B1} becomes very low, of the order of a few ohms. When $r_{B1} = r_s$ [the (I_V, V_V) point of Fig. 9-9*e*], the regenerative decrease in r_{B1} becomes limited because the lifetime of the injected current carriers is shortened by the high concentration of holes and electrons in the r_{B1} region. In other words, recombination of the injected holes and electrons becomes more probable when the concentrations are high. Beyond the valley point (V_V and I_V of Fig. 9-9*e*) the emitter current becomes an approximately linear function of emitter voltage, and the resistance becomes positive. The actual operating points of various UJT circuits and a table summarizing UJT nomenclature will be presented in the next two sections.

9-5-3 NOMENCLATURE, CHARACTERISTICS, AND STRUCTURES

Table 9-2 summarizes the UJT nomenclature. Most of the terms have been mentioned previously and are self-explanatory. The *intrinsic standoff ratio* η is a new term which is defined as the base-resistor voltage-divider ratio, so that

$$\eta = \frac{r_{B1}}{r_{B1} + r_{B2}}$$

Values of η range from 0.4 to 0.9, depending on the type of unijunction. The ratio η determines the peak-point (firing) voltage V_P and is related to V_P by the following formula:

$$\begin{aligned} V_P &= \eta V_{B2B1} + V_D = V_X + V_D \\ &\approx \eta V_{BB} + V_D \qquad \text{(if } R_1 \text{ and } R_2 \ll r_{BB}) \end{aligned} \tag{9-10}$$

TABLE 9-2 UJT nomenclature

Symbol	Definition
I_P, V_P	Emitter current and voltage at the peak point of the emitter characteristics where the negative resistance region starts. These levels must be reached before the UJT triggers.
I_V, V_V	Emitter current and voltage at the valley of the emitter characteristics. This point separates the negative resistance and saturation regions.
r_{BB}	Interbase resistance between B_2 and B_1. $r_{BB}=r_{B1}+r_{B2}$.
V_{B2B1}	Voltage between B_2 and B_1, where V_{BB} is the power-supply voltage.
V_D	Emitter-diode forward voltage drop.
η	Intrinsic standoff ratio. $\eta = r_{B1}/(r_{B1} + r_{B2})$ or $(V_P - V_D)/V_{B2B1}$.
I_{EO}	Emitter reverse leakage current measured between E and B_2 with B_1 open.
V_{OB1}	Base 1 peak pulse voltage with a specific R_1 and C_T.
$V_{EB1,\text{sat}}$	Also termed $V_{E,\text{min}}$. Emitter saturation voltage or minimum voltage from E to B_1 at a specified I_E and V_{B2B1}.

(See Fig. 9-9*c*.) If V_{B2B1} is constant, with the UJT not fired, Eq. (9-10) merely states that the voltage at point X is determined by η, and when the emitter voltage exceeds the diode drop V_D plus V_X, the UJT fires. Hence η determines the firing voltage, the other quantities in Eq. (9-10) being essentially constant. η is also a weak function of temperature, life, interbase voltage, and the particular unit within a batch.

Although the bar- or alloy-type UJT structure portrayed by Fig. 9-9*a* is useful in explaining the basic operation of the UJT, alloy unijunctions are rapidly being replaced by less-expensive annular and planar structures. The planar process offers significant improvements over the older alloying technique, mainly higher input impedance, lower required triggering power, lower emitter leakage current for higher accuracy in timing circuits, and higher base-1 voltage for triggering thyristors.

Excerpts from typical high- and low-standoff-ratio planar UJT data sheets are given in Fig. 9-10. Notice that when the UJT emitter characteristics are plotted on a log scale, as in Fig. 9-10*b*, the negative leakage current is not shown. Considering the UJT parameters that are temperature-dependent, notice that

1. From Fig. 9-10*c*, r_{BB} increases directly with temperature at $\approx 1.0\,\%/^\circ\text{C}$.
2. The emitter reverse leakage current I_{EO} has roughly the same magnitude and slope as most reverse-biased silicon diodes ($< 1.0\ \mu\text{A}$ at room temperature and doubling about every 8°C).
3. η is negligibly affected by temperature, except that the diode voltage drop V_D decreases by the usual 2.5 mV/°C as temperature increases.

Owing to item 3 above and the fact that V_P is related to V_D by Eq. (9-10), V_P also varies inversely with temperature. The change of V_{OB1} with temperature is shown in Fig. 9-10*c*.

	PARAMETER	TEST CONDITIONS	2N3980 MIN	2N3980 MAX	2N4947 MIN	2N4947 MAX	2N4948 MIN	2N4948 MAX	2N4949 MIN	2N4949 MAX	UNIT
r_{BB}	Static Interbase Resistance	$V_{B2\text{-}B1} = 3$ V, $I_E = 0$	4	8	4	9.1	4	12	4	12	kΩ
α_{rBB}	Interbase Resistance Temperature Coefficient	$V_{B2\text{-}B1} = 3$ V, $I_E = 0$, $T_A = -65°$C to 100°C,	0.4	0.9	0.1	0.9	0.1	0.9	0.1	0.9	%/deg
η	Intrinsic Standoff Ratio	$V_{B2\text{-}B1} = 10$ V,	0.68	0.82	0.51	0.69	0.55	0.82	0.74	0.86	
$I_{B2(mod)}$	Modulated Interbase Current	$V_{B2\text{-}B1} = 10$ V, $I_E = 50$ mA	12		12		12		12		mA
I_{EB2O}	Emitter Reverse Current	$V_{EB2} = -30$ V, $I_{B1} = 0$		−10		−10		−10		−10	nA
		$V_{EB2} = -30$ V, $I_{B1} = 0$, $T_A = 125°$C		−1		−1		−1		−1	μA
I_P	Peak-Point Emitter Current	$V_{B2\text{-}B1} = 25$ V		2		2		2		1	μA
$V_{EB1(sat)}$	Emitter — Base-One Saturation Voltage	$V_{B2\text{-}B1} = 10$ V, $I_E = 50$ mA		3		3		3		3	V
I_V	Valley-Point Emitter Current	$V_{B2\text{-}B1} = 20$ V	1	10	4		2		2		mA
V_{OB1}	Base-One Peak Pulse Voltage		6		3		6		3		V

(a)

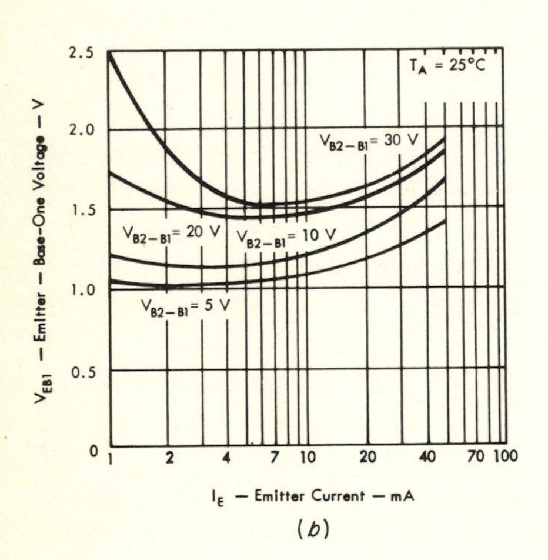

(b)

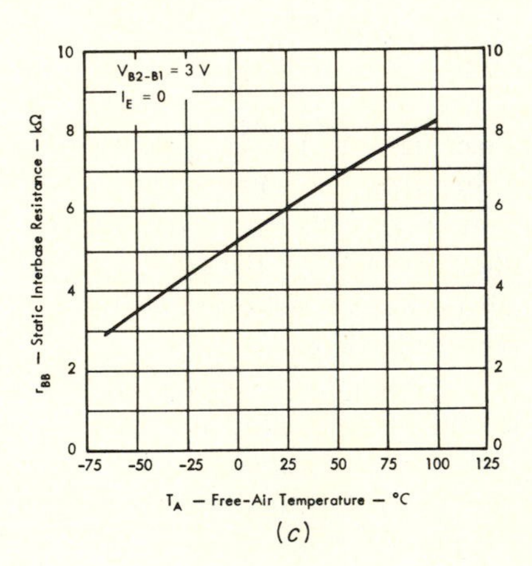

(c)

Fig. 9-10 Typical planar-UJT characteristics: (*a*) ratings and characteristics; (*b*) V_E versus log I_E; (*c*) r_{BB} and V_{OB1} versus temperature. *Texas Instruments.*

9-5-4 THE UJT RELAXATION OSCILLATOR

The basic UJT relaxation oscillator is shown in Fig. 9-11*a*. The operation of this circuit is as follows. When power is applied, C_T charges exponentially through R_T toward the supply voltage V_{BB}. When the emitter voltage reaches the UJT peak-point voltage V_P, the emitter becomes forward-biased, and we can consider the ideal emitter switch to have closed since the emitter diode now conducts. At this point r_{B1} drops to a very low value, accompanied by a heavy flow of I_E. C_T can now discharge through the emitter and low $r_{B1} + R_1$ path, and a positive voltage spike thus results at B_1. A similar, but negative, spike is simultaneously available at B_2. The steady-state and pulse magnitudes of V_{B1} and V_{B2} depend on the voltage divider formed by resistors R_1, R_2, and r_{BB}, as shown in Fig. 9-11*b* and *c*. Figure 9-11*c* shows that there are three UJT outputs available and that the emitter ceases conduction at $V_{E,\,\min}$, and the cycle is repeated. Pulses are available at either base, and an exponential waveform at the emitter. The frequency of oscillation can be varied by adjusting the values of the timing components R_T and/or C_T.

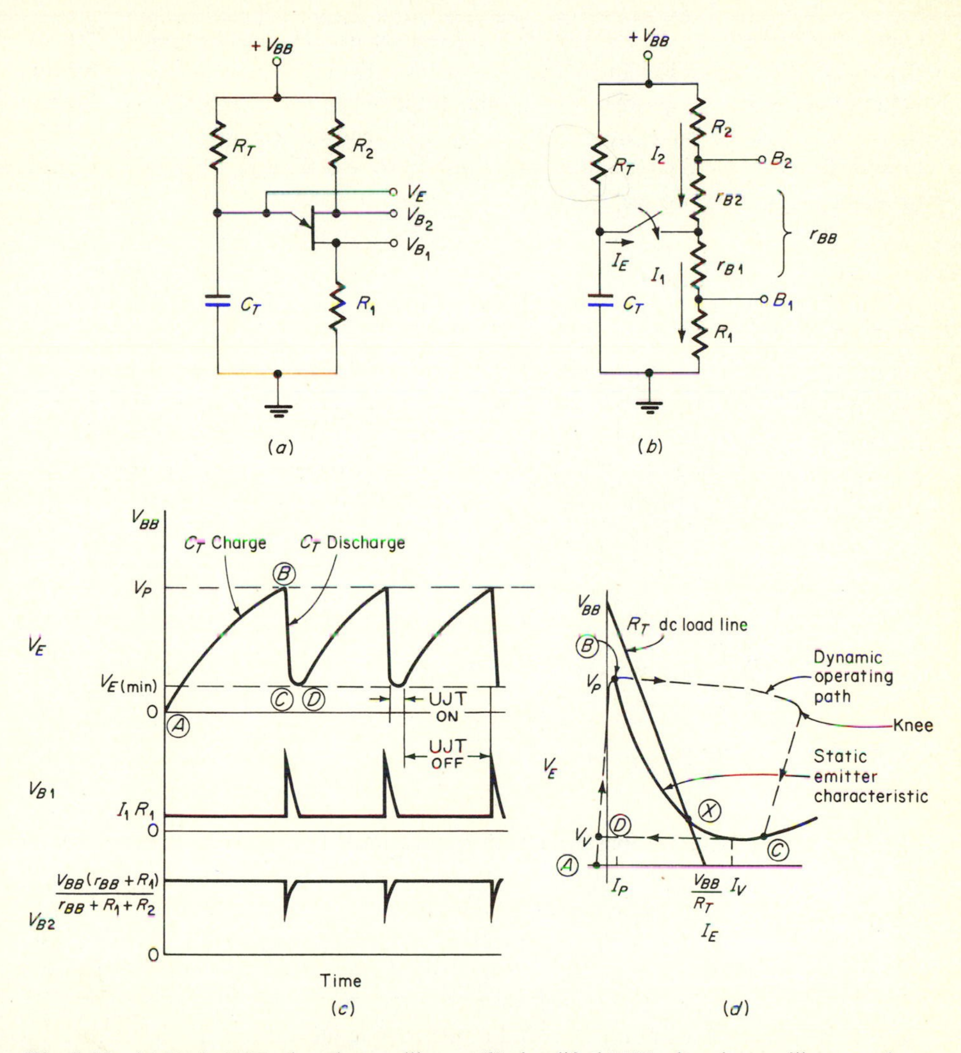

Fig. 9-11 (*a*) Basic UJT relaxation oscillator; (*b*) simplified UJT-relaxation-oscillator equivalent circuit; (*c*) UJT-relaxation-oscillator waveforms; (*d*) UJT emitter characteristics with dc load line and dynamic operating path.

The UJT relaxation oscillator is thus an extremely simple, versatile, and inexpensive oscillator requiring as few as four components. This circuit is the basis of all UJT timing, sawtooth, and pulse-generation circuits.

As an example of the reduction in number of components inherent to designing a relaxation oscillator with unijunction rather than bipolar transistors, consider a conventional-transistor relaxation oscillator. The parts count is nearly four times that required for the UJT relaxation oscillator of Fig. 9-11*a*.

The UJT relaxation oscillator shown in Fig. 9-11*a* has a rather simple set of biasing criteria, but these must be satisfied to ensure oscillation. Figure 9-11*d* shows the optimum UJT relaxation-oscillator dc load line with intersections at the V_E and I_E axes of V_{BB} and V_{BB}/R_T, respectively. To obtain this load line, the value of R_T must fall between two limits. First, when the charge on C_T reaches V_P, there must be an emitter current $\geq I_P$ to ensure triggering of the UJT. Obviously if R_T were large enough to limit I_E to less than I_P, triggering would never occur. Hence, we see that

$$R_{T,\max} < \frac{V_{BB} - V_P}{I_P}$$

where $V_{BB} - V_P$ is the voltage across R_T at the instant of triggering. Considering the lower limit of R_T, we see that if R_T were small enough to allow the emitter current to exceed the UJT's valley current, the UJT would never turn OFF. Therefore, the dc load line must intersect the static emitter characteristics in the negative-resistance region:

$$R_{T,\min} > \frac{V_{BB} - V_V}{I_V}$$

Combining these limits, we see that R_T must fall between the two limits

$$\frac{V_{BB} - V_P}{I_P} > R_T > \frac{V_{BB} - V_V}{I_V} \qquad (9\text{-}11)$$

The actual dynamic operating path for a typical UJT relaxation oscillator is shown in Fig. 9-11*d*. The sequence begins at point *A* as C_T starts charging toward the supply voltage. No emitter current flows at this point because the UJT has not yet triggered. At point *B*, $V_E = V_P$ and the UJT fires with effective closing of the switch, as shown in Fig. 9-11*b*. Since the voltage across a capacitor cannot change instantaneously but the current through it can, the dynamic operating path moves away from point *B* with a considerable increase in I_E. However, the supply of current from C_T and the reduction of r_{B1} are limited, and the dynamic operating point becomes *C*. The actual shape of the dynamic operating path from *B* to *C* is determined by the magnitude of C_T, R_1, and r_{B1}. As C_T is decreased, less I_E is available, and the knee moves to the left. The slope just prior to *C* increases as R_1 increases. V_E seeks point *C* because S_1 (Fig. 9-11*b*) is now closed, and the capacitor attempts to discharge as far as possible. From point *C* one might expect the dynamic path to go to point *X*, which is the intersection of the dc load line and the static-emitter characteristic. Operation must move to a lower emitter-current level because C_T cannot continue to supply I_V, but it is impossible for V_E to increase from point *C* to *X*. To do this, the capacitor would have to gain charge; however C_T is discharging, so this is clearly impossible. Therefore, the dynamic operating path follows an approximately constant V_E line to point

D, where the cycle is repeated. We thus see that the UJT relaxation oscillator turns OFF mainly because the capacitor in the emitter circuit does not allow point C or X to be a stable operating point.

The waveforms shown in Fig. 9-11c can be explained using the simplified equivalent circuit of Fig. 9-11b. Figure 9-11c also gives an indication of the relative time required for the point-to-point transitions of Fig. 9-11d. Notice the relatively long C_T charge time, as compared to its discharge time. This results because C_T charges through R_T, which is generally considerably greater than r_{B1} (during triggering) and R_1 ($< 100\ \Omega$), through which C_T discharges. Before the UJT fires, a dc interbase current equal to I_1 ($=I_2$) flows down the voltage divider formed by R_2, r_{BB}, and R_1. After firing, S_1 is effectively closed, and r_{B1} decreases drastically. The increased emitter current then adds to I_1, since $I_1 = I_E + I_2$. This increase in I_1 causes V_{B1} to increase because R_1 is fixed while I_1 increases.

The frequency of oscillation of the UJT relaxation oscillator can be approximated by

$$f_{osc} = \frac{1}{T} \approx \frac{1}{R_T C_T} \tag{9-12}$$

A more exact formula for the frequency of oscillation is given by (Ref. 1):

$$f_{osc} = \frac{1}{T} = \frac{1}{R_T C_T \ln[(V_{BB} - V_V)/(V_{BB} - V_D - \eta V_{B2B1})]} \tag{9-13}$$

Since $V_{BB} \gg V_V$ or V_D, and $V_{BB} \approx V_{B2B1}$,

$$f_{osc} \approx \frac{1}{R_T C_T \ln[1/(1 - \eta)]}$$

The reason why Eq. (9-12) provides adequate accuracy for most applications is demonstrated by Fig. 9-12. UJT intrinsic standoff ratios (η) vary from 0.4 to 0.9 with an average at ≈ 0.65. This number coincides closely with the definition of one RC time constant, which is the time for the exponential waveform to reach

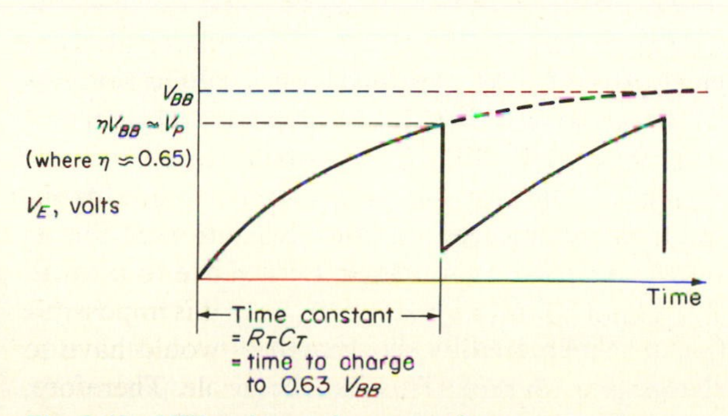

Fig. 9-12 UJT emitter voltage waveform.

63 percent of the final voltage. In other words, the UJT fires at ηV_{BB} or ≈ 0.65 V_{BB}, and the point of triggering determines the oscillation period, which is approximately equal to $R_T C_T$. The errors involved in such a simplification include the spread of η values above and below 0.65, the UJT ON and OFF times, and the fact that after the first cycle the UJT's emitter voltage does not return to zero as shown in Fig. 9-12. In addition, the frequency as determined by Eq. (9-12) differs from the more exact one only by the ln $[1/(1-\eta)]$ multiplier, or only a few percent.

As an example of UJT relaxation-oscillator design, consider the circuit of Fig. 9-11*a* with $V_{BB} = +12$ V, a 2N4948 UJT (Fig. 9-10*a*), and a desired frequency of oscillation of 1 kHz. Using Eq. (9-10) and typical η values, we have

$$V_P \approx (0.7)(12 \text{ V}) + 0.7 \text{ V} \approx 9.1 \text{ V}$$

For $C_T = 0.1\ \mu$F and a frequency of oscillation of 1 kHz, Eq. (9-12) gives

$$R_T = \frac{1}{f_{\text{osc}} C_T} = \frac{1}{(10^3)(10^{-7})} = 10^4 = 10 \text{ k}\Omega$$

Equation (9-11) can be used to check this value of R_T. From Eq. (9-11) and Fig. 9-10*a*,

$$\frac{12 \text{ V} - 9.1 \text{ V}}{2 \times 10^{-6} \text{ A}} > 10.0 \text{ k}\Omega > \frac{12 \text{ V} - 3 \text{ V}}{2 \times 10^{-3} \text{ mA}}$$

$$1.45 \text{ M}\Omega > 10.0 \text{ k}\Omega > 4.5 \text{ k}\Omega$$

In applications requiring a change in frequency with light or heat, a photoresistor (Sec. 7-1-2) or a thermistor can be substituted for R_T.

The basic UJT relaxation-oscillator waveforms of Fig. 9-11*c* show that some improvements are still desirable. These include the following:

1. To maximize the swing of V_E, it would be preferable to have $V_{E,\text{min}}$ closer to zero voltage.
2. The V_E ramp linearity is inadequate for most ramp-generator applications.
3. The frequency stability with temperature is also inadequate for many precise oscillator applications.
4. Excessive loading and performance degradation are sometimes encountered when low-impedance circuits are connected to the relaxation-oscillator emitter output.

The maximum swing of V_E can be increased by shorting the timing capacitor C_T to ground during its discharge (Fig. 9-11*c*). This can be accomplished by adding a transistor switch in parallel with C_T, as shown in Fig. 9-13*a*.

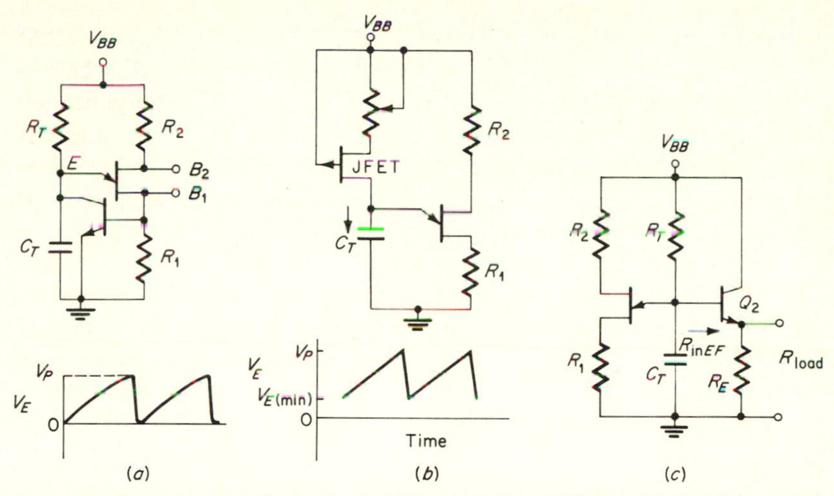

Fig. 9-13 Improvements in the basic UJT relaxation oscillator: (*a*) transistor switch used to short the timing capacitor and reduce voltage offsets; (*b*) using a JFET to improve the linearity of the UJT'S emitter voltage waveform by acting as a constant-current source for charging C_T; (*c*) emitter follower used to couple the UJT emitter to a load to reduce loading effects.

The linearity of the emitter sawtooth waveform can be significantly improved by charging C_T from a constant-current source rather than through R_T. Figure 9-13*b* shows the use of a junction FET to provide a constant-current source for C_T. The drain current of the JFET approximates a constant-current source in that this current is only slightly dependent on the device's output voltage, and a constant-current source provides a capacitor charging voltage which is linear with time, compared with the exponential shape obtained if R_T is used.

To utilize the UJT as a sawtooth generator, it is generally necessary to couple the UJT's emitter to the load via a follower circuit in order to decrease the effects of loading on the oscillator. Such an addition is shown in Fig. 9-13*c*. Notice that no coupling network is required because the UJT's $V_{E,\,\text{min}}$ is generally >1.0 V, which is a large enough V_B to keep a bipolar transistor ON. In order that the emitter follower not alter the effective R_T or provide a leakage path for C_T, β_{Q2} and R_E should be as large as possible to ensure adequate emitter-follower input resistance. The bipolar transistor could also be replaced with an FET or operational amplifier.

The typical UJT data and discussion of Sec. 9-5-3 showed that practically all the UJT parameters are temperature-dependent. Of primary concern are the variations of V_P and $V_{E,\,\text{min}}$, because any variation in these characteristics would cause a change in the frequency of oscillation. To visualize this, consider that V_P has decreased due to an increase in temperature. According to the UJT emitter waveforms of Fig. 9-11*c*, this would mean that C_T would charge to the lower V_P in less time, and thus the oscillation period would decrease (frequency would increase). This temperature dependence of V_P and oscillation frequency can be significantly reduced by the addition of a small-value fixed resistor R_2 in series with B_2, as shown in Fig. 9-14. To analyze this temperature-compensation scheme, consider Eq. (9-10):

$$V_P = \eta V_{B2B1} + V_D$$

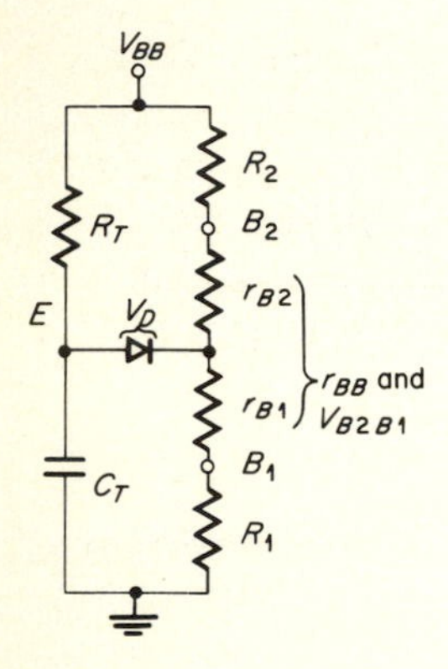

Fig. 9-14 Temperature compensation with an external resistor R_2.

We have seen that V_D and η decrease with temperature (negative temperature coefficient); therefore V_P does likewise. If we neglect R_1 and derive V_{B2B1} from the voltage divider of Fig. 9-14*a*, we have

$$V_{B2B1} = V_{BB} \frac{r_{BB}}{r_{BB} + R_2}$$

Substituting this into Eq. (9-10) yields

$$V_P = \eta \left[V_{BB} \frac{r_{BB}}{r_{BB} + R_2} \right] + V_D \tag{9-14}$$

Since V_D and η both decrease with increasing temperature, if the bracketed term in Eq. (9-14) were to increase with temperature, V_P might be stabilized. This is in fact the case, since R_2 remains constant and r_{BB} increases with temperature, as shown by Fig. 9-10*c*. Therefore, the positive temperature coefficient of r_{BB} can be made to approximately cancel the negative temperature coefficient of η and V_D, thereby making the coefficient of V_P approach zero. Considering the rates of change of V_D, η, and r_{BB} with temperature, the empirical formulas for calculating the optimum value of R_2 are

For the annular structure (Ref. 2): $\quad R_{2,\text{opt}} \approx 0.015 V_{BB} r_{BB} \eta$

For the bar structure (Ref. 3): $\quad R_{2,\text{opt}} \approx \dfrac{0.7\, V r_{BB}}{\eta V_{BB}} \quad$ if $R_2 \ll r_{BB}$

(9-15)

9-5-5 UJT THYRISTOR TRIGGER CIRCUITS

The UJT is an excellent thyristor trigger device. (See Chap. 8 for a discussion of the various thyristors.) The level of thyristor gate voltage needed for reliable triggering is a direct function of the thyristor power rating, the type of load, and the thyristor's anode-to-cathode voltage (Sec. 8-2-3).

There are many ways of obtaining gate trigger pulses, but the UJT offers one of the least expensive, simplest, most compact, and least power consuming, especially for a dc system. The simplest form of the UJT thyristor trigger circuit is shown in Fig. 9-15*a*. In this case the UJT circuit is the familiar relaxation oscillator whose frequency depends on the values of R_T and C_T. As shown in Fig. 9-15*a*, the UJT base-1 pulses are used to provide the thyristor gate trigger.

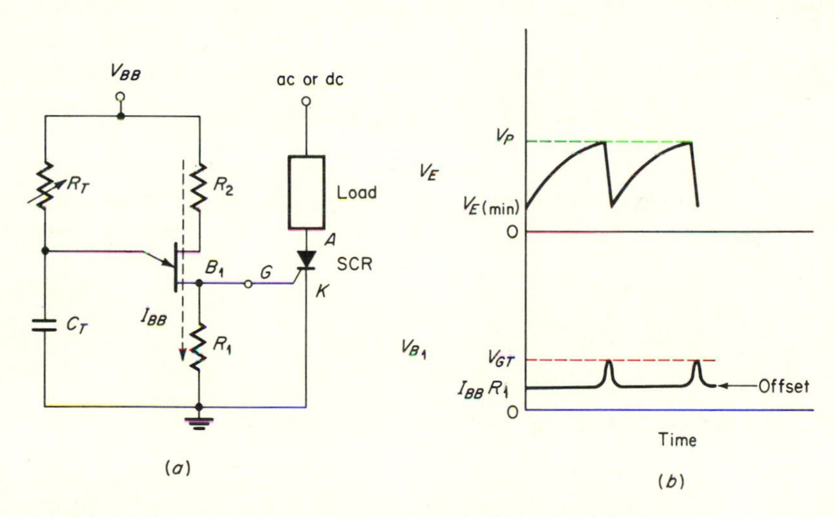

Fig. 9-15 (*a*) UJT thyristor trigger circuit; (*b*) voltage waveforms.

This is why it is important to choose an adequate V_{BB} supply and a UJT with sufficient V_{OB1} pulse capability. The resistor in series with the B_1 terminal is included to provide a path for the UJT interbase current I_{BB}, which flows prior to UJT triggering. If this current were allowed to flow into the thyristor's gate, it would probably turn the thyristor ON all the time. There is one disadvantage associated with the inclusion of R_1; the maximum swing of V_{B1} is then limited by the $I_{BB} R_1$ offset as shown in Fig. 9-15*b*. The value of R_1 is chosen to reduce V_{GK} below the minimum gate trigger voltage, which is generally a fraction of a volt depending on the particular thyristor. From Fig. 9-15*a* we see that $R_{1,\max}$ is limited to

$$R_{1,\max} \leq \frac{V_{GK,\min}}{I_{BB}} \tag{9-16}$$

Since $r_{BB} \gg R_1 + R_2$, $I_{BB} \approx V_{BB}/r_{BB}$, and substituting into Eq. (9-16) gives

$$R_{1,\,\max} \approx \frac{V_{GK,\,\min}\, r_{BB}}{V_{BB}}$$

For example, if $V_{BB} = 25$ V, $V_{G,\,\min} = 0.2$ V, and $r_{BB} = 4$ kΩ,

$$R_{1,\,\max} = 32\ \Omega$$

When the thyristor conducts, its resistance from anode to cathode becomes very low, allowing power to flow through the load.

In many thyristor applications a separate dc supply is not available, and the only source of power is the ac line. The type of application we are referring to involves controlling the amount of power delivered to a motor, lights, or a heater. This is generally accomplished by phase control, a technique used extensively to control the turn-ON of gas-filled thyratrons and SCRs, as shown by the ac phase-control circuits of Chap. 8.

9-5-6 COMPLEMENTARY UNIJUNCTION TRANSISTORS (CUJTS)

The complement of the conventional *n*-type UJT is actually an integrated circuit which retains many desirable UJT properties while substantially reducing many of the undesirable ones. The more important advantages of the complementary UJT (CUJT) compared to the conventional UJT are:

1. Intrinsic standoff-ratio spread reduced by a factor of >4
2. Factor-of-3 lower saturation resistance, which results in a similar increase in the maximum base output voltage
3. Operation at approximately one-third the supply voltage needed by UJTs (more compatible with integrated circuits)
4. Simpler and more predictable temperature compensation (10 times better frequency stability)
5. Higher frequency operation (up to 100 kHz)
6. Factor-of-10 (or more) reduction in emitter reverse leakage current
7. Offers a *complement* to the conventional UJT

Figure 9-16 gives the symbol, circuit, emitter characteristics, and equivalent circuit for the CUJT. For comparison purposes some of the conventional UJT information is repeated in the primed figures of Fig. 9-16. Figure 9-16*a* shows the CUJT emitter arrow emanating from the base of the CUJT, indicating a *p*-type base. Since this is the opposite of the conventional UJT, the CUJT emitter characteristics are inverted as shown.

Figure 9-16*d* shows that the CUJT is a simple integrated circuit consisting of two transistor elements and two diffused resistors. The *ratio* of these two

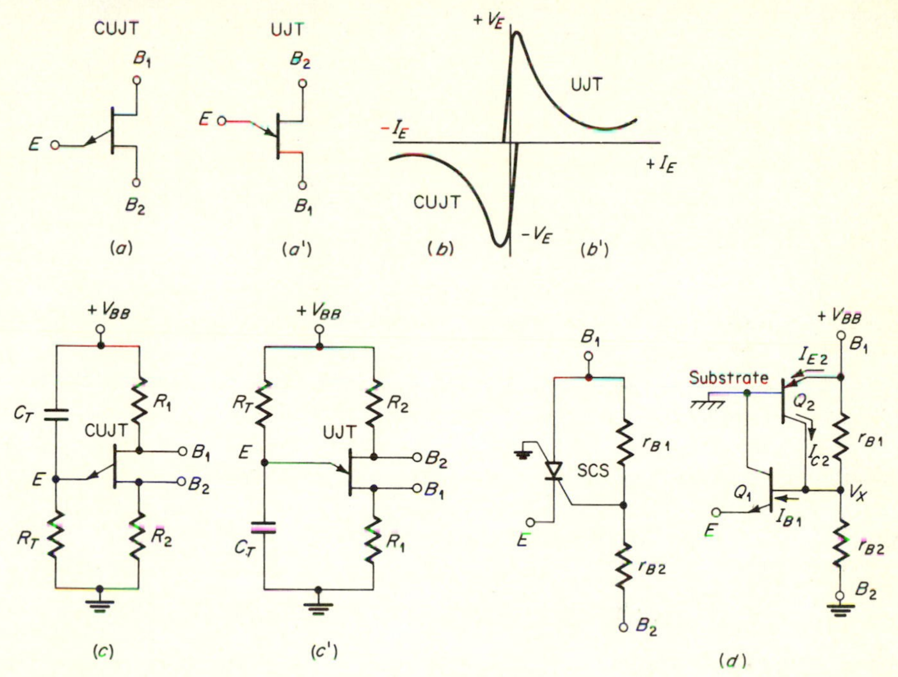

Fig. 9-16 Comparison of conventional and complementary UJT's (all primed figures are for the conventional UJT, and all unprimed are for the complementary UJT): (*a*) symbol; (*b*) emitter characteristics; (*c*) relaxation oscillator; (*d*) complementary-UJT equivalent circuits.

integrated-circuit resistors can be held to tight tolerances (better than ± 5 percent), and therefore the CUJT spread in η is small. Also, although such diffused resistors are temperature-dependent, temperature will have little effect on η if the diffused-resistor temperature coefficients are the same. The active part of the CUJT is also referred to as a silicon-controlled switch (SCS of Sec. 8-1).

The two-transistor equivalent circuit of Fig. 9-16*d* will be used to explain the operation of the CUJT. As the emitter of the CUJT is made more negative than V_X, Q_1 will begin to conduct. This turns Q_2 ON, and regenerative action is started. The SCS anode-to-cathode resistance becomes very low, effectively shorting r_{B1}. The SCS thus serves the same function as the emitter diode in the conventional *n*-type UJT (Sec. 9-5-3); that is, it causes part of r_{BB} to decrease with increasing emitter current. Whereas this was accomplished in the conventional UJT by conductivity modulation, in the CUJT the diffused r_{B1} resistor is actually shorted out by the low SCS ON resistance. Notice, in Fig. 9-16*c*, that the B_1 and B_2 terminals (and R_T and C_T) are all reversed from that of the conventional *n*-type UJT. This upside-down image is used because the CUJT polarities are reversed from the *n* type.

Figure 9-17 gives the relaxation-oscillator circuit and waveforms for the CUJT. Notice again that the CUJT circuit and emitter voltage waveform are reversed from the conventional UJT. In the CUJT, C_T charges from $+V_{BB}$

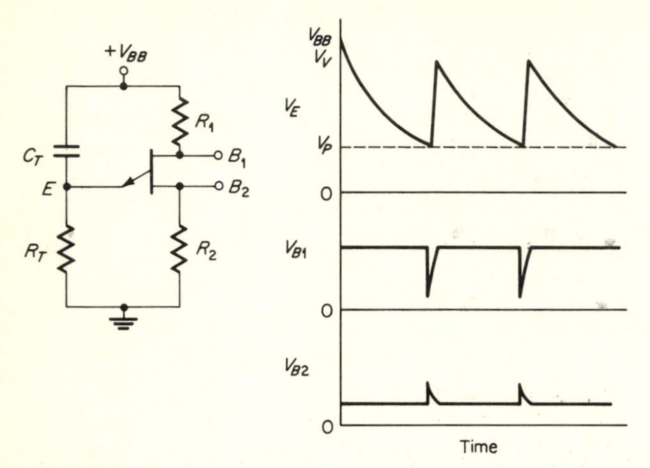

Fig. 9-17 Complementary-UJT relaxation-oscillator circuit and waveforms.

toward ground, and triggering occurs at V_P. The CUJT B_1 pulse is of the wrong polarity to trigger most thyristors, and the B_2 output is of insufficient magnitude. Therefore, the CUJT circuit must be modified as shown in Fig. 9-18, so that a large positive spike is available when the CUJT fires. This pulse originates from ground potential and can directly trigger the gate of an SCR.

Since the CUJT stability with temperature is approximately an order of magnitude better than that of the conventional UJT, it becomes much more useful in precision oscillators, timers, voltage sensing, and divider circuits.

9-5-7 PROGRAMMABLE UNIJUNCTION TRANSISTORS (PUTS)

Another improved UJT is the *programmable* UJT or PUT. It is similar to the complementary unijunction in that it is not really a UJT but rather a thyristor or *pnpn* device whose characteristics can be made similar to those of a UJT. However, in the case of the PUT the base resistors are not included in the package but are added externally. This capability of selecting the effective r_{BB} is the basis

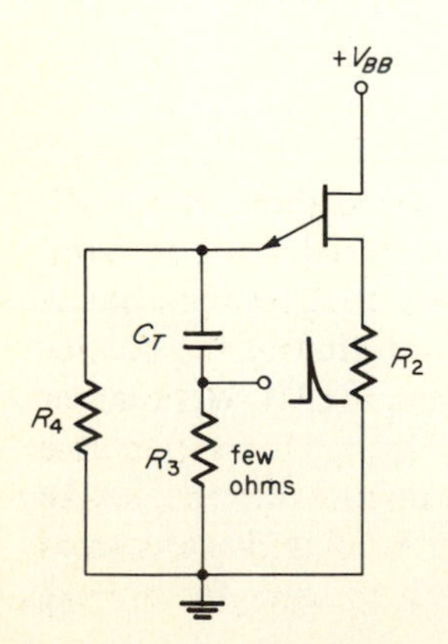

Fig. 9-18 Modification required for CUJT thyristor triggering. *From W. Spofford, "Complementary Unijunction Transistors," General Electric Appl. Note 90.72, Feb., 1968.*

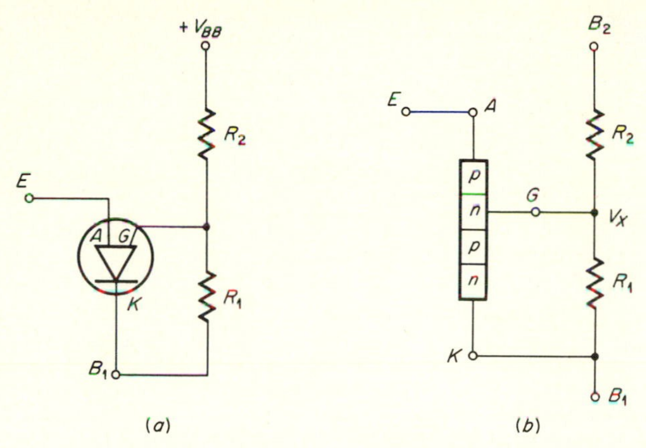

Fig. 9-19 (*a*) Programmable-UJT symbol; (*b*) PUT four-layer equivalent circuit.

of the PUT's programmability. Figure 9-19 shows the basic PUT symbol with external R_1 and R_2 base resistors and the equivalent circuit. Notice that the base resistors are external to the device itself and are necessary for proper operation.

The PUT *pnpn* device is not the same as the dual-gate silicon-controlled switch in the CUJT. The PUT terminals are labeled anode, gate, and cathode. To see the similarity of the PUT's operation to that of the conventional UJT, recall that the conventional UJT emitter diode (Fig. 9-9*c*) becomes forward-biased when the emitter voltage exceeds V_X by ≈ 0.7 V, and r_{B1} decreases drastically with emitter current. The PUT equivalent circuit of Fig. 9-19*b* also has a *pn* junction connected to an R_2-R_1 voltage divider, and in this case the values of R_1 and R_2 determine V_X. As with most *pnpn* devices, when the gate diode conducts, regeneration takes place, and the anode-to-cathode resistance decreases to a very low value. When the PUT thyristor is in this ON condition, R_1 is paralleled with $\approx 10\ \Omega$; thus R_1 is effectively "modulated" just as r_{B1} was in the conventional UJT. The emitter characteristics (really the anode characteristics in this case) are very similar to those of the conventional UJT, except that the forward-current values are much lower.

9-5-8 COMPARISON OF UJTS

Now that at least four different types of unijunction transistors—bar UJT, planar UJT, CUJT, and PUT—have been introduced, it is appropriate to compare the various types and thereby note some of the inherent advantages of the PUT. Figure 9-20 compares the various types of unijunction transistors. Notice that the effective interbase resistance of the PUT is higher than that of the other UJTs, and its peak and valley currents are considerably lower. Its respectable reverse leakage current and extremely low firing and operating currents make the PUT ideal for long-interval timers, in which large values of timing resistors necessarily limit the available triggering current, and high emitter leakage

Characteristic	*Bar UJT*	*Planar UJT*	*CUJT*†	*PUT*†	*Prefer*
Emitter reverse current, nA	20 to 2,000 (V_{EB1} = 30 V)	1.0 to 10	0.1 to 1.0 (V_{EB1} = 5 V)	1.0 to 10 (anode to anode gate at V_G = 40 V)	Low for accurate timing applications
η range (at V_{BB} = 10 V)	0.45 to 0.80	0.55 to 0.85	0.58 to 0.62	Adjustable	Depends on the application
r_{BB} (typical at I_E = 0), kΩ	7.0	6.5	8.0	6–30 (equivalent)	Generally high for low *off*-power consumption
I_P, μA	9.0	3.0	10.0 (max) (V_{BB} = 10 V)	0.1 to 1.0 (V_G = 10 V, R_G = 10 kΩ)	Low for ease of triggering
I_V, mA	8.0 (V_{BB} = 20 V)	2.0	2.0 (V_{BB} = 10 V)	0.1 (V_G = 10 V, R_G = 1 MΩ)	Depends on the application
Minimum operating voltage, V	10	5	5	4	Low
Peak base pulse voltage, V	3.0	4.0 (V_{BB} = 20 V)	3.5 (V_{BB} = 12 V)	10.0 (V_{BB} = 20 V)	High to provide thyristor trigger
Emitter saturation voltage (at I_E = 50 mA), V	4.5	3.5	1.1	1.5	Low so base pulse high
Max emitter reverse voltage, V	60	30	≈10	GA = 40 GK = 5	Depends on the application
Max emitter current, mA	70	50	150	150	High
Max dissipation at 25°C, mW	450	300	200	300	High

† These are lower voltage units; therefore operation and specifications are at lower levels than for the other UJTs.

Fig. 9-20 Comparison of bar, planar, complementary, and programmable UJT characteristics.

currents cause intolerable timing errors. These advantages must be weighed against the inability to accurately temperature-compensate the peak-point voltage of the PUT. The PUT has a much higher peak-pulse output and frequency response, which is advantageous in thyristor triggering applications. The PUT also shares the CUJT advantages of very low operating-voltage capability. In addition, the parameters η, I_P, I_V, and r_{BB} are programmable by adjusting the values of the external resistors, R_1 and R_2. Lastly, all the PUT breakdown voltages except V_{GK} are $\approx$40 V; hence, by proper choice of R_1 and R_2, the PUT can be made suitable for 120-V line operation.

Figure 9-21 shows a PUT 10-min delay timer. Time delays of up to 2 h are possible by using a PUT and large values of R_T and C_T. However, low-leakage capacitors must be used, or such leakages will deprive the PUT of adequate I_P. In such a failure mode, C_T will charge partway and stay there, and the PUT will not trigger. By using precision components in the PUT relaxation oscillator, the necessity for frequency-trimming potentiometers can be eliminated because its important characteristics are precisely determined by the precision of R_1 and R_2.

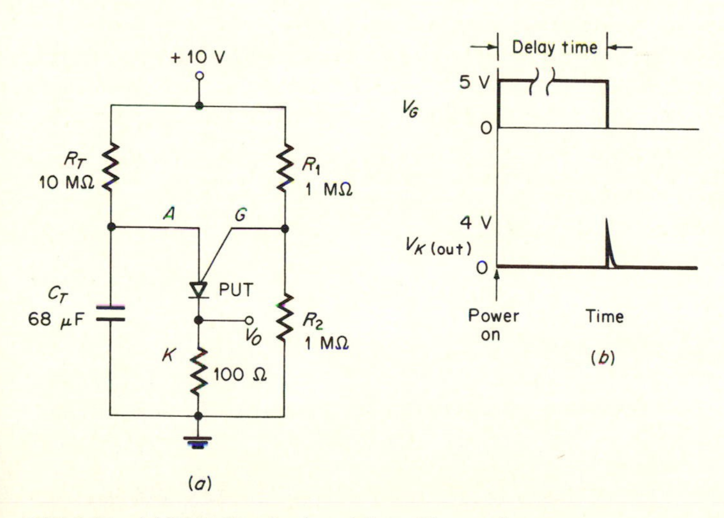

Fig. 9-21 (*a*) PUT 10-min time delay; (*b*) waveforms.

PROBLEMS

9-1 If a 1-V peak-to-peak sine wave is applied to the circuit in Fig. 9-1*a*, what will be the peak value of the output if a silicon diode is used? What will it be if a germanium diode is used? Assume that silicon and germanium diodes have $V_T = 0.6$ and 0.15 V, respectively.

9-2 Give two advantages of using a center-tapped transformer (secondary winding) in a rectifier circuit, rather than a non-center-tapped transformer.

9-3 Explain why the output current capability of a zener regulator cannot be increased by paralleling two zeners as in Fig. Prob 9-3. Give a better method.

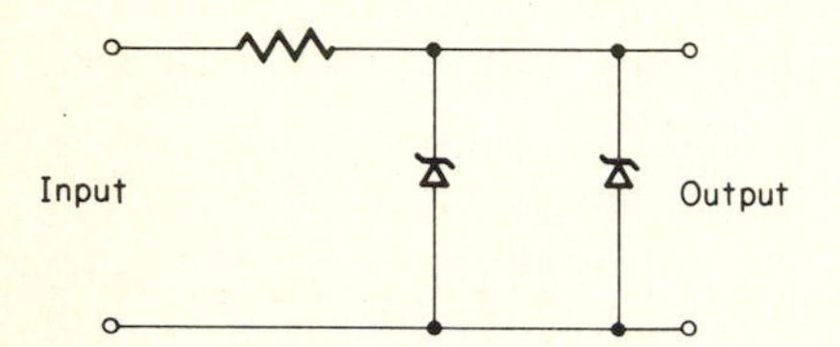

Fig. Prob. 9-3

9-4 A zener regulator of the type shown in Fig. 9-4*a* uses a nominal 20-V diode. The unregulated input can vary from 30 to 40 V, and the load current from 10 to 20 mA. Assume the zener impedance is constant at 5 Ω, and the minimum current required in the diode for zener action is 1 mA. Find the value required for the series resistor, its maximum power dissipation, and the maximum fluctuation in the output voltage.
9-5 Draw two different circuit configurations for a zener regulator with an operational-amplifier isolator for

(*a*) Variable output voltages smaller than the zener voltage
(*b*) Variable output voltages greater than the zener voltage

9-6 Give three differences between avalanche and zener diodes.
9-7 Distinguish between the dynamic and static resistance of a zener diode. Calculate both for the plot in Fig. 9-3*b*.
9-8 How does a VVC diode work? Give three advantages VVCs have over mechanically varied capacitors. Give one disadvantage.
9-9 In a tuned circuit, a VVC diode capacity changes by 5 percent. By how much will the resonant frequency of the tuned circuit change?
9-10 A VVC has equivalent series resistance and capacitance of 3.2 Ω and 1000 pF. Find its Q at 1 MHz.
9-11 Briefly explain the operation of a conventional unijunction transistor.
9-12 What is meant by negative resistance?
9-13 (*a*) Name five UJT applications.
(*b*) Briefly describe the UJT features or characteristics which enable it to perform many functions with fewer components than are required by bipolar-transistor design.
9-14 Describe the operation of the basic UJT relaxation oscillator. Use figures, and include appropriate waveforms.
9-15 Referring to the UJT relaxation oscillator of Fig. Prob. 9-15, calculate the

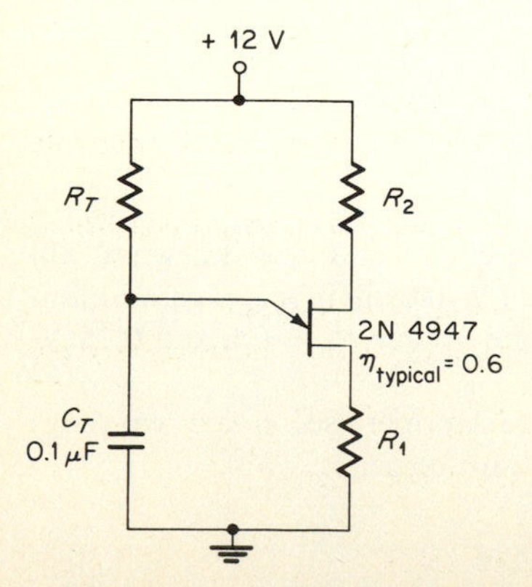

Fig. Prob. 9-15

approximate value of timing resistor R_T to obtain a frequency of oscillation of (*a*) 500 Hz, (*b*) 1 kHz, and (*c*) 10 kHz. Make sure your value of R_T meets the load-line criteria. Refer to Fig. 9-10 for the UJT data, and assume $I_P = 0.25\ \mu A$.

9-16 (*a*) Briefly explain the two temperature effects that tend to cancel each other when a fixed R_2 is inserted in a UJT relaxation oscillator.

(*b*) Calculate the approximate value of R_2 for a 2N4949 UJT and a supply voltage of 24 V. (See Fig. 9-10.)

9-17 Calculate the frequency of oscillation for the UJT sawtooth generator of Fig. Prob. 9-17. (See Fig. 9-10.)

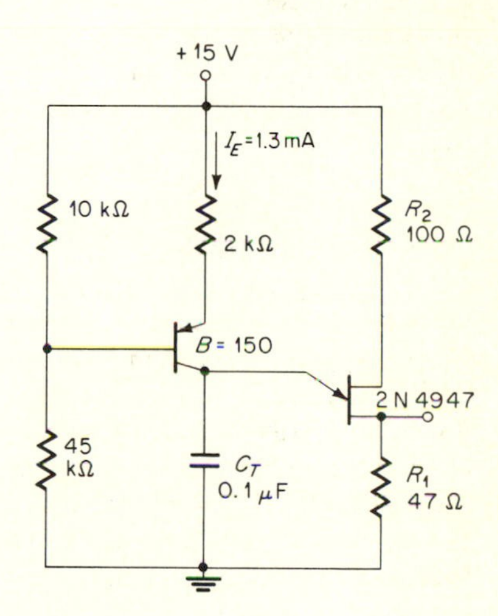

Fig. Prob. 9-17

9-18 Referring to Fig. 9-15*a*, calculate the maximum value of R_1 to ensure that the thyristor is not falsely triggered by I_{BB}. Assume the thyristor has a 0.4-V minimum gate trigger level, $V_{BB} = 25$ V, $R_2 = 220\ \Omega$, and the UJT is a 2N4947 (Fig. 9-10). What is the magnitude of V_{EB1} if $I_E = 10$ mA?

9-19 Briefly describe the CUJT, including its operation.

9-20 (*a*) List four advantages of the CUJT over the conventional UJT.

(*b*) List three applications in which the inherent advantages of the CUJT provide performance superior to that of conventional UJTs.

9-21 (*a*) Calculate the value of R_T necessary for a 1-h time delay using a conventional UJT. Assume $C_T = 100\ \mu F$.

(*b*) Describe another technique for obtaining such a long time delay.

9-22 (*a*) Name two major differences between programmable and complementary UJTs.

(*b*) What does one do to program the PUT?

9-23 (*a*) What thyristor does the *pnpn* device in the PUT most resemble?

(*b*) Briefly explain the operation of the PUT.

9-24 List five advantages of the PUT over other types of UJTs.

9-25 Referring to the PUT circuit of Fig. Prob. 9-25, calculate the values of R_1 and R_2 for an effective r_{BB} of 50 kΩ and an η of 0.6.

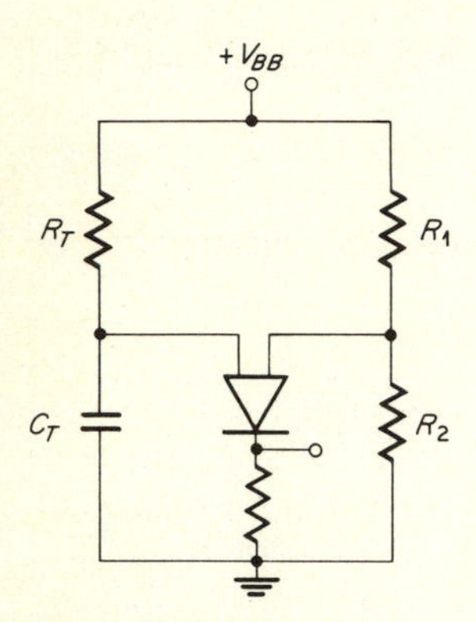

Fig. Prob. 9-25

9-26 Calculate the equivalent gate resistance R_G and gate voltage V_G of Fig. Prob. 9-26.

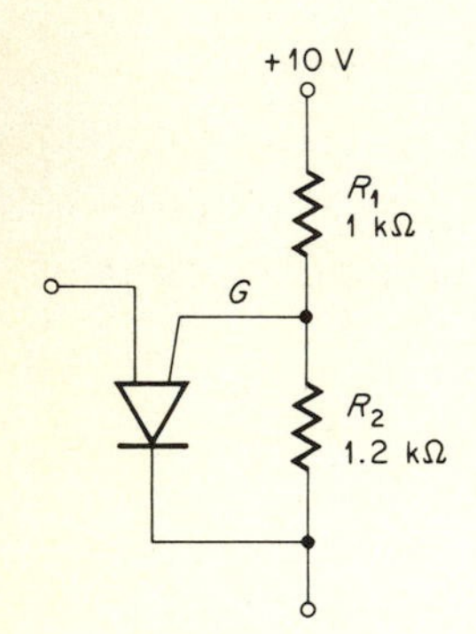

Fig. Prob. 9-26

REFERENCES

1. Bergersen, T.: Unijunction Transistor Timers and Oscillators, *Motorola Appl. Note AN*-294, p. 3; and Sylvan, T.: The Unijunction Transistor, Characteristics and Applications, *General Electric Appl. Note* 90.10, p. 37, May, 1965.
2. Bergersen, T.: Unijunction Transistor Timers and Oscillators, *Motorola Appl. Note AN*-294, p. 11.
3. Jones, D.: Unijunction Temperature Compensation, *General Electric Appl. Note* 90.12, p. 2, April, 1962.

Appendix

Bipolar-Transistor Circuit Analysis Using the Hybrid-Pi Equivalent Circuit

Bipolar transistor circuits are commonly analyzed using models which are characterized by such quantities as r and h parameters. There is another type of equivalent circuit which is of great value for two reasons. First, it requires a minimum of information from a data sheet, and second it is useful from dc to very high frequencies. This equivalent circuit is called the *hybrid* π.

A-1 HYBRID-PI EQUIVALENT CIRCUIT

Although originally intended as a model for high-frequency use, the hybrid-π equivalent circuit is useful from near dc to high frequencies. The circuit derives its name from its configuration, which is like the Greek letter π, as can be seen in Fig. A-1.

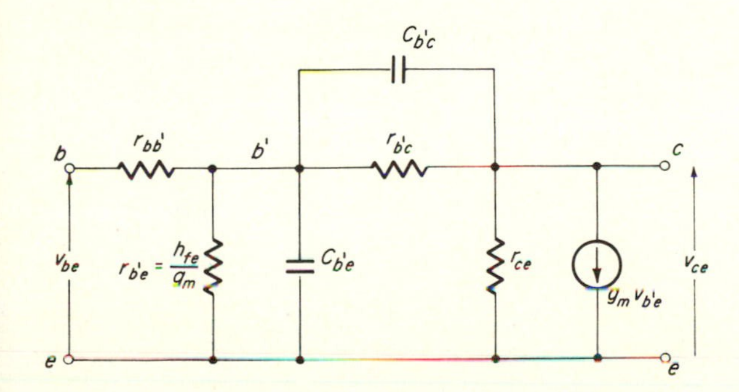

Fig. A-1 Hybrid-π equivalent circuit.

A feature of the hybrid-π model is that it includes the high-frequency characteristics of the transistor. It is worth stressing the difference between the hybrid-π model and other models such as the h-parameter model in this respect. The hybrid π contains elements such as $C_{b'e}$ and $C_{b'c}$ (on Fig. A-1), which account for the high-frequency performance of the transistor. However, neither the values of these two elements, nor those of any of the other elements, vary with frequency. Frequency effects are accounted for by the fact that capacitor reactances vary as frequency varies. In contrast, the h parameters are themselves frequency-dependent. One cannot draw an h-parameter equivalent circuit, insert numerical values such as $h_{ie} = 1$ kΩ and $h_{fe} = 50$, and then use these values at any frequency. h_{ie} and h_{fe} are only 1 kΩ and 50 at low frequencies. They change values at high frequencies. Thus, the designer has a problem if the manufacturer does not give the h parameters as a function of frequency. In the hybrid-π model we might have $r_{bb'} = 50\ \Omega$, $C_{b'e} = 1000$ pF, etc. Because of RC effects, the voltages and currents are functions of frequency as we would expect, but the *values of the elements themselves do not change with frequency.*

Another feature of the hybrid-π model is that it is *not difficult to obtain values for the elements.* This can be done by combinations of simple calculations, using information supplied by the manufacturer, or by measurement. Very often, reasonably accurate results can be obtained without data sheets or measurements.

Finally, it is important to realize that the hybrid-π model is useful at low as well as high frequencies. Thus, it competes very well with the h-parameter model even at low frequencies, especially since the hybrid-π model in Fig. A-1 can be drastically simplified at low frequencies if certain assumptions are made. (This is done in Fig. A-3.)

Now let us consider the structure of the hybrid-π model shown in Fig. A-1. There is an inaccessible terminal, designated b', which plays a very important role in an understanding of the hybrid π. b' is separated from the input base b by the base-spreading resistance $r_{bb'}$. This is a function of the geometry and resistivity of the base region. Values of $r_{bb'}$ vary from 10 to 1000 Ω and are substantially independent of the transistor operating point.

$r_{b'e}$ is the incremental (ac) base-emitter junction resistance and is approximately equal to h_{fe}/g_m, where h_{fe} is the CE low-frequency ac current gain, and g_m is the transistor transconductance. Transconductance is defined in the usual way as the ratio of an output current change to the input voltage change. In this case $g_m = i_c/v_{b'e} \approx i_e/v_{b'e}$. Notice that since the incremental resistance of the base-emitter diode (which is often called r_e) is $v_{b'e}/i_e$, we can write $g_m = 1/r_e$. This relationship is of great value to us because r_e is a function of the dc emitter current I_E:

$$r_e = \frac{kT}{qI_E} \tag{A-1}$$

where k = Boltzmann's constant
T = absolute temperature
q = charge on an electron

At room temperature $kT/q = 0.026$ V, so that

$$g_m = \frac{1}{r_e} = \frac{I_E}{kT/q} = \frac{I_E}{0.026} = (38.4)[I_E\,(\text{mA})] \quad \text{ma/V} \tag{A-2}$$

From Eq. (A-2), $g_m = 38.4$ mA/V at $I_E = 1$ mA.

Thus, we can obtain a value for g_m knowing only the value of the dc emitter current. Also,

$$r_{b'e} = \frac{h_{fe}}{g_m} = \frac{h_{fe}}{(38.4)[I\,(\text{mA})]} \qquad \text{k}\Omega \tag{A-3}$$

To account for certain losses that occur in practice, the 38.4 in Eq. (A-3) is often replaced by 35 as in Table A-1.

TABLE A-1 Hybrid-π circuit elements

Parameter Represents	Formula	Typical Value*
$r_{bb'}$, base-spreading resistance	$h_{ie} - h_{fe}/g_m$	10–1000 Ω
$r_{b'e}$, incremental base-emitter junction resistance	h_{fe}/g_m	2.8 kΩ for $h_{fe} = 100$ and $I_E = 1$ mA
$r_{b'c}$, feedback from output to input	$\cong h_{fe}/g_m h_{re}$	>1 MΩ
r_{ce}, output resistance	$1/(h_{oe} - g_m h_{re})$	>200 kΩ
$C_{b'e}$, artificial capacitor used to account for decrease in h_{fe} with frequency	$g_m/2\pi f_T$	122 pF at $I_E = 1$ mA if $f_T = 50$ MHz
$C_{b'c}$, capacitance of reverse-biased collector-base junction	$C_{ob}\left(\frac{V'_{CE}}{V_{CE}}\right)^3$ (V'_{CE} = voltage at which C_{ob} was measured.)	40 pF at 5 V
g_m, transconductance	$(35)[I_E(\text{mA})]$ mA/V	35 mA/V at $I_E = 1$ mA
$g_m v_{b'e}$, output-circuit current generator	Use g_m from above times $v_{b'e}$	. . .

* Typical values obtained for a transistor with:
$f_T = 50$ MHz (= frequency at which $h_{fe} = 1$)
$C_{ob} = 5$ pF measured at $V'_{CE} = 10$ V
$h_{fe} = 100$
$I_E = 1$ mA

The fact that $r_{b'e}$ equals the current gain divided by the transconductance can be checked for the frequent situation in which $r_{b'e} \gg r_{bb'}$. $r_{b'e}$ is then the input resistance. Notice that

$$\frac{h_{fe}}{g_m} \approx \frac{i_c/i_b}{i_c/v_{b'e}} \approx \frac{v_{b'e}}{i_b} \approx \frac{v_{be}}{i_b} = \text{input resistance}$$

$r_{b'c}$ is a feedback resistor linking the output and input circuits. Usually it is sufficiently large to be neglected.

r_{ce} is the output resistance, and though not as large as $r_{b'c}$, it is also often large enough to be neglected.

$C_{b'e}$ is an artificial capacitance inserted into the model to account for decreasing current gain and increasing phase shift with frequency. It is a function of frequency and of the incremental base-emitter resistance. To clarify we shall list the parameters of which $C_{b'e}$ is a function. They are

$$r_e = \text{emitter diode resistance} = kT/qI_E = 1/g_m$$

$$r_{b'e} = h_{fe} r_e = h_{fe}/g_m$$

f_β = frequency at which h_{fe} is 3 dB down from its low-frequency value

f_T = frequency at which $h_{fe} = 1$

= gain-bandwidth product

= (f_β)(low-frequency value of h_{fe})

$C_{b'e}$ is artificially chosen to have a value such that $C_{b'e}r_{b'e}$ is the time constant of the 3-dB point on the h_{fe} versus frequency curve (see Fig. A-2). Therefore,

$$f_\beta = \frac{1}{2\pi C_{b'e} r_{b'e}}$$

In terms of f_T, which is more commonly given on a data sheet,

$$\frac{f_T}{h_{fe}} = \frac{1}{2\pi C_{b'e} r_{b'e}}$$

Hence,

$$C_{b'e} = \frac{h_{fe}}{2\pi f_T r_{b'e}} = \frac{g_m}{2\pi f_T} \tag{A-4}$$

Frequently, f_T is not given directly on a data sheet. Instead, the current gain at some specified frequency is stated, and this enables f_T to be found. For example, if h_{fe} has a low-frequency value of 100 and falls to 10 at 10 MHz as in Fig. A-2 the gain-bandwidth product f_T is (10)(10 MHz) = 100 MHz.

It is important to realize that we designate current gain h_{fe} in the hybrid-π model only because this nomenclature is so entrenched from the h-parameter model. In an h-parameter model, h_{fe} varies with frequency as in Fig. A-2. In the hybrid-π model, h_{fe} *does not vary with frequency*, the variation in performance of the transistor being accounted for by other means. We use Fig. A-2 only to

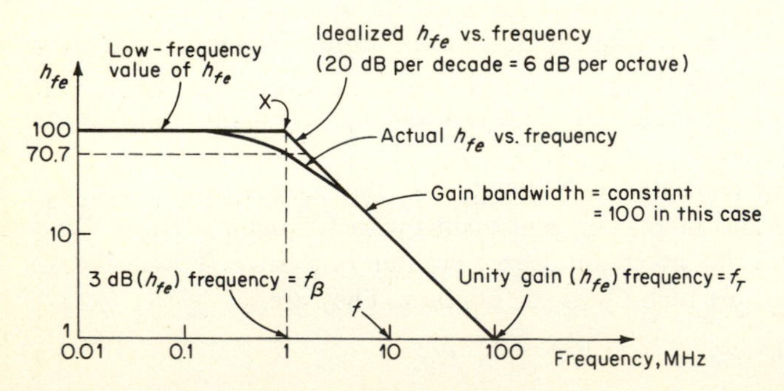

Fig. A-2 Variation of h_{fe} with frequency.

obtain f_T, which we need to calculate $C_{b'e}$ in Eq. A-4. If f_T is given directly, we do not even have to bother with Fig. A-2 or gain-bandwidth-product calculations.

$C_{b'c}$ is the capacitance of the reverse-biased collector-base junction. It is a function of junction area and the dc collector-base voltage. It is essentially equal to C_{ob}, which is usually given on a data sheet. $C_{b'c}$ is related to V_{CE} as follows:

$$C_{b'c} = \frac{K}{V_{CE}^{\;n}} \tag{A-5}$$

where $n \approx 2$ for an abrupt junction and $n = 3$ for a graded junction. $n \approx 3$ is most common, so if the data sheet gives C_{ob} at $V_{CE} = 10$ V, we can find $C_{b'c}$ at other voltages (since $V_{CE} \approx V_{CB}$) from

$$C_{b'c} = C_{ob}\left(\frac{10}{V_{CE}}\right)^3 \tag{A-6}$$

Thus, if $V_{CE} = 5$ V and $C_{ob} = 5$ pF at $V_{CE} = 10$ V,

$$C_{b'c} = (5)(10/5)^3 = 40 \text{ pF}$$

The last element in our model is the current generator $g_m v_{b'e}$. g_m is the transconductance as defined in Eqs. (A-1) and (A-2). $v_{b'e}$ is the signal voltage from our inaccessible terminal b' to the emitter. In many cases $v_{b'e}$ is essentially equal to v_{be}, the base-to-emitter input voltage.

Table A-1 lists the various circuit elements used in the hybrid π and gives numerical values for a typical small-signal transistor.

Table A-1 may appear to be somewhat complicated, but this is very deceptive. The table contains information that is often very difficult to get *at all* from data sheets. In fact, it is a feature of the hybrid-π model that numerical values for Table A-1 can be obtained with a minimum of information from a data sheet. This is especially true if we use the hybrid-π model at low frequencies and make a couple of simplifying assumptions. At low frequencies we can neglect $C_{b'e}$ and $C_{b'c}$. Let us also assume that $r_{b'c}$ and r_{ce} are large enough to be neglected. The hybrid π then reduces to the circuit in Fig. A-3*a*.

In many cases Fig. A-3*a* can be simplified even further. This is because if $r_{bb'}$ is small compared with h_{fe}/g_m, we can eliminate $r_{bb'}$ from the circuit and obtain Fig. A-3*b*. Notice that the only information required from a data sheet for Fig. A-3*b* is h_{fe}, since we obtain g_m from Eq. (A-2).

Let us use the model in Fig. A-3*b* to find the voltage gain A_V of a common-emitter amplifier operated with an emitter current of 2 mA and a collector load $R_L = 5$ kΩ. The output current i_c will be that from the output current generator, or $g_m v_{b'e}$. i_c will flow through R_L, which is connected from c to e in Fig. A-3*b*, to produce the output voltage v_{ce}. So we can write

$$v_{ce} = i_c R_L = -g_m v_{b'e} R_L$$

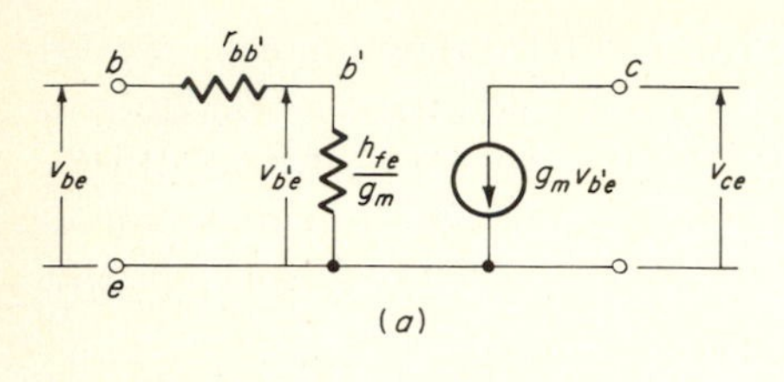

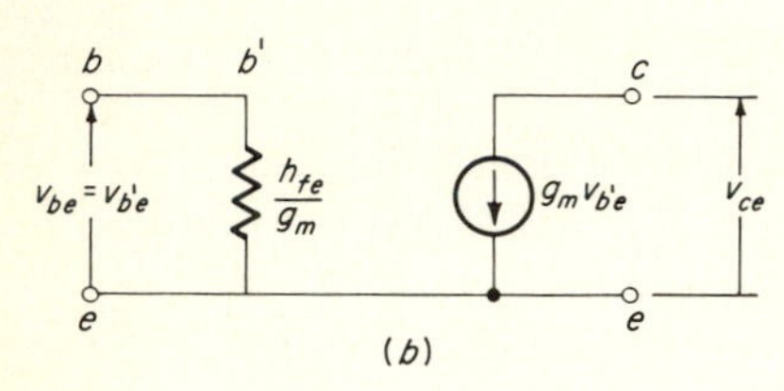

Fig. A-3 Simplified hybrid-π circuit: (*a*) including $r_{bb'}$; (*b*) excluding $r_{bb'}$.

Since $v_{b'e} = v_{be}$ in Fig. A-3*b*, we have

$$A_V = -\frac{v_{ce}}{v_{be}} = -g_m R_L \tag{A-7}$$

From Eq. (A-2), $g_m = 38.4 I_E = (38.4)(2 \text{ mA}) = 76.8 \text{ mA/V}$. Therefore,

$$A_V = -(76.8 \text{ mA/V})(5 \text{ k}\Omega) = -384$$

A-2 SOME USEFUL APPROXIMATIONS

Table A-2 lists some formulas which are of value in rapidly evaluating the performance of a bipolar-transistor circuit. Only the basic circuits are shown, but the formulas are readily applied to circuits with bias resistors, additional loads, and source resistors. For example, if the CE amplifier drives another stage, R_L in the voltage-gain formula would be the first-stage collector resistor in parallel with the input impedance of the following stage. Similarly, to find the input impedance of the CE amplifier in Table A-2 if there are base resistors present, R_{in} from the formula in Table A-2 must be paralleled with the impedance of the base-bias string. The use of these techniques is best illustrated by a numerical example.

Consider the three-stage amplifier shown in Fig. A-4. The input stage is a CE amplifier with its emitter resistor unbypassed, presumably to maintain a reasonably high input impedance. The second stage is a similar CE amplifier, but with its emitter resistor bypassed, so that we can consider this the high-gain stage. The third stage is a CC (emitter follower), no doubt employed to lower the circuit output impedance. Let us find the input impedance, voltage gain, output impedance of the amplifier, and the upper and lower cutoff frequencies. Assume that the three capacitors are ac short circuits at the frequency we are considering, and that all three transistors have $h_{fe} = 250$.

TABLE A-2 Approximate formulas

Circuit	Quantity	Formula*	Simplified Formula	Conditions for Simplification
CE	A_V	$-R_L/(R_E + 1/g_m)$	$-R_L/R_E$	If $R_E \gg 1/g_m$
			$-g_m R_L$	If $R_E \ll 1/g_m$ or R_E bypassed
	A_i	h_{fe}	. . .	. . .
	R_{in}	$h_{fe}(R_E + 1/g_m)$	$h_{fe}R_E$	If $R_E \gg 1/g_m$
			h_{fe}/g_m	If $R_E \ll 1/g_m$ or R_E bypassed
	R_{out}	R_L	. . .	. . .
CB	A_V	$R_L/(R_S + 1/g_m)$	R_L/R_S	If $R_S \gg 1/g_m$
			$g_m R_L$	If $R_S \ll 1/g_m$
	A_i	-1	. . .	. . .
	R_{in}	$R_E/(1 + g_m R_E)$	$1/g_m$	$R_E \gg 1/g_m$
	R_{out}	R_L	. . .	. . .
CC (emitter-follower)	A_V	$g_m R_E/(1 + g_m R_E)$	1	If $g_m R_E \gg 1$
	A_i	$-h_{fe}$	. . .	. . .
	R_{in}	$h_{fe}(1/g_m + R_E)$	$h_{fe}R_E$	If $R_E \gg 1/g_m$
	R_{out}	$1/g_m + R_S/h_{fe}$	$1/g_m$	If $R_S h_{fe} \ll 1/g_m$
			R_S/h_{fe}	If $R_S/h_{fe} \gg 1/g_m$

* h_{fe} is a positive number.
$g_m \approx 35\ [I_E(\text{mA})]$ mA/V, $r_{bb'}$ assumed negligible.
See Sec. A-2 for application of formulas.

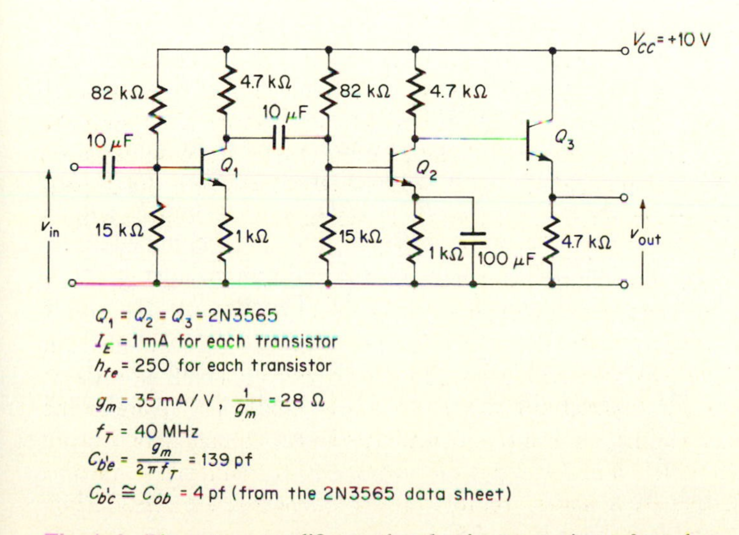

Fig. A-4 Three-stage amplifier analyzed using approximate formulas.

Many of the formulas in Table A-2 have g_m in them, so let us find g_m for each transistor.

We are using high-current-gain transistors ($h_{fe} = 250$), so we can, as a first approximation, neglect base currents. The voltage at the base of Q_1 is therefore

$$V_{B1} = \frac{15\ \text{k}\Omega}{15\ \text{k}\Omega + 82\ \text{k}\Omega} 10\ \text{V} = 1.55\ \text{V}$$

The emitter of Q_1 will be 0.5 to 0.6 V below V_{B1}, say at 1.0 V. The emitter current of Q_1 is therefore 1 V/1 kΩ = 1 mA. Q_1 therefore has a g_m of (35)(1 mA) = 35 mA/V from Table A-1.

Q_2 is biased to the same dc operating point as Q_1, since the coupling and emitter bypass capacitors do not affect the dc bias. The g_m of Q_2 is therefore also 35 mA/V. The 1 ma into the 4.7-kΩ load of Q_2 sets Q_2 collector voltage at 10 − (1 mA)(4.7 kΩ) = 5.3 V. The base of Q_3 is at this same potential, so its emitter will be at about 4.7 V. Since Q_3 emitter resistor is 4.7 kΩ, its emitter current is 4.7 V/4.7 kΩ = 1 mA.

Thus all three transistors are biased at 1 mA and all have a g_m of 35 mA/V. We shall also need $1/g_m$, which is

$$\tfrac{1}{35}\ \text{V/mA} = \tfrac{1}{35}\ \text{k}\Omega = 28.6\ \Omega$$

Now let us find the input impedance of the circuit. The input impedance of Q_1 is, from Table A-2,

$$\begin{aligned} R_{\text{in}} &= h_{fe} R_E \qquad \text{since } R_E \gg 1/g_m\text{, that is, } 1\ \text{k}\Omega \gg 28.6\ \Omega \\ &= (250)(1\ \text{k}\Omega) = 250\ \text{k}\Omega \end{aligned}$$

The input impedance of the bias string is 15 kΩ||82 kΩ, which is 12.7 kΩ. The *circuit* input impedance is 12.7 kΩ||250 kΩ, which is 12.1 kΩ.

Now let us find the circuit voltage gain. This is the product of the gains of the three stages, so first we shall find the gain of the first stage. Since $R_E \gg 1/g_m$ (1 kΩ ≫ 28.6 Ω), we can use $A_V = -R/_L R_E$. R_E is 1 kΩ, so let us find R_L. R_L is the total load "seen" by Q_1 collector and consists of three parallel impedances. One is the 4.7-kΩ collector load of Q_1. The second is the impedance of Q_2's base-bias string, or 15 kΩ||82 kΩ, which is 12.7 kΩ. The third is the input impedance of transistor Q_2. Since R_E for Q_2 is bypassed, the second-stage R_{in} is h_{fe}/g_m or 250/35 mA/V = 7.14 kΩ. The total collector load for Q_1 is therefore 4.7 kΩ||12.7 kΩ||7.14 kΩ, which is 2.32 kΩ. The gain of the Q_1 stage is therefore

$$\frac{-R_L}{R_E} = \frac{-2.32\ \text{k}\Omega}{1\ \text{k}\Omega} = -2.32$$

The voltage gain of the Q_2 stage is $-g_m R_L$ since R_E is bypassed in this case. R_L consists of the 4.7-kΩ collector load in parallel with the input impedance of Q_3. Q_3 is an emitter follower with an input impedance of $h_{fe} R_E$ or (250)/(4.7 kΩ) = 1175 kΩ. For all practical purposes 4.7 kΩ‖1175 kΩ ≈ 4.7 kΩ. The voltage gain of the Q_2 stage is therefore $-g_m R_L = -(35\text{ mA/V})(4.7\text{ k}\Omega) = -165$.

The voltage gain of the Q_3 stage is ≈ +1 since it is an emitter follower with $g_m R_E \gg 1$. The total circuit voltage gain is therefore

$$A_V = (-2.32)(-165)(+1) = +383$$

The circuit output impedance is that of emitter follower Q_3. The source impedance R_s for the emitter follower is the output impedance of Q_2, which is ≈4.7 kΩ. The circuit output impedance is, from Table A-2,

$$R_{\text{out}} = \frac{1}{g_m} + \frac{R_s}{h_{fe}} = 28.6 + \frac{4700}{250} = 28.6 + 18.8 = 47.4\ \Omega$$

Now let us find the lower and upper cutoff frequencies for the amplifier in Fig. A-4.

The low-frequency cutoff will be due to the effects of the two 10-μF coupling capacitors and the 100-μF emitter bypass capacitor. These capacitors, with resistive elements in the circuit, form time constants which limit the low frequency response.

We can quickly determine that it is the time constant involving the 100-μF capacitor that will predominantly set the low-frequency cutoff. The input 10 μF will see the circuit input resistance of 12.1 kΩ, so the input time constant is $(10)(10^{-6})(12.1)(10^3) = 0.121$ s.

The interstage 10 μF will see a resistance consisting of the output resistance of Q_1, the 4.7-kΩ collector load of Q_1, the bias string of Q_2, and the input resistance of Q_2, which is $h_{fe}/g_m = 250/35 = 7.14$ kΩ. Before calculating the exact net resistance, let us note that it will be around a few kilohms, giving a time constant of perhaps a quarter of that due to the input, which is 0.121 s. Therefore we can expect around 0.03 s due to the interstage 10 μF.

The 100-μF capacitor sees the output resistance of Q_2 emitter in parallel with the 1-kΩ emitter resistance. Q_2 emitter output resistance is $1/g_m + R_S/h_{fe}$ from Table A-2. R_S is 4.7 kΩ in parallel with 82 kΩ in parallel with 15 kΩ, or 3.43 kΩ. Q_2 emitter output resistance is therefore 28 + 3430/250 = 42 Ω; 42 Ω in parallel with the 1-kΩ emitter resistance is 40 Ω. The time constant associated with the emitter of Q_2 is therefore $(40)(100)(10^{-6}) = 0.0040$ s. This time constant is appreciably smaller than either of the other two time constants and will therefore predominantly determine the low-frequency cutoff. The lower 3-dB point of the amplifier will be $\approx 1/2\pi RC$, where $R = 40\ \Omega$ and $C = 100\ \mu$F, or 39.8 Hz.

The high-frequency cutoff will be due to the shunting effects of $C_{b'e}$ and $C_{b'c}$ inside the transistors. If Q_1 is fed from a voltage source, any input shunt capacity in Q_1 will have no effect. Also, Q_1 has considerable emitter feedback, and this will reduce the effective base-to-emitter capacitance and minimize the Miller capacitance since the stage voltage gain is low, being less than 3. Also, because the gain-bandwidth product is fixed for a given transistor, the use of feedback to reduce gain increases the bandwidth.

Q_3 is an emitter follower, so again the effective input capacities are minimized, and the frequency response will be extended by the 100 percent feedback.

The amplifier bandwidth will be predominantly limited by the high input capacity of the second stage. This stage has the highest voltage gain, so it will also have the highest Miller capacity. Figure A-5*a* shows the relevant parts of the

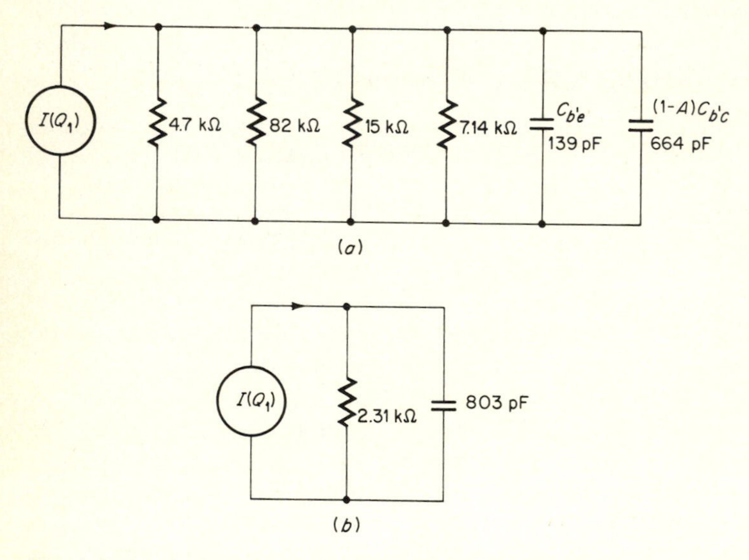

Fig. A-5 Equivalent circuits for amplifier in Fig. A-4: (*a*) for upper 3-dB cutoff frequency; (*b*) simplified for upper 3-dB cutoff frequency.

equivalent circuit. Current generator Q_1 drives a current into several resistors and capacitors. The resistors are the 4.7-kΩ collector load of Q_1, the 15- and 82-kΩ bias string of Q_2, and the input resistance of Q_2, which is 7.14 kΩ. These resistors in parallel are equivalent to a resistor of 2.31 kΩ. The two input capacitors for Q_2 are $C_{b'e}$, which is 139 pF from Fig. A-4, and the Miller capacitor, which is $(1 - A)(C_{b'c})$. The voltage gain A of stage Q_2 has already been found to be -165, and $C_{b'c} = C_{ob} = 4$ pF from the manufacturer's data sheet. The Miller capacity is therefore $[1 - (-165)](4) = 664$ pF. The total input capacity is therefore $139 + 664 = 803$ pF. The equivalent resistance and capacitance may be combined with the current generator as in Fig. A-5*b*. This circuit has a 3-dB cutoff frequency of

$$\frac{1}{2\pi RC} = 85.8 \text{ kHz}$$

As a check, the circuit in Fig. A-4 was analyzed on a computer and the effects of *all* resistors and capacitors in Figs. A-1 and A-4 were included in the analysis. The results of the computer analysis are compared with the measured and calculated results in Table A-3. Measurements were made on a breadboard, in which the 2N3565s were not selected for $h_{fe} = 250$, as used in the calculations. The h_{fe} spread for the 2N3565 is 120 to 750 for $I_C = 1$ mA and $V_{CE} = 10$ V. All transistors in Fig. A-4 were operated with $V_{CE} < 10$ V, and resistors and capacitors were ± 10 percent.

The authors would like to thank Professor Thomas T. L. Chou of California State Polytechnic College, San Luis Obispo, California, for performing the computer analysis of the circuit in Fig. A-4.

TABLE A-3 Comparison of data for amplifier in Fig. A-4 obtained by computer analysis, rough calculations and by measurement.

	Computer	Rough Calculations	Measured
Gain (mid-band)	353	383	340
Input resistance	12.0 kΩ	12.1 kΩ	13 kΩ
Output resistance	47 Ω	47.4 Ω	53 Ω
Low-frequency 3-dB point	35 Hz	39.8 Hz	33 Hz
High-frequency 3-dB point	90 kHz	85.8 kHz	81 kHz

Answers to Selected Problems

CHAPTER 1

1-7 1.22 kΩ, 25 kΩ
1-10 (*a*) 0 (*b*) 2 mA (*c*) 0.5 mA
1-11 (*a*) 0 (*b*) 800 μmhos (*c*) 400 μmhos
1-12 20 kΩ, 1 mA, 1000 μmhos
1-13 (*a*) −10, 10 MΩ, 10 kΩ
(*b*) +0.952, 10 MΩ, 952 Ω
1-14 60 pF, 5.25 pF
1-18 87.1 μV rms
1-20 (*a*) $V_G = +2$ V (*b*) $V_G = -3$ V
1-21 4 inputs are required
1-24 −206

CHAPTER 2

2-5 $R_s = 4.33$ kΩ
2-6 +0.897, 448 Ω
2-11 $R_F = 10^{11}$ Ω
2-16 (*a*) $R_S = 1.5$ kΩ, $R_D = 1$ kΩ
(*b*) $V_{DS} = +15$ V
(*c*) Gain = −10
2-17 (*a*) Gain = −10/11
(*b*) V_{DS} is unchanged at +15 V

CHAPTER 3

3-1 150 Ω per square
3-2 $R_4 = 9.22$ kΩ
3-3 $I_D = .100$ mA
3-4 $R_1 = R_2 = 70.92$ kΩ
Collector voltages are still at +10 V
3-9 $V_1 = 0$ Hz (dc), $V_2 = 40$ kHz, $V_3 = 10$ kHz
3-10 $R_1 = 10$ kΩ, $R_2 = 4.4$ kΩ, $R_3 = 2.2$ kΩ, $R_4 = R_5 = 5$ kΩ

CHAPTER 4

4-1 −1, 2
4-2 −4 V
4-3 −1 V
4-4 $-j$ 15.9 kΩ
4-6 1.01 V/°C
4-18 The end of the cable connected to the operational amplifier, not the end connected to the load
4-20 Inverting: $A_F = -100$, loop gain = 10^4, $R'_{in} = 1$ kΩ, $R'_o = 0.1$ Ω. Noninverting: $A_F = 101$, loop gain = $10^6/101$, $R'_{in} = 10^{11}/101$, $R'_o = 0.101$ Ω
4-22 Loop gain

$$= \frac{A_o}{(1 + R_2/R_1 + R_2/R_{in})}$$

4-23 0.11 Ω
4-25 $e_5 = -1.206$ V

CHAPTER 5

5-2 Binary = 1101111, BCD = 0001 0001 0001
5-7 After $J = K = 0$, no change
After $J = 1$, $K = 0$, no change

After $J = 0$, $K = 1$, $Q = 0$ and $\bar{Q} = 1$
After $J = K = 1$, $Q = 1$ and $\bar{Q} = 0$

5-10 NOR
5-14 00011 00011
00111 11111
01000 00000
5-26 For a $X = 1$, $Y = 0$ address, the output = 0.
5-28 For a three input only the "3" AND gate has all inputs high.
5-30 For $A = 1$ and $B = 0$: $C = 0$, $D = 0$, and $E = 1$ $(A > B)$
5-32 (*a*)

```
  01100
+ 01001
  10101 = 16 + 4 + 1
        = 21
```

5-33 (*a*)

```
  111
- 101
  010 = 2
```

(*b*)

```
   0011
 × 0110
   0000
  0011
 0011
0000
10010 = 16 + 2 = 18
```

5-34

```
   1010
 + 1011   (comp. of 4)
(1) 0101
 →   +1   (end-around carry)
   0110 = 6
```

5-41 10 bits

CHAPTER 6

6-4 1.37 eV, 2.25 eV, and 0.367 eV respectively
6-7 100 cm^2
6-9 0.714 MSCP and 88.2 h (or 75 h via Fig. 6-4*b*)
6-10 3.43 MSCP and 986 s
6-11 1.85 mW/cm^2
6-12 1.11 mW/cm^2
6-13 (*a*) 0.69 mA
6-14 56.4 mW/cm^2
6-17 886 nm
6-18 0.267%
6-20 0.18 mW
6-21 0.186 mW/cm^2
6-22 1.44 mW/cm^2
6-24 107 Ω
6-26 67.4°

CHAPTER 7

7-2 (*a*) 517 nm (*b*) 729 nm
7-6 battery = 10 V
7-8 $R_2 = 24.7$ kΩ and $V_1 \simeq 14$ V
7-9 R_1 and $R_3 = 90$ kΩ
7-10 $R_{E,\max} = 330$ Ω, $R_{2\min} = 3.6$ kΩ
7-13 $\simeq 10$ MΩ
7-17 At 1 Hz bandwidth $I_{shot} = 1.39 \times 10^{-14}$ A
7-18 (*a*) 38.9 to 1
7-20 79 MHz neglecting stray capacitances
7-21 (*a*) 4.0 μA (*b*) −4.0 V (*c*) −1.0 mV
7-22 104 kΩ
7-23 25 mA
7-24 −2.02 V
7-27 500 cm^2
7-28 (*a*) 28.8 mW
7-29 (*c*) 50 parallel strings of 30 series cells each
7-30 $R_E = 3.57$ kΩ
7-31 1.07 mA
7-35 617 kΩ
7-39 potentiometer = 1 MΩ, $V_{out} = 30.0$ V
7-45 87.5 mA (worst case)
7-47 (*a*) 170 Ω

CHAPTER 8

8-4 (*a*) A 400 V C35D (*b*) A C35B
8-5 50°C
8-6 (*b*) 0.53 μF
8-7 59 W
8-8 3 V and 80 mA (double to insure triggering)
8-11 164.6° max and 126.9° min
8-14 32 μH and 0.32 μF
8-15 21.4 kΩ
8-17 $R_{1(\max)} = 300$ kΩ, $C_1 \geq 1.5$ μF
8-19 710 Ω
8-20 Assuming a linear charge, V_{C_1} must charge 0.025 V/cycle, $R_{1(\max)} = 8.11$ MΩ, $C_1 \simeq 7$ μF
8-27 At 80°C, 8.0 A rms

CHAPTER 9

9-1 0 V, 0.35 V, 0.11 V
9-4 476 Ω, 0.84 W, 0.155 V
9-9 −2.41 percent for an increase in capacity and +2.47 percent for a decrease in capacity.
9-10 50
9-15 $R_T = 10$ kΩ
9-17 1.53 kHz
9-18 $R_{1(\max)} \approx 64$ Ω, $V_{EBI} \approx 1.5$ V
9-21 $R_T \approx 36$ MΩ
9-25 $R_1 = 30$ kΩ, $R_2 = 20$ kΩ
9-26 $R_{G,\ \mathrm{THEV}} = 545$ Ω, $V_G = 5.45$ V

Index